## Praise for P. W. Singer's *Wired for War*

"Likely to be the definitive work on this subject for some time to come. . . . Riveting and comprehensive, encompassing every aspect of the rise of military robotics, from the historical to the ethical."
                                                                    —*Financial Times*

"War by remote control is here and is brilliantly depicted in Peter W. Singer's latest bestseller, *Wired for War*. . . . Singer's book is a fascinating, well-researched, page-turning account of where technologies are taking us next in the wars of the future."
                                                                    —*The Washington Times*

"Genuinely provocative."
                                                                    —*Book Forum*

"As Singer explores the issues raised by military robotics—meeting with entrepreneurs, engineers, and operators, ethicists, and pundits—his enthusiasm becomes infectious. With its informal style and cultural references, and because of its topic, *Wired for War* is a book of its time: this is strategy for the Facebook generation."
                                                                    —*Foreign Affairs*

"Addresses some ominous and little-discussed questions about the military, technology, and machinery."
                                                                    —*Harper's*

"An important and influential book, [*Wired for War*] explains not only what we see in today's headlines, but where it's all leading. It's nothing short of a civilization game changer. . . . Singer has picked up the strands of known information from the military, the laboratories, the private companies currently producing robotics, and military people in Iraq, Afghanistan and in the distant cubicles where the wars are also being fought. He's blended all of this into a highly readable, nontechnical, thoroughly convincing picture of today's realities—along with a UAV-like over the horizon image of what's ahead for the future of warfare."      —*U.S. Politics Today*

"Heed Singer's suggestion of a serious dialogue before the robots walk among us."
                                                                    —*The Brooklyn Rail*

"Splendid . . . riveting, important . . . It is the only book in my reading experience that quotes Immanuel Kant and Biggie Smalls with equal enthusiasm. . . . An intoxicating, encyclopedic trip—made intensely readable by all the colorful characters Singer salts along this story."
                                                                    —*San Francisco Chronicle*

"An in-depth and at times frightening look at the growing use of robotics by the military."
                                                                    —*Mother Jones*

"Informative and useful to interested parties of all ideological inclinations . . . *Wired for War* is a portrait of highly competent military embedded within a prosperous society adapting to a difficult situation through technological innovation. But it's also a depiction of a country that has lost its way and an alarming military-industrial complex that isn't helping us find our way back."

—*American Conservative*

"A vivid picture of the current controversies and dazzling possibilities of war in the digital age."

—*Kirkus Reviews*

"If you shrug off *Terminator* and *Battlestar Galactica* as never-gonna-happen impossibilities, P. W. Singer has news for you. His spine-tingling book *Wired for War*, carefully explains the robotics revolution that's gripped our military since 9/11."

—Gizmodo.com

"*Wired for War* is a sprawling, eye-opening, important look at an evolving technology that promises to change the future in profound ways. Read it. Be prepared."

—Military.com

"An excellent book."

—*The Daily Times* (Pakistan)

"A comprehensive look at how robots have become integral to the modern military."

—*The Japan Times*

"A compelling work."

—General Norton Schwartz, chief of staff of the U.S. Air Force

"A superb book . . . If you read *Wired for War* you'll actually get a sense for the complexities that we are creating. We're not making a simpler world with these robots I don't think at all; I think we're making a more complex world, and that is something I got from this great book."

—General James Mattis, USMC, NATO Supreme Allied Commander
for Transformation and the Commander of U.S. Joint Forces Command

"P. W. Singer has fashioned a definitive text on the future of war around the subject of robots. In no previous book have I gotten such an intrinsic sense of what the military future will be like."

—Robert D. Kaplan, author of *Imperial Grunts: The American Military on the Ground*

"P. W. Singer's *Wired for War* is a rare chance to glimpse history before it happens. If you're an armchair warrior—and your armchair happens to be a time machine—this book will enthrall."

—Greg Bear, Hugo and Nebula award–winning science fiction author of
*Darwin's Radio, Forge of God,* and *Blood Music*

PENGUIN BOOKS

# WIRED FOR WAR

P. W. Singer, Director of the 21st Century Defense Initiative at the Brookings Institution, has worked in the Pentagon, as well as consulted for the departments of Defense and State, the CIA, and Congress. The author of two previous books, *Corporate Warriors* and *Children at War,* he has also written for publications such as the *The New York Times* and *Foreign Affairs.* For further information, visit www.pwsinger.com.

# [WIRED FOR WAR]

The Robotics Revolution and Conflict in the Twenty-first Century

## P. W. SINGER

PENGUIN BOOKS

PENGUIN BOOKS

Published by the Penguin Group

Penguin Group (USA) Inc., 375 Hudson Street, New York, New York 10014, U.S.A.

Penguin Group (Canada), 90 Eglinton Avenue East, Suite 700, Toronto,
Ontario, Canada M4P 2Y3 (a division of Pearson Penguin Canada Inc.)

Penguin Books Ltd, 80 Strand, London WC2R 0RL, England

Penguin Ireland, 25 St Stephen's Green, Dublin 2, Ireland (a division of Penguin Books Ltd)

Penguin Group (Australia), 250 Camberwell Road, Camberwell,
Victoria 3124, Australia (a division of Pearson Australia Group Pty Ltd)

Penguin Books India Pvt Ltd, 11 Community Centre,
Panchsheel Park, New Delhi – 110 017, India

Penguin Group (NZ), 67 Apollo Drive, Rosedale, North Shore 0632,
New Zealand (a division of Pearson New Zealand Ltd)

Penguin Books (South Africa) (Pty) Ltd, 24 Sturdee Avenue,
Rosebank, Johannesburg 2196, South Africa

Penguin Books Ltd, Registered Offices:
80 Strand, London WC2R 0RL, England

First published in the United States of America by The Penguin Press,
a member of Penguin Group (USA) Inc. 2009
Published in Penguin Books 2010

1   3   5   7   9   10   8   6   4   2

ISBN 978-1-59420-198-1 (hc.)
ISBN 978-0-14-311684-4 (pbk.)
CIP data available

Printed in the United States of America
DESIGNED BY NICOLE LAROCHE

*This is your last chance. After this, there is no turning back. You take the blue pill—the story ends, you wake up in your bed and believe whatever you want to believe. You take the red pill—you stay in Wonderland and I show you how deep the rabbit-hole goes. Remember that all I am offering is the truth. Nothing more.*

—Larry and Andy Wachowski, *The Matrix*, 1999

# [CONTENTS]

# WHY A BOOK ON ROBOTS AND WAR?

*Those people who think they know everything are a great annoyance to
those of us who do.*

—ISAAC ASIMOV

Because robots are frakin' cool.

That's the short answer to why someone would spend four years researching
and writing a book on new technologies and war. The long answer is a bit more
complex.

As my family will surely attest, I was a bit of an odd kid. All kids develop their
hobbies and even fixations, be it baseball cards or Barbie dolls. Indeed, I have yet to
meet a six-year-old boy who did not have an encyclopedic knowledge of all things
dinosaur. For me growing up, it was war. I could be more polite and say military
history, but it was really just war. In saying the same about his childhood, the great
historian John Keegan wrote, "It is not a phrase to be written, still less spoken with
any complacency." But it is true nonetheless.

Perhaps the reason lies in the fact that the generations before me had all served
in the military. They left several lifetimes' worth of artifacts hidden around the
house for me to pilfer and play with, whether it was my dad's old military med-
als and unit insignia, which I would take out and pin to my soccer jersey, or the
model of the F-4 Phantom jet fighter that my uncle had flown over Vietnam, which
I would run up and down the stairs on its missions to bomb Legoland.

But the greatest treasure trove of all was at my grandparents' house. My grandfather passed away when I was six, too young to remember him as much more than the kindly man whom we would visit at the nursing home. But I think he may have influenced this aspect of me the most.

Chalmers Rankin Carr, forever just "Granddaddy" to me, was a U.S. Navy captain who served in World War II. Like all those from what we now call "the Greatest Generation," he was one of the giants who saved the world. Almost every family gathering would include some tale from his or my grandmother's ("Maw Maw" to us grandkids) experiences at war or on the home front.

It's almost a cliché to say, but the one that stands out is the Pearl Harbor story; although, as with all things in my family, it comes with a twist. On December 7, 1941, my grandfather was serving in the Pacific Fleet on a navy transport ship. For three months after the Pearl Harbor attack, the family didn't hear any word from him and worried for the worst. When his ship finally came back to port (it had actually sailed out of Pearl Harbor just two days before the attack), he immediately called home to tell his wife (my grandmother) and the rest of his family that he was okay. There were only two problems: he had called collect, and that side of my family is Scotch-Irish. No one would accept the charges. While my grandfather cursed the phone operator's ear off, in the way that only a sailor can, on the other end the family explained to the operator that since he was calling, he must be alive. So there was no reason to waste money on such a luxury as a long-distance phone call.

Granddaddy's study was filled with volume after volume of great books, on everything from the history of the U.S. Navy to biographies of Civil War generals. I would often sneak off to this room, pull out one of the volumes, and lose myself in the past. These books shaped me then and stay with me now. One of my most prized possessions is an original-edition 1939 *Jane's Fighting Ships* that my grandfather received as a gift from a Royal Navy officer, for being part of the crew that shipped a Lend-Lease destroyer to the Brits. As I type these very words, it peers down at me from the shelf above my computer.

My reading fare quickly diverged from that of the other kids at Myers Park Elementary School. A typical afternoon reading was less likely to be exploring how Encyclopedia Brown, Boy Detective, cracked *The Case of the Missing Roller Skates* than how Audie Murphy, the youngest soldier ever to win the Medal of Honor, went, as he wrote in his autobiography, *To Hell and Back*. War soon morphed over into the imaginary world that surrounds all kids like a bubble. Other kids went to Narnia, I went to Normandy. While it may have looked like a normal Diamondback dirt bike, my bicycle was the only one in the neighborhood that mounted twin .50-caliber machine guns on the handlebars, to shoot down any marauding Japanese Zeros that dared to ambush me on my way to school each morning. I still remember my mother yelling at me for digging a five-foot-deep foxhole in our backyard when I was ten years old. She clearly failed to understand the importance of setting up a proper line of defense.

I certainly can't claim to have been a normal kid, but in my defense, you also have to remember the context. To be so focused on war was somewhat easier in that period. It was the Reagan era and the cold war had heated back up. The Russians wouldn't come to our Olympics and we wouldn't go to theirs, the military was cool again, and we had no questions about whether we were the good guys. Most important, as a young Patrick Swayze and Charlie Sheen taught us in *Red Dawn*, not only were the Commies poised to parachute right into our schools, but it was likely us kids who would have to beat them back.

What I find interesting, and a sign of the power of Hollywood's marketing machine, is that usually some artifact from science fiction is in the background of these memories, intertwined with the history. For example, when I think back to my childhood bedroom, there are the model warships from my grandfather's era lined up on display, but also Luke, Leia, Han, and Chewbacca peeking up from my *Star Wars* bedsheets.

As most of science fiction involved some good guy battling some bad guy in a world far, far away, the two memes of my fantasy world went together fairly well. In short, your author was the kind of little boy to whom a stick was not a mere piece of wood, but the makings of a machine gun or a lightsaber that could save the world from both Hitler and Darth Vader.

## WAR! WHAT IS IT GOOD FOR?

I look back on these memories with some embarrassment, but also guilt. Of course, even then, I knew that people die in war and many soldiers didn't come home, but they were always only the buddy of the hero, oddly enough usually from Brooklyn in most World War II movies. The reality of war had no way of sinking in.

It was not until years later that I truly understood the human costs of war. I remember crossing a jury-rigged bridge into Mostar, a town in Bosnia that saw some of the worst fighting in the Yugoslav civil war. I was there as part of a fact-finding mission on the UN peacekeeping operation. Weeks of back-and-forth fighting had turned block after block of factories and apartments on the riverfront into a mass of hollowed-out hulks. The pictures of World War II's Stalingrad in an old book on my grandfather's shelf had sprung up to surround and encompass me. The books never had any smell other than dust, but here, even well after the battles, a burnt, fetid scent still hung in the air. Down the river were the remnants of an elegant 500-year-old bridge, which had been blasted to pieces by Serb artillery. The people, though, were the ones who drove it home. "Haunted" is the only adjective I can think of to describe the faces of the refugees.

The standout memory, though, was of a local provincial governor we met with. A man alleged to have orchestrated mass killing and ethnic cleansing campaigns for which he would soon after be indicted, he sat at an immense wooden desk, ominously framed by two nationalist paramilitary (and hence illegal) flags. But he banally talked about his plans to build up the tourism industry after the war. He explained that the war had destroyed many of the factories and cleaned out whole villages. So on the positive side, the rivers were now clear and teeming with fish. Forget the war crimes or the refugees, he argued, if only the United States and United Nations would wise up and give him money, the package tourists would be there in a matter of weeks.

This paradox between the "good" wars that I had fought in my youth and the seamy underside of war in the twenty-first century has since been the thread running through my writing. During that same trip, I met my first private military contractors, a set of former U.S. Army officers, who were working in Sarajevo for a private company. Their firm wasn't selling widgets or even weapons, but rather the very military skills of the soldiers themselves. This contradiction between our ideal of military service and the reality of a booming new industry of private companies leasing out soldiers for hire became the subject of my first book, *Corporate Warriors: The Rise of the Privatized Military Industry*. During the research, I was struck by another breakdown of the traditional model of who was supposed to be at war. In West Africa, the main foes of these new private soldiers were rebel bands, mostly made up of children. Many of these tiny soldiers had been abducted from their schools and homes. For me as a child, war had merely been a matter of play; for these children, war was the only way to survive. My next book, *Children at War*, tried to tell their story, in a way that didn't just tug at heartstrings, but also explained the causes and effects of child soldiers, such that we might finally act to end this terrible practice.

This contradiction of war as we imagine it to be, versus how it really is, isn't just the matter of a young boy growing up and putting his lightsaber away. It is part of something bigger that has haunted humanity from its very start.

One of the original sins of our species is its inability to live at peace. From the very beginning of human history, conflicts over food, territory, riches, power, and prestige have been constant. The earliest forms of human organization were clans that first united for hunting, but soon also for fighting with other clans over the best hunting grounds. The story of the dawn of civilization is a story of war, as these clans transformed into larger tribes and then to city-states and empires. War was both a cause and effect of broader social change. From war sprung the very first specializations of labor, the resulting stratification into economic classes, and the creation of politics itself.

The result is that much of what is written in human history is simply a history of warfare. It is a history that often shames us. And it should. War is not just merely human destruction, but the most extreme of horror and waste wrapped together. Our great religions view war as perhaps the ultimate transgression. In the Bible, for example, King David was prohibited from building his holy Temple, because, as God told him, "You are a warrior who has shed blood" (1 Chronicles 28). The ancient prophets' ideal vision of the future is a time when we "will learn warfare no more" (Isaiah 2:4). As one religious scholar put it, "War is a sign of disobedience and sinfulness. War is not intended by God. All human beings are made in the image of God and they are precious and unique."

The same disdain for war was held by our great intellectuals. Thucydides, the founder of both the study of history as well as the science of international relations, described war as a punishment springing from man's hubris. It is our arrogance chastised. Two thousand years later, Freud similarly described it as emanating from our Thanatos, the part of our psyche that lives out evil.

Yet for such a supposed abomination, we sure do seem to be obsessed with war. From architecture to the arts, war's horrors have fed the heights of human creativity. Many of our great works of literature, arts, and science either are inspired by war or are reactions to it, from the founding epics of literature like *Gilgamesh* and the *Iliad* to the great painters of surrealism to the very origins of the fields of chemistry and physics.

War then appears in many more guises than the destruction and waste of human talent that we know it to be. War has been described as a testing ground for nobility, the only true place where man's "arête" (excellence) could be won. In the *Iliad*, the master narrative for all of Western literature, for example, "fighting is where man will win glory." From Herodotus to Hegel, war is described as a test of people's vitality and even one culture's way of life versus another. War is thus often portrayed in our great books as a teacher—a cruel teacher who reveals both our strengths and faults. Virtues are taught through stories of war from Homer to Shakespeare, while evils to avoid are drawn out by war in stories ranging from Aeschylus to Naipaul.

War is granted credit for all sorts of great social change. Democracy came from the phalanx and citizen rowers of the ancient Greeks, while the story of modern-day civil rights would not be the same without Rosie the Riveter or the African American soldiers of the Red Ball Express in World War II.

War then is depicted as immoral, yet humanity has always found out-clauses to explain its necessity and celebration. The same religions that see violence as a sin also licensed wars of crusade and jihad. And it is equally the case in politics. We repeatedly urge war as the means to either spread or defeat whatever ideology is in vogue at the time, be it enlightenment, imperialism, communism, fascism, democracy, or even simply "to end all wars."

This paradox continues in American politics today. Avoidance of war has been a traditional tenet of our foreign policy. Yet we have been at war for most of our nation's history and many of our greatest heroes are warriors. We are simultaneously leaders of weapons development, being the creators of the atomic bomb, and the founders of arms control, which seeks its ban.

We are repulsed by the idea of war, and yet entranced by it. In my mind, there are two core reasons for humankind's seeming obsession with war. The first is that war brings out the most powerful emotions that define what it is to be human. Bravery, honor, love, leadership, pity, selflessness, comradeship, commitment, charity, sacrifice, hate, fear, and loss all find their definitive expressions in the fires of war. They reach their ultimate highs and lows, and, in so doing, war is almost addictive to human culture. As William James put it, "The horror is the fascination. War is the strong life; it is life in extremis."

The other reason that war so consumes us is that for all humanity's advancement, we just can't seem to get away from it. After nearly every war, we cite the immense lessons we learned that will prevent that calamity from repeating itself. We say over and over, "Never again." Yet the reality is "ever again."

## "THE FUTURE AIN'T WHAT IT USED TO BE"

If humanity's fascination comes down to how war reveals its best and worst qualities, this book comes from wrestling with a new contradiction in warfare that humanity finds itself hurtling toward. We embrace war, but don't like to look to its future, including now one of the most fundamental changes ever in war.

Mine is a generation that, as one analyst put it, is "producing more history than it can consume." With all the focus over orange alerts and Iraq, it is tough to take a step back and notice some of the tidal shifts we are living through. For example, in my lifetime, computers went from an oddity to omnipresent. I still remember when my dad first took me to the local science museum at the age of eight to see what a computer looked like. You could only communicate with it through an obtuse "Basic" language, a sort of evil technologic shorthand. As I recall, the only useful thing we could do with that early computer (I think it was a Texas Instruments) was design a smiley face made out of hundreds of the letter M, which we then printed on one of those old spool printers that you had to tear the paper off the edges. Today, the last thing my wife does before she goes to bed is check her e-mail on a handheld computer linked up wirelessly to a shared global server, all the while brushing her teeth. In the blink of an eye for history, something revolutionary happened.

Computers seem overwhelming enough, but over the last few years I became more and more convinced that my generation was living through something perhaps even more momentous. From the robot vacuum cleaner that patrols my floors to the unmanned planes that my friends in the air force use to patrol Iraq, humanity has started to engineer technologies that are fundamentally different from all before. Our creations are now acting in and upon the world without us.

I certainly cannot claim to be the only one to see these changes. Bill Gates, the world's richest man and perhaps most responsible for the spread of computers, for instance, describes robotics today as being where the computer industry was around 1980, poised to change the way we think about what technology can do for us. "As I look at the trends that are now starting to converge, I can envision a future in which robotic devices will become a nearly ubiquitous part of our day-to-day lives.... We may be on the verge of a new era, when the PC will get up off the desktop and allow us to see, hear, touch and manipulate objects in places where we are not physically present."

By the end of 2007, a United Nations report found that there were 4.1 million robots around the world working in people's homes (as vacuum cleaners and the like). That is, there were more robots than the entire human population of Ire-land. The same study found that this "personal" robotics industry had a current market value around $17 billion. What is more important than the raw numbers is the trajectory of the growth. In 2004, the number of personal robots in the world was estimated at 2 million. By 2007, it had doubled. Another 7 million more were expected to be bought by the end of 2008.

Looking forward, many see the numbers expanding at higher rates. By 2010, one technology research group predicts there will be 55.5 million personal robots in the world. This would be just the start. Indeed, in South Korea (human population 49 million), the Ministry of Information and Communication has announced plans to put a robot in every home by 2013. Here in the United States, it will likely take a little longer. One industry leader projects 2014 as the year by which 10 percent of the Ameri-can population will have some form of personal robot in their household.

Robots are also showing up at work, from the more than forty-five hundred drones doing crop-dusting on Japanese farms to what many males would perhaps find the most disturbing example of mechanized outsourcing, the robot that han-dles security access at the offices of Victoria's Secret. Indeed, assembly-line factory robotics is an $8 billion a year industry, growing at a 39 percent pace in the United States. This is not good news for everyone, of course, as it has put many blue-collar workers out of work, most notably at carmakers. Roughly one of every ten workers in automobile manufacturing is now a robot, and Toyota has announced a plan to eventually automate all its factories.

These trends project an industry that many analysts believe is poised for a breakout. Future Horizons, a technology research group based in Kent, England, describes how "the electronics industry is on the cusp of a robotics wave." Many even think that by 2025, the robotics industry might rival the automobile and com-puter industries in both dollars and jobs. *BusinessWeek* summed up the future of the industry as "A Robotics Gold Mine."

To put it another way, the robots that had once only populated my action figure collection were now becoming all too real. Science fiction appeared to be turning into science reality.

## THE PARADOX OF THE FUTURE AND OF WAR

If robots were starting to appear in almost every aspect of life, I began to won-der how they would matter for war and politics. It is with some trepidation that

I made such a seeming leap of reason. Indeed, people have long peered into the future and then gotten it completely and utterly wrong. My favorite example took place on October 9, 1903, when the *New York Times* predicted that "the flying machine which will really fly might be evolved by the combined and continuous efforts of mathematicians and mechanicians in from one million to ten million years." That same day, two brothers who owned a bicycle shop in Ohio started assembling the very first airplane, which would fly just a few weeks later.

Similarly botched predictions frequently happen in the military field. General Giulio Douhet, the commander of Italy's air force in World War I, is perhaps the most infamous. In 1921, he wrote a best-selling book called *The Command of the Air*, which argued that the invention of airplanes made all other parts of the military obsolete and unnecessary. Needless to say, this would be news both to my granddaddy, who sailed out to another world war just twenty years later, and to the soldiers slogging through the sand and dust of Iraq and Afghanistan today.

The result is yet another paradox. It is completely normal to look forward into the future in realms like science, business, or even the weather. But forecasts of the future and, even more important, serious explorations of the changes that might result from such a future are generally avoided in the study of war. People play it safe and the gatekeepers of the field often try to knock down anything that feels too unfamiliar.

My own first experience with this was when I began my research on private military firms. A senior professor thereupon informed me that I would do well to quit graduate school and instead "go become a screenwriter in Hollywood," for thinking to waste his time on such a fiction as companies providing soldiers for hire. I still wonder how he squares this worldview with the 180,000 private military contractors now deployed in Iraq. A similar thing happened when I first presented my early research on the problem of child soldiers. A professor at Harvard University told me that she didn't believe child soldiers existed and that I was "making it up." Today there are some 300,000 children at war around the globe, fighting in three out of every four wars.

The irony is that while we accept change in other realms, we resist trying to research and understand change in the study of war. For example, the very real fear about what the environment will look like as far away as 2050 has driven individuals, governments, and companies alike to begin (belatedly) changing their practices. Yet we seem willing to stay oblivious to the changes that will come well before then for war, even though, just like the changes in global climate, we can already see the outlines of the transformation under way.

Each time I perused the Sharper Image catalog or read a report mentioning a drone taking out a terrorist camp in Afghanistan, I felt myself living at the time of the most important weapons development since the atomic bomb. One could even argue that the rise of these digital warriors is more significant, in that robotics alters not merely the lethality of war, but the very identity of who fights it. The end of humans' monopoly on war surely seemed something momentous, which historians would talk about centuries from now, if humankind is so lucky to still be around.

Yet for something so seemingly important, no one was talking about it. Time and again, I was struck by this disconnect. For example, as I describe later in the book, I once went to a major conference in Washington, D.C., on "the revolution in military affairs." The speakers included many of the most notable scholars in the field, as well as several key political and military leaders. And yet, over the course of several hours of pontificating on what was supposedly new and exciting in security issues today, not one mention was made of these new technologies, not even a single word.

Another time I arrived at the airport, facing a long flight but having forgotten a book to read on the plane. So I picked up one of those potboiler paperback novels at the bookstore. It turned out to be a courtroom drama about the suspicious murder of a beautiful scientist. About halfway through the flight, I read one of the characters describe the scientist's work. "Genetics, nanotechnology, and robotics.... It has the capacity to replace the NBCs of the last century—nuclear, biological and chemical. In its own way the potential is much more insidious. There is always a downside. The other side of the coin of progress. Some people don't want to take the chance. You can see why. The question is: How do you stop it? How do you put the genie of knowledge back in the bottle?"

And with that, I thought, a fictional character in a cheesy crime novel had just put more thought into the future of war than pretty much the entire set of real-world university political science departments, think tanks, and the foundations that fund them.

This lack of study began to disturb and fascinate me more and more. A failure to research, understand, and weigh the changes going on around us could only lead to bad results for our politics and policies. Yet some of the most important changes in the wars of today and tomorrow were either not talked about at all or merely tossed aside, as one military expert put it, to the categories of "science fiction and futurism."

And that didn't seem right. My fear also began that while all this change was incredibly exciting, it was also somewhat terrifying. We seemed to be repeating past cycles of only dealing with a huge change after the fact, when the genie was already out of the bottle. And thus somewhere along the way from reading military books in my grandfather's study to playing with lightsabers in the backyard, I decided that the serious questions that surround robotics in war and what happens when humankind's monopoly over it is broken were worthy of study. *Wired for War* is the result.

## YOUR MISSION, SHOULD YOU CHOOSE TO ACCEPT IT

When we look back at history, one thing that stands out is how the truly momentous events were often missed. When Gutenberg invented the printing press, no one held a parade. Likewise, when Hitler decided to give up his painting career, no one thought to convince him to give selling crappy watercolors just one more try, in lieu of a bid for world domination. This is all the more difficult in our present chaotic news environment of talk shows, blogs, webcasts, and so forth. As one writer put it, "The true watersheds in human affairs are seldom spotted amid the tumult of headlines broadcast on the hour."

If my growing sense was that we are in the midst of something important, maybe even a revolution in warfare and technology that will literally transform human history, then my aim in the book's research quickly became to capture that incredible moment. Imagine, I thought, if we had been able to wrestle with the great changes that atomic bombs brought to politics while they were being invented, rather than waiting to puzzle our way through their implications years later. Moreover, as I set off on my research path, I quickly learned that what was impossible in 1945 is possible now. This revolution is not occurring in secret desert test facilities, but playing out right in front of us.

My goal then became to write a book built on careful scholarship, resting on hard-core research, not speculation or exaggeration. I hoped to offer not only an entry point into this exciting and unnerving change, but also a 360-degree view of what was going on, a resource that could prove useful to leaders and public alike, both today and tomorrow. And yet I also hoped to bring readers the same sense of wonder and amazement that originally drove me on this journey.

If you haven't noticed by now, this will be a book somewhat different from the normal look at either war or technology. It's a product of who I am and the forces that shape me. I am the kid who played with Transformers who now consults for the military. I am a scholar who studied under Sam Huntington, one of the most distinguished political scientists of the twentieth century, and yet I am shamefully addicted to watching *The Real World*. The eloquence and brilliance of authors like John Keegan and Jared Diamond inspire me, and yet the writer I read most religiously is Bill Simmons, the irreverent "Sports Guy" columnist for ESPN, who blogs on the finer points of the NBA Draft and *The Bachelor* dating show.

Fortunately, the topic I am wrestling with is located where warfare, history, politics, science, business, technology, and popular culture all come together. So unlike most books on war or politics, this one isn't aimed at just one audience. The issues of robotics and war are so compelling and important that people with all sorts of interests and backgrounds can and should dig into them. In other words, you will find references of both sorts. There will be references of the scholarly type, pointing to the sources of the data, and there will be references of the pop culture type, pointing to parallels and lessons in mass media.

As such, the research for this book involved a melding of methodologies. I spent nearly four years seeking out anything and everything useful that I could find on the topic, whatever the source. I checked out musty old history books that hadn't left the library in years. I scoured the last twenty years of each U.S. military branch's professional journals, printing table-high stacks of any article relevant to the topics of war, technology, leadership, and change. I searched the online archives of all the major technology journals. Indeed, I even spent several wonderful days cruising through the "Wookipedia," the Internet hub for all things *Star Wars*.

The comedian Stephen Colbert famously said, "I don't trust books. They're all fact, no heart." So I also made sure to interview anyone I could find who brought an important or unique perspective to the issues. I sought out the ideas of robot scientists and weapons developers, professors and journalists, human rights activists and science fiction writers, as well as the men and women now using these new technologies to fight. Rank mattered less to me than what they could bring to the issue. I quizzed four-star generals and secretaries of the army, navy, and air force, as well as nineteen-year-old specialists at the bottom of the chain of command. I met with pilots of robotic drones, who had never left the United States, and special operations soldiers just back from missions in Iraq and Afghanistan. Robots have no one nationality, so my interviews also became quite international.

I gathered the views of everyone from German army officers and an Indian news editor to a set of Iraqi insurgents. Where possible, these individuals are identified, but sometimes the interviewees chose not to be named, which I have respected in the citations.

Sometimes these interviews would take place in person and sometimes via e-mail or phone. The research took me from robot factories and military bases around the world to one interview of an Arab general from the backseat of his BMW 7 series luxury car, discussing robot strike scenarios as we tooled about town. A hotel conference room oddly enough turned out to be the most dangerous of all these research locales. While I was observing a meeting of robotics developers and their military counterparts, a rogue robot tried to run me over. A demonstration model brought in by one of the developers, the little bugger was programmed to patrol around the room but avoid contacting any humans. But it just kept on coming and coming and nearly broke my foot. In my mind, it makes me a bit like the nerdy version of John Connor, hunted down by a machine obviously sent to keep you from reading about our robotic future.

## THE END OF THE BEGINNING

As you work your way through the chapters, then, you will likely note a few things. Some admittedly may be a bit different from the traditional book that originates at a public policy "think tank."

The first is that there is a mix of traditional hard data (numbers, statistics, and the like) as well as a heavy dose of untraditional anecdotes. In turn, the chapters tend to weave together scores of "characters," rather than following just one throughout the book. My research and interviews brought in an amazing and colorful cast of people at work in this field. In the translation of research to book, I didn't want to lose this human side. The reason is that, ironically for a topic on nonhuman changes in war, the stories and personalities are what tell us the most about where we are today, as well as where we are heading tomorrow. Or as one scientist described, "These robots are extensions of us."

The use of vignettes, personalities, and anecdotes also has a methodological rationale. It is not just a more effective way of giving readers a true "feel" of what is going on and capturing our historic moment in time. It is also an echo of the strategy used by ethnographers, who collect individual stories and anecdotes to discern broader trends and conclusions. Indeed, so much of what I learned, as well as how we tend to communicate to each other, came via storytelling that it seems only fitting to share many of these stories in the context of the issues. Indeed, one story may be an anecdote, but a collection of them is data.

Second, the book deals with the future and thus has to be, in part, predictive or conceptual. As the earlier examples showed, this is no easy task. The prognostications of nonscientists often fail because they frequently don't pay close attention to what is technically feasible or not. In turn, scientists' predictions tend to overstate the positive, especially when it comes to war. Franklin, Edison, Nobel, and Einstein, for example, all thought that their inventions would end war. They knew the science, but not the social science. Both groups tend to disregard how social change can intervene and intertwine with technology, yielding not one definite future, but rather many possible futures.

Researchers have found that these three problems can be diminished by relying on actual facts rather than hopes or fears, having a firm technical and social science footing, making conclusions built on sound reasoning, and being sure not to ignore the doubts of the skeptics. This book follows those lessons. For instance, you'll read here only about technologies either operating now or already at the prototype stage. I steer clear of the imaginary ones fueled by the Klingon power packs, dragon's blood, or the hormones of teenage wizards.

Third, for a book supposedly on the future, there is a lot of history. My sense is not that history repeats itself, but that there are patterns and lessons that we can draw from, a key way to ground any look forward. There will be much change in the future of war, but also much continuity.

Fourth, nothing in this book is classified information. I only include what is available in the public domain. Of course, a few times in the course of the research I would ask some soldier or scientist about a secretive project or document and they would say, "How did you find out about that? I can't even talk about it!" "Google" was all too often my answer, which says a lot both about security as well as what AI search programs bring to modern research.

Fifth, the book makes many allusions to popular culture, not something you normally find in a research work on war, politics, or science. Some references are obvious and some are not (and thus the first reader to send a complete list of them to me at www.pwsinger.com will receive a signed copy of the book and a Burger King *Transformers* collectible). It is also, as far as I know, the first book to come out of a think tank with a recommended music playlist, designed to get into the vibe of the research results, also available on the Web site.

The reason for this different approach is not simply to break the mold, or rather mould, of scholarly style or to give heart attacks to the old guard with my generation's manner of thinking and writing, even on important issues like war. Rather, as much as we pointy-headed scholars hate to admit, this is how people process information most efficiently. Humankind has long best understood and digested things that are new by flavoring them with stories of personal experience ("There was this one time, in band camp, where we...") as well as by allusions to what is already culturally familiar, especially icons, symbols, and metaphors ("It's just like when..."). And, whether we like it or not, our twenty-first-century folklore is that of the popular movies, TV shows, music, gadgets, and books that shaped us growing up.

You have now been dutifully warned of what may come. I hope you find the results of this journey simultaneously interesting, educational, and maybe even a bit scary. In other words: frakin' cool.

# PART ONE

# THE CHANGE
# WE ARE CREATING

# ONE

# INTRODUCTION:
# SCENES FROM A ROBOT WAR

*We are building the bridge to the future while standing on it.*

—U.S. ARMY COLONEL

There was little to warn of the danger ahead. The Iraqi insurgent had laid his ambush with great cunning. Hidden along the side of the road, the bomb looked like any other piece of trash or scrap metal. American soldiers call these jury-rigged bombs "IEDs," official shorthand for improvised explosive devices. The team hunting for the IED is called an Explosive Ordnance Disposal, or EOD, team; they are the military's bomb squads.

Before Iraq, the EOD teams were not much valued by either the troops in the field or their senior leaders. They usually deployed to battlefields only after the fighting was done, to defuse any old weapons caches or unexploded ammunition that might be found. It was dangerous work, but not one that gained the EODs much acclaim. But in Iraq, the IED quickly became the insurgents' primary way of lashing back at U.S. forces. In the first year of the fighting, there were 5,607 road-side bomb attacks. By 2006, the insurgents were averaging nearly 2,500 a month.

Cheap and easy to make, IEDs took a grievous toll, becoming the leading cause of casualties among American troops as well as Iraqi civilians. They also limited the ability of U.S. forces to move about safely and carry out their missions, such that the commanding general quickly determined that among all the myriad problems in Iraq, "IEDs are my number one threat." In response, the Pentagon soon was spending more than $6.1 billion to counter IEDs in Iraq.

The EOD teams were tasked with defeating this threat, roving about the battlefield to find and defuse the IEDs before they could explode and kill. The teams went from afterthought to, as one journalist put it, "one of the most important assignments on the battlefield." In a typical tour in Iraq, each team will go on more than six hundred IED calls, defusing or safely exploding about two IEDs a day. Perhaps the best sign of how critical the EOD teams became is that the insurgents began offering a rumored $50,000 bounty for killing an EOD team member.

Unfortunately, this particular mission would not end well. By the time the soldier had advanced close enough to see the telltale explosive wires of an IED, it was too late. There was no time to defuse the bomb and no time to escape. The IED erupted in a wave of flame.

Depending on how much explosive the insurgent has packed into an IED, a soldier must be as far as fifty yards away to escape death and even as much as half a mile away to escape injury from the blast and bomb fragments. Even if you are not hit, the pressure from the blast can break your limbs. This soldier, though, had been right on top of the bomb. Shards of metal shrapnel flew in every direction at bullet speed. As the flames and debris cleared and the rest of the team advanced, they found little left of the soldier. Hearts in their throats, they loaded the remains onto a helicopter, which took them back to the base camp near Baghdad International Airport.

Writing home after such an incident may be the toughest job for a leader. That night, the team's commander, a U.S. Navy chief petty officer, did his sad duty. The effect of this explosion had been particularly tough on his unit. They had lost their most fearless and technically savvy soldier. More important, they had also lost a valued member of the team, a soldier who had saved the others' lives many times over. The soldier had always taken the most dangerous roles, scouting ahead for IEDs and ambushes. Despite this, the other soldiers in the unit had never once heard a complaint.

This being a war in the age of instant communication, there was no knock on the door of some farmhouse in Iowa, as is always the case in the old war movies. Instead, the chief's letter was sent via e-mail. In his condolences, the chief noted the soldier's bravery and sacrifice. He apologized for his inability to change what had happened. But he also expressed his thanks and talked up the silver lining to the tragedy. As the chief wrote, "When a robot dies, you don't have to write a letter to its mother."

## CLEANING FLOORS AND FIGHTING WARS

The destination of the e-mail was a gray, concrete, two-story office building located in a drab corporate park just outside Boston. On the corner of the building is the sign for the soldier's maker, a company called iRobot.

The complex is an outgrowth of the Burlington Shopping Mall, so across the street are a Men's Wearhouse discount suit store and a Macaroni Grill, a faux Italian restaurant chain known less for its pasta than the fact that it allows you to crayon on your tablecloth. It may seem an odd cradle for the future of war, but then again, no one standing outside a bicycle shop in Dayton, Ohio, some hundred years back thought, "Ah yes, this must be the home of a new era of package tourism, lost luggage, and strategic bombing."

The inside of iRobot is just like any other office building, with drab colors on the walls and mind-numbing rows of cubicles filled with staff punching away at their keyboards. The difference at iRobot is that the obligatory corporate boardroom doubles as a small museum of robots and every so often a loud thump comes from a robot crashing into the wall. When I arrive for my visit, some of the employees are testing out a tracked robot, driving it down the hallway with a jury-rigged Xbox video game controller. Think *Office Space* crossed with Asimov.

iRobot was founded in 1990 by three MIT computer geeks, Colin Angle, the CEO, Helen Greiner, the chairman of the board, and Rodney Brooks, their former professor, who doubled as chief technical officer. Brooks was already considered one of the world's leading experts on robotics and artificial intelligence, Greiner would eventually be named one of "America's Best Leaders" by *U.S. News & World Report*, and Angle's work would become so influential that his undergrad thesis paper ended up at the Smithsonian. iRobot, though, was no sure thing at the start. There was no real market for robots, the company's first home was Angle's living room, and their CEO's only previous job had been as a summer camp counselor.

iRobot the company took its name from *I, Robot* the Isaac Asimov science fiction novel (later made into a Will Smith movie). Asimov laid out a vision in which humans of the future share the world with robots. His fictional robots carry out not only mundane chores but also make life-and-death decisions.

The real-world firm started out slowly with some small-scale government contracts and several attempts at robotic toys. Its first robot was Genghis, a tiny bot designed to scramble across the surfaces of other planets for NASA. On the toy front, it tried to sell a doll that laughed when tickled and a robot dinosaur, a velociraptor inspired by the movie *Jurassic Park*. None of their products made a splash. "We were the longest overnight success story ever," said Greiner.

In time, iRobot developed two products that would make its mark on the world. The first is Roomba, the first mass-marketed robotic vacuum cleaner. Roomba is a disc-shaped robot thirteen inches in diameter and just over three inches high. It is basically a large Frisbee that roams about the floor, automatically vacuuming it clean. Roomba's sensors figure out the size and shape of your room, and with the push of the "clean" button it goes to work. Indeed, Roomba is smart enough to avoid falling down the stairs and even knows how to return to its charger when the power is running low. Roomba actually evolved from Fetch, a robot that the company designed in 1997 for the U.S. Air Force. Fetch cleaned up cluster bomblets from airfields; Roomba cleans up dust bunnies under sofas. Released in 2002, Roomba became a media darling, appearing in everything from the Sharper Image catalog to the *Today* show, and soon was one of the most sought-after gadgets to give at Christmas.

iRobot's other breakout product was PackBot, the "soldier" blown up by that IED in Iraq. PackBot came out of a contract from the Defense Advanced Research Projects Agency, or DARPA, in 1998. Weighing forty-two pounds and costing just under $150,000, PackBot is about the size of a lawn mower. It is typically controlled via remote control, although it can drive itself, including even backtracking to wherever it started its mission. PackBot moves using four "flippers," essentially tank treads that can rotate on one axis. These allow PackBot not only to roll forward and backward like regular tank tracks, but also to flip its tracks up and down to climb stairs, rumble over rocks, squeeze down twisting tunnels, and even swim in under six feet of water. The tracks are made of a hard rubberlike polymer that iRobot patented. They are specially designed to be used on any surface, ranging from the mud of a battlefield to the tiled floor of an office building.

The designers at iRobot look at their robots as "platforms." PackBot has eight separate payload bays and hookups that allow its users to swap in whatever they need: mine detector, chemical and biological weapons sensor, or just extra power packs. The EOD version of the PackBot that served in Iraq comes with an extendable arm on top that mounts both a head, containing a high-powered zoom camera, and a clawlike gripper. Soldiers use these to drive up to IEDs, peer at them closely, and then, using the gripper, disassemble the bomb, all from a safe distance.

PackBot made its operational debut on the fateful day of September 11, 2001. With all air traffic grounded after the destruction of the World Trade Center, engineers from iRobot loaded their robots into cars and drove down to help in the rescue and recovery efforts at Ground Zero. A *New York Times* article entitled "Agile in a Crisis, Robots Show Their Mettle" described them as "rescuers [that] are unaffected by the carnage, dust and smoke that envelop the remains of the World Trade Center. They are immune to the fatigue and heartbreak that hang in the air."

Soon after, PackBot went to war. As U.S. forces deployed to Afghanistan, troops came across massive cave complexes that had to be scouted out, but were often booby-trapped. The only specialized tool the troops had were flashlights, and they had to crawl through the caves on hands and knees. Usually, the GIs would send their local Afghan allies down into the caves first, but as one soldier put it, "We began to run out of Afghans." iRobot was then asked by the Pentagon to send help. Just six weeks later, PackBots made their debut in a cave complex near the village of Nazaraht, in the heart of Taliban territory. iRobot was now at war.

With both the Roomba and PackBot becoming hits (the first test robots sent to Afghanistan were so popular with the troops, they wouldn't let the company take them back), the business that had started in a living room took off. In the next five years, the company's revenue and profits grew by a factor of ten. By 2007, more than three million Roombas had been sold at over seven thousand retail stores. On the military side, the war robot business grew by as much as 60 percent a year, culminating in a $286 million Pentagon contract in 2008 to supply as many as three thousand more machines. The PackBot was in such demand that the space reserved for it in iRobot's museum was empty when I visited the offices. The display model had been deployed to Iraq.

With these successes under its belt, iRobot was ready for the big time. It entered the stock market, with its IPO underwritten by two of the most prestigious investment houses in the world, Morgan Stanley and J.P. Morgan. On the first day of trading, iRobot's public value hit $620 million. At the market's close, a PackBot rang the bell at the New York Stock Exchange, the first robot ever to do so.

## THE iROBOT WAY

iRobot's business model splits its sales effort between a consumer division that targets robots for the home and a government and industrial robots division that mainly targets the military. The military business currently makes up about a third of revenue, but market analysts are "really excited by it" and predict it will soon become about half the company's revenue. iRobot also has a vibrant research team led by Andrew Bennett, who was part of the team that raced to New York on 9/11. This group lays the groundwork for future advances, and has some fifty patents either approved or pending.

This split in iRobot's customer base can make for some amusement. iRobot may be the only company that sells at both Pentagon trade shows and Linens 'n Things. In the customer testimonials section of its Web site, the chief's letter about his robot in Iraq is just below one from "Janine," a housewife from Connecticut. While he talked about how his robot saved lives in battle, she thanked the company because "I have four boys and two cats and this little 'robot' keeps my rugs and hardwood floors dirt and hair free!"

The firm plans to continue to advance the frontiers of cleaning floors and fighting wars. It has followed up the Roomba with Scooba, which washes and scrubs floors, and the Dirt Dog, a heavy-duty cleaner designed for sucking up nuts and bolts in workshops and off factory floors. The online advertisements for the robots tell potential buyers, "You've done enough; leave the cleaning to a robot."

"One of our challenges is getting people who aren't familiar with the product and who haven't really thought about robots being real before to give it a shot," says Colin Angle, iRobot's CEO. "This is all very new stuff. We're continually trying to find new ways of helping people get over the skepticism to really imagine how robots in their lives could really be helpful." Indeed, with less than 1 percent of U.S. households owning cleaning robots, "the demographics of our purchasers suggest we're just scratching the surface of what's possible," says Angle. iRobot recently launched a multimillion-dollar advertising campaign called "I Love Robots" that shows people talking about their robots and the work they do.

On the military side, iRobot has similar dreams of growth. It has new and improved versions of the PackBot as well as a host of plans to convert any type of vehicle into a robot, be it a car or ship, using a universal control unit that you plug into the engine and steering wheel. One new robot that iRobot's designers are especially excited to show off is the Warrior. Weighing about 250 pounds, the Warrior is essentially a PackBot on steroids. It has the same basic design, but is about five times bigger. Warrior can run a four-minute mile for five hours, while carrying 100 pounds. Yet it is agile enough to fit through a doorway and go up stairs. iRobot built the robot, even though, as one designer put it, "there are no clear buyers yet...we don't know yet just who will use it." The firm is essentially using the *Field of Dreams* model: if they build it, the buyers will come.

Warrior is really just a mobile platform, with a USB port on top. USB ports are the universal connectors used to plug anything into a computer, from your mouse to a printer. With the USB on Warrior, users can hook up whatever they want to their robot, whether it be sensors, a gun, and a TV camera for battle, or an iPod and loudspeakers for a mobile rave party. The long-term strategy then is that other companies will focus on the plug-in market, while iRobot corners the market for the robotic platforms. What Microsoft did for the software industry, iRobot hopes to do for the robotics industry.

With this kind of grand vision, its rapid growth, and immense financial backing, iRobot may well be on its way to becoming the Ford or GE of the twenty-first century. Indeed, Asimov's book that inspired its name tells the fictional history of a small company, "U.S. Robotics," that becomes the largest corporation in the world within fifty years of its founding.

It sure sounds exciting, but also comes with a catch. iRobot, the company, may well be ignoring the warnings of *I, Robot* the book. Isaac Asimov is remembered not merely for his vision of the future, but also for his "Three Laws of Robotics" that supposedly guided robots' development in his fictional world. The laws are so simple, yet so complex in their implications, that ethicists now teach them at colleges in the real world. Asimov's first and most fundamental law is: "A robot may not harm a human being, or, through inaction, allow a human being to come to harm."

It is hard to square the fictional rules with the present reality of a company at war. Some argue that Asimov would definitely not approve of the latest plug-in accessory for the PackBot, a shotgun. The folks at the company think such people are missing the point; the firm is leading a thrilling technologic revolution. When Helen Greiner is asked how Asimov might react to iRobot, she responds, "I think he would think it's cool as hell."

## ENGINEERING THE COMPETITION

Just a twenty-minute drive from iRobot's offices outside the Burlington Mall is an old industrial park in Waltham, Massachusetts. Here, in a complex of brown concrete-block buildings dating back to the 1950s, is the headquarters of the Foster-Miller company.

Like iRobot, Foster-Miller was founded by MIT graduates. Eugene Foster and Al Miller were engineers who shared an office at MIT and did consulting work on the side. After graduating, Al Miller left, never to be heard from again. His office replacements were Charles Kojabashian and Edward Nahikian. Foreign-sounding last names weren't a big selling point in 1955, so the trio continued the work under the name of Foster-Miller Associates. A year later, Foster-Miller opened its shops in Waltham.

Foster-Miller makes the PackBot's primary competitor, the Talon, which first hit the market in 2000. The Talon looks like a small tank, driven by two treads that run its length. Weighing just over a hundred pounds, it is a bit bigger than the PackBot. It too has a retractable arm with a gripper, but mounts its main sensors on a separate antennalike "mast" sticking up from the body and carrying a zoom camera. Talon can go up to speeds of about 5.5 mph, the equivalent of a decent jog on a treadmill, a pace it can maintain for five hours.

Like the PackBot, the Talon helped sift through the wreckage at Ground Zero and soon after deployed to Afghanistan. And like iRobot, Foster-Miller has boomed, doubling the number of robots it sells every year for the last four years. The company received an initial $65 million in orders for Talons in the first two years of the insurgency in Iraq. By 2008, there were close to two thousand Talons in the field and the firm had won a $400 million contract to supply another two thousand. Under an additional $20 million repair and spare parts contract, the company also operates a "robot hospital" in Baghdad. Foster-Miller now makes some fifty to sixty Talons a month, and repairs another hundred damaged systems.

The similarities between the two firms end there. iRobot was started by researchers, focused on invention. iRobot's facilities are mostly cubicles in a large office building, as it outsources much of the manufacture of its robots to factories in the Midwest and China. In its lobby, the name of each visitor is displayed on a flat-screen television mounted on the wall of the reception area.

Foster-Miller was founded by engineers, focused on the practical end. Foster-Miller's headquarters are a complex of more than two hundred thousand square feet of offices, labs, and machine shops. It makes most of its products on site. In its lobby, visitors are greeted by an elderly receptionist, who then announces your arrival on one of those old intercom microphones that I last saw at my elementary school.

In the back of the Foster-Miller complex is a large warehouse that you enter from an employee parking lot. In the corner, men tinker at various high tables loaded with machinery. A large American flag hangs from the ceiling. It all looks like Santa's workshop for robots, crossed with a car company advertisement.

When I toured Foster-Miller's shop in the autumn of 2006, more than twenty-five Talons were lined up on the floor. In one row were shiny new Talons ready for shipment to Iraq. In a second row were dinged-up robots back from the war for repair, their arms a bit mangled and with burnt scars on various parts. I noticed some scorched paper stuck to one of the bots. Edward Godere, a vice president at Foster-Miller, explained, "The soldiers have started taping *Playboy* centerfolds to the side of the robots. It's the twenty-first-century version of the pin-up art on the bomber planes during World War II."

The two companies also see the world quite differently. iRobot, as its research team leader Andrew Bennett puts it, "is all about robotics." Still a research firm at heart, it has little interest in other industrial sectors and turns down opportunities that it sees as "boring." As one of the researchers put it, "We don't build Buicks."

Helen Greiner goes even further. "These robots are on a mission and so are we: to bring robots into the mainstream.... We can make robots to do a better job than humans in some cases." The result is a fairly unique corporate mission statement: "Have Fun, Make Money, Build Cool Stuff, Deliver a Great Product, and Change the World."

Whereas iRobot just works on robotics, Foster-Miller makes everything from armor for tanks to air conditioners for gold mines. At Foster-Miller, the motto is, "We engineer ideas into reality," and there is no interest in trying to change the world via inventions. For example, Foster-Miller's Talon gets its night vision from simply slipping a soldier's night vision goggles over the robot's camera and drives on tank treads that originally came from a snowmobile. By contrast, iRobot's PackBot drives on specially designed tracks that took nine months to develop and originally came with so much artificial intelligence software that the army actually asked iRobot to make PackBot dumber by stripping out some of the programs.

Foster-Miller is also noticeably more comfortable in its relationship with the Pentagon than iRobot seems to be. It is, as its vice president Bob Quinn puts it, "a defense firm at heart," with roughly 90 percent of its business defense and security related. Or, as one Foster-Miller executive put it frankly, "We're industrialists looking for needs to meet. You gotta follow the money."

As the market takes off, however, Foster-Miller is finding more and more of its business identity in the robotics sector. Moreover, it is bleeding over to other jobs. For example, the company has a long history of engineering for the navy. The navy wants to reduce the number of personnel on its ships because one fewer sailor on board saves $150,000 a year in operating costs. So Foster-Miller came up with a design for an "automated galley" that would go into the latest warship. The system starts with sailors ordering their meal in advance by computer. A management system then allocates the food onboard to their preferences and the meal is transferred from storage by a robot. It is then cooked largely by automation and sent down via a "hot food" robot to a service station, where each sailor picks it up.

In looking at the potential futures of the two firms, it is also notable that they have vastly different ownership structures. iRobot is publicly held, meaning anybody with an online account can buy a slice of its future. Many believe this will drive it more and more toward widening its role in consumer products to balance out the defense growth. By contrast, Foster-Miller is privately held and expresses no interest in consumer robotics. Indeed, in 2004 it was bought by QinetiQ for $163 million. QinetiQ is a multibillion-dollar partnership between the Defence Evaluation and Research Agency (the British government's defense labs, which were privatized in 2001) and the Carlyle Group.

Carlyle is one of those quietly influential firms that conspiracy theorists love. It is the only large private equity firm located in Washington, D.C., and oversees some $44 billion in equity capital. Its members and advisers include former secretary of state James A. Baker III, former secretary of defense Frank C. Carlucci (who was also the college wrestling partner of then secretary of defense Donald Rumsfeld), former White House budget chief Richard Darman, former British prime minister John Major, and former president George H. W. Bush. This "who's who" is certainly enough fodder for the conspiracy theorists. Raising eyebrows even further, the bin Laden family was one of the Carlyle Group's investors. They made out pretty well; the *Wall Street Journal* reported the family got 40 percent annual returns on its investments in Carlyle. In one of those stranger-than-fiction moments, on the very morning the hijacked planes smashed into the World Trade Center, the Carlyle Group was holding its annual investor conference, with Shafiq bin Laden, the brother of Osama bin Laden, in attendance.

The two companies feel a keen sense of competition with each other, and with their close distance, tensions are certainly there. At iRobot, researchers describe their rivals as thinking, "We hear that robots are trendy, so let's do that." At Foster-Miller, they retort, "We don't just do robots and we don't suck dirt."

The two companies have even become locked in a bit of a marketing war. If robots were pickup trucks, Foster-Miller represents the Ford model, stressing how the Talon is "Built Tough." Its promotional materials describe the Talon as "The Soldier's Choice." They repeatedly mention its ruggedness, and even make a point to highlight an e-mail from a marine in Iraq, who wrote of his unit's Talon, "I wouldn't use anything else over here."

The executives at Foster-Miller love to recount tales of how the Talon has proven it "can take a punch and stay in the fight." One Talon was riding in the back of a Humvee while the truck was crossing a bridge. The unit was ambushed and an explosion blew the Talon into the river. After the battle ended, the soldiers found the damaged control unit and drove the Talon right out of the river. Another Talon serving with the marines was once hit by three rounds from a .50-caliber heavy machine gun (meaning the robot was actually a victim of friendly fire), but still kept working. The repair facility in Waltham has even worked on one Talon that was blown up on three separate occasions, each time just giving it new arms and cameras.

The iRobot team bristles at the idea that their systems are "agile but fragile." They insist that the PackBot is tough too, but being more science-oriented, cite various statistics on how it can survive a 400 g-force hit, what they describe as the equivalent of being tossed out of a hovering helicopter onto a concrete floor. They are most proud of the fact that their robots have a 95 percent out-of-the-box reliability rate, higher than any others in the marketplace, meaning that when the soldiers get them in the field, they can trust the robot will work as designed.

Beneath all the difference and rancor that divides the companies, they are similar in one telling way. The hallways and cubicles of both their offices are covered with pictures and thank-you letters from soldiers in the field. A typical note from an EOD soldier reads, "This little guy saved our butts on many occasions."

## THE KILLER APP

For all its talk of eschewing new inventions in lieu of simple solutions, Foster-Miller is where matters get even more revolutionary. Just down from the workshop repair room of Talons sits what *Time* magazine called one of the "most amazing inventions of the year." In technology circles, new products that change the rules of the game, such as what the iPod did to portable music players, are called "killer applications." Foster-Miller's new product gives this phrase a literal meaning.

Like the PackBot, the Talon comes in all sorts of different versions, including EOD, reconnaissance, and a hazmat (hazardous materials) robot. The real killer app, though, is its SWORDS version. This robot's name comes from the acronym for Special Weapons Observation Reconnaissance Detection System. SWORDS is the first armed robot designed to roam the battlefield.

The SWORDS is basically the Talon's pissed-off big brother, with its gripping arm replaced with a gun mount. Akin to a Transformers toy made just for soldiers, SWORDS is armed with the user's choice of weaponry. The robot's mount can carry pretty much any weapon that weighs under three hundred pounds, ranging from an M-16 rifle and .50-caliber machine gun to a 40mm grenade launcher or an antitank rocket launcher. In less than a minute, the human soldier flips two levers and locks his favorite weapon into the mount. The SWORDS can't reload itself, but it can carry two hundred rounds of ammunition for the light machine guns, three hundred rounds for the heavy machine guns, six grenades, or four rockets. One report on SWORDS declares that "with this increased firepower, soldiers and their robots will be able to wreak absolute havoc on the battlefield."

Unlike the PackBot, SWORDS has very limited intelligence on its own, and is remote-controlled from afar by either radio or a spooled-out fiber optic wire. The control unit comes in a suitcase that weighs about thirty pounds. It opens up to reveal a video screen, a handful of buttons, and two joysticks that the soldier uses to drive the SWORDS and fire its weapons. At the time of my visit, Foster-Miller was exploring replacing the controller with a Nintendo Game Boy–style controller, hooked up to virtual reality goggles.

The operator sees what SWORDS sees through five cameras mounted on the robot: a target acquisition scope linked to the weapon, a 360-degree camera that can pan and tilt, a wide-angle zoom camera mounted on the mast, as well as front and rear drive cameras. With these various views, the operator can not only see as if they have eyes in the back of their head, but farther than had previously been possible when shooting a gun. As one soldier put it, "You can read people's nametags at 300 to 400 meters, whereas the human eye can't pick that up. You can see the expression on his face, what weapons he is carrying. You can even see if his [weapon's] selector lever is on fire or on safe." The cameras can also see in night vision, meaning the enemy can be fired on at any hour and in any climate. This capability has gained added appeal in current operations; during the 2003 invasion of Iraq, three days of sandstorms shut down U.S. forces.

The inspiration for the SWORDS is generally credited not to a scientist, but to a soldier, army sergeant first class David Platt. Platt first used the Talon while sifting through the wreckage at the World Trade Center and later at EOD tasks. His thinking behind giving the robot a gun was fairly straightforward: "It's small. It's quiet, and it goes where people don't want to be."

In keeping with Foster-Miller's philosophy, converting the Talon to SWORDS was a "bootstrap development process." It only took six months and less than $3 million to make the first prototype. As one of the developers put it, "It is important to stress that not everything has to be super high tech. You can integrate existing componentry and create a revolutionary capability." Guided by that ethic, these lethal little gunslingers cost just $230,000.

Napoleon once said, "There are but two powers in the world, the sword and the mind. In the long run, the sword is always beaten by the mind." The invention of the SWORDS might one day invalidate his statement. In an early test of its guns, the robot hit the bull's-eye of a target seventy out of seventy tries. In a test of its rockets, it hit the target sixty-two out of sixty-two times. In a test of its antitank rockets, it hit the target sixteen out of sixteen times. A former navy sniper summed up its "pinpoint precision" as "nasty."

The robot's zoom lens not only extends the shooter's sight, but matches it exactly to the weapon's. Rather than trying to align their eyes in exact symmetry with the gun in their hand, it is as if the soldier's eagle eye was the gun. The weapon also isn't cradled in the soldier's arms, moving slightly with each breath or heartbeat. Instead, it is locked into a stable platform. As army staff sergeant Santiago Tordillos says, "It eliminates the majority of shooting errors you would have."

The robot can be set to fire either one bullet at a time or in bursts of eight bullets. Since it is a precisely timed machine pulling the trigger, the "one shot" mode means that any weapon, even a machine gun, can be turned into a sniper rifle. Finally, it makes no difference to the robot whether it is at the shooting range or in the middle of a firefight; the situation does not affect its accuracy. "The SWORDS doesn't care when it's being shot at. Indeed, it would like you to shoot at it," says Sergeant Platt. "That way we can identify you as a valid target and engage you properly."

The arming of SWORDS has opened up a host of new roles for robotic systems on the battlefield beyond just bomb disposal. Missions so far for what Fox News called the "G.I. of the 21st century" include street patrols, reconnaissance, sniping, checkpoint security, as well as guarding observation posts. It is especially attuned for urban warfare jobs, such as going first into buildings and alleyways where insurgents might hide. SWORDS's inhuman capabilities could well result in even more intrepid missions. For example, the robot can drive through snow and sand and even drive underwater down to depths of one hundred feet, meaning it could pop up in quite unexpected places. Likewise, its battery allows it to be hidden somewhere in "sleep" mode for at least seven days and then wake up to shoot away at any foes. Described one report of the robotic gunner, "They have been a hit with the soldiers."

## THE WAR BEYOND BOSTON

The PackBot, Talon, and SWORDS are only a few of the many new unmanned systems that are operating in war today. When U.S. forces went into Iraq, the original invasion had zero robotic systems on the ground. By the end of 2004, the number was up to 150. By the end of 2005, it was up to 2,400. By the end of 2006, it had reached the 5,000 mark and growing. It was projected to reach as high as 12,000 by the end of 2008.

The unmanned systems roaming about Iraq come in all sorts of shapes and sizes. One of the smallest, but most commonly used, is the MARCBOT (Multi-Function Agile Remote-Controlled Robot). MARCBOT looks like a toy truck with a video camera mounted on a tiny antennalike mast. Costing only $5,000, the tiny bot is used to scout out where the enemy might be and also to drive under cars and search for hidden explosives. Many soldiers are so used to driving remote-controlled cars growing up that it typically takes less than an hour to learn how to use the system. MARCBOT isn't just notable for its small size. The little truck actually drew first blood on the battlefield, even before SWORDS. One unit of soldiers jury-rigged their MARCBOTs to carry a Claymore antipersonnel mine. Whenever they thought an insurgent was hiding in an alley, they would send a MARCBOT down first, not just to scout out the ambush, but to take them out with the Claymore. Of course, each insurgent found meant $5,000 worth of a blown-up robot's parts, but so far the army hasn't billed the soldiers.

All told as of 2008, some twenty-two different robot systems were operating on the ground in Iraq. As one retired army officer put it, "The Army of the Grand Robotic is taking place."

The world of unmanned systems at war doesn't end at ground level. They have also taken to the air. One of the most notable is the Predator. The Predator is a UAV (unmanned aerial vehicle), or drone, that "looks like a baby plane." At twenty-seven feet in length, it is just a bit smaller than a Cessna, and is powered by a "pusher" propeller in the back. Unlike most planes, the Predator lacks a cockpit and its tail wings are canted downward, instead of the normal sideways; one observer even said it looked like "a flying meat fork." Since it is made of composite materials instead of metals, the Predator weighs just 1,130 pounds. Perhaps its best quality is that it can spend some twenty-four hours in the air, flying at heights of up to twenty-six thousand feet.

Each Predator costs just under $4.5 million, which sounds like a lot until you compare it to the cost of other military planes. Indeed, for the price of one F-22, the air force's latest jet, you can buy eighty-five Predators. More important, the low price and lack of a human pilot means that the Predator can be used for missions where it might be shot down, such as traveling low and slow over enemy territory. About a quarter of the cost of the Predator actually goes into the "Ball," a round mounting under the nose of the drone. The rotating Ball carries two variable-aperture TV cameras, one for seeing during the day and an infrared one for night, as well as a synthetic-aperture radar that allows the Predator to peer through clouds, smoke, or dust. The exact capabilities of the system are classified, but soldiers say they can read a license plate from two miles up. It also carries a laser designator to lock on to any targets that the cameras and radar pick up.

Predators are flown by what are called "reachback" or "remote-split" operations. While the drone flies out of bases in the war zone, the pilot and sensor operator for the plane are physically located seventy-five hundred miles away, connected with the drone plane only via satellite communications. Their control panels look a bit like one of the 1980s' two-player video games you used to see at arcades, each sitting behind three TV screens (one screen has a video feed of what the drone is seeing, one displays technical data, and the third is the navigation map, akin to the GPS display in a car).

The Predator has thus introduced not only tactical but also organizational changes in the units that use them. The mechanics and ground crew go with the plane to the battle zone, usually an "undisclosed location," shorthand for a base in an allied state in the Persian Gulf. The pilots flying the planes remain in the United States, working out of a set of converted single-wide trailers. Most of these trailer parks for robots are located at Nellis and Creech air force bases, just outside of Las Vegas and Indian Springs, Nevada. But as trailer parks tend to do, these drone bases are multiplying. There are plans to start up Predator operations at bases in Arizona, California, New York, North Dakota, and Texas.

Predators originally were designed for reconnaissance and surveillance, flying over enemy territory to scout for targets and monitor the situation. The prototypes were first used in the Balkan wars, but truly entered their own after 9/11. Indeed, in the first two months of operations in Afghanistan, some 525 targets were laser-designated by Predators. The generals, who had once had no time for such systems, couldn't get enough of them. Tommy Franks, the commander of all U.S. forces in the region at the time, declared, "The Predator is my most capable sensor in hunting down and killing Al Qaeda and Taliban leadership and is proving critical to our fight."

"Our major role is to sanitize the battlefield," says Service Airman Medric Jones. "We . . . make sure our own guys aren't walking into danger." A typical reconnaissance operation in Iraq involves the Predator circling over a city like Baghdad from five miles up. The pilots communicate with commanders, flight coordinators, intelligence teams (who might be located in the region or back in the States), and even troops on the ground via e-mail or radio. Sometimes, they send out the Predator's live feed via "Rover," a remote video system that transmits what the Ball is seeing to Panasonic notebook computers carried by the troops on the ground.

If the enemy is spotted, the Predator can also orchestrate the attack, pointing its laser at targets and even warning the troops if there are any "squirters," bad guys running away. "I can watch the rear of a building for a bad guy escaping when troops go in the front, and flash an infrared beam on the guy that our troops can see with their night-vision goggles," said U.S. Air Force Major John Erickson. Erickson's experience is illustrative of the changes. He had been an F-16 pilot, but when he tells his grandkids about what he did in the Iraq war, it will be about the eighteen months he spent flying a Predator, never leaving the ground.

Predators don't just watch from afar, but have also begun to kill on their own. The backstory of how this happened is one of the sad "what ifs?" of what could have been done to prevent the 9/11 attacks. Over the course of 2000 and 2001, Predators operated by the CIA sighted Osama bin Laden in Afghanistan many times, usually when he was driving in a convoy between his training camps. But the unmanned spy planes were toothless and could only watch as bin Laden soon disappeared.

The idea then arose to arm the drone by mounting laser-guided Hellfire missiles on the wings. Since the Predator already could direct missiles at targets with its lasers, the only difference is that the drone would carry its own, instead of having to rely on the kindness of strangers to blow up those below. The Predator would truly become a predator.

The plan made sense but quickly got mired in bureaucratic politics, as the CIA and air force argued over who would have control over the now-armed drones and, most important, whose budget would be stuck with the $2 million in costs. It seems like a small amount in retrospect, a "shoestring operation," according to the air force general in charge of the effort. But, as he now laments, "it was a big problem, I hate to say." With the two agencies at loggerheads, a senior White House official was needed to cut through the dispute. But terrorism was not at the top of the priority list of the new Bush administration. The issue of how to deal with bin Laden and the growing warnings of an attack inside the United States was tabled until everyone got back from their summer vacations. "It saddens me to know we could have done a heck of a lot more," says the officer.

After 9/11 and the more than three thousand people killed, the issue of $2 million became null and void. The CIA armed its Predators and the air force decided that it couldn't be left behind. In the first year, armed Predators took out some 115 targets in Afghanistan on their own. Many commented on the oddity of a war where many of the forces still rode to battle on horses, and yet robotic drones were flying above. In the words of one U.S. officer, it was "the *Flintstones* meet the *Jetsons*."

Predators continue to operate over Afghanistan today. Frequently, the drones carry an American flag onboard, which is then given to the family of a soldier killed in the fighting. In one case, a Predator carrying a flag for a family actually took out the same group of Taliban that had killed their son.

With the precedent set in Afghanistan, the Predator also joined the fight in Iraq. Among its first missions was to help take down the Iraqi government's television transmissions, which broadcast the infamous "Baghdad Bob" propaganda. In the days and weeks that followed, the Predator struck at everything from suspected insurgent safe houses to cars being prepped for suicide attacks.

The ugly little drone has quickly become perhaps the busiest U.S. asset in the air. From June 2005 to June 2006, Predators carried out 2,073 missions, flew 33,833 hours, surveyed 18,490 targets, and participated in 242 separate raids. Even with this massive effort, there is demand for more. Officers estimate that they get requests for some 300 hours of Predator imagery a day, but that there are only enough Predators in the air to supply a little over 100 hours a day. The result is that the Predator fleet has grown from less than 10 in 2001 to some 180 in 2007, with plans to add another 150 over the next few years.

Besides the Predator, there are many other drones that fill the air over Iraq and Afghanistan. At some forty feet long, Global Hawk could be described as the

Predator's big brother. Others uncharitably say it looks like "a flying albino whale." The Global Hawk was originally conceived as the replacement for the U-2 spy plane, which dates back to the 1950s. Besides not putting a human pilot in harm's way (the U-2 is perhaps most famous for the crisis over the downing of pilot Francis Gary Powers at the height of the cold war), "physiological factors" limited the amount of time that the U-2 pilots could fly missions (that is, they would pass out from fatigue, boredom, or a buildup in their kidneys). In contrast, Global Hawk can stay in the air up to thirty-five hours. Powered by a turbofan engine that takes it to sixty-five thousand feet, the stealthy Global Hawk carries synthetic-aperture radar, infrared sensors, and electro-optical cameras. Working in combination, these sensors can do a wide-area search to look at an entire region, or focus in on a single target using the "high-resolution spot mode." The link of the sensors with the long flight time means that the drone can fly some three thousand miles, spend twenty-four hours mapping out a target area of some three thousand square miles, and then fly three thousand miles back home. In other words, Global Hawk can fly from San Francisco, spend a day hunting for any terrorists in the entire state of Maine, and then fly back to the West Coast.

Like the Predator, the Global Hawk is linked back to humans on the ground, but it mainly operates autonomously rather than being remotely piloted. Using a computer mouse, the operator just clicks to tell it to taxi and take off, and the drone flies off on its own. The plane then carries out its mission, getting directions on where to fly from GPS (Global Positioning System) coordinates downloaded off a satellite. Upon the return, "you basically hit the land button," describes one retired air force officer.

With such capability, the Global Hawk is not cheap. The plane itself costs some $35 million, but the overall support system runs over $123 million each. Even so, the U.S. Air Force plans to spend another $6 billion to build up the fleet to fifty-one drones by 2012.

At the smaller end of the scale in Iraq and Afghanistan are unmanned planes flown not out of air force bases back in the United States, but rather launched by troops on the ground. The big army units fly Shadow, which looks like the sort of radio-controlled planes flown by model plane hobbyists. Just over twelve feet long, it takes off and lands like a regular plane. Compared to a Predator or Global Hawk, however, it is underpowered, only able to stay up five hours and fly seventy miles. Driven by a propeller, it has a distinctive noise that sounds like a weed-whacker flying overhead. Most of the Shadow's UAV pilots are enlisted soldiers, such as Private First Class Ryan Evans, who explains why he volunteered to fly robotic planes in lieu of performing his normal army duties. "It is more of a rush that you are in control of something in the sky."

The most popular drone, though, is one of the smallest. The Raven is just thirty-eight inches long and weighs four pounds. In a sort of irony, soldiers launch the tiny plane using the same over-the-shoulder motion that the Roman legionnaires used in war two thousand years ago, just tossing a robot instead of a javelin. The Raven then buzzes off, able to fly for ninety minutes at about four hundred feet. Raven carries three cameras in its nose, including an infrared one. Soldiers love it because they can now peer over the next hill or city block, as well as get their own spy planes to control, rather than having to beg for support from the higher-ups. "You throw the bird up when you want to throw it. You land it when you want to land," says Captain Matt Gill, a UAV company commander with the 82nd Airborne Division. The other part of the appeal is that the pilots of the Raven are just regular soldiers; a cook from the 1st Cavalry is actually considered among the best. In just the first two years of the Iraq war, the number of Ravens in service jumped from twenty-five to eight hundred.

A veritable menagerie of unmanned drones now circles above the soldier in Iraq, reporting back to all sorts of units. The small UAVs like Raven or the even smaller Wasp (which carries a camera the size of a peanut) fly just above the rooftops, sending back video images of what's on the other side of the street. The medium-sized ones like Shadow circle over entire neighborhoods, at heights above fifteen hundred feet, and are tasked out by commanders at division headquarters to monitor for anything suspicious. Reporting back to pilots thousands of miles away, the larger Predators roam above entire cities at five thousand to fifteen thousand feet, combining "reconnaissance with firepower." Finally, sight unseen, the Global Hawks every so often zoom across the entire country at some sixty thousand feet, monitoring anything electronic and capturing reams of detailed imagery for intelligence teams to sift through. Because they rarely see the Global Hawks, officers in the field joke that these pictures are mainly used to fill the PowerPoint briefings for the generals back in D.C. Added together, by 2008, there were 5,331 drones in the U.S. military's inventory, almost double the amount of manned planes. That same year, an air force lieutenant general forecast that "given the growth trends, it is not unreasonable to postulate future conflicts involving tens of thousands."

The reach of unmanned systems also extends to the sea. REMUS, the Remote Environmental Monitoring Unit, is helping to clear Iraqi waterways of mines and explosives. Shaped like a torpedo, REMUS is about six feet long, weighs eighty-eight pounds, and costs $400,000. It was originally built by the Woods Hole Oceanographic Institute to carry out automated surveys of coasts, reefs, and shipwrecks, but the navy soon modified it for military uses, purchasing more than 140 of the undersea robots by 2008.

It is another modification of an unmanned system originally designed for sea that may be the most novel to hit the battlefield in Iraq and Afghanistan. Second behind the threat of IEDs is that from mortars and rockets. Insurgents will often set up a mortar or rocket launcher in a residential neighborhood, quickly pop off a few rounds at an unsuspecting U.S. base, and then get out of the area before any response can be made. Although most miss their targets, plenty of damage and many casualties have been caused by lucky shots.

Enter the Counter Rocket Artillery Mortar technology, or CRAM for short. The navy has long equipped many of its ships with the Phalanx 20mm Gatling gun, capable of firing up to forty-five hundred rounds per minute. The radar-guided gun is mounted in a cylindrical shell that tilts and moves in circles, such that the sailors affectionately call it "R2-D2," after the little robot in *Star Wars*. The gun was designed as a "last chance" defense against antiship missiles that skim just above the waves. R2-D2 automatically tracks and shoots down any missiles that have gotten past all other defenses and are too quick for humans to react to.

CRAM is basically R2-D2 taken off the ship and crammed (mounted) onto a flatbed truck. Its software was modified to target mortar shells and rockets instead of missiles, with the idea that it would essentially put up a wall of bullets to protect bases. Tests showed that CRAM had a 70 percent shootdown capability. By December 2007, at least twenty-two of them were deployed to Afghanistan and Iraq.

Not all has gone perfectly with the CRAM. The original, naval version of R2-D2 used bullets made of depleted uranium. As they were intended to fall into the middle of the sea, no one worried much about what happened to the shells after they fired. In an urban environment, thousands of bullets filled with radioactive dust falling from the sky is more of a concern. So the shells had to be altered to incendiary rounds that blow up in midair, but are less effective. Also, R2-D2 apparently once mistook an American helicopter flying over Baghdad for the Emperor's Death Star. It locked in on the chopper to shoot it down, as if it were a rocket with some funny rotors spinning on the top. So CRAM had to be reconfigured to avoid any "blue on blue" friendly fire incidents. Finally, R2-D2 does not come cheap. Once you count in all the radar and control elements, the CRAM required a congressional earmark of $75 million in funding.

## THE NEW WARRIOR AT HOME

The "war on terrorism" hasn't just taken place on battlegrounds far far away. The result has been the creation of immense bureaucracies and massive spending dedicated to this war at home, or what we now call "homeland security."

A few numbers illustrate the vast industry that has been built around homeland security. In 1999, there were nine companies with federal contracts in homeland security. By 2003, there were 3,512. In 2006, there were 33,890. The business of protecting buildings, borders, and airports and preparing to respond to disaster generates $30 billion a year and is projected to reach $35 billion by 2011. As one report on the homeland security industry put it, "Thank you, Osama bin Laden!"

This money has not just been spent on the Einsteins who seize your shampoo at airports, but also on new technology research for homeland security. *Popular Science* reported that it "reached heights not seen since the Sputnik era." In 2003, $4 billion of the newly formed Department of Homeland Security's budget went to technology research programs.

The outcome is that unmanned systems have also started to serve on the front lines of the war at home. One of the early scares in the war on terrorism was the rash of letters carrying deadly anthrax powder sent to prominent officials and media. Some of the powder also leaked out inside post offices. Since those attacks, some one thousand robots have been installed to sort parcels, with the U.S. Postal Service planning to add as many as eighty thousand more.

With the war on terror involving the need to protect everything from airports to office buildings, industry analysts also foresee a booming market for "sentry-bots." These systems can guard entrances, automatically patrol perimeters, check IDs, and even use facial recognition software to know who should or shouldn't be allowed into the area. Examples of such systems range from the Guard Robo made by Sohgo Security Services, which looks like Rosie, the maid from *The Jetsons*, to the Robot Guard Dog made by Sanyo. It looks, well, like a robot guard dog, just with a video camera for eyes and a mobile phone mounted inside to call for help whenever it finds intruders. An executive I met at one robotics conference predicted that "we will sell tens of thousands of them to everything from military bases to power plants."

With America under threat, robots haven't just hidden out in post offices or passively stood guard. They also have taken flight to guard the nation's borders. While the Predator was originally designed for the military to find enemy missiles and tanks, the federal government quickly became interested in its potential for another role. Through most of 2005 and 2006, the Department of Homeland Security flew a Predator drone over the U.S.-Mexico border. The robot border-cop helped arrest 2,309 people and seize seven tons of marijuana.

In 2008, DHS presented plans to Congress to buy eighteen drone planes to patrol the U.S. border. Of course, all realize that the drones are actually focused on stopping a different type of border crosser than al-Qaeda agents—illegal immigrants. "But the acceptability of using these systems for border surveillance has increased dramatically since terrorism became such a real, in-our-backyard threat," says Cyndi Wegerbauer of General Atomics, which sold the Predator drone to the Border Patrol.

Indeed, in the war to defend against would-be immigrants, robots have also gone to work not only for the government, but also for the private border patrols, or "militias," as some have called themselves. One example is the "Border Hawk" drones serving with the American Border Patrol, a private organization operating in Cochise County, Arizona.

Some have accused the American Border Patrol of racism. Its founder, Glenn Spencer, is certainly a controversial figure. He describes illegal immigration as "The Second Mexican-American War" and Latin America as "a cesspool of a culture" that threatens the "death of this country."

Spencer may sound like a sad throwback to the 1950s or even 1350s, but his group's technology is twenty-first century. They operate three drones that carry video and infrared cameras. The drones are launched by radio control and then automatically fly a patrol pattern using GPS, staying at four hundred feet, just below what the government requires for certification. While in the air, they search out any illegal immigrants crossing the border and record the images to TiVo for playback and review. The group doesn't arrest the illegal aliens themselves, but passes on the information to the United States Border Patrol as well as loads its robots' footage onto the Internet using a satellite connection, or, as the group describes, "broadcasting the invasion live on the internet."

Besides battling terrorists and would-be immigrants, the war at home also involves responding to disaster. In the aftermath of 9/11, brave little PackBots and Talons joined the search for survivors. In the aftermath of Hurricane Katrina, Silver Fox UAVs searched for survivors in flooded areas of New Orleans, while two tiny robotic helicopters from the Center for Robot-Assisted Search and Rescue at the University of South Florida worked on the Mississippi coast. Many think robotic systems will have an even wider role in future disasters. For example, after Katrina, cell phone towers went out because of storm damage and a lack of power, which hampered both residents on the ground as well as rescue efforts. During the next disaster, the plan is to use a UAV as an "aerial cell tower."

## THE REST OF THE STORY

PackBot, Talon, SWORDS, Predator, Global Hawk, and all their digital friends are the first signs that something big is going on. Man's monopoly of warfare is being broken. We are entering the era of robots at war.

It sure sounds like science fiction to claim such a wild thing. But we have to remember that pretty much everything we now take for granted sounded like fiction at some point, whether it was the fantastic dreams of mechanical flying beasts to the absurd concept of talking to someone on the other side of the world.

What follows is an effort to understand this change, to travel through this new world of unmanned war and unwrap just what it might mean. Part 1 attempts to capture this moment of great change, to understand the changes that we are creating. In order to assess what is going on in technology, robotics, and war today, it will explore such key issues as the history of robots, how these new technologies work, what is coming in the next wave, who is working on them, and what inspires them. Then, part 2 of this book will explore what all this change is creating for us. It will cover everything from the resulting shifts in how wars are fought and who is fighting them to important questions that our new machine creations are starting to raise in politics, law, and ethics. War just won't be the same.

# SMART BOMBS, NORMA JEANE, AND DEFECATING DUCKS: A SHORT HISTORY OF ROBOTICS

*The further backward you look, the further forward you can see.*

—SIR WINSTON CHURCHILL

"Perhaps the most wonderful piece of mechanism ever made" is how the famous Scottish engineer Sir David Brewster would describe it some one hundred years after it was invented. By contrast, the great poet Johann Wolfgang von Goethe called it "most deplorable...like a skeleton [with] digestive problems." The two men were talking about Vaucanson's duck, the mechanical wonder of its age, or, as present-day scientists call it, "the Defecating Duck."

Jacques de Vaucanson was born in Grenoble, France, in 1709. At the age of twenty-six, he moved to Paris, then the center of culture and science during the Age of Enlightenment. Inspired by Isaac Newton's idea of the universe as a great clock that had been set in motion by the Creator, the Deist philosophers of the time saw the world as guided by mechanical forces. They believed that everything, from gravity to love, could be understood if you could just scientifically reason it out.

Arriving in this cauldron of rationality gone wild, Vaucanson became fascinated with the concept of using reason and mechanics to reproduce life itself.

More important, needing funds, the young engineer hit upon the idea of "getting assistance by producing some machines that could excite public curiosity." So he did what any other enterprising young man would do: he built a duck.

Vaucanson's duck was no ordinary duck; it was actually an intricate mechanical creation modeled after a sculpture in the gardens of the Palais des Tuileries, a cultural center at the time, now more famous as one of the sites of *The Da Vinci Code*. While the duck looked lifelike from the outside, the true amazement was that it could stand up, sit down, preen, waddle, quack, eat pellets of corn, drink water, and then, wonder of wonders, defecate. Claiming that he had made the duck with methods "copied from Nature," Vaucanson presented the mechanical fowl at the court of King Louis XV. The duck then became the talk of all the Paris salons, as the nation's leaders debated how it worked and just what it signified for politics, philosophy, and life itself.

Once the duck was placed on public display, people came from far and wide, paying an admission fee equivalent to a week's wages. Also accompanying the bird were mechanical mandolin, flute, and piano players, who tapped their feet, moved their heads, and seemed to breathe as they played music. But it was the duck and, most important, the inexplicable fact that it could do number two that was the star attraction. The duck seemed to show that the incomprehensible processes of life could be re-created.

Vaucanson became a rich man and soon thereafter was given the highest possible honor for a scientist, election to the esteemed Académie des sciences, joining such luminaries as Descartes, Colbert, and Pascal. The duck was then sent out on tour (where the German poet Goethe would meet it some years later, showing its age like all great stars do when they've been on the road for too long), and Vaucanson would become the director of the French government's silk mills. In 1745, he would invent the world's first automated loom, which used a system of cards with holes punched in them to repeatedly create patterns in silk. Centuries later, these punch cards would inspire the early developers of computers.

It wasn't until four decades after its invention that the duck's secret was discovered. It was, in fact, unable to digest food. The corn that was seemingly eaten and then digested was instead stored in a pod hidden in the back of the duck's throat, initiating a timer that would then, after a suitable pause, release another hidden container of "artificial excrement." The duck, and its waste, were both frauds. But if Vaucanson's duck was a hoax at re-creating life, it was a remarkably intricate one. The blueprints for the mechanical bird show it to involve hundreds of moving, interlocking parts and scores of inventions, all for the sole purpose of simulating the most routine part of life's daily business.

## THE QUEST FOR ARTIFICIAL LIFE

Vaucanson's duck is relevant today because it illustrates how humankind's attempts to use technology to mimic and replace life go further back than we often think. The robots searching for IEDs in Iraq didn't just spring out from nowhere. They have a past that shapes their present and future.

The idea of creating mechanical beings to replace the work of humans is at least as old as ancient Greek and Roman mythology. For example, the god of metalwork (Hephaestus to the Greeks, Vulcan to the Romans) had a host of mechanical servants that he made out of gold. Occasionally, he also gave out his creations to the mortals, one example being Talos, a huge statue that protected the island of Crete by throwing huge boulders at any ships that came nearby. If any stranger made it ashore, Talos would heat up his metallic arms to a red-hot glow and then give the intruder a deadly welcome hug. Talos was later the name for an Apple computer operating system, as well as the first computer-guided missiles on U.S. Navy ships.

These myths were not just stories, but became inspirations for both real-world philosophers and inventors. Indeed, it was in this period that Aristotle (384–322 B.C.), one of the founding philosophers of Western thought, would describe his vision of the ultimate free world: "If every tool, when ordered, or even of its own accord, could do the work that befits it...then there would be no need either of apprentices for the master workers or of slaves for the lords."

Likewise, the engineers of ancient times made advances that were often well beyond what we might think possible. Around 350 B.C., the Greek mathematician Archytas of Tarentum built "the Pigeon," a mechanical bird that was propelled by steam. Besides building what was likely the world's first model airplane, Archytas used it to carry out one of the first studies of flight. Perhaps most remarkable was the "Antikythera computer." In A.D. 1900, a Greek sponge diver found a wreck of an ancient Greek sailing ship that had sunk off the island of Antikythera near Crete around 100 B.C. In the wreck was a small box about the size of a laptop computer. It contained thirty-seven gears that, when a date was entered, worked to calculate the position of the sun, moon, and other planets. Many credit it as the first known mechanical analog computer.

This fascination with mechanical creations subsided during the Dark Ages, but would rise again in the Renaissance, perhaps most famously with Leonardo da Vinci. Among his many sketches is a mechanical knight. Like most of his flashes of brilliance, such as his plans for helicopters and planes, the design was ahead of its time. If built, this sixteenth-century version of the SWORDS, armed with a sword, would have been able to sit up and move its arms and legs. The fascination with such systems, though, was not limited to Europe. In feudal Japan in the 1600s, several craftsmen were noted for having made automated dolls that served tea.

For all the wonder of these early mechanical creations, though, it is important to note that they were not actually what we now think of as robots. The devices typically did the same thing every time they were activated, rather than moving about or responding to any changes in the environment. That is, they were automated, but not robotic. Moreover, many turned out to be hoaxes, either elaborate ones like Vaucanson's duck that actually pushed the frontiers of technology in the pursuit of fakery, or more traditional ones. The most famous of this latter type may have been "The Turk," later referenced in the Terminator series. This was a "chess automaton" in the shape of a Turkish-looking figure on top of a cabinet, made by Wolfgang von Kempelen in what is now Slovakia. Preceding IBM's chess-playing supercomputer Deep Blue by almost two hundred years, "The Turk" consistently beat humans at chess, including even Napoleon. It turned out, though, that von Kempelen had hidden a dwarf chessmaster inside.

Ducks and Turks aside, most of the research to develop technologies that replicated human powers was frequently intertwined with war. Archimedes, for example, may have been the most influential scientist in ancient history, shaping the future development of the fields of mathematics, physics, engineering, and astronomy. In his era, though, he was best known for his various inventions used in the defenses of the city of Syracuse. These ranged from a "death ray" (supposedly using mirrors to light ships afire) to a huge "claw" (a large crane that grabbed ships). Similarly, the field of modern chemistry was largely founded by Antoine-Laurent Lavoisier, who served Louis XVI as one of the directors of the French national commission on gunpowder.

In no area was this link greater than in the first calculating machines, the forerunners of computers. Charles de Colmar is credited with inventing the first mechanical calculator, which he called the Arithmometer, in 1820. The machine was as big as a desk. His first customers were the French and British militaries, which used it for navigation and plotting the trajectory of cannonballs. Similarly,

the Royal Navy hired Charles Babbage, the man generally credited with designing the first programmable computer. Babbage's 1822 machine, called a "difference engine," was designed of some twenty-five thousand parts. In a foretaste of the
innovators of today, Babbage was also a bit of an oddball. He once baked himself in
an oven for four minutes, just "to see what would happen."

## ROBOTS GO TO WAR

Ultimately, technology caught up with ambition around the turn of the twentieth
century. Science finally had advanced to create machines that could be controlled
from afar and move about on their own. The robotic age was getting closer, and
robots' link with war would become even more closely intertwined.

The first real efforts started with Thomas Edison and Nikola Tesla, two rival
scientists and the first of what we now would call electrical engineers. While working on various ways to transmit electricity, Edison and Tesla both experimented
with radio-control devices. Because of his eccentric personality and lack of a good
public relations team like Edison, Tesla would not gain the same place in history as
his rival, the "Wizard of Menlo Park," and died penniless.

Tesla, though, did perhaps the most remarkable work at the time with remote-
control devices. He first mastered wireless communication in 1893. Five years later,
he demonstrated that he could use radio signals to remotely control the movements
of a motorboat, holding a demonstration at Madison Square Garden. Tesla tried to
sell this first remotely operated vehicle, along with the idea of remote-controlled
torpedoes, to the U.S. military, but was rejected. As Tesla recounted, "I called an
official in Washington with a view of offering the information to the government
and he burst out laughing upon telling him what I had accomplished." Tesla would
not be the last inventor to find out that what was technically possible mattered
less than whether it was bureaucratically imaginable. Two brothers from Dayton,
Ohio, had the same experience a few years later when they first tried to sell their
invention of manned flight.

The foundations then were laid for remote-controlled vehicles and weapons
just as the First World War began. World War I proved to be an odd, tragic mix
of outmoded generalship combined with deadly new technologies. From the
machine gun and radio to the airplane and tank, transformational weapons were
introduced in the war, but the generals could not figure out just how to use them.
Instead, they clung to nineteenth-century strategies and tactics and the conflict was

characterized by brave but senseless charges back and forth across a no-man's-land of machine guns and trenches.

With war becoming less heroic and more deadly, unmanned weapons began to gain some appeal. On land, there was the "electric dog," a three-wheeled cart (really just a converted tricycle) designed to carry supplies up to the trenches. A precursor to laser control, it followed the lights of a lantern. More deadly was the "land torpedo," a remotely controlled armored tractor, loaded up with one thousand pounds of explosives, designed to drive up to enemy trenches and explode. It was patented in 1917 (appearing in *Popular Science* magazine) and a prototype was built by Caterpillar Tractors just before the war ended. In the air, the first of what we would now call cruise missiles was the Kettering "Bug" or "aerial torpedo." This was a tiny unmanned plane that used a preset gyroscope and barometer to automatically fly on course and then crash into a target fifty miles away. Few of these remote-controlled weapons were bought in any numbers and most remained prototypes without any effect on the fighting.

The only system to be deployed in substantial numbers was at sea. Here, the Germans protected their coast with FL-7s, electronically controlled motorboats. The unmanned boats carried three hundred pounds of explosives and were designed to be rammed into any British ships that came near the German coast. Originally, they were controlled by a driver who sat atop a fifty-foot-high tower on shore, steering through a fifty-mile-long cable that spooled out of the back of the boat. Soon after, the Germans shifted the operator from a tower onto a seaplane that would fly overhead, dragging the wire. Both proved unwieldy, and in 1916 Tesla's invention of wireless radio control, now almost two decades old, was finally deployed in warfare.

Perhaps reflecting the fact that they were outnumbered in both these wars, the Germans again proved to be more inclined to develop and use unmanned systems when fighting began again in World War II. The best known of their weapons, akin to the land torpedo, was called the Goliath. About the size of a small go-cart and having a small tank track on each side, the Goliath of 1940 was shaped almost exactly like the Talon that Foster-Miller makes over six decades later. It carried 132 pounds of explosives. Nazi soldiers could drive the Goliath by remote control into enemy tanks and bunkers. Some eight thousand Goliaths were built; most saw service as a stopgap on the Eastern Front, where German troops were outnumbered almost three to one.

In the air, the Germans were equally revolutionary, deploying the first cruise missile (the V-1), ballistic missile (V-2), and jet fighter (Me-262). The Germans were also the first to operationally use remotely piloted drones. The FX-1400, known as the "Fritz," was a 2,300-pound bomb with four small wings, tail controls, and a rocket motor. The Fritz would drop from a German plane flying at high altitude. A controller in the plane would then guide it into the target using a joystick that steered by radio. The Fritz made a strong debut in 1943, when the Italian battleship *Roma* was trying to defect to the Allies. Not knowing of the Fritz, the Italian sailors saw a German bomber plane, but didn't worry too much as it was at a distance, height, and angle from which it couldn't drop a bomb on top of them. A Fritz launched from the bomber and then flew into the *Roma*, sinking it with more than a thousand sailors lost.

The Allies were behind the Germans in these technologies, but they were no less futuristic in some of the things they sought to develop. In the United States, the focus of research was on aerial weapons and actually led to another of the great "what ifs?" of recent history. In 1944, "Operation Aphrodite" was launched in Europe. The idea was to strip down bomber planes and load them up with twenty-two thousand pounds of Torpex, a new explosive discovered to be 50 percent more powerful than TNT. A human crew would fly the plane during takeoff, arm the explosives in midair, and bail out. A mothership flying nearby would then take remote control of the bomber and, using two television cameras mounted in the drone's cockpit, steer the plane into Nazi targets that were too well protected for manned bombers to hit.

On August 12, 1944, the naval version of one of these planes, a converted B-24 bomber, was sent to take out a suspected Nazi V-3, an experimental 300-foot-long "supercannon" that supposedly could hit London from over 100 miles away (unbeknownst to the Allies, the cannon had already been knocked out of commission in a previous air raid). Before the plane even crossed the English Channel, the volatile Torpex exploded and killed the crew.

The pilot was Joseph Kennedy Jr., older brother of John Fitzgerald Kennedy, thirty-fifth president of the United States. The two had spent much of their youth competing for the attention of their father, the powerful businessman and politician Joseph Sr. While younger brother JFK was often sickly and decidedly bookish, firstborn son Joe Jr. had been the "chosen one" of the family. He was a natural-born athlete and leader, groomed from birth to become the very first Catholic president. Indeed, it is telling that in 1940, just before war broke out, JFK was auditing classes at Stanford Business School, while Joe Jr. was serving as a delegate to the Democratic National Convention. When the war started, Joe Jr. became a navy pilot, perhaps the most glamorous role at the time. John was initially rejected for service by the army because of his bad back. The navy relented and allowed John to join only after his father used his political influence.

When Joe Kennedy Jr. was killed in 1944, two things happened: the army ended the drone program for fear of angering the powerful Joe Sr. (setting the United States back for years in the use of remote systems), and the mantle of "chosen one" fell on JFK. When the congressional seat in Boston opened up in 1946, what had been planned for Joe Jr. was handed to JFK, who had instead been thinking of becoming a journalist. He would spend the rest of his days not only carrying the mantle of leadership, but also trying to live up to his dead brother's carefree and playboy image.

The Aphrodite program was not the only remotely controlled weapons program that the Allies devised in World War II. The Brits, for example, developed what they darkly called "bombing without knowledge of path, place, or time" that used radio signals from afar to guide bombers in the dark. In the Pacific theater, more than 450 VB-1 Azons, a 1,000-pound radio-controlled glider bomb, were used to destroy targets in Burma, mainly bridges of the sort made famous in the movie *The Bridge over the River Kwai*.

The most widely produced unmanned plane in World War II, however, was used for training rather than combat. It was called the OQ-2 Radioplane, or sometimes the "Dennymite" after its maker, Reginald Denny. Denny was a British pilot during World War I, who then moved to Hollywood to become an actor. With his dashing looks and aristocratic accent, his career took off. Over the next forty years, he would appear in 172 films. The high point was his starring role opposite Greta Garbo in 1935's *Anna Karenina*, the low point perhaps his final role as "Commodore Schmidlapp" in 1966's *Batman: The Movie*.

While horsing around on set, Denny became a hobbyist of radio-controlled model airplanes. He saw a business opportunity in other fans, and so in 1934 opened Reginald Denny Hobby Shops, a model plane store located on Hollywood Boulevard. As war grew closer, Denny got the idea that cheap radio-controlled planes would make perfect targets to give more realistic training to antiaircraft gunners. In 1940, he pitched the idea of the planes, which he marketed to hobbyists as the "Dennymite," for use as a target drone. The army signed a contract for fifty-three. Then Pearl Harbor happened. Over the next five years, the army would buy another fifteen thousand drones, making the Dennymite the first mass-produced unmanned plane in history.

To build so many drones, Denny had to move his manufacturing out of Hollywood and into a plant at the Van Nuys Airport. In 1944, army photographer David Conover was sent to this factory for a magazine shoot about women contributing to the war effort. He spotted a buxom woman spraying the drones with fire retardant. It was not the most sexy of settings but he thought this woman had potential as a model and sent his photos on to a friend at a model agency. Norma Jeane Dougherty soon dyed her mousy brown hair to platinum blond and changed her name to Marilyn Monroe. After the war, the Northrop company bought out Denny, meaning that the icon of the blonde bombshell and the Global Hawk drone both were born in the same place.

More advancement was made during this period with computers and other automated systems, though, than with remote-controlled ones that went out into the world on their own. The most widely used of these automatic systems was the Norden bombsight.

Carl Norden was a Dutch engineer who moved to the United States in 1904. In 1920, he developed an analog computer that could calculate the trajectory of how a bomb would fall off a plane in flight. In a plane moving faster than three hundred feet per second, the human's reaction time was too slow to use the computer's calculation effectively, so the system automatically released the bomb at just the right time when it was sighted on a target. Norden's bombsight could even be linked to the plane's autopilot, taking over the flight controls on the final bombing run.

While it was advertised as being able to "put a bomb in a pickle barrel from twenty thousand feet," the reality was that in combat conditions, the system was a little less accurate, typically hitting targets within one hundred to one thousand feet. Even so, the Norden was far more accurate than anything before it, and was used in all the U.S.'s heavy bombers during World War II. The device was considered so valuable that it was taken out of the plane and put in a safe after each mission. If their plane was about to crash, the crew was to shoot the bombsight with a thermite gun that would melt the computer.

The cost of the Norden program was $1.5 billion, almost the same as the Manhattan Project to make the first atomic bomb. Like many of the inventors, though, the "cranky" and "irascible" Norden was a bit of an oddball and never profited to the extent he might have. He didn't like how the U.S. Army Air Corps had treated him when he had tried to sell them unmanned planes during World War I. So to get back, he sold his sight to the army's greatest nemesis, not the Japanese or the

Germans, but the U.S. Navy, for the grand price of one dollar. Throughout World War II, then, the U.S. Army had to buy its bomber sights from the U.S. Navy.

By the end of the war, the early B-17 and B-24 planes that Norden had equipped were being replaced by the far more sophisticated B-29 Superfortress. Besides the automated bombsight, the B-29 was the first plane to have a computer-controlled firing system, made up of twelve .50-caliber machine guns mounted in electric turrets, all remotely fired using an analog computer called the "Black Box." It was a B-29, the *Enola Gay*, that would use a Norden bombsight to drop the first nuclear bomb on Hiroshima.

The real breakthrough was in computers that stayed off the battlefield. The first that used programming as we now understand it was Colossus, built at the top-secret codebreakers' lab at Bletchley Park, England. Weighing a ton, Colossus had fifteen hundred electronic valves to crank out the complex mathematics needed to break the Enigma code used by the Germans.

Colossus, however, used physical switches to store data, so the first truly electronic computer was ENIAC, the Electric Numerical Integrator and Computer. Built at the University of Pennsylvania in 1944, it weighed twenty-seven tons and took up eighteen hundred square feet of floor space. While it was an unwieldy system that required the wires to be reset for each different problem, ENIAC could crunch out equations in thirty seconds that took a human engineer with a slide rule more than twenty hours. It was put to work on everything from shell trajectories to the development of the hydrogen bomb. In 1951, the first commercial version was released, and it was soon put to use at such things as predicting election results. Officially, it was termed the UNIVAC, but the media called it the "Giant Electronic Brain."

## A COLD WAR AND A COLD MARKET

This dichotomy of robotics and computers continued into the cold war. On one hand, the work on unmanned vehicles and weapons stagnated. Indeed, the new U.S. Air Force (formed from the army's Air Corps) was so uninterested in drones and guided missiles that their further development was left to the army and navy ordnance departments.

Computers, though, continued to take off, with the military at the center of their funding and development. Among the early pioneers in this period was "Amazing" Grace Hopper. Hopper was a U.S. naval officer who worked on the development of the Harvard Mark I computer made by IBM. The Mark I, which was fifty-one feet in length and had some five hundred miles of wire, is credited by many as being the first digital computer that could store numbers and automatically calculate them. The challenge for these early computer pioneers was that all the instructions for the computer had to be written out in binary code. Hopper was part of the team that developed software known as a "compiler," which essentially turned each machine's codes into something universal. This early common language was called COBOL (Common Business Language). With the U.S. military as the largest buyer in the market, it became the standard. It was a huge breakthrough. By employing the same programming language, computers were no longer limited to computer scientists, and all sorts of machines could now communicate with each other. Hopper would retire as a rear admiral. Today, she is the only mathematician in history to have a navy ship named in her honor, the guided-missile destroyer U.S.S. *Hopper*.

Many remember Hopper for something else. These early computers were difficult to use, and being so big, developed all sorts of problems in the hardware as well as software. One of the most vexing was with the Harvard Mark I's replacement (the Mark II), which kept inexplicably crashing. Eventually, Hopper's team figured out that the reason was not faulty programming, but a moth that had become trapped between two relays inside the computer. From then on, computer program glitches have been called "bugs."

These programs laid the groundwork for what we now call the Internet. In 1965, a new employee named Bob Taylor joined what was then ARPA (Advanced Research Projects Agency; soon after, "Defense" was added to the agency's name, changing it to DARPA). Taylor inherited an office that had three different computer terminals sitting in it, each linking up to a different computer in Cambridge, Massachusetts, Santa Monica, California, and Berkeley, California. Taylor asked, "Why don't we just have a network such that we have one terminal and we can go anywhere we want?"

It only took twenty minutes to convince Taylor's boss that this was a great idea. Four years and half a million dollars later, the system was ready to launch. What was especially novel is that the system didn't just link each computer directly to all the others, but instead figured out that messages could be passed on between computers by using interface message processors. Each of these processors, which we now commonly call "routers," weighed half a ton.

In October 1969, the new "Darpanet" was ready to go, with the first message remotely connecting with a computer from UCLA. Prefacing the future way that technology doesn't always work as planned, the world's first computer network crashed on the "g" of the "login" command. Soon, however, all was fixed, and by 1973 a "formal set of instructions" for how the different networks could communicate to each other, called a Transmission Control Protocol, was created. It was in one of these papers that the term "Internet" was first used to describe the network of computer networks that was building.

Computers were delivering well past expectations, with the military seeking to integrate them in all possible manners. But robotics wasn't completely dormant. In 1956, the world's first robot company, Unimation (Universal Automation), was founded, and in 1962, the first industrial robot, Unimate, was placed on a production line at General Motors. In 1973, the first industrial robot controlled by a computer was installed by Unimation's only competitor, the Cincinnati Milacron Corp. It was called T3 (The Tomorrow Tool). The first real mobile robot, not bound to an assembly line or lab, came in 1968. Shakey was built at the Stanford Research Institute and was novel for being able to move down a hallway without bumping into the walls. By 1976, robotics had reached the point that they served onboard both the Viking 1 and 2 space probes. Numerous advancements were being made, though not on the scale of the computing revolution.

## A SLOW START

It is around this point in robotics that Robert Finkelstein entered the business. Today, Finkelstein is president of Robotic Technology Inc. and a consultant for numerous government agencies, including DARPA. Among his current projects is one on "how to get the driver out of the car and save lives." Finkelstein is optimistic that this will happen "certainly in the range of 2015–2030."

Wearing glasses, a black shirt, and a paisley tie, Finkelstein certainly looks the part of an early military robotics pioneer. Indeed, he is so excited by robots that at conferences he briefs audiences on the history and future of unmanned systems using a PowerPoint presentation. Many people do such presentations in industry and academia; few of them, however, show a presentation with 321 slides.

But Finkelstein is a straight shooter. His greatest frustration is with uninformed policies that hold up the development of unmanned systems in both the military and civil markets. He minces no words, telling story after story about "friggin brain-dead" bureaucrats "who have no vision." What was technically possible long ago, he believes, could already have saved thousands of lives. "The sad thing is that many useful systems could have been fielded years ago."

Finkelstein was into physics growing up and then joined the army. After he left, he joined MIT as an engineer working on the Apollo space program. "Then Nixon pulled the plug." Finkelstein moved on to working on space and defense projects at several "beltway bandits" before eventually landing at the MITRE company, a research center that mainly does work for the Defense Department. Finkelstein became part of a project researching how to defeat enemy air defenses using unmanned drones, at the time called RPVs, short for remotely piloted vehicles. The concept was that they would create extra targets, forcing the enemy to use up their ammunition on the unmanned planes, rather than shooting down American pilots.

Finkelstein was hooked by the idea of robotics and became part of the very first RPV society in 1977. He remembers that "the gift wrap industry was larger than the robotics industry and all the engineers in it could fit into one room.... Everyone was filled with such angst. 'Why don't they like us?' we would ask all the time. You constantly had to justify to the bosses on why robotics made any sense at all."

Despite the poor prospects, in 1985 Finkelstein formed his own company, Robotic Technology Inc. He explains, "One decision criterion of mine is the minimization of regret. You don't want to be someday sitting in your rocking chair, in your shawl drooling, wishing you had taken your shot." Among its first proposals was a control system that sought to allow a QF-4 Phantom fighter jet, which had already been converted into an unmanned target drone, also to engage in air-air combat training simulations against real human pilots. But with the new software, the program threatened not merely to train the pilots, but also occasionally to beat them. So the views of it changed; it went from being a potentially useful training tool to a potential professional threat. Tells Finkelstein, "The air force was terrified of unmanned planes. You know, the whole silk scarf mentality. Pilots are what become generals, not anyone else.... So, the front office came back to us and said, 'Great project, but we now need it for a submarine to use instead.' "

Stories like Finkelstein's abound across the military robotics field when discussing the cold war years. It isn't that the systems weren't getting better, but that the interest, energy, and proven success stories necessary for them to take off just weren't there. The only substantial contract in this period was one that the Ryan aeronautical firm received in 1962 for $1.1 million to make an unmanned reconnaissance aircraft. The drone that came out of it, the Fire Fly, flew 3,435 missions in Southeast Asia. Overall, though, the Vietnam experience was as bad for robotics as it was for the broader U.S. military. Most of the uses of unmanned systems were classified and thus there was little public knowledge of their relative successes, as well as no field tests or data collection to solve the problems they incurred (16 percent of the Fire Flys crashed). As Finkelstein points out, "It took decades for UAVs to recover from Vietnam misperceptions."

The next big U.S. military spending on unmanned planes didn't come until 1979, with the army's Aquila program. The Aquila was to be a small propeller-powered drone that could circle over the front lines and send back information on the enemy's numbers and intentions, much like the Predator of today. Soon, though, the army began to load up the plane with all sorts of new requirements. It now had to carry night vision and laser designators, spot artillery fire, survive against enemy ground fire, and so on. Each new requirement came at a cost. The more you loaded up the drone, the bigger it had to be, meaning it was both heavier than planned and an easier target to shoot down. The more secure you wanted the communications, the lower the quality of the images it beamed back. The program originally planned to spend $560 million for 780 Aquila drones. By 1987, it had spent over $1 billion for just a few prototypes. The program was canceled and the cause of unmanned vehicles was set further back, again more by policy decisions than the technology itself.

Work continued, but mainly on testing various drones and ground vehicles, which were usually regular vehicles jury-rigged with remote controls. During this period, most of the ground systems were designed to be tele-operated, that is, using long fiber optic wires to link the robot to the controller. Any enemy with a pair of scissors could take them out. One of the few to be built from the ground up to drive on its own was Martin Marietta's eight-wheeled "Autonomous Land Vehicle." Unfortunately, the weapon had a major image problem. It was shaped like an RV, what your grandparents would use to drive cross-country to see the Grand Canyon. This killed any chance of convincing the generals of its use for warfighting.

Another significant program that didn't take off in this period was a 1980 army plan for a robotic antitank vehicle. The idea was to take a commercial all-terrain vehicle, rig it for remote control, and load it with missiles. Congress thought that ATVs, while certainly fun for country kids to ride around behind trailer parks, were a bit too small to be taking on Soviet tanks. So the program was canceled. But a mistaken belief soon grew in the military that the real thing Congress had objected to was weaponizing unmanned systems. "So," as Finkelstein says, "misinterpretation kept weapons off for almost a decade."

Those working on military robotics during this period learned that a major problem was generating what is called "customer pull." Too often, they were developing new projects based on "technology push," focusing purely on technological research in all sorts of directions, rather than having the customers' needs direct them. What was possible mattered less than what the military wanted. At the same time, they learned that "support from the top is essential." They needed buy-in from the generals and the politicians.

Despite these setbacks, the military robotics community didn't waver in its belief in the utility of its work. Helping to keep the faith was seeing other nations begin to gain some success with unmanned systems, which could be used to build support in the United States. Most significant was the growing Israeli experience with drones. In 1982, the Israelis carried out strikes on Syrian-occupied areas in the Bekaa Valley that decimated the Syrian air defenses (which were using the latest-model Soviet technology), with no Israeli losses. The secret to their success was a stratagem of first using UAVs to gather the electronic frequencies of the Syrian radars. Then a swarm of UAVs flew over the area, sending out fake signals. The Syrians, thinking it was the real attack, fired off their missiles. While they reloaded, a second wave of Israeli jets flew in and took out the entire defense system, using missiles that homed in on the radars that the drones had unmasked.

## THE RISE OF "SMART" BOMBS

By the time of the 1991 Persian Gulf War, unmanned systems were gradually making their way into the U.S. military, but in very small numbers. The army had a handful of M-60 tanks converted into unmanned land-mine clearers, but they were left behind in the famous "left hook" invasion force that drove across the desert into Iraq. The air force flew just one UAV drone.

The only notable success story was the navy's use of the Pioneer drone. The Pioneer was an unmanned plane (almost exactly like the planned Aquila) that the navy had bought secondhand from the Israelis. It flew off of World War II–era U.S. battleships that had been taken out of mothballs in the 1980s and updated for use in pounding ground targets with their massive sixteen-inch guns. The guns fired shells that weighed 2,000 pounds and could leave a crater the size of a football field. The little drones, which the Iraqis took to calling "vultures," would fly over targets and spot where the shells were landing. "The Iraqis came to learn that when they heard the buzz of a Pioneer overhead, all heck would break loose shortly thereafter because these 16-inch rounds would start landing all around them," said Steve Reid, an executive at the Pioneer's maker, AAI. In one case, a group of Iraqi soldiers saw a Pioneer flying overhead and, rather than wait to be blown up by a 2,000-pound cannon shell, waved white bedsheets and undershirts at the drone. It was the first time in history that human soldiers surrendered to an unmanned system.

The real stars of the Gulf War were not unmanned systems in the way we think of them now, but new, guided missiles and bombs, commonly referred to as "smart bombs." There were two main types that caught attention, laser-guided bombs and cruise missiles.

Laser-guided weapons came out of the earlier experiences with the television-guided glider weapons from World War II. The concept was the same, except now a human didn't have to steer the bomb, but rather just illuminate, or "paint," a target with a laser. The bomb or missile would then guide itself in. The air force was not interested in such devices, so the first research was actually done by the army in 1962. It wasn't until microchips became small and cheap enough to go on weapons that the device truly became useful. These guided bombs made their debut at the very end of the Vietnam War, where in May 1973 they were used to destroy the Thanh Hoa Bridge, a heavily defended site that had survived over eight hundred previous attacks by unguided bombs and missiles.

As one navy admiral put it, " 'Smart bombs' are really only 'pretty obedient bombs.' " A human finds and designates the target and the bomb just goes where it is told. On the early models, the human also had to keep the target continually painted with the laser, exposing themselves to danger. Later models had memory capabilities built in, so the human pilot could fly off while the bomb automatically stayed on target. Of course, a key weakness of the system is that the weather had to be clear enough for the laser to go through, meaning dust, haze, or smoke could make it useless. The early models were also fairly expensive, sometimes costing far more than the targets they took out. The pilots called such bombing runs "dropping a Cadillac."

Cruise missiles were a bit more advanced. With the various World War I "aerial torpedoes" and World War II Aphrodite bomber planes as their forebears, these were missiles that flew themselves, using either preset coordinates or recognition software to find their target. The one used most in the Gulf War was the Tomahawk, which flew under radar by hugging the earth at low altitudes that would be unsafe for human pilots. Still, such systems relied on the target's being decided before it took off, and had to go over terrain that had already been mapped out or photographed. It could not react to change.

Back home, a massive PR campaign was built around the guided weapons as the "heroes" of the short hundred-hour Gulf War. The only problem was that they weren't. Only 7 percent of all the bombs dropped were guided; the rest were "dumb."

The most influential technology was not the sexy smart weapons, but the humble desktop computer. By 1990, the U.S. military had bought into the idea of digitizing its forces and was spending some $30 billion a year on applying computers to all its various tasks. The Gulf War was the first war in history to involve computers to a significant extent, doing everything from organizing the movement of hundreds of thousands of troops to sorting through reams of satellite photos to find targets for the missiles to hit. Computers even gamed out potential Iraqi responses to American battle plans; indeed, they came up with more effective battle plans than the Iraqis ended up using in reality. Calling it a "technology war," the victorious commanding general, "Stormin'" Norman Schwarzkopf, said, "I couldn't have done it all without the computers."

Over the rest of the 1990s, the systems became ever more capable. But the "magic moment," as one retired air force officer put it, occurred in 1995, when unmanned systems were integrated with the GPS. "That's when it really came together."

The GPS is a constellation of military satellites that can provide the location, speed, and direction of a receiver, anywhere on the globe. It allowed unmanned systems (and their human operators) to automatically know where they were at any time. With GPS, as well as the advance of the video game industry (which the controllers began to mimic), the interfaces became accessible to a wider set of users. The drones began to be far more intuitive to fly, while the information they passed on to the generals and troops in the field became ever more detailed. Drones like Predator and Global Hawk made their debut in the Balkan wars a few years later, gathering information on Serb air defenses and refugee flows.

The programs also began to pass some key hurdles of acceptability. The various military services had long resisted buying any unmanned systems, but slowly they began to see their use. In 1997, for example, the air force chief of staff, General Ronald R. Fogleman, instructed his planners that his service could "no longer…spend money the way we have been," and mandated that they begin to think "outside the box," including on new technologies such as UAVs.

This key step for the air force actually came out of a good old-fashioned turf war. As it saw other services showing interest in unmanned planes in the early

1990s, "it was threatened by other services' infringement on what it saw as traditional Air Force missions." So, akin to what happened with ballistic missiles, where the air force only became interested in them when it saw the army getting into the business of space, the air force began several of its own drone programs.

Early air force plans had civilians piloting the drones, as it didn't feel it was worth shifting its own pilots over to such missions. However, the company that they originally hired was run by a retired U.S. Navy admiral, and so most of the pilots he brought in were ex-navy. Many cite this as another key step in advancing the user base inside the military. The air force decided that maybe having its own pilots fly the drones was a better outcome than "running an after-retirement jobs program for the squids."

## "FIRE IT INTO THE HEAVENS"

By the start of the twenty-first century, technology was starting to mature, each year getting better and easier to use. In turn, whether it was the UAVs in the Kosovo war or NASA sending out robotic explorers to Mars, unmanned systems were collecting a portfolio of success stories to show that they could be useful.

More important, these technologic developments began to coincide with changing political winds. As the cold war ended, the U.S. military was getting smaller, shrinking by more than 30 percent in the 1990s. At the same time, leaders began to think that the public tolerance for military risk had dramatically shifted, with expectations newly set by the relatively costless victory in the Gulf War. This was soon followed by the rapid withdrawal of American troops from Somalia after the Black Hawk Down disaster in 1993 and the unwillingness to send in ground troops during the genocides in the Balkans and Rwanda in 1994, for fear of casualties. As Major General Robert Scales argued, the new era of warfare was one in which "dead soldiers are America's most vulnerable center of gravity." With this, an added reason for investing in unmanned systems began to grow, centering on the new nature of foreign policy in the post-Vietnam, post–cold war era.

At a congressional hearing on February 8, 2000, it finally all came together for military robotics on the "demand" side. Senator John Warner from Virginia, the powerful chairman of the Senate Armed Services Committee, laid down a gauntlet, mandating into the Pentagon's budget that by 2010, one-third of all the aircraft designed to attack behind enemy lines be unmanned, and that by 2015, one-third of all ground combat vehicles be driverless.

Warner, who had once been navy secretary, was more popularly known for being the seventh husband of the actress Elizabeth Taylor. In his third decade in the Senate, however, he brooked no challenges and knew how to shake up the system. Yet Warner had shown no earlier obsession with robotics or technology. His insistence on pushing unmanned systems to the next level had nothing to do with what was possible with robotics at the time.

The first factor was his concern over what the growing intolerance for human casualties meant for U.S. foreign policy. "When you look at the history of casualties, beginning with almost half a million killed in World War II, over 35,000 killed in Korea, and more than 50,000 killed in Vietnam, and zero combat deaths in Kosovo, in my judgment this country will never again permit the armed forces to be engaged in conflicts which inflict the level of casualties we have seen historically," Warner explained. "So what do you do? You move toward the unmanned type of military vehicle to carry out missions which are high risk in nature....The driving force is the culture in our country today, which says, 'Hey! If our soldiers want to go to war, so be it. But don't let any of them get hurt.'"

The second factor was Warner's belief that the military needed a new way to convince youth to enlist. The more the military was using futuristic technology, Warner explained, "the more likely we'll attract quality men and women because they're interested in learning high tech in the military and then moving on and using those skills in the civilian community."

The military had no choice but to follow through on Warner's mandate. As chairman of the Armed Services Committee, he controlled the schedule of all hearings and bills on defense matters. Most important, when it came to approving both military budgets and officer promotions, he was the boss. And you don't cross the boss. As Warner summed up the episode, "Every now and then somebody like me has to take out their shotgun and fire it into the heavens to get somebody's attention."

Warner's goals were entered into law as part of the National Defense Authorization Act of fiscal year 2001. And then came September 11.

## "THE ROBOT IS OUR ANSWER TO THE SUICIDE BOMBER"

In the wake of 9/11, the shackles came off not only the use of force and willingness to send American troops around the world, but also the amount being spent on the military in general and robotics in particular.

From 2002 to 2008, the annual national defense budget has risen by 74 percent, to $515 billion. This figure does not include the several hundred billion dollars additionally spent on the cost of operations in Afghanistan and Iraq, which have been funded in separate budget supplementals. If you include these, the total Pentagon budget is at its highest level in real (inflation-adjusted) terms since 1946, the last budget to reflect World War II–related spending, and $36 billion and $126 billion (in 2008 dollars) more than the peak spending during the Korean and Vietnam wars (though the percentage of GDP is far lower). Research and development (R&D) and procurement costs, what it takes to design and build new weapons systems, have thus experienced an equivalent boom, or what one analyst described as "unchecked growth."

This is what we know. In addition, there is the "black budget," the Pentagon's classified budget for buying and researching what it wants to keep secret. For obvious reasons, the black budget is not released to the public, but it is estimated by the Center for Strategic and Budgetary Assessments to be around $34 billion in 2009, up roughly 78 percent since 9/11.

A core part of this massive post-9/11 research and buying spree has been new technologies, with a particular focus on anything unmanned. The amounts spent on ground robots roughly doubled each year, while the amounts on drones grew by around 23 percent each year. As an industry report commented, unmanned systems may have had a long history, going well back to World War I, but 9/11 was when things finally took off. "Prior to 9/11, the size of the unmanned vehicle market had been growing, but at an almost glacial pace. Thanks to battlefield successes, governments are lavishing money on UAV programs as never before." "Make 'em as fast as you can" is what one robotics executive recounts being told by his Pentagon buyers after 9/11.

As noted earlier, the number of unmanned ground systems in the U.S. military inventory went from almost zero in 2001 to five thousand by the end of 2006 and twelve thousand by the end of 2008. Whereas 93 percent of the bombs and missiles dropped on Iraqi forces in 1991 were unguided "dumb" bombs, 70 percent of the bombs and missiles dropped in the 2003 air campaign were precision "smart" bombs.

And with this change in military mentality, money, and use, the groundwork was finally laid for a real military robotics industry. As one report put it, "The undertaking has attracted not only the country's top weapons makers but also dozens of small businesses... all pitching a science-fiction gallery of possible solutions." Robert Finkelstein recalled a time when he personally knew most of the engineers working on military robotics. Today the Association for Unmanned Vehicle Systems International (AUVSI) has fourteen hundred member companies. Almost four thousand people showed up at its last annual meeting.

This trend seems only just the beginning. A historic comparison is often made to how certain technologies were jump-started by war. As one DARPA report put it, in 1908, there were just 239 Model T Ford cars sold. Ten years later, the figure was over one million. They predicted the same for robotics. "Just as World War I accelerated automotive technology, the war on terrorists will accelerate the development of humanoid robot technology."

The scientists are not the only ones who see this future. The Teal Group is a defense consultancy firm that specializes in forecasting the financial trends for war. As clients turn to Teal for investment ideas, it is not known for zany thinking. And yet Teal describes unmanned planes with zeal, as "the most dynamic growth sector of the world aerospace industry." Its experts expect the global spending on unmanned planes and computer-guided missiles over the next decade to yield $103.7 billion by 2016.

Teal's estimates consider only airborne unmanned systems; they don't even begin to include those on the ground or the sea. Scientists such as Finkelstein think the ground sector could prove to be even bigger, noting that "ground vehicles are just now on the edge of that same sort of acceptance in major use." Second, this spending base won't just be coming from the U.S. taxpayers, but will include those around the globe. Teal's numbers, for example, have Europe spending 20 percent of the worldwide total, followed closely by the Asia-Pacific region and the Middle East.

And no one expects these trends to abate anytime soon. As U.S. Navy researcher Bart Everett, another early robotics pioneer, cites, two crucial factors will continue the growing demand for military robots: "One, the technology has finally matured to the point where reasonably affordable robots can actually do something useful. And, two, the world situation has changed for the worse in terms of the variety, sophistication, and lethality of the various threats we now face in the free world.... To me, the robot is our answer to the suicide bomber."

## YOU ARE THE WEAKEST LINK

In the background of how the demand for military robotics finally caught up to the long history behind them has been one more important development. The more the military used unmanned systems, the more people came to believe that machines brought certain advantages to the battlefield. "They don't get hungry," says Gordon Johnson of the Pentagon's Joint Forces Command. "They're not afraid. They don't forget their orders. They don't care if the guy next to them has just been shot. Will they do a better job than humans? Yes."

Robots proved attractive for roles that fill what people in the field call the "Three Ds" ("Dull, Dirty, or Dangerous"). An irony is that many military missions can be incredibly boring as well as physically taxing. For example, with aerial refueling, spy planes are now able to stay in flight for as long as twenty hours or more. And yet the air force has found that humans lose effectiveness after ten to twelve hours. They simply wear down physically and psychologically from doing the same task that long. Unmanned systems, by contrast, don't need to sleep, don't need to eat, and find monitoring empty desert sands as exciting as partying at the Playboy Mansion. As one unmanned plane advertisement put it, "Can you keep your eyes open for thirty hours without blinking?"

Scientists are also finding that certain tasks also take incredibly high concentration for humans. But keeping at that level of intense concentration for lengthy periods of time is quite difficult, so people need to pause between tasks to collect themselves and gear back up. For example, detecting land mines is obviously a job for which a person needs to be at the top of their game. So they will pause and recollect themselves every so often. An unmanned system doesn't need that. Even using the same mine-detecting gear as a human, current robots can do the same task in about a fifth of the time, with greater accuracy.

Unmanned systems can also operate in dirty environments, such as a battle zone filled with biological or chemical weapons, where a human would have to wear a bulky suit and protective gear. But say you just happen to stumble into a war where the claimed threat of WMD turns out to be false. Even then, there are other kinds of dirty that robots may be more apt for. As an air force captain comments, things as simple as "inclement weather, smog, and smoke can hinder pilot visibility. How is this different between a manned and unmanned aircraft? The UAV has EO/IR/SARS [electro-optical, infrared, and synthetic-aperture radar sensors] to rely on. The pilot has the Mark I Eyeball."

Beyond just the factor of putting humans into dangerous environments, technology does not have the same limitations as the human body. For example, it used to be that when planes made high-speed turns or accelerations, the same gravitational pressures (g-forces) that knocked the human pilot out would also tear the plane apart. But now, as one study described of the F-16, the machines are pushing far ahead. "The airplane was too good. In fact, it was better than its pilots in one crucial way: It could maneuver so fast and hard that its pilots blacked out."

If, as an official at DARPA observed, "the human is becoming the weakest link in defense systems," unmanned systems offer a path around those limitations. They can fly faster and turn harder, without worrying about that squishy part in the middle. Looking forward, a robotics researcher notes that "the UCAV [the unmanned fighter jet] will totally trump the human pilot eventually, purely because of physics." This may prove equally true at sea, and not just in underwater operations, where humans have to worry about small matters like breathing or suffering ruptured organs from water pressure. For example, small robotic boats (USV) have already operated in "sea state six." This is when the ocean is so rough that waves are eighteen feet high or more, and human sailors would break their bones from all the tossing about.

Working at digital speed is another unmanned advantage that's crucial in dangerous situations. Automobile crash avoidance technologies illustrate that a digital system can recognize a danger and react in about the same time that the human driver can only get to mid-curse word. Military analysts see the same thing happening in war, where bullets or even computer-guided missiles come in at Mach speed and defenses must be able to react against them even quicker. Humans can only react to incoming mortar rounds by taking cover at the last second, whereas "R2-D2," the CRAM system in Baghdad, is able to shoot them down before they even arrive. Some think this is only the start. One army colonel says, "The trend towards the future will be robots reacting to robot attack, especially when operating at technologic speed. . . . As the loop gets shorter and shorter, there won't be any time in it for humans."

Robots also offer quicker learning curves. Computers not only speak the same language, but can be connected directly via a wire or a network, which means they have sharable intelligence. If one soldier learns French or marksmanship, he cannot pass on that knowledge easily. Barring a Vulcan mind meld, his squadmates would have to learn it in much the same painful way. And, no matter how hard all tried, there would be many differences and inconsistencies in their final skills. A computer, by contrast, can share that skill or knowledge with another computer or robot in only the time that it takes to download the software file.

Finally, many robotics salesmen are starting to sell an undervalued advantage that comes with the fact that robots simply aren't human: they don't carry all our wonderful "human baggage." They don't show up at work red-eyed from a night of drinking, they don't think about their sweethearts back home when they are supposed to be on mission, and they don't get jealous when a fellow soldier gets a promotion. One executive tells how his primary selling point for robotic sentries at warehouses was not their technologic capabilities or cost advantages, but simply that "robots don't participate in 'inside jobs.' "

Overall, there are a variety of reasons and motivations for why the military has become more and more interested in buying unmanned systems. But they all come down to one basic aspect. As military analyst and Bush administration adviser Eliot Cohen says, "The military is deciding that in the long run we can do more with machines than it can do without them."

## THE FUTURE IS SO BRIGHT

Just six years after Senator Warner fired his shotgun into the sky, Congress revisited the issue of military robotics. This time, the changed attitude was encapsulated by a new mandate in the Senate Armed Service Committee's version of the Defense Department budget. Congress ordered the Pentagon to show a "preference for joint unmanned systems in acquisition programs for new systems, including a requirement under any such program for the development of a manned system for a certification that an unmanned system is incapable of meeting program requirements." If the U.S. military was going to buy a new weapon, it would now have to justify why it was *not* a robotic one.

In a certain way, then, the history of robots had come full circle. Jacques de Vaucanson had impressed the most powerful leaders of his time with a futuristic vision of a world filled with artificial creations. Some 250 years later, President George Bush, the first president of the twenty-first century, saw the world turning out to be much the same way, just without the duck. "Now it is clear the military does not have enough unmanned vehicles," he said. "We're entering an era in which unmanned vehicles of all kinds will take on greater importance—in space, on land, in the air, and at sea."

It had taken a long time, but the field of robotics was now set to deliver on its great promise, most notably through its relationship with the military. Unmanned systems had started out as abnormal, limited in their use and acceptance. As the twenty-first century began, they were becoming the new normal.

# THREE

# ROBOTICS FOR DUMMIES

*Like a robot, sometimes I just know not.*

—EMINEM

"The ROBOTs are dressed like people. Their movements and speech are laconic. Their faces are expressionless and their eyes fixed..."

This very first mention of the word "robot" was in the stage directions for *R.U.R (Rossum's Universal Robots)*, a 1921 play by Karel Čapek, a writer living in what was then Czechoslovakia. The play opens in a fictional factory with posters on the walls that say things like: "Tropical Robots—A New Invention—$150 a Head." Business must be good at Rossum's, as the play opens with the company's general director sitting in a fancy office, dictating a letter about an order of fifteen thousand "robots."

All is not well, however, and by the end of the play these new machines have revolted against their human makers. As he watches his robots go out into the world, their human designer closes the play by quoting from the Bible: "And God said unto them, Be fruitful, and multiply, and replenish the earth, and subdue it: and have dominion over the fish of the sea, and over the fowl of the air, and over every living thing that moveth upon the earth."

The Czech word *robota* described the work that a peasant owed a landowner and also had a second meaning of "drudgery." A *robotnik* is a peasant or a serf, and *rabota* was the old Slavic word for slave. So the new word used to describe these machine characters actually came packed with all sorts of added meanings to Czech audiences. A few years later, Čapek's play made it onto the stages of New York and the word "robot" entered the English language.

## WHAT IS A ROBOT?

Unfortunately, a Czech actor with an expressionless face isn't all that helpful a definition for understanding robots. What follows is a broad guide to robots and how they work, not enough to create your own R2-D2, but enough to understand the basics.

Robots are machines that are built upon what researchers call the "sense-think-act" paradigm. That is, they are man-made devices with three key components: "sensors" that monitor the environment and detect changes in it, "processors" or "artificial intelligence" that decides how to respond, and "effectors" that act upon the environment in a manner that reflects the decisions, creating some sort of change in the world around a robot. When these three parts act together, a robot gains the functionality of an artificial organism.

If a machine lacks any of these three parts, it is not a robot. For example, the difference between a computer and a robot is the former's lack of effectors to change the world around it. Interestingly, a machine's sophistication has nothing to do with whether it is a robot. Just like biologic life might range in intelligence from bacteria and Paris Hilton to *Homo sapiens* and Albert Einstein, man's artificial creations too show wide levels of complexity.

Despite the seeming simplicity of this definition, it is still subject to some debate. For example, some scientists say that in order to be a robot, the machine has to be mobile. Yet this forgets that movement is just one way to change the world around you (as the world now has you in a different location). Defining only mobile systems as robots would not only exclude robots that work on factory lines, but would also be akin to defining paraplegics out of the human race.

## INTERFACE: THE MAN AND THE MACHINE

Our machines are designed to work for us, so an important part of understanding them is the user interface. The interface is the way the human receives information from the robot and, in turn, sends information to it, including orders that seek to control what the robot does.

The controls for many robots are often like those of any home computer: a screen, a keyboard, a mouse, and so forth. The Crusher robot, a six-ton truck mounting a .50-caliber machine gun, can even be controlled remotely with an Apple iTouch music player.

But as systems get more and more capable, there is a bit of a paradox going on with user interface. In the words of a sergeant just back from Iraq, sometimes there is just "too much technology.... It can be overwhelming." The major problem is the ever-growing amount of data that robots send to the user. As artfully described in *National Defense* magazine, it is like "the TV episode of *I Love Lucy* where Lucy and Ethel are at the chocolate factory and the chocolate just gets out of control, and you never get back in gear." An iRobot engineer confesses, "User interface is a big, big problem."

Military researchers are now trying to solve the interface problem by "playing to the soldiers' preconceptions." And with young males today, that means video games. Greg Heines, who runs the marines' Dragon Runner (a small ground robot) project, explains, "We modeled the controller after the PlayStation because that's what these 18-, 19-year-old Marines have been playing with pretty much all of their lives."

By using video game controllers, the military can piggyback on the billions of dollars that game companies have already spent designing controllers and training up an entire generation in their use. Yet their use in war doesn't always transfer over perfectly. The first problem is that the systems have to be ruggedized. A PlayStation controller, for example, only has to survive being thrown across the room in frustration when you give up that crucial first down in *Madden NFL*. The ones that the soldiers use have to be able to survive desert heat, sand, or even explosions.

Secondly, such video game controllers often require a ridiculously complicated series of moves to do anything complex. Anyone who has played the game *Mortal Kombat* recalls that you had to press Away, Toward, Toward, Down, and 3 in the span of a second to perform Sub-Zero's renowned finishing move of relieving a foe of his spinal cord. In real-world mortal combat, you have neither the time nor focus for that kind of fancy fingerwork.

Technology waits for no game, and the direct handheld joysticks and controllers like in the Xbox or PlayStation are already being replaced in popularity by Nintendo's Wii controller. The new feature of the "Wiimote" is motion sensing capability. Instead of just responding to punched buttons and twisted joysticks, the system also reacts to how the human moves the overall controller. The Wiimote recognizes movements like pointing a gun or swinging a golf club and then registers the same movement in the game.

Such advances may prove to be only a stepping-stone to eliminating joysticks and other remote controls altogether. In Steven Spielberg's movie *Minority Report*, for instance, Tom Cruise wears gloves that turn his fingers into a virtual joystick/mouse, allowing him to call up and control data, including even video, without ever touching a computer. He literally can "point and click" in thin air. Colonel Bruce Sturk, who runs the high-tech battle lab at Langley Air Force Base, liked what he saw in the movie. "As a military person, I said, 'My goodness, how great would it be if we had something similar to that?'" So the defense contractor Raytheon was hired to create a real version for the Pentagon. Bringing it full circle, the company then hired John Underkoffler, the technology guru who had first proposed the fictional idea to Spielberg. The result is the "G-Speak Gestural Technology System," which lets users type and control images on a projected screen (including even a virtual computer keyboard projected in front of the user). Movie magic is made real via sensors inside the gloves and cameras that track the user's hand movements. Taking it to the next level, researchers are integrating such systems into immersive virtual environments, taking the human operator of a UAV out from behind a computer and into a 3-D virtual world like Second Life, controlling as many as eight real world UAVs at the same time.

Even these sophisticated systems may not be enough to convey all that the soldier or their robot needs. A defense industry rule of thumb is that "if it takes more than two clicks to get the information, you are wasting your time." At each stage, as much information as possible must be packed into the interface. Yet the main interfaces to our technologies are through sight and sound. We read the information typed out on a screen or see it visually represented, and sometimes hear it via warning alerts. Relying only on these two senses limits how much can be digested and controlled; they are also difficult to use in the chaos and noise of battle.

One way to pack in more control and information is to tap as many of our other senses as possible. The Pentagon is pursuing "haptics," technologies that use the body's sense of touch as another portal for interfacing, akin to how the blind read Braille or people set their cell phones on vibrate. Given the different ways that our senses of feel and touch work, haptics multiply the amount of information our bodies can take in. The new controller programs range from the simple, such as buzzers that would let a soldier know there is danger in a certain direction, to variant thermal and pressure switches that might be placed on everything from biceps to toes.

For example, instead of having to look down to check if ammunition is running low, a soldier might get a quick pinch on their bicep when ten rounds are left in the gun. Or if a squadmate is wounded, a patch on their back might go ice cold.

The most advanced haptic projects in research right now are "symbiotic systems," such as suits designed for pilots to wear that let them "feel" parts of the plane. Explains their designer, "If there's an overload in one wing, the pilot will feel a vibration, or heat, in his corresponding arm." In turn, the plane monitors the pilot, even, for example, knowing when they are in the deepest part of their sleep cycle, so that it wakes them at just the perfect moment during a long mission. "It will really make a complete fusional relation between the plane and the pilot."

Another promising interface program is voice recognition software. In 2004, for example, an unmanned UCAV fighter jet was controlled on a war-game mission via voice commands. Some covert missions, though, require quiet. So there is also work on "subvocalization" systems, which allow the computer to register the human's voice commands solely by movement of the tongue and jaw; a command is mouthed but not actually said out loud.

Much like the haptic sensors, the controllers might be mounted on any part of the body that can move. For example, the Florida Institute for Human and Machine Cognition has even tested out a controller that is essentially a strip of red plastic tape that you wear on your tongue. It contains 144 microelectrodes that create various tingling sensations when activated. Navy divers have used it to interface with sonar systems, to help them find underwater obstacles and mines in dark or muddy water. One navy veteran said using the technology felt just like "Pop Rock candies." Just don't drink Coke afterward.

## ALL JACKED UP

Kevin Warrick is the head of cybernetics and robotics at Reading University in the United Kingdom. While watching an episode of *Star Trek: The Next Generation*, he became fascinated by the Borg, a species on the show obsessed with bettering themselves through the assimilation of other species and their technology into their bodies. In the show the Borg enslave or kill trillions of species, but for Warrick they were a revelation. "In terms of evolution, humans have gone as far as we can go."

Warrick began work on what he saw as the next evolutionary step in human-machine interfaces. He experimented with placing various controllers of technology inside his body. For example, he put a computer chip in his arm that was linked to a robotic arm via the Internet. When he moved his arm, the robotic arm moved. "To all intents and purposes, I was a Borg."

Such implants, Warrick believes, are the only way for humans to keep up with the machines we have made. "We know that machines will have phenomenal memory and speed of processing, so I say 'why can't I have a bit of that?' Machines aren't really limited to three dimensions the way we are. I'd love to be able to think in 20 dimensions.... Although I was born a human, I will die a cyborg, a very, very enhanced being."

For all the potential of French-kissing your robot's sonar controller, the best portal for communicating with our machines is the big mushy computer inside our skulls. When the neurons in our brains fire to communicate with each other, each signal beams out on a different frequency called "brain waves." Already in electrical form, these waves portray our thoughts and intent most rapidly and directly.

The challenge is transforming these electrical signals into something that can usefully connect to a machine. The signals intended to control the machine must be isolated from all the trillions of other signals going in and out of the brain. Likewise, the brain's instructions have to be decoded and converted into digital software that a machine understands.

There are noninvasive ways to tap into these brain waves from the outside. An electroencephalograph, or EEG, for example, is basically a cap covered with electrodes that listen in on the electrical signals that leak out through your skull. An EEG wearer has been able to move a cursor around a computer screen, much like using a computer mouse by hand. Such systems, however, remain limited by the fact that the technology is not directly connected to the body and is exposed to interference. Describes one researcher, "It's a blurry vision of what the brain is doing. It's like watching TV through waxed paper."

For that reason, most of the cutting-edge interface research is focused on direct links. The idea of such a "jack" into the brain comes out of science fiction, most notably William Gibson's 1984 novel *Neuromancer*. In the book, computer hackers of the future plug wires into their brains to link up with a virtual world of computers, which Gibson entitled "cyberspace" (another idea coined in the book). If picking up the waves through the skull is like watching TV through wax paper, a direct link through a jack, explains the researcher, is "like watching a high-definition plasma screen."

This idea of human brain jacks remained theoretic until Matthew Nagle, a young man from South Weymouth, Massachusetts, was paralyzed from the neck down in 2001. "Every other day I wanted to die," Nagle recalled. "I had nothing in my life to look forward to." With doctors telling him he had no hope of moving again, or even breathing without a ventilator, Nagle turned to BrainGate technology. A computer chip was implanted into his head. The goal was to isolate the signals leaving Nagle's brain whenever he thought about moving his arms or legs, even if the pathways to those limbs were now broken. The hope was that Nagle's intent to move could be enough; his brain's signals could be captured and translated into a computer's software code.

A mere three days into what was supposed to be a twelve-month research study, there was a breakthrough; just by thinking about it, Nagle moved a cursor on a computer screen. And with the ability to move a cursor, a new world opened up. He could move a robotic hand, surf the Web, send e-mail, draw, and even play video games, "just by imagining it." He even had the scientists link his brain to his TV's remote control, allowing him to change the channels just by thinking it. Nagle remained physically paralyzed, but through technology, he was changing the world around him.

Originally, the mind-machine interface worked by Nagle's thinking about moving the position of the cursor on the computer screen as if he were pushing it with his hand, with the computer translating his intent into action. Soon, he found it easier and more "natural" to just think about moving the cursor directly, akin to ESP. As he gained experience, he became a multitasker, able to talk while also playing a video game via his thoughts. "I do feel like it was a part of me," Nagle said. "They plugged me in and it was go, go, go. It was cool, man."

This ability to link up to a computer directly opens up some wild new possibilities for war, which is why the Pentagon's DARPA helped pay for the research. Its Brain-Interface Project is "the most lavishly funded of nearly all the DARPA bioengineering efforts." A project run out of the National Institutes of Health took it to the next step, where two severely disabled patients played a video game against each other, both controlling their sides solely by thought. "It's as if the first flight at Kitty Hawk has gone a few hundred feet," program director Joseph Pancrazio described of the possibilities that might follow.

In the world of war, in which microseconds are the difference between life and death, such thought interfaces infinitely speed up reaction time. Many scientists make the comparison to the movie *Firefox*, in which Clint Eastwood pilots a thought-controlled plane to easy victory over regular planes. In high-speed air battles, the ability to maneuver just by thinking, versus having to jerk a joystick around while fighting g-forces, can be huge. Even more, the ability will allow humans to fight virtually from "inside" unmanned systems, combining the advantages of manned planes with unmanned systems. Not only could a remote operator fly a UAV by thought (something already done in 2009 with brain interfaces to Honda's Asimo robot), but that system might also beam back images directly into their brain. This would allow the operator to sense what the robot is sensing, such as "seeing" in infrared or thermal.

Experiments with such virtual interfaces are finding that people also begin to develop a sort of sixth sense that further links them to the machine they are interfacing with. Professor Warrick (the Borg fan), for example, used chips implanted in his arms to communicate with tiny robots that signaled him whenever they made contact with something. Warrick described that it wasn't so much that he could "feel" as sense the presence of what the robots were touching. It's much like how when you first learn to drive a car, you have no natural sense of where the vehicle begins and ends. As you gain experience, soon you can squeeze it into tight parking spaces without much thought at all. The machines, like that car or even a pen when you write, become an unconscious extension of the body.

These brain jacks are also evolving from the ugly vision of tubes stuck into the back of your skull, much like in *The Matrix*, to ever smaller implants. At Duke University test subjects have been connected via electrodes as thin as a human hair, while a program at Emory University has developed implants the size of a grain of rice.

Much as Internet connections have gone wireless, so might the implanted brain chips one day. This development will give soldiers in the field access to all sorts of new capabilities beyond just controlling their robots by thought. For example, when I couldn't remember who starred in *Firefox*, I punched in a search on my desktop computer's Internet browser. Imagine instead being able to do such searches inside our heads. One researcher explains that the ability to directly connect to the Internet "is going to be my mental prosthesis. Everything I want to know, I can look up. Everything I can forget, I can find. I'm going to get old, but it won't matter. I won't have to remember anything."

If our brains are connected to machines, it also means they can be connected to each other. As with any other computer file, if a thought can be transformed into computer code, nothing prevents that file from being accessible to someone else. For warfare, this means that a soldier may one day not need to radio his buddies of an ambush ahead or snap a picture and e-mail them, but instead could just pass it on by thought. Robotics scientist Robert Finkelstein is quite excited by this technology's prospects. "We would all share information in an instant process.... What I see, you see.... It could be very potent, for good or ill. If al-Qaeda is still around, it could be very scary." What scientists are talking about, says one, is "network-enabled telepathy." It sounds otherworldly, but the U.S. government's National Science Foundation envisions such communication to be possible within the next two decades.

But even all this direct interface will still not be enough. The systems will still be producing overwhelming amounts of data, only now dumping it directly into our heads. It may sound great, for example, to be able to fly a Global Hawk by thoughts alone and see what the drone is able to see. However, that drone can fly for thirty-five hours and cover an area the size of Maine with its sensors. So no matter how jacked up we get, our poor little monkey brains will still need some computer to help us control the drone, as well as make sense of what it is finding. This disparity between ourselves and our machines is what drives research into "autonomy" and "artificial intelligence."

## AUTONOMY: WHEN A ROBOT DECLARES INDEPENDENCE

That a machine can make a decision on its own, with or without a human in the loop, does not define whether or not it is a robot. The relative independence of a robot is merely a feature called "autonomy." Autonomy is measured on a sliding scale from direct human operation at the low end to what is known as "adaptive" at the high end.

The potential autonomy of a robotic spy plane can illustrate. Direct human operation is having some guy behind a computer control all the plane's operations from the ground. Human-assisted is when the pilot on the ground takes off and lands the plane, but can let the plane fly itself while in the air. In human delegation, the pilot just has to instruct the plane to take off and land and give it waypoints to fly to. In human-supervised, the operator is no longer really a pilot, but just monitors what information it sends back. In mixed-initiative, the human might give a robotic plane a mission to accomplish, but doesn't need to oversee it. The machine is given a mission file to complete or even loosely told to be "curious" and only report back when it finds something interesting. In fully autonomous mode, the machine decides on its own what to report and where to go. Finally, a machine is adaptive when it can learn; it can update or change what it should search out, even evolving to gather information in new ways.

Autonomy is thus about more than simply whether the human is in control or not, but also about how it relates to the world. Can the robot build its own model of the world? Can it operate in the world on its own using that model? Can it change and update that model on its own? And finally, can it decide to throw that old model out and find a new way to figure out what to do? Autonomy, then, relates to many of the same questions that we usually use to define a human being's maturity.

When thinking about all this in the context of war, it is easy to see the attraction of building increasing levels of autonomy into military robots. The more autonomy a robot has, the less human operators have to support it. As one Pentagon report put it, "Having a dedicated operator for each robot will not pass the common sense test." If robots don't get higher on the autonomy scale, they don't yield any cost or manpower savings. Moreover, it is incredibly difficult to operate a robot while trying to interpret and use the information it gathers. It can even get dangerous as it's hard to operate a complex system while maintaining your own situational awareness in battle. The kid parallel would be like trying to play *Madden* football on a PlayStation in the middle of an actual game of dodgeball.

With the rise of more sophisticated sensors that better see the world, faster computers that can process information more quickly, and most important, GPS that can give a robot its location and destination instantaneously, higher levels of autonomy are becoming more attainable, as well as cheaper to build into robots. But each level of autonomy means more independence. It is a potential good in moving the human away from danger, but also raises the stakes of the robot's decisions. As one defense analyst succinctly put it, "The autonomy thing is f'ing hard. All the little decisions build up, especially in a chaotic situation like war."

## "INTELLIGENCE IS INTELLIGENCE"

Wrapped up in the idea of autonomy, essentially the robot's level of independence and maturity, is something even more complex: "intelligence." This is perhaps the most important aspect of a robot, which processes information and decides what to do with it. As one military analyst argued, "Forget about whether the intelligence is carbon-based like humans or silicon-based like machines. Intelligence is intelligence and must be respected."

The dictionary definition of intelligence is "an ability to act appropriately (or make an appropriate choice or decision) in an uncertain environment." That seems simple enough, but when that ability is located in a man-made creation, not a living thing, it gets complex. The definition of "artificial intelligence" (often called AI) is actually in dispute, not only because of all the technical baggage that comes with determining what is an "appropriate choice" or not, but also because the definition of AI links to broader debates about what it means to be human or not.

For some, the definition of a machine's intelligence depends on a comparison with a human. They argue that a machine is artificially intelligent if it can do a task that requires some measure of intelligence for a human to do.

This is a difficult way to go about determining intelligence, as it completely depends on the specific task at that specific moment in time, not the machine or the human performing the task. Computers, for instance, outperform us in tasks that involve numbers, calculations, and searching for stored information. They can remember literally trillions of points of data, whereas most of us have a hard time remembering even the PIN number to our ATM at the bank.

However, in other ways, computers have so far proven to be "ridiculously stupid." As one science writer describes, they may be able to "calculate faster than any human being, but they lack the common sense of a two-year-old." The reason is that computers are limited to the world of numbers in both their language and processes. But sometimes this logical, mechanical manner of intelligence is hard to translate to activities in the real world. Moreover, while a computer can churn through the numbers behind a single or a few problems with ease, the human brain is massively parallel. It may process far slower than a digital computer, but the mushy gray blob inside our head can do a hundred trillion computations at the same time. We still crush computers when it comes to matching patterns with memories and applying knowledge to current contexts.

A good illustration is what some describe as the Apple-Tomato test. For a computer to tell the difference between an apple and a tomato is actually quite tricky. It could resort to all sorts of visual analyses, comparing the size, shape, and color. But soon the machine would find that in certain cases there would be overlap, so any and all tests, no matter how rapid, would be inconclusive. It could next proceed to taking samples, such as capturing its chemical makeup via a smell test, and then comparing the data to other known test subjects. Ultimately, it could only be sure beyond a reasonable doubt with a DNA sample, which would occupy a massive part of its processing power.

By comparison, pretty much any two-year-old human instantly "knows" that an apple is not a tomato, without any calculation. At the same time, that toddler can pick his nose, kick a ball, and realize that it is raining outside. Thus, the toddler may not be able to count to infinity, but they blow the computer out of the water when it comes to pattern recognition and multitasking, which is 95 percent of what we ask our brains to think on. "If you think it's easy for you to do, most of the time it's very difficult for robots to do," says Takeo Kanade, director of Carnegie Mellon University's Robotics Institute. So it seems unfair then to compare machines to humans in this type of intelligence.

What should matter more in defining intelligence is simply whether there is some use made of information in order to achieve the task. For our purposes, then, when talking about machine intelligence it seems most apt to use the definition that leading roboticists (robot scientists) all center around. As Sebastian Thrun, director of the Artificial Intelligence Laboratory at Stanford University, explains, artificial intelligence is the ability of a machine to "perceive something complex and make appropriate decisions."

This also recognizes that there are various types of intelligence that we value. Some intelligence is reactive, basically sensing and acting upon information. Some is predictive, anticipating what will happen from prior information and acting beforehand. And some is creative, recognizing patterns of information and inventing new solutions to problems. To put it another way, if the task is to figure out what to do when it rains, the reactive answer would be to get under a tree when you feel water on your head, the predictive answer would be to check the weather and avoid the rain before it even starts, while the creative answer would be to invent the umbrella. Each requires intelligence, just of a different type. For this reason, researchers are at work on all sorts of advanced AI, such as "expert systems" that organize behavior into millions of rules to follow, to "evolutionary" or self-educating AI, such as neural networks that mimic the human brain, to genetic algorithms that continually refine themselves.

AI learning is thus the key to robots' continual expansion and usefulness in the real world. As Lynne Parker, head of the Distributed Intelligence Laboratory at the University of Tennessee, describes, "Simply put, we can't know or predict everything that a robot might encounter in performing its task. As the common adage states, 'Only change is certain.' So, to deal with all of these issues, robots must be able to learn and adapt to changes in their environment."

## AI GETS STRONG

Today, there are all sorts of artificial intelligence that appear in our daily lives, without our even thinking of them as AI. Anytime you check your voice mail, AI directs your calls. Anytime you try to write a letter in Microsoft Word, an annoy-ing little paper-clip figure pops up, which is an AI trying to turn your scribbles into a stylistically sound correspondence. Anytime you play a video game, the characters in it are internal agents run by AIs, usually with their skill levels graded down so that you can beat them. Indeed, almost any movie made today that has a crowd shot is actually filmed by populating the scene with AIs. Starting with the acclaimed battles in the *Lord of the Rings* trilogy, production companies have found it cheaper to build tiny avatars that interact in the scenes than actually pay tens of thousands of real human extras to stand around for the day wearing orc armor. Overall, the size of the AI market was estimated by the Business Com-munications Company to be roughly $21 billion in 2007 with annual growth of 12.2 percent.

It is this part of robots that makes decisions, artificial intelligence, that may be the part most important to their impact on war. Up until today, each of the func-tions of war took place within the human body and mind. The warrior's eyes saw the target, their brain identified it as a threat, and then it told their hands where to direct the weapon, be it a sword or rifle or missile. Now each of these tasks is being outsourced to the machine. For this reason, the U.S. military funds as much as 80 percent of all AI research in the United States. Thus, while firms like Microsoft or Google lead and the military follows in other parts of the information technology world, the military sets the agenda in AI.

Both the complexity of AI and its application to the tasks that go into warfighting are clearly growing. As Helen Greiner at iRobot says, "We are not close to having AI on a human level. Nobody is. But if you take a particular mission, like vacuuming a floor, we are able to provide the intelligence to accomplish that mission. On the mili-tary side, going and doing perimeter security autonomously, going into a building and doing a full coverage operation, looking for terrorists or weapons caches, we can do that autonomously right inside the robot."

The GT Max, for example, is an unmanned helicopter project at Georgia Tech that is sponsored by DARPA and has been tested out at nearby Fort Benning. Max not only can fly on its own, but can think for itself how best to do it. Just as a pilot's brain has to fly the plane as well as react to changing weather or enemy fire, Max's software can handle multiple, unexpected challenges. That is, the "UAV is able to learn as it flies." GT Max has been able to automatically plan its way through obstacles, fly via onboard camera rather than GPS navigation, maneuver aggressively as a human pilot might, and even reconfigure itself when accidents happen, such as staying in the air even when the primary flight control systems fail, something that a human pilot would find incredibly difficult to pull off.

Various programs are pushing AI (and the robots that will be guided by it) well past the capabilities of the PackBot, Roomba, Predator, SWORDS, or even the Max. One example at the Air Force Research Laboratory is based on the research of Stephen Thaler. Better known as the inventor of the Oral-B electronic toothbrush, Thaler has created the "Device for the Autonomous Generation of Useful Information." Also called "The Creativity Machine," it is a neural network AI program with two extra features. The first constantly introduces new information, or "noise," to help jumble together new and old ideas. The second is a filter process that measures the new ideas against old knowledge and preferences. By generating new ideas based on old ones, the machine has done everything from write catchy pop music (by learning from what kind of songs make it on the top ten lists), design soft drinks, discover substances harder than diamonds, optimize missile warheads, and search the Internet for terrorist communications. Recently, the air force lab contracted Thaler to marry up his AI software with robotic hardware to create "Creative Robots."

Another program at the University of Reading in England, is looking at how robots can learn to interact and even develop "personalities" without human guidance. A group of robots, each with the brainpower equivalent to a snail (roughly fifty neurons), are programmed to avoid each other, based on a system that rewards or punishes different types of contact. They are then placed in an enclosed space and monitored. The wrinkle is that the robots are able to factor in past experience, meaning each can develop different lessons over time. Despite all the robots having the same initial software, the researchers are seeing the emergence of "good" robots that cooperate and "bad" robots that constantly attack each other. There was even one robot that became the equivalent of artificially stupid or suicidal, that is, a robot that evolved to constantly make the worst possible decision.

This idea of robots, one day being able to problem-solve, create, and even develop personalities past what their human designers intended is what some call "strong AI." That is, the computer might learn so much that, at a certain point, it is not just mimicking human capabilities but has finally equaled, and even surpassed, its creators' human intelligence.

This is the essence of the so-called Turing test. Alan Turing was one of the pioneers of AI, who worked on the early computers like Colossus that helped crack the German codes during World War II. His test is now encapsulated in a real-world prize that will go to the first designer of a computer intelligent enough to trick human experts into thinking that it is human. So what is the reward for inventing what some hope will be the real-world equivalent of Data from *Star Trek*, but others worry will be Skynet from *The Terminator*? One hundred thousand dollars. It almost seems not worth the bother.

## SENSORS AND SENSIBILITY

If the processors behind AI make decisions about the world and how to respond to it, "sensors" are the part of robots that define just what this world is. They collect information about the environment in which a robot is located. Instead of being encapsulated into what we describe as the "five senses" (and yes, there are electronic taste and smell sensors that can even identify wines and cheeses as well as most sommeliers), robotic sensors are generally categorized into two categories, passive or active.

Passive sensors sense merely by receiving information. An example would be infrared sensors that collect surrounding heat source emissions. Active sensors first send some form of information or energy out into the world, so as to collect even more information. One of the most common forms of these is called Laser Detection and Ranging (LADAR). It sends both laser beams and radar waves out widely, which then bounce back and help create a map of the obstacles around the robot.

Sensors are then linked back to AI processors to create "perception," understanding the meaning of the object detected in the context of the environment. The levels of perception range from simply sensing an obstacle (a big thing is in the way), to recognizing shapes (the big thing is a rectangle), to being able to categorize the shape and thus identify it (rectangles of this shape are tanks), to understanding the significance of the object (tanks should be reported back to base, shrubs should not). The goal is to be able to perceive things like the difference between friend and foe, rate their importance, and decide potential responses. Donald Verhoff, vice president of technology at the Oshkosh Truck Corporation, which builds robotic army trucks, says, "If it's a child, you want to stop. If it's a guy with an RPG-7 [a rocket-propelled grenade launcher], you want to run him over."

These tasks are actually quite difficult in the chaos of the real world. It is hard enough for a computer to tell the difference between a tomato and an apple. What about when driving at fifty miles per hour? Likewise, how can the robot tell the difference between humans and mannequin decoys made to look like them? Because of these challenges, Sebastian Thrun of Stanford University says that "understanding the environment is the Holy Grail for artificial intelligence."

As in AI, there are all sorts of amazing developments going on with sensors to make robots ever more capable. One of the most useful advances may be in millimeter-wave radiation. These sensors work like the X-ray machine at your doctor's office, but gather more detail. They detect not only the shape of something on both the outside as well as inside, but also the different materials that make up whatever is being looked at. A scan of a person hiding a gun would not only pull out the shape of the gun from their skeleton, but also assign it a different color than the cell phone in their other pocket. Already, such sensors are being used to scan trucks in the Chunnel that links Britain and France.

When various sensors are brought together, the results are truly powerful. In the battle against IEDs, the U.S. Army is now working on deploying UAVs that can do what is known as hyperspectral imagery. Besides gathering regular visual information from a drone's stereo cameras and infrared detectors, the system can distinguish such things as the color of an object or whether any objects are hidden inside it. It also has a "Bloodhound mode" that uses the sensors to search out specific spectral signatures. So if the military gets a report of a suspicious "black pick-up truck driven by two men," the UAV flying high over Baghdad will be able to hunt it down.

Unfortunately, better sensors don't always yield more useful information. More data collected means more data to process and often more time required to make decisions. It is like trying to walk through the world looking through a telescope. Sensors also can capture only what is going on in that moment in time, not the context or undertone of what they are seeing. Imagine a robot carrying out surveillance of a bar for suspected terrorists. Its multispectral sensors detect a man entering the bar. Its sophisticated eavesdropping devices hear him ask for a glass of water. The bartender pulls a gun. The man says "Thank you" and leaves the bar. While we might get the trick that the man had hiccups and the bartender was just scaring him, could the robot? Or would it just assess the bartender as a threat and terminate him?

## TAKING EFFECT

If sensors gather data, and processors decide how to react to it, "effectors" are the part of a robot that create the desired change in the environment. They translate intent into action. Describes David Bruemmer of the Idaho National Laboratory, "As opposed to a computer that can just hold data,...effectors are what allow robots to take part in the drama of the real world."

The most obvious type of effectors are those that allow movement; that is, the propulsion system. On the ground, movement effectors range from wheels and tracks to legs. So far, roboticists are finding good legs to be difficult to build. A kindergartner still possesses more speed and agility than most legged robots. The challenge that legs present for robots is that they mix the task of movement with that of balance.

One effort to solve this challenge is the conversion of the infamous Segway into a robotic platform. Originally marketed as a secret invention that would "revolutionize society," the Segway turned out only to be a two-wheeled scooter that you stand on to drive. George W. Bush famously tumbled off of his while tooling around his Texas ranch. The backers hoped it would sell millions when it was released to great fanfare in 2001, but it has since sold only six thousand units, mainly due to its high cost. However, its ability to self-balance is giving the Segway a second life. Pentagon-funded university research efforts have converted the human scooter into a robotic platform that can do such things as carry supplies, patrol a building, and even carry a stretcher.

Propulsion systems vary for robots, depending on their environment. In the air, UAVs have been powered by everything from propellers to jet engines. Recently there has been much work on rotary wings, which allow the robot to take off and land from small spaces, as well as hover in place, allowing it to identify and inspect targets. One example is the U.S. Army's Autonomous Rotorcraft Sniper System (ARSS), an unmanned helicopter that mounts a sniper rifle. It is designed to fly above a messy urban battlefield and take out foes one by one. The system is guided using a modified Xbox 360 video game controller. While rotary wings are typically helicopters, there is also work on tilt-rotor unmanned aircraft, such as the Eagle Eye system. These take off and land like a helicopter, but then the propellers flip down and the drone can fly like a plane, ideally getting the best of both worlds. Propellers are also the primary propulsion system among unmanned vehicles at sea. However, there is recent U.S. Navy work on a system called the "water strider," essentially thin mechanical legs that skip across the water, allowing the robot to move like a water bug.

Just as our limbs do more than just move us, robot effectors do more than just propulsion. For example, manipulators are a robot's arms, which can touch, grip, or pick up objects. But, as with your own arms, the end effector that truly shapes how the robot interacts with the world is whatever tool is at the end of the manipulators. It might be fingerlike grippers or a "diamond-tipped" buzz saw (actually proposed in a 2008 U.S. Navy report). But of course, the robotic effectors that create the most "drama in the world" are weapons.

Given the four thousand years that humankind has spent perfecting its tools of war, weapons come in all sorts and sizes. The most common are those that use some sort of chemical propellant to shoot a projectile. We know these better as guns or rockets, and as the SWORDS illustrates, robots already can carry as many as their human counterparts.

Besides their ability to carry heavier weights, robots have one more fundamental difference. They come with their own power systems. This opens up new possibilities, as the robot can now power its own weapon. One example is called Metal Storm and has been tested out on iRobot's Warrior. Metal Storm, originally invented by an Australian grocery store worker, is a gun that uses electricity rather than gunpowder to shoot out stacks of bullets. The switch from chemical to electric power allows it to fire far faster, as many as a million rounds per minute. Thus, instead of shooting at one target with one bullet, Metal Storm can do such things as deconstruct a target, by shredding it apart bullet by bullet, or put up an actual wall of bullets in the air to protect against incoming missiles. The makers also note that this electric machine gun is good for "crowd control."

For some strange reason, a few people have concerns about super-smart robots carrying machine guns that can shred entire buildings. Many believe that if a robot is going to have a weapon, it should be a nonlethal one. These are weapons not designed to destroy and kill, but to incapacitate without causing any permanent damage. As Steven Metz, a professor at the U.S. Army War College, says, "The combination of robotics and nonlethality could be incredibly important. Instead of 'killing them all and letting God sort them out,' you could have systems that just zap them and let the police then come in."

There are all sorts of ways to incapacitate a person or machine, so nonlethal weapons come in almost as many shapes and sizes as lethal weapons. Acoustic weapons use sound waves in lieu of bullets. Perhaps the most noted of these is the Long Range Acoustic Device (LRAD). If regular electronics make what is known as "white noise," the LRAD puts out what soldiers affectionately call "the brown sound." That is, it sends acoustic waves in such frequencies that they overwhelm the human body and even make a targeted person defecate upon themselves. The devices have a range of up to one kilometer.

The LRAD actually made its first combat debut not with the military but on a vacation cruise ship. In 2005, one of the Seaborne line's luxury ships was attacked off Somalia by pirates armed with machine guns and rockets. Instead of fighting them off with shuffleboard sticks, the crew used LRAD sonic blasters to chase them away. There are also smaller, handheld acoustic weapons that can send out "sonic bullets." These bursts of sound last but a few seconds, but are so powerful (150 decibels, the equivalent of standing in front of a jet engine or guitar speaker set at "11") that they can literally knock people off their feet.

Another category uses some form of chemical to incapacitate enemies without hurting them permanently. In Britain, for example, defense researchers have built a gun that shoots blobs of compressed glue (so the robot then can ensnare the target just like Spider-Man does), while others shoot out supercharged versions of stink bombs.

The final category of nonlethal weapons targeted for robot use is those that emit various forms of directed energy. One already tested out in Iraq is the Active Denial System. Sometimes called the "pain ray," the system shoots out waves like those used to heat up frozen pizza in a microwave oven. The rays, which have a range of over five football fields, penetrate the top sixty-fourth layer of skin (even if you are wearing clothes over the skin) and heat up the water inside. The ray doesn't permanently hurt the person, or even cause a sunburn. But the sensation is excruciating, enough to make test subjects feel like their skin was catching on fire. If the ray is turned off, or the person moves out of its focus, the pain instantly ends.

Other systems send out various forms of radio-wave beams. They can screw up enemy machines by disrupting mechanical signals, or create "an artificial fever" by heating up the core body temperature of any human target, which instantly knocks the person out. It could even be modulated to hit all the people in an entire building. Of course, the danger is that, just like overcooking that pizza in your microwave, being off just a few degrees could kill them all instead of just giving them a fever. The line from nonlethal to lethal is a fuzzy one.

The Pulsed Energy Projectile fires balls of plasma that can disrupt the functions of nerve cells. Depending on the tuning, it could cause a shock, create a stunning effect, or temporarily disable a person. Another prototype is the tetanizing beam weapon. Just as lightning occurs when storm clouds build enough electronic potential difference to cause ionization between the ground and the clouds, the weapon creates an artificial ionization course between the weapon and the target. Artificial lightning then strikes whoever is unlucky enough to be on the other end.

So far, the system has worked out to two hundred meters, but its designers think it might soon have a range of over a mile.

Lasers, though, are where most of the robot energy effector action is headed. Lasers basically work by exciting certain types of atoms so that they emit particles of light, called photons, in one direction, along one wavelength.

The idea of lasers first came in H. G. Wells's famous 1898 story *The War of the Worlds*. For the next century, lasers remained mostly in science fiction. They didn't find much use in war except as targeting devices. During the Reagan administration, that changed. Edward Teller, the father of the hydrogen bomb, successfully pushed the idea of using powerful lasers to shoot down enemy missiles from space. The project was officially called the Strategic Defense Initiative, or SDI, but soon got tagged as "Star Wars." The name was originally given by opponents who meant to mock the idea as only useful for a galaxy far, far away. But the supporters soon turned it around and began referring to the weapons plan as "Star Wars" in official government documents in 1985. Hollywood was not amused and George Lucas sued the U.S. government for trademark infringement.

With more than $1 billion a year spent on its research since, laser weapons technology has advanced. While it is still not set for wars in space, by 2003 the Zeus system was ready for deployment to Iraq. Zeus, named after the Greek god who threw thunderbolts, is a ground-based laser, small enough to fit into a Humvee, which shoots a beam capable of blowing up IEDs and land mines from a hundred feet away. As the systems have gotten more effective, there are now plans to fit the R2-D2-like CRAM system with a laser (to accompany the machine gun) that can shoot down rockets or mortar rounds over five miles away.

There is immense excitement among soldiers, scientists, and sci-fi geeks alike about the possibilities offered by lasers, to the point that one study called it the "Holy Grail" of weapons. Besides being incredibly accurate (hence the phrase "laserlike precision"), lasers can also be controlled even after the trigger is pulled, and even change direction or modulate the power level, from a mere dazzling effect to deep fry. Weapons makers get almost into a frenzy when they talk about all the sorts of roles for laser weapons. Tanks and other land systems could use lasers to shoot down incoming rockets. Planes could zap targets on the ground or defend against incoming missiles, much like the old World War II bombers could protect themselves with machine-gun turrets. Dan Wildt, Northrop Grumman's Directed Energy Director of Business Development, even pitched lasers' use as a useful defense against "terrorists on Jet-Skis."

The appeal of lasers lies not just in their flexibility, but also in the belief that they would have an immense psychological effect. In 2006, the First Marine Expeditionary Force, then stationed in Iraq, issued an "urgent operational need" request for the development of a "Precision Airborne Standoff Directed Energy Weapon (PASDEW)." This would be a helicopter- or drone-mounted laser that could, in the words of the request, create "instantaneous burst-combustion of insurgent clothing, a rapid death through violent trauma, and more probably a morbid combination of both." The request went on to describe how the laser "can be compared to long range blow torches or precision flame throwers, with corresponding psychological advantages."

Many of the more powerful lasers right now remain quite bulky and require huge amounts of energy. This limits the variety of robots on which they can be mounted. One solution is the Tactical Relay Mirror System, in which the robot (such as a Predator drone, or even an unmanned blimp) carries not the weapon itself, but instead a mirror that redirects a beam fired by a ground laser down on unsuspecting targets. Still, as with electronics in general, lasers are getting smaller and smaller, with companies now working on prototypes that would be small enough for a person-sized robot to carry. One laser weapon design is even small enough to fit in a woman's purse, making it the ultimate accessory.

## CAN'T FIGHT THE POWER

Flying over cities and shooting precision flamethrowers at unsuspecting terrorists on Jet-Skis can certainly be tiring, so robots have one more critical need. "Power" sources are what supply a robot with the energy it needs to operate.

A battery of some type is typically used for smaller robots. Most robots are trending toward using rechargeable batteries in some way, such as through an electrical hookup, or even through an infrared beam, as the Roomba uses. A novel program at Tel Aviv University has even built a nanobattery as thin as a strand of hair, which can recharge faster than the common lithium-based batteries.

If the robot is a large or converted manned system (like a car or plane), it will often draw its main power supply from the vehicle's normal engine source, such as gasoline. But robot power need not be limited to the climate-changing fossil fuels that bind us to Middle Eastern potentates. For example, ethanol is just a form of drinkable grain alcohol. So one inventive research group at St. Louis University built a robot powered by various beverages that could be sourced at any college mixer. In the words of one grad student, robots prefers the hard stuff: "It didn't like carbonated beer and doesn't seem fond of wine, but any other [alcohol] works."

The future will likely see a range of novel energy sources powering unmanned systems, especially in the air, where payload and space are at a premium. In the 1950s, for example, there was immense research on using nuclear power to fly bomber planes, which would then never need to land. The obvious concerns over what might happen if one of these planes crashed, as well as how to get the crew on and off, ended the idea. Today, there is new research on using nuclear isomers, such as the element hafnium. These release energy through a triggered decay, rather than atomic fission or fusion. The U.S. Air Force is presently exploring how isomers might power long-duration unmanned aircraft. Another UAV, the Global Observer, is hydrogen-powered and can remain in the air for up to a week.

Just as AI research often takes its lead from how the brain works, so too is research on power looking to biology. There are all sorts of projects on how to harness the same processes that sustain living organisms, such as photosynthesis or other biochemical energy mechanisms. One example is Chew-Chew, the "gastrobot." Built at the University of South Florida, Chew-Chew is a twelve-wheeled robot that runs on a microbial fuel cell. In a sort of reverse Atkins diet, bacteria in the fuel cell break down whatever carbohydrates Chew-Chew eats and convert the released energy into electricity. Ultimately, such fuel cells could be used on such systems as robotic lawn mowers that draw power from the grass they cut.

As the bulging waistlines of our Fast Food Nation demonstrate, green leafy stuff is not as apt as the fleshy red stuff for efficiently packing in the calories. This is where things begin to get truly weird. A contemporary of Chew-Chew's at the University of the West of England is a gastrobot that powers itself by eating slugs. At the University of Texas, researchers have built a tiny fuel cell that draws electricity from the glucose-oxygen reaction in human blood. It is called a "vampirebot." A group of Japanese scientists working on a similar project found that such systems could draw about 100 watts, equal to a bright lightbulb, from the blood of one human being.

Finally, there is the EATR (Energetically Autonomous Tactical Robot), discussed at a 2006 military robotics conference. The EATR concept is the pairing of one of the oldest types of power devices, the nineteenth-century Stirling engine, with a twenty-first-century autonomous robot. First designed by the Scottish clergyman Robert Stirling in 1816, Stirling engines convert heat into mechanical work, so EATR would power itself by scavenging about for anything organic that burns. At one presentation I witnessed, a scientist (a little too nonchalantly) described the EATR's potential battlefield diet as "grass, wood, broken furniture, [and] dead bodies."

## ROBOTS TAKE FORM

When we think of robots, we tend to imagine them to look like metal versions of people, as most of the classic Hollywood robots were usually just an actor sweating it out in a metal suit. Yet robots in reality come in all shapes and sizes. As Bill Gates says, "Although a few of the robots of tomorrow may resemble the anthropomorphic devices seen in *Star Wars*, most will look nothing like the humanoid C-3PO."

Robots are made up of sensors, processors, effectors, and power sources. The variety of each of these parts means that there is almost a limitless array of possible combinations to give a robot its appearance. Yet in many ways, the form that a robot takes is most dependent on the effectors. As one robot technician puts it, "The tool has to fit the task and the robot has to fit the tool."

Many robots are actually just vehicles that have been converted into unmanned systems. As robotics writer Daniel Wilson describes, "Every vehicle is a robot waiting to happen." Examples of these range from KITT, the brilliant but snide car that drove David Hasselhoff around in *Knight Rider*, to the Israeli military's Caterpillar bulldozers, which have been converted to remote control so that they can bash through obstacles in urban combat zones.

The simplest robot conversions involve plugging in a robot control unit (RCU) and sensors. These do the computations needed to fill the role of the human driver and "slave" the rest of the vehicle to its commands. The first notable robot-convertible car was the Navlab 5, built by the Carnegie Mellon University Robotics Institute. Nicknamed "Ralph," the Navlab 5 was actually a 1990 Pontiac minivan, outfitted with a GPS and a camera mounted on the dashboard, as well as a chain that linked the system to the steering wheel, to move it instead of the driver's hands. Everything was powered via a plug into the cigarette lighter.

In 1995, Ralph went on the "No Hands Across America" drive, in which the Navlab 5 drove from Pittsburgh to Los Angeles. The whole drive took place on normal roads (and without notifying any authorities). Two humans sat in the front seats, to take over in case anything went wrong, as well as not to freak out any fellow drivers. Ralph drove 2,800 of the 2,850 total miles on its own, 98.2 percent of the way, and even stopped at nearly every tourist site on the way, from the Indianapolis Motor Speedway to the Hoover Dam. Its journey ended at *The Tonight Show with Jay Leno*.

The advantage of such systems is that they can enhance the life and utility of a normal vehicle for a surprisingly cheap amount. For instance, it costs just $70,000 to convert a military Humvee into an unmanned system. Many expect that over the coming decades, all sorts of vehicles originally designed just for human drivers will be converted to add the option of unmanned control. For example, the F-35 Joint Strike Fighter is to be the U.S. Air Force's twenty-first-century mainstay. It is yet to even finish its testing and design, but its maker, Lockheed Martin, is already exploring how it might be modified to fly without a pilot inside. Many defense experts believe it may be the first of such convertible fighter jets, or the last of the manned ones, depending on your point of view.

## HUMANOIDS: THE HIGHEST FORM OF ARROGANCE

For all of humankind's progress in making various vehicles to move us from place to place, nothing yet beats our own effectors made for walking. Wheeled vehicles can only operate on 30 percent of the Earth's land surface, tracked vehicles on roughly 50 percent, while legs can tackle nearly 100 percent. Moreover, almost all the adjustments we have made to that surface to make it of value to us, our cities and buildings, were designed for those with legs.

The result is that while our image of robots as metal humans may come from a mix of Hollywood movies and arrogance, this "humanoid" form of two arms and two legs may well be a necessary design for many roles, especially in war. In 2004, DARPA funded a study of optimal military robot forms. It found that "humanoid robots should be fielded—the sooner the better."

The human form is just a shape that robots might take. There is no limit on its size. Asimo, the robot that Honda has spent over $100 million developing, is roughly the size of a person, while Chroino, a robot from the University of Kyoto, stands just a foot high. Then there are "mechas," basically giant robots. The word "mecha" comes from the Japanese abbreviation *meka*, shorthand for all things mechanical. Mechas are a staple of video games like *Metal Gear Solid* and Japanese *manga* comics in which the Tokyo of the future is filled with giant robots that work at construction, policing, and, of course, fighting wars. In Western science fiction, mechas have appeared as huge, building-sized robots, such as in the film *The Iron Giant,* or as just slightly bigger than human robotic suits, such as the one famously driven by Sigourney Weaver in *Aliens.*

With these inspirations in mind, many organizations have taken to making mechas real. Toyota Motor Corp. has developed the i-Foot. It is a 200-kilogram (440-pound) robot that stands on two legs and can climb stairs. The most popular military mecha designs borrow liberally from the world of science fiction. Saka-kibara Kikai Co., for example, makes the Land Walker, which is effectively the AT-ST All Terrain Scout Walker from the world of *Star Wars* made real (this was the machine that the Ewoks took on in *Return of the Jedi*). Standing on two legs, it is eleven feet high, and mounts two cannons.

The advantage of such mecha designs is that, just like with humans, legs give such giants the means to step over any obstacles that might limit where a truck or tank could go. However, the legs are also the major weakness. Robotic legs remain incredibly complex and expensive, and less capable the bigger they get. Moreover, being tall may allow the mecha to look down on opponents, but it also means that every enemy out there can see it. And even if those enemies are as unsophisticated as the stupid, despicable little Ewoks (who along with Jar Jar Binks are to blame for the ruination of the *Star Wars* franchise), all they have to do is take out the legs to ruin the mecha's day.

For similar reasons many disparage the humanoid design, for robots big or small. When most of us look in the mirror, we have to admit that our bodies are not perfect from an efficiency standpoint. For example, our visual sensors (our eyes) are really quite badly situated, create bad periphery, give us multiple blind spots, can't see in multiple spectrums, and are blind in the dark. Rodney Brooks at iRobot even goes so far as to describe human eyes as "badly designed," and cites octopus eyes as far more elegant and efficient.

The result then is that while humanoid robots are a central type of robot form, they will not be the only one. The same DARPA study that extolled the future of humanoid soldiers also found that two legs are not necessarily the optimal form. As Brooks predicts, "In the next 10–20 years, we will get over our *Star Wars–Star Trek* complexes and build truly innovative robots."

## BIO-BOTS: STEALING FROM NATURE

One source of inspiration for roboticists is the world around us. Taking lessons from nature goes deeper than just the broad outlines of what a robot looks like. As Georgia Tech professor Ronald Arkin notes, "Every aspect of robotics is touched by biology....It's a pervasive influence. A more appropriate question might be what aspects of biology have not had an influence—even seemingly esoteric fields such as ontogeny, immunology, and endocrinology have had their impact in the robot-ics research community." DARPA even employs a self-described "combat zoologist," who describes his job as "getting robots to jump, run, crawl, do things that nature does well. We're evolving our machines to be more like animals."

When you think about the history of evolution, this makes perfect sense. The ani-mal world has a huge head start on our human-designed technology. For instance, while humans only started flying about a century ago, insects have been doing it for three hundred million years. The result is that designers often take inspiration from biology in many ways, from animal capabilities and their patterns of movement to how animals act in the wild, as well as their overall form.

For instance, as they fly over large areas in search of prey, eagles' eyes can focus on distant objects without losing an overall wide-angle perspective. In essence, the bird's eyes come equipped with nature's version of the "picture in a picture" option now available on many big-screen TVs. The middle of their retina has a much higher concentration of light-sensitive cells than the surrounding areas. Thus, they see the center of whatever they look at in incredibly sharp focus without losing the wide-angle picture. A California company called Nova Sensors copied this to create a "detection tracking algorithm," giving such "eagle eyes" to unmanned planes. Another project, the Stickybot, copied how geckos climbed walls; the research was also used by the Pixar film company to help design their animated characters in *A Bug's Life*.

Some robots take broad ideas from nature in their overall design, but look nothing like them. One example is marsupial robots, in which one robot carries another inside. Just as Kanga carried little Roo in the *Winnie-the-Pooh* stories, so does REV, the Robotic Evacuation Vehicle, carry REX, the Robotic Extraction Vehicle. REV is a robot version of an ambulance, while REX is a robotic stretcher bearer that zips in and out to drag soldiers to safety.

Nature, as Darwin argued, can be quite cruel when it comes to design competition, and thus is as much of a guide of what not to build. For example, a DARPA survey of scientists and military officers found that while there was great interest in bipeds, they also expected to see future warbots with three legs (the extra leg for stability) and four legs (which the military officers found to be the most likely). What was interesting is that not one scientist or officer called for six-legged robots. As one said, "Fact of nature: There are no large land creatures with six legs and there never have been."

Designs that find their inspiration in living organisms are known as "biomi-metic" ("bio" from "biology" and "mimetic" meaning to "mimic" or "copy"). Perhaps the best known of these in military circles is a four-legged robot made by Boston Dynamics. The "Big Dog" (others call it the "Robot-Ass," but that name hasn't stuck for marketing reasons) is designed to serve as a modern-day packhorse, following after soldiers with their backpacks and other gear. The current prototype is the shape and size of a mule. The four legs differ from a mule's in having three joints and springs built into them that can change length, much like a tent pole. These joints and springs readjust five hundred times a second to balance the robot. The system guides itself with a stereo camera and a laser scanner. The firm hopes that a fully autonomous Big Dog will be "unleashed" around 2014.

## BREAKING THE MOLD

Evolution does not always lead to perfection or even the most efficient design. How else do we explain the platypus, Pauly Shore, or the fact that there are no wheels in nature? When vehicles, the human form, and nature don't provide the perfect solution, robot designers must then get truly innovative.

In some cases, robot engineers might mix and match different elements from the various forms to create a "hybrid." For example, the Office of Naval Research (ONR) is working on a robot that has a humanoid torso mounted on a Segway bottom. The army's Future Combat Systems (FCS) program is building an unmanned helicopter shaped like a floating fan, affectionately known in the community as the "UFO-Buttplug Hybrid." Finally, DARPA's survey on robotics futures found that many experts expect a future military use for "centaurs," robots having four legs but a humanoid body, like the mythic ancient Greek creatures.

These are just the stopgaps, however. Professor Shigeo Hirose at the Tokyo Institute of Technology is one of the true pioneers of robotics, having made the first "snake-bot" in the late 1970s. Since then, he has created everything from a "Ninja robot" that uses suction pads to crawl up the walls of high-rise buildings to a construction and rescue robot that is basically a seven-ton bulldozer equipped with four huge robotic spider legs. He believes that future robots will look like nothing before: "I have so many dreams about robots that have still not been realized."

The ultimate step in robot shapes and forms are systems able to change their shape to suit the mission. Such "self-transforming" or mighty "morphing" robots will range from changing slightly between a few designs like the Transformers to ones that could recast themselves into hundreds of forms like the T-1000 robot in the movie *Terminator 2*.

At the most simple level are robots with morphing effectors that alter to allow more efficient movement in different domains. An example of this is the RHEX made by Boston Dynamics. It has legs that can transform into flippers, allowing it to walk on land or swim underwater. The Naval Postgraduate School in Monterey, California, similarly built a plane the size of a small bird that can both fly and crawl. It originally came out of a request by the special forces for robotic planes that could do such things as fly up to a windowsill and then creep inside.

As systems advance, scientists expect to see a bevy of shape-shifting or self-reconfigurable robots. They might use legs to scramble up a hill and convert to wheels to roll down it. As more and more individual parts become capable of morphing, soon can the whole, making the idea of the T-1000 not just Hollywood fantasy. Indeed, scientists in Palo Alto have already made the Polybot, which uses hinged cubes to shape its entire body into all sorts of forms, such as shifting from a snake into a spider, while a team at MIT (working for DARPA's "Programmable Matter" project) has built "self-folding origami" machines that can fold themselves into "virtually any three-dimensional object" (swans included). Such transforming robots do have a key vulnerability. Any opaque molecule can block the reformation and communication pathways. It would have made a far less exciting movie, but all that Sarah Connor needed to do to defeat the T-1000 Terminator was pour maple syrup on it.

With all these possibilities, one day we may even see the very definition of robot turned on its head. Some argue that having a robot in one piece is old news and that the future is in distributed systems. These are where the robot is broken down into many pieces. At Carnegie Mellon University, for example, researchers are at work on "claytronic" robots, prototype, pocket-sized machines that use electromagnetic forces to move, communicate, and even share power. They can each act independently or all attach together into one big robot, akin to the Constructicons from *Transformers*.

The descriptor of "robot" started out as being just some human characters in a play, yet soon took on greater meaning. But as our mechanical creations "become more and more common," predicts Bill Gates, "it may be increasingly difficult to say exactly what a robot is. Because the new machines will be so specialized and ubiquitous—and look so little like the two-legged automatons of science fiction— we probably will not even call them robots."

# FOUR

# TO INFINITY AND BEYOND:
# THE POWER OF EXPONENTIAL TRENDS

*The saddest aspect of life right now is that science gathers knowledge*
*faster than society gathers wisdom.*

—ISAAC ASIMOV

"I decided I would be an inventor when I was five. Other kids were wondering what they would be, but I always had this conceit. And I was very sure of it and I've never really deviated from it."

Ray Kurzweil stuck to his dreams. Growing up in Queens, New York, he wrote his first computer program at the age of twelve. When he was seventeen, he appeared on the game show *I've Got a Secret*. His "secret" was a song composed by a computer that he had built.

Soon after, Kurzweil created such inventions as an automated college application program, the first print-to-speech reading machine for the blind (considered the biggest advancement for the visually impaired since the Braille language in 1829), the first computer flatbed scanner, and the first large-vocabulary speech recognition system. The musician Stevie Wonder, who used one of Kurzweil's reading machines, then urged him to invent an electronic music synthesizer that could re-create the sounds of pianos and other orchestral instruments. So Kurzweil did. As his inventions piled up, *Forbes* magazine called him "the Ultimate Thinking Machine" and "rightful heir to Thomas Edison." Three different U.S. presidents have honored him, and in 2002 he was inducted into the National Inventors Hall of Fame.

Kurzweil has found that the challenge isn't just inventing something new, but doing so at just the right moment that both technology and the marketplace are ready to support it. "About thirty years ago, I realized that timing was the key to success....Most inventions and predictions tend to fail because the timing is wrong."

Kurzweil has now founded a business that centers on figuring out this timing issue. Guessing the future seems a task for a psychic, not a prolific inventor. But Kurzweil comes with a pretty good batting average. In the early 1980s, he made the seemingly absurd forecast that a little-known project called the Arpanet would become a worldwide communications network, linking together humanity in a way previously impossible. Around the same time, he made the equally ridiculous claim that the cold war, which had just heated up with the Soviet invasion of Afghanistan, was going to end in just a few years. The Internet and the fall of the Berlin Wall made Kurzweil look like a clairvoyant.

"You'll often hear people say that the future is inherently unpredictable and then they will put up some stupid prediction that never came to bear. But actually the parameters of it are highly predictable," says Kurzweil. He isn't arguing that he can see into the future exactly and his business plan isn't to pick lottery numbers. Rather, he argues that the overall flow of the future can be predicted, especially when it comes to technologic change, even if the individual components cannot. He makes a comparison to thermodynamics. Imagine a kettle of water being put on a stove. What each individual molecule of water does as it heats up is inherently unpredictable. But the overall system is predictable; even if we don't know which water molecule will turn to steam first, we know the kettle will ultimately whistle.

An example of how Kurzweil's business makes money predicting the future happened in 2002. His research group looked at all the various technology trends and predicted that a pocket-sized reading device would be possible within four years. Such a prediction seemed a bad investment, as the technology wasn't even invented yet. But they positioned a project to be ready to deliver in 2006, just as the advancing technology made it workable. As he describes, "We use predictions to catch the moving train of technology at the right time."

To have such a business model, you have to have an immense faith in science. For Kurzweil, this even covers how he plans to extend his own life. Each day, the sixty-year-old takes a mix of some 250 dietary supplements. "I've slowed down aging to a crawl," he says. "By most measures my biological age is about forty, and I have some hormone and nutrient levels of a person in his thirties." With life spans advancing and technologic breakthroughs happening every day, Kurzweil believes that if he can just hold out long enough, he may even be able to live forever. It sounds crazy, but then again, this is a guy whom Bill Gates described as a "visionary thinker" and to whom thirteen universities have given honorary degrees.

Kurzweil gets a reported $25,000 for every speech he gives on the future of technology. At many of these speeches, he shares the stage with "Ramona." She is an AI programmed to be his alter ego (that is, if he looked like a twenty-five-year-old female rock star) and projected onto a screen behind him. The presentations, with both him and Ramona interacting with the audience, are considered so revolutionary and creative that they even inspired the 2002 movie *S1m0ne*, in which Al Pacino played a filmmaker who creates a Ramona-like AI to be the perfect actress.

With such a backstory, Kurzweil can sound a bit like the twenty-first-century technology version of "Professor" Harold Hill from *The Music Man*. But some serious folks are counting on his understanding of how the future is unfolding. Wall Street investors are pouring money into his FatKat (Financial Accelerating Transactions from Kurzweil Adaptive Technologies), the first hedge fund to make investment decisions using AI predictions. He is also one of five members of the U.S. Army's Science Board and has thrice given the keynote speech at the army's annual conference on the future of war.

Kurzweil describes the robots we now see in Iraq and Afghanistan, like the Predator or PackBot, as "only an early harbinger" of greater trends. Just around the curve is a moment for robotics and AI, which will "create qualitative change and social, political, and technological change, changing what human life is like and how we value it." He expounds, "In just 20 years the boundary between fantasy and reality will be rent asunder."

Kurzweil recalls that it was in 2002 when he first shared such visions with the army on the future of technology and war. His discussion of AI and robotics becoming the norm in war "was seen as amusing, even entertaining." Now his predictions of the future are "very much at the mainstream."

## EXPONENTIAL POWER

Kurzweil doesn't just pull his vision of the future from a crystal ball, but rather from a historic analysis of technology and how it changes the world around us. As opposed to gaining in a linear fashion, he argues that "the pace of change of our human-created technology is accelerating and that its powers are expanding at an exponential pace."

When something is moving at an "exponential" pace, it grows faster and faster each time it gets bigger. A familiar example is the idea of compound interest. Imagine a genie offers you the choice of either $1 million today or a magic penny that doubles in value every day for one month. The obvious choice would seem to be to take the $1 million. But that would actually be the sucker's play. Because of the exponential growth, the penny would be worth almost $11 million at the end of that month.

The challenge of exponential change is that it can be deceptive, as things often start out at a seemingly slow pace. Halfway into the month, the penny would have produced only $300. It's only as the pace goes up the exponential curve that the change truly accelerates.

Kurzweil's favorite example to illustrate how understanding exponentials can prove tricky even for scientists is the Human Genome Project. When it started in 1990, it had a fifteen-year goal of sequencing the more than three billion nucleotides that go into our complete DNA. The problem was that at the start of the project, only one ten-thousandth of the genome was mapped. "Skeptics said there's no way you're gonna do this by the turn of the century." Indeed, by year seven, close to the planned halfway point of the project, only 1 percent was complete. Kurzweil says, "People laughed and thought it would take another 693 years to complete. But they didn't account for the exponential." By that point the project was doubling its pace every year. "If you double from 1 percent every year over seven years, you get 100 percent. It was right on schedule." Here again, Kurzweil proved right and the project was completed in time.

Exponential change is most evident perhaps in technology products. A quick look at your cell phone should be persuasive enough. The first commercial mobile phone was the Motorola DynaTAC, which came out in 1983. It cost $3,500 and weighed two and a half pounds; it was nicknamed "The Brick." By 1996, Motorola was selling the Startac, which cost $500. Today, cell phones fit in your pocket and have gone from a luxury item to commonplace; some two billion people around the world have them and many are even thrown in for free when you purchase long-distance plans.

When it comes to computer technology, exponential progress is encapsulated in "Moore's law." In 1965, Gordon Moore, the cofounder of Intel, noticed that the number of transistors on a microchip was roughly doubling every two years. This realization was actually more exciting than it sounds, as each time you doubled the number of transistors on a microchip, you also cut the space between them by half. This meant that the time needed by electric signals to move between them was also cut in half. As companies crammed more and more transistors onto a chip, each and every year, Moore foresaw it would lead to faster and faster chips. Moore predicted that this simple doubling factor would spur everything from more powerful computers to automated cars.

Moore's prediction of microchip transistor doubling has held true in the four decades since, and has even sped up, now doubling every eighteen months. Showing how far we have come, Tradic, the first computer using transistors, built in 1955, had eight hundred. Almost sixty years later, Moore's old company Intel released the Montecito, which has 1.72 billion transistors on just one chip. Computers powered by these microchips have gotten more and more capable, again in an exponential way rather than an additive one. For example, the circa 2005 Dell computer I typed this book out on is already antiquated, but it has roughly the same capacity and power as all the computers that the entire Pentagon had in the mid-1960s combined.

But personal computers only tell part of the story of where Moore's law has taken us. An average PC today works in the scale of megaflops, being able to do millions of calculations per second. This is pretty impressive-sounding. But a present-day supercomputer, such as Purple that runs tests of nuclear weapons at Lawrence Livermore National Labs, can calculate 100 teraflops—100 million million calculations per second. Purple can do calculations in six weeks that would have taken supercomputers ten years ago, like the ones that first beat the human chessmasters, over five thousand years. But today's supercomputer is tomorrow's Commodore 64. The Department of Energy has already contracted IBM to build a next-generation supercomputer able to do 1,000 trillion calculations per second, or one petaflop, equivalent to the power of ten Purples.

The corollary to Moore's law is not just that microchips, and the computers powered by them, are getting more and more powerful, but that they are also getting cheaper. When Moore first wrote on the phenomenon in 1965, a single transistor cost roughly five dollars. By 2005, five dollars bought five million transistors. With lower exponential costs comes greater exponential demand. In 2003, Intel made its one billionth microchip after thirty-five years of continuous production. Only four years later, it had made its next one billion chips. The same changes have happened with the ability to store data. The cost of saving anything from the military's Predator drone footage of Iraqi insurgents to your old Depeche Mode songs is going down by 50 percent roughly every fifteen months.

Moore's law explains how and why we have entered a world in which refrigerator magnets that play Christmas jingles have more computing power than the entire NORAD nuclear defense system had in 1965. Exponential change builds upon exponential change and advancements in one field feed advancements in others. And lower prices in one field help feed new development in others. A good example is how advancements in microchips made portable electronics accessible to consumers. As more and more people bought such items as video and then digital cameras, it dropped the cost of equipping robots with the same kind of cameras (their electronic vision systems) by as much as 75 percent. This eliminated the barriers to entry for robots to be used across the marketplace, further dropping costs for robots as a whole, as more people could buy them. Rodney Brooks at iRobot calls this kind of cross-transfer "riding someone else's exponentials."

## AN EXPONENTIAL WORLD

Historic data shows exponential patterns beyond just Moore's law, which referred just to semiconductor complexity. For example, the annual number of "important discoveries" as determined by the Patent Office has doubled every twenty years since 1750. Kurzweil calls this pattern of exponential change in our world "The Law of Accelerating Returns."

This convergence of exponential trends is why technologic change, especially for electronics, comes not only quicker, but in bundles, rather than staying within one category. While microchip performance is now doubling roughly every eighteen months and storage every fifteen months, we are also seeing similar acceleration in categories far and wide. Wireless capacity doubles every nine months. Optical capacity doubles every twelve months. The cost/performance ratio of Internet service providers is doubling every twelve months. Internet bandwidth backbone is doubling roughly every twelve months. The number of human genes mapped per year doubles every eighteen months. The resolution of brain scans (a key to understanding how the brain works, an important part of creating strong AI) doubles every twelve months. And, as a by-product, the number of personal and service robots has so far doubled every nine months.

The darker side of these trends has been exponential change in our capability not merely to create, but also to destroy. The modern-day bomber jet has roughly half a million times the killing capacity of the Roman legionnaire carrying a sword in hand. Even within the twentieth century, the range and effectiveness of artillery fire increased by a factor of twenty, antitank fire by a factor of sixty.

These changes in capabilities then change the way we fight. For instance, exponentially more lethal weapons helped lead to equivalent exponential "stretching" of the battlefield. In antiquity, when you divided the number of people fighting by the area they would typically cover, on average it would take a Greek hoplite and five hundred of his buddies to cover an area the size of a football field. This is why in movies like *Spartacus* or *300* you can see the entire army during a battle. By the time of the American Civil War, weapons had gained such power, distance, and lethality that roughly twenty soldiers would fight in that same space of a football field. By World War I, it was just two soldiers in that football field. By World War II, a single soldier occupied roughly five football fields to himself. In Iraq in 2008, the ratio of personnel to territory was roughly 780 football fields per one U.S. soldier.

The same exponential change in how we fight has also gone on in the short time that war has taken place in the air. During World War II, roughly 108 planes were needed to take out a single target. By the time of the airstrikes over Afghanistan in 2001, the ratio had flipped; each plane was destroying 4.07 targets on average per flight.

Connectivity is also expanding at an exponential rate, allowing new technologies to change human society quicker and quicker. For example, the wheel first appeared in Sumer around 8500 B.C. But it took roughly three thousand years for the wheel to be commonly used in animal-drawn carts and plows. So the agricultural revolution that made possible human cities, and what we now know as "civilization," played out over several millennia. By the eighteenth century, communication and transportation had sped up to the point that it took only just under a century for the steam engine to become similarly widespread, launching the Industrial Age. Today, the spread of knowledge is nearly instantaneous. The Internet took roughly a decade to be widely adopted (and Internet traffic doubles every six months). And now that it is in place, an invention is shared across the world in nanoseconds.

And yet this change happened so quickly that we often forget how new it all is. In less than a decade, over a billion people, whether it was soldiers, terrorists, or grandmothers in Peoria, went from (1) never having heard of the Internet, to

(2) having heard of it, but never having used it (I still recall my mother asking, "What is this new 'Inter-web' thing?" soon to be followed by her asking about sending an "electronic letter"), to (3) trying it out, such as sending their first e-mail (when I was in college, e-mail was primarily used for sending out "Your momma so fat" joke lists), to (4) using it on a regular basis, to (5) that same soldier, terrorist, or grandmother not being able to professionally or *socially* succeed without it. And with the rise of three-dimensional "virtual worlds" like Second Life, that massive change is already old news.

When Kurzweil did a historic analysis of overall technologic change (measuring its advancement, complexity, and importance to human society), he found that the doubling period of this convergence of invention, communication, and progress happened just about every ten years. Individual technologies certainly move in fits and starts, but the overall flow for the aggregate of technologic change has clocked in at a fairly steady 7 percent annual rate of growth. This means that for the period up to the Industrial Age, the overall weight of technologic change was so slow that no one would significantly notice it within their lifetime. A Roman legionnaire or knight of the Middle Ages could go their entire life with maybe one new technology changing the way they lived, communicated, played, or fought. By the late 1800s, change was playing out over decades and then years, fast enough that people began to call it the "Golden Age of Invention."

But this change period was just the start of an acceleration up that exponential curve. The current rates of doubling mean that we experienced more technologic change in the 1990s than in the entire ninety years beforehand. To think about it another way, technology in 2000 was roughly one thousand times more advanced, more complex, and more integral to our day-to-day lives than the technology of 1900 was to our great-grandparents. More important, where they had decades and then years to digest each new invention, ours come in ever bigger bundles, in ever smaller periods of time.

## "THE SINGULARITY IS NEAR"

"We often say things like, 'No way this will happen in a hundred years!' But we are talking in about a hundred years at the current rate of progress," Ray Kurzweil points out. "If we are using today's rate, the twentieth century only had about twenty years of progress."

If Moore's law continues to play out, some pretty amazing advancements will happen that will shape the world of robots and war. By 2029, $1,000 computers would do twenty million billion calculations a second, equivalent to what a thousand brains can do. This means that the sum total of human brainpower would be less than 1 percent of all the thinking power on the globe.

Likewise, the trends for storing information are leading toward the same direction. Hugo de Garis, the head of the StarBrain AI project, has written a cheerily titled article on this entitled "Building Gods or Building Our Potential Exterminators?" In it he writes, "Within a single human generation, it will very probably be possible to store a single bit of information on a single atom." If this proves true, an object the size of a disc then would be able to hold a trillion trillion (a 1 with twenty-four zeros after it) bits of information. By comparison, the human brain is created from a genome of roughly twenty-three million bits of information. If computers can match this almost incomprehensible processing speed with such amazing memory, the advantage that human brains have of being so parallel starts to fall by the wayside. Moreover, many of the latest AI research projects, including StarBrain, are modeled after human brains. So they can build this parallelism into their own programs, nullifying our advantage.

With the ability to think faster and source more data, more and more becomes potentially possible for machines. The question then becomes, will computers ever be able to match the human brain in its thinking ability and then surpass it? Think of it this way: if a computer can process and store information billions or trillions times faster than a human, what research could it then accomplish? Would it be inconceivable that it could think up things just a thousand times faster, or even better? Could it even become so advanced as to become self-aware? This is the essence of what the scientists call "strong AI" or what science fiction writers call "HAL." If you project the current trends even further, Kurzweil claims, we are on track to experience "about twenty thousand years of progress in the twenty-first century, one thousand times more than we did in the twentieth century."

At a certain point, things become so complex we just don't know what is going to happen. The numbers become so mind-boggling that they simply lose their meaning. We hit the "Singularity."

## A SINGULAR SENSATION

In astrophysics, a "singularity" is a state in which things become so radically different that the old rules break down and we know virtually nothing. Stephen Hawking, for example, describes black holes as singularities where "the laws of science and our ability to predict the future would break down."

The historic parallel to singularities is "paradigm shifts," when some concept or new technology comes along that wipes out the old way of understanding things. Galileo's proof that the Earth rotated around the sun and not the other way around would be an example for astronomy, much as Einstein's theory of relativity was for physics. The key is that someone living in a time before a paradigm shift would be unable to understand the world that follows.

An example that many scientists cite is if you asked monks living in 1439 to predict advances in the future. They might predict such slight changes as better quills or ink for their illustrated manuscripts, or how a new well might be built. But they would likely not be able to conceive of how a rickety contraption made that year by Johannes Gutenberg, a German goldsmith, would become what *Time* magazine called "the most important invention of the millennium." Before the creation of the printing press and the singular break it created for society, it would have been simply impossible for those monks to imagine such things as mass literacy, the Reformation, or the *Sports Illustrated* swimsuit issue.

The idea of a singularity in relation to computer technology first came from Vernor Vinge. Vinge is a noted mathematician and computer scientist, as well as an award-winning science fiction writer. His most recent novel, *Rainbows End: A Novel with One Foot in the Future*, is set in 2025. He describes a world in which people "Google all the time, everywhere, using wearable computers, and omnipresent sensors." Vinge doesn't dedicate the book to his wife or parents or cat. Instead, perhaps sucking up to our future owners, he dedicates it to "the Internet-based cognitive tools that are changing our lives—Wikipedia, Google, eBay, and the others of their kind, now and in the future."

In 1993, Vinge authored a seminal essay. The title he chose, "The Coming Technological Singularity: How to Survive in the Post-Human Era," pretty much says it all. Vinge described the ongoing explosion in computing power and projected that "within thirty years, we will have the technological means to create superhuman intelligence. Shortly thereafter, the human era will be ended." Once superhuman intelligence gets involved, argued Vinge, the pace of technological development would accelerate even further than the doubling we have gone through for the last generations. There would be a constant feedback loop of artificial intelligence always getting better by improving itself, but with humankind now outside the equation. This would be the "point where our old models must be discarded and a new reality rules."

Vinge presented the paper at a NASA colloquium, arguing, "We are on the edge of change comparable to the rise of human life on Earth.... Developments that before were thought might only happen 'in a million years' (if ever) will likely happen in the next century." His idea became incredibly influential in both science and science fiction (writers such as Neal Stephenson and William Gibson wrestled with what it would mean for humanity, and such movies as *The Matrix* are set in a post-Singularity future).

Vinge's concept underlies how Kurzweil and other futurists envision the coming decades. If the present trends in technology continue, then the current exponential growth is picking up such steam that we hit a paradigm shift. The old models of understanding the world and what is and isn't possible will no longer hold true. "It's a future period," writes Kurzweil, "during which the pace of technological change will be so rapid, its impact so deep that human life will be irreversibly transformed." When we look at supercomputers and robots, we may well be like those monks seeing Gutenberg's printing press for the first time, trying to wrap our heads around what such a metal contraption really does signify.

But it is called *the* Singularity for a reason, as proponents of the idea see the change with AI and robotics as different from all the other paradigm shifts that have come before. Robert Epstein, a psychologist who has also worked on AI, explains, "It's not merely a technology that will change how we act, but it is a technology that is akin to a new species. It will change everything. Indeed, more than we can imagine because the new entity will be doing the imagining."

Vinge was ambivalent about whether this Singularity with an uppercase S was a good or bad outcome. He thought it could play out in a way that "fits many of our happiest dreams: a time unending, where we can truly know one another and understand the deepest mysteries." Or it could lead to the "physical extinction of the human race." You win some, you lose some.

By contrast, Kurzweil is the ultimate optimist. "At the onset of the twenty-first century, humanity stands on the verge of the most transforming and the most thrilling period in its history. It will be an era in which the very nature of what it means to be human will be both enriched and challenged, as our species breaks the shackles of its genetic legacy and achieves inconceivable heights of intelligence, material progress, and longevity."

Whether and when this all happens is an issue of debate. Kurzweil thinks the Singularity will become possible in the 2020s, but projects some lag time might be built in. Even then, the potential changes that he projects will occur soon (before most of us pay off the mortgages on our houses) sound pretty stunning. If the current rates of change hold up, by 2045, he writes, "the non-biological intelligence created in that year will be one billion times more powerful than the sum of all human intelligence today." Another way of thinking about it is that Kurzweil and others are arguing that my generation will be the last generation of humans to be the smartest thing on the planet. "Generation X" takes on a whole new meaning.

## QUESTIONING THE RAPTURE

Of course, not everyone buys such projections, or even the idea of the Singularity. Some argue it isn't possible, and others just mock it. The most stinging may be those who call the Singularity "The Rapture for Nerds."

That said, an amazing array of people have begun to weigh in on the side of the Singularity. Bill Joy, the cofounder of Sun Microsystems, and thus one of the Internet's godfathers, is very much a believer. "By 2030 we are likely to be able to build machines a million times as powerful as the personal computers of today." He then projects that "once an intelligent robot exists, it is only a small step to a robot species—to an intelligent robot that can make evolved copies of itself." In turn, while doing research for this book, I interviewed a U.S. special operations forces officer, just back from hunting the terrorist Abu Musab al-Zarqawi in Iraq. Our discussion was supposed to be on how his team uses unmanned systems, but at the end of the discussion, he added, "By the way, Joy's thesis is spot-on."

The economist Jeremy Rifkin, named by the *National Journal* as one of the 150 most influential people in shaping U.S. government policy, agrees as well. "Never before in history has humanity been so unprepared for the new technological and economic opportunities, challenges, and risks that lie on the horizon. Our way of life is likely to be more fundamentally transformed in the next several decades than in the previous 1,000 years. By the year 2025, we and our children may be living in a world utterly different from anything human beings have ever experienced in the past." The Singularity was even the subject of a 2007 U.S. Congress study by the Joint Economic Committee, entitled "The Future Is Coming Sooner Than You Think."

Rodney Brooks at iRobot acknowledges that the idea of the Singularity seems too futuristic to be true, but then describes a pattern he has noticed again and again. Incredibly bright people often draw a line in the sand on what computers will "never be able to do." But then technology continually forces them to erase that line and draw a new one. He likes to cite the story of Hubert Dreyfus as instructive for those who doubt the potential of technology.

Dreyfus is a noted philosopher at the University of California–Berkeley, located in the heart of Singularity fandom. In 1967, he famously predicted that no computer would ever beat him at chess. It turns out he wasn't the greatest of players and lost to a computer in his first and only match soon after. Dreyfus, who went on to author the 1972 book *What Computers Can't Do*, was undeterred. He revised his prediction to say that a computer would never be able to beat a skilled chess player, a nationally ranked player. A computer soon did. When that happened, he revised his prediction again (as well as his book title, which in 1992 was reissued as *What Computers Still Can't Do*), claiming that while computers may be able to beat most humans, they would never be able to beat the very best, such as the world champion chessmaster. Of course, this then happened in 1997 with IBM's Deep Blue.

Psychologist and AI expert Robert Epstein, a Singularity proponent who administers the Turing test program, acknowledges that "some people, smart people, say I am full of crap. My response is that someday you are going to be having that argument with a computer. As soon as you open your mouth, you've lost. In that context, you can't win. The only person able to deny the changes occurring around us is the one who hides, the one who has their head in the sand."

## THE MILITARY AND THE SINGULARITY

The question as to whether the Singularity will come and when depends on whether the same sort of exponential growth that happened in the past will continue in the years ahead. Does an exponential past necessarily mean an exponential future?

Between now and the Singularity (or not), all sorts of things could happen, from an asteroid hitting the Earth to World War III (then again, wars tend to spur technologic change to go even faster). More pertinently, it would seem that Moore's law can't stay true forever. At a certain point, around 2020 in the projections, the number of transistors packed onto a microchip must move down into the atomic level; that is, there may be no space left between the atoms themselves for electric signals. Overheating is another problem at this density, as the electric currents have to run through ever more tightly packed transistors.

Yet, again, technology may well leap over and around the problem. In 2007,

IBM and Intel found a way to use hafnium (the same isomer used for novel UAV nuclear power systems) to build a next generation of microchips with circuits as small as 45 nanometers, about one two-thousandth the width of a human hair. In 2008, scientists at the National Institute for Materials Science in Japan designed a three-dimensional "duroquinone" that was only seventeen molecules large but could simultaneously carry out sixteen times more operations than a normal computer transistor. Other breakthroughs have been made in subatomic circuitry. Instead of switching an electric current on and off, to create the 0s and 1s that make up binary language, these take electricity out of the equation and use magnets to control the direction in which electrons spin. Not only is there no overheating, but it also means the chip can work for as long as it keeps its magnetic charge. That is, while an electric charge needs to be linked to some power source, a magnet keeps its charge even after you pull the plug. Here again, the credit goes to the military, with DARPA pouring more than $200 million into such quantum research.

For this reason people like Microsoft founder Bill Gates are uniformly optimistic that each of the various hurdles to robotics will be knocked out in the coming years. "The challenges facing the robotics industry are similar to those we tackled in computing three decades ago." Or as military robotics developer Robert Finkelstein puts it, finding the solutions needed to take robots to the next level and beyond "doesn't require us to try to discover new laws of physics, antimatter, or cold fusion. It's just a matter of proper funding and dedication." Which brings us back to the military.

Some believe the military is an integral part of bringing the Singularity into being, because of the massive investments it has made in R&D for things like artificial intelligence and sensors, as well as the immense marketplace it has created for hardware. I asked an executive at one defense contractor whether he agreed with the crazy ideas being bandied about on singularities and robots becoming as smart as humans. He replied, "If this war keeps going on a few more years, then yes."

Robert Epstein sees the military's role as more than simply funding the Singularity. It is the most likely integrator needed to bring it all together. He describes how there are all sorts of research programs and companies around the globe, working on various technologies, from pattern recognition software and robotic sensors to artificial intelligence and subatomic microchips. "When you marry all that up with the strategic planning that the military brings to the table, you will end up with a qualitative advance like no other. At that point prediction of what comes next becomes difficult.... That's when you hit the Singularity, where all the rules change, in part because we are no longer making the rules."

In the end, we don't now know yet whether computer, AI, and robotics development will reach a singularity or the Singularity. Indeed, this could be the one prediction that Kurzweil and his cohort simply get wrong. We do know, however, that major shifts are already going on in computing power and machine intelligence. And if the trends for the future do hold true even at the most minimal level, then things are going to get real interesting in the not too distant future.

# FIVE

# COMING SOON TO A BATTLEFIELD NEAR YOU: THE NEXT WAVE OF WARBOTS

*They're going to sneak up on us.... They're going to do more and more of the toting. They're going to do more and more of the surveilling. And when they start fighting, no organized force could stand against them.*

—JOHN PIKE, GlobalSecurity.org

"William James once said, 'We are literally in the midst of an infinite.' Today, there is an infinite going on in the world of war.... The challenge is that there are fewer things to look for and more information. The needle in the haystack is at the essence of counterinsurgency. Machines can filter down what we need to see. Instead of us telling machines where to go, it is increasingly machines telling us."

Noah Shachtman is the new breed of war correspondent. He's quoting the nineteenth-century philosopher William James, but doing so while talking about the next generation of robots, as we sit in a chic Manhattan bar filled with rap stars and models. Describing his beat as "technology, national security, politics, and geek culture," Shachtman writes for the *New York Times* and is a contributing editor at *Wired*, the digital world's most popular magazine. He also runs *Danger Room*, the blog focusing on "what's next in national security."

In the course of his reporting, Shachtman has done everything from sneaking into the Los Alamos nuclear lab to riding out on missions in Iraq with an EOD team and their robots. Based on these experiences, he is emphatic that we've only seen the start of the robotics trend in war. "In both war and police actions, you will see more and more of robots in all shapes and sizes.... There is a huge growth curve, with no signs of slowing down. To see having one [robot] in every squad isn't all that crazy. And that is before you get into the sexy, futuristic stuff." For military robotics in the next decade, "there is zero chance of the field not increasing exponentially."

## THE COMING WARBOTS BY LAND

The systems just rolling out or already in prototype stage are far more capable, intelligent, and autonomous than ones now in Iraq and Afghanistan. But even they are just the start. As one robotics executive put it at a demonstration of new military prototypes in 2007, "The robots you are seeing here today I like to think of as the Model T. These are not what you are going to see when they are actually deployed in the field. We are seeing the very first stages of this technology."

Charles Shoemaker, who runs the Robotics Program Office at the Army Research Laboratory, agrees. "It is really, really hard" to create military robots that fight on land, more officially called UGVs (unmanned ground vehicles), and especially ones that can operate independent of a human controller. "But I'm convinced that we're going to develop systems that work for a whole range of tactical missions.... We could be at the dawn of a golden age of military UGVs."

To make such visions come true, the Pentagon's Joint Robotics Program is currently developing twenty-two different prototypes of intelligent ground vehicles. They range in size from tiny eight-pound robots to the world's current biggest robot, a 700-ton robotic dump truck capable of hauling 240 tons of earth at a time, which also served as the model for the character Long Haul in the *Transformers* series. In addition, there are various programs to convert existing manned vehicles into UGVs. Robert Finkelstein thinks that the conversion of supply trucks to unmanned vehicles will actually be among the first major uses. A converted Humvee has already driven around military bases at an average of thirty-five miles per hour and never veered from its planned route by more than eight inches. Describing how supply convoys of such unmanned trucks would cut losses in Iraq, he exclaims, "It's already been done. The kits are available. We can save lives!"

In addition to these plug-in kits, the next wave of new robots to be deployed on land will mostly be "new and improved" versions of existing platforms. For example, iRobot's original PackBot just had a digital camera that sent back views of what the robot was seeing, making it essentially a mobile pair of binoculars. Now most PackBots perform EOD roles with fairly simple effector arms and grippers.

But as new add-ons are developed, the same robot will be able to take on a wider and wider set of battlefield roles. For example, the company has already tested out an armed PackBot. For iRobot's first weapon, it eschewed the variety that Foster-Miller had for the SWORDS and instead chose a good old-fashioned shotgun, because it is "so versatile." The robot can now fire a variety of ammunition, including nonlethal rubber bullets, rounds that can blow down a door, and even more powerful "elephant killer" bullets.

Another version is called the REDOWL (Robotic Enhanced Detection Outpost with Lasers), which uses lasers and sound detection equipment to find any sniper who dares to shoot at the robot or accompanying troops, and then instantly targets them with an infrared laser beam. "You'll actually see the sniper before the smoke disappears from the shot," said retired admiral Joe Dyer, who leads the military programming at iRobot. He adds that in tests, it's been 94 percent accurate and is smart enough that "it can tell the difference between a 9 millimeter pistol and an AK-47 or an M-16."

Foster-Miller has similar plans to upgrade its current generation of ground robots. For example, the first version of the armed SWORDS needed the remote human operator to be situated within a mile or two, which can still put the human in danger. Vice President Robert Quinn describes how the company plans to vastly extend the range of communications to get ground robot operators completely off the battlefield. "It is not an insurmountable problem. It is nothing that money and time can't solve." The SWORDS itself is being replaced by a new version named after the Roman god of war. The MAARS (Modular Advanced Armed Robotic System) carries a more powerful machine gun, 40mm grenade launchers, and, for nonlethal settings, a green laser "dazzler," tear gas, and a loudspeaker to warn any insurgents that resistance is futile.

As these systems evolve, we will also soon see entirely new unmanned combat vehicles hit the battlefield. One such prototype was the Gladiator. Described as the "world's first multipurpose combat robot," it came out of a partnership between the Marine Corps and Carnegie Mellon University. About the size of a golf cart, the vehicle was controlled by a soldier wielding a PlayStation video game controller, but software plug-ins will allow it to be upgraded to semiautonomous and then fully autonomous modes. Fully loaded, it costs $400,000 and carries a machine gun with six hundred rounds of ammunition, antitank rockets, and nonlethal weapons. "It is just fucking nasty," raves journalist Noah Shachtman.

Not all ground robots will take on combat roles. For instance, medics have long had one of the most dangerous jobs on the battlefield. A former army special forces officer explains how this is generating a pull for robotic solutions. "If you can avoid unnecessary situations where you expose them [medics] to fire and you end up with two dead guys, then we have a responsibility to the American people to avoid that."

An early entry into the "medbot" field is yet another improved version of the PackBot, known as the Bloodhound. Whenever a soldier is hurt, an alert will go out and the robot will find the wounded soldier on its own. Then the robot's human controller, who might be located anywhere in the world, will check out the soldier via the video link and treat them using the robot's onboard medical payload, which will include a stethoscope (likely very cold, with no one to breathe on it), liquid bandages, and even automatic syringes to dispense morphine or antidotes.

The next step will be specially designed medbots, such as the previously mentioned marsupial pair of REV and REX. REV, the Robotic Evacuation Vehicle (a robot version of an ambulance), carries REX, the Robotic Extraction Vehicle, a tiny stretcher bearer that zips out to drag soldiers into the safety of the ambulance. REX has an arm with six joints to drag a soldier to safety, while REV has a life-support pod that even comes with a flat-screen TV facing the wounded soldier's face so that operators can see and communicate with the human on the other end if they are conscious.

Ultimately, REV will be configured so that complex surgeries can occur inside the medbot. DARPA has already spent more than $12 million on developing such a remote "trauma pod" (originally called a "crechepod" in Frank Herbert's *Dune* novels) that will automatically diagnose and treat a wounded soldier. The soldier will be loaded up into the protected pod and sped away to safety, all the while being scanned from head to toe, given oxygen, and their information processed to remote doctors, who might even perform surgery. The system is based on the da Vinci robotic surgical system, a commercialized technology that is already used at some three hundred facilities around the world. The maker, SRI International, thinks that such a system "could be operational on the battlefield in ten to fifteen years." As described by Russell Taylor, an engineering professor at Johns Hopkins University, these robotic systems don't just allow a surgeon to do their work remotely, but to do it with far greater machine-enabled dexterity than before. "The average surgeon will become as good as the star surgeon, and the star will have superhuman capabilities."

Of course, robots will have a hard time replicating the compassion of a real-life medic. As one special forces soldier says, "The last thing I want to see if I'm about to die is a robot coming for me. I want to see a human." On the other hand, that robot may be able to go where humans could not, so their lack of a bedside manner is seen as an acceptable trade-off.

All these various robots were originally supposed to come together in the army's $340 billion Future Combat Systems (FCS) program. Begun in 2003, the FCS concept was to transform the army of the twenty-first century into smaller, lighter, and more lethal units of manned and unmanned components, joined together by a massive computer network.

The FCS was certainly ambitious. Its original plan involved everything from replacing the army's twenty-eight thousand armored vehicles with a new genera-tion of manned and unmanned vehicles to writing some thirty-four million lines of software code to link them all together. Starting in 2011, the army hoped to start spiraling a few new FCS technologies at a time into the force. By 2015, it planned to reorganize its units into a new revolutionary model of organization. Each bri-gade is actually to have more unmanned vehicles than manned ones (a planned 330 unmanned to 300 manned). These would include the Multifunction Utility/ Logistics and Equipment Vehicle, or MULE. Made by Lockheed Martin, it is about the size of a small car. The aptly named MULE can do anything from carrying equipment and supplies to mounting its own weapons, such as a machine gun or rockets. The runt of the FCS litter is the Small Unmanned Ground Vehicle. This is essentially a smaller, but souped-up version of the Packbot. iRobot has received a contract of $51 million to make the first run of 3,600 of these robots. Each unit would also have its own unmanned air force of over a hundred drones, ranging in size from a fifteen-pounder that fit into soldiers' backpacks to a twenty-three-foot-long robotic helicopter.

In addition to vehicles, FCS also started the development of a variety of unmanned ground sensors. One example is the Sensor Dart, a small missile packed with sensors that will be carried on a drone and then dropped behind enemy lines to report back on what's going on. In the air it will have wings, and then transform into an earth-penetrating dart. In the testing so far, soldiers have been enthusi-astic. One commander said that if his unit had the systems during their previous deployment to Iraq, "it would have saved an NCO's life, his squad leader's legs, and his team leader's hand."

The FCS plan soon ballooned in cost, however. Military robots expert Robert Finkelstein described it as "the largest weapons procurement in history . . . at least in this part of the galaxy," while former army officer Ralph Peters jokes that FCS spending has gotten so out of control that "it's the system that ate the army."

With the Congressional Budget Office projecting that FSC would cost as much as $170 billion more than the original plan, the Obama Pentagon moved to trim FCS. Interestingly, this actually lead to the purchasing of more robots rather than less. While unmanned systems made up roughly half of the new FCS vehicles to be bought in the original plan, they represented only 15 percent of the planned costs. Similarly, the majority of technical hurdles remaining for the program (twenty-seven identified by the CBO) were on the manned vehicles rather than the unmanned ones. So, in 2009, the Pentagon cut FCS. Or, more specifically, it canceled all the plans to buy new manned vehicles and committed just to buying the next generation of unmanned ones.

## WET AND WILD, ROBOT STYLE

A broad new set of robots is also being introduced for war at sea, where the main differentiation is whether they are designed to operate on the surface, like a boat, or underwater, like a submarine.

Robots of the first type, unmanned boats, are called USVs (unmanned surface vessels). They actually have a great deal in common with the simpler land robots, as they both primarily operate in a two-dimensional world. Many basic USVs merely entail taking sensors and a remote control unit and plugging them into a boat.

However, many think the sea is actually a far more difficult environment for robots than land. "Everything's working against you," says Robert Wernli of the Ocean Systems Division of the Space and Naval Warfare Systems Center (SPA-WAR) in San Diego. Waves and currents can pull a boat off course. Visibility is lower, and sometimes communications are more difficult. Plus, robots can get seasick; the constant motion and corrosive effects of salt water cause mechanical breakdowns much more rapidly than on land.

So far, the prototype USVs tend to be smaller boats than large navy ships. One example is a thirty-six-foot robotic motorboat called the Spartan Scout, which the navy has spent some $30 million developing. Guided by a GPS navigation system, the boat can be on its own for up to forty-eight hours, and speed up to fifty miles per hour. Filled with sensors (including day and night video cameras), Spartan Scout is designed to carry out surveillance, patrol harbors, and inspect any suspicious ships that might be trying to pull another U.S.S. *Cole*–type attack by sneaking up on a navy vessel. If it finds something fishy, the robot boat is also packing a .50-caliber machine gun. Spartan Scout got its first real-world use in the Iraq war in 2003, inspecting small civilian boats in the Persian Gulf without risking sailors' lives. The boat also mounts a loudspeaker and microphone, so an Arab linguist back on the "mothership" would interrogate any suspicious boats that the Spartan Scout had stopped. As one report put it, "The civilian sailors were somewhat taken aback when they were interrogated by this Arab speaking boat that had no one aboard."

The other type of navybots are UUVs (Unmanned Underwater Vehicles). These are designed for underwater roles such as searching for mines, the cause of most naval combat losses over the last two decades. Many UUVs are biologically inspired, like the "Robo-lobster," which operates in the choppy waters close to shore. But others are converted torpedoes, like the REMUS, which was used to clear sea mines in Iraq, or even mini-submarines, which are launched from manned submarines to hunt down the enemy.

The sea will also prove to be a new platform for robots to fly from. The navy plans to equip many of its ships with the MQ-8 Fire Scout, a sister version of the robotic helicopter used in the army FCS plans. Able to take off from and land autonomously from any warship with a small deck, the Fire Scout can fly more than six hours. It packs thermal imagers, radar, high-powered video cameras, and a laser designator that can target for the mothership's weapons or fire its own rockets. With a range of over two hundred miles, the robotic chopper can take the ship captain's eyes farther than ever before, including even inland.

The most novel of the drones at sea may be the Cormorant, DARPA's design for a submarine-launched flying drone. Operating a plane off a submarine may sound new, but it actually dates back to World War II; indeed, the very first air attack on the mainland United States was in 1942, when a submarine-launched Japanese plane bombed Brookings, Oregon. What is novel about the Cormorant is not only that it would be unmanned, but also that it would be able to be both launched and recovered while the submarine stays hidden under the water. Having wings like a seagull, the drone would be squeezed into a missile launch tube. Whenever the sub commander wants to scout above or launch a surprise air attack, the drone would be fired from the tubes, float to the surface, and then launch into the air using converted rocket boosters. The drone would then fly back to a rendezvous location on its return. It lands in the water, sinks back down, and the submarine scoops the robot plane back inside.

## TOP (UNMANNED) GUNS

As with ground robots, the next wave of robot planes, also known as "unmanned aerial vehicles" or "systems" (UAVs or UASs), will be a mix of upgraded current systems, converted manned vehicles, and brand-new designs. For example, the Predator drone today does surveillance and also some ground attack missions. New versions are being reconfigured for electronic warfare, submarine hunting, and even air-to-air combat. Thomas Cassidy, a former navy fighter pilot (so respected that he even had a cameo in the movie *Top Gun*) and now CEO of Predator's manufacturer, General Atomics, declares, "I want to see a Predator coming back here with MiG kills painted on its side; and that will happen soon."

The next generation of the Predator is the even more menacing-sounding Reaper, an Air Force drone about four times bigger and nine times more powerful. Among its improvements is a Microsoft Windows software package that has "automatic man-made object detection" and "coherent change detection." Not only can the plane come close to flying itself, but its sensors can recognize and categorize humans and human-made objects. It can even make sense of changes it is watching, such as being able to interpret and retrace footprints or even lawn mower tracks. As of 2009, twenty-eight Reapers were in service, with many deployed to Afghanistan, "standing alert somewhere in case a certain high-priority target pops his head out of his cave."

In turn, just as the Reaper hit the battlefield in 2009, Cassidy's company turned out a prototype of its successor, called the Avenger. Powered by a jet engine, it can fly twice as fast as the turboprop-powered Reaper, and more than three times as quick as the Predator, while carrying over three thousand pounds of weapons. It is also specially designed to be stealthy, with radar absorbing materials, a swept wing, and internal bomb bay. Of note, the prototype also came equipped with a tailhook, potentially allowing it to land on aircraft carriers at sea.

As new prototypes of unmanned planes hit the battlefield, the trend will be for the size extremes to be pushed in two directions. Among the planes being made at the military's flight test center near Groom Lake, Nevada, better known as Area 51, is the Lockheed Martin "Polecat." Described as looking like "a B-2 bomber's chick," the bomber drone is made of only two hundred parts that are glued, rather than riveted, together to increase its stealthiness. It will be rigged up with "a fully autonomous flight control and mission-handling system," meaning it will be able to carry out its mission from takeoff to landing without any human instruction. Lockheed Martin claims its studies show Polecat to be five times more survivable and mission-effective than the air force's plans for a manned bomber version of its new F-22 fighter jet.

Not having pilots who need to be replaced every ten hours or so will also allow unmanned planes to have greater endurance and become far bigger than any created so far. For example, Boeing is at work on a glider powered by solar energy and liquid hydrogen that could stay aloft for seven to ten days. It would have a wingspan almost the length of a football field. The next step is DARPA's plan announced in 2007 for a "VULTURE" (Very-high-altitude, Ultra-endurance, Loitering Theater Unmanned Reconnaissance Element) drone, which the agency hopes will be able to stay aloft for as long as five years.

We may even see the return of blimps to warfare. Lockheed Martin has been given $150 million to design and build a robotic "High Altitude Airship" twenty-five times larger than the Goodyear blimp. Such huge, long-endurance blimps open up a whole new range of roles not normally possible for planes. For example, airships could literally be "parked" in the air, as high as one hundred thousand feet up, for weeks, months, or years, serving as a communications relay, spy satellite, hub for a ballistic missile defense system, floating gas station, or even airstrip for other planes and drones.

At the other end, there will also be more of what Noah Shachtman describes as "itty-bitty, teeny-weeny UAVs." Some even think that small, pilotless planes will make up as much as 75 percent of the military's future air forces, mainly because they are cheap, easy to use, and perhaps most suitable for the clogged urban battlefields of the twenty-first century.

Any plane that is smaller than fifteen centimeters is technically known as a "micro-unmanned aerial vehicle." As far back as the 1970s, the CIA experimented with a "bio-inspired" drone the size of a dragonfly. The problem during testing was that, as one scientist describes, "It was tough to track on film and easy to lose in the grass." Today, the exact nature of this program is classified. But the military's belief in what is possible is illustrated by a contract let by DARPA in 2006. It sought an insectlike drone that weighs less than 10 grams, is smaller than 7.5 centimeters, has a speed of 10 meters per second, a range of 1,000 meters, and can hover in place for at least a minute. The agency has also given Lockheed Martin a $1.7 million contract to build the SAMERAI drone. As Shachtman says, this drone is "similar in size and shape to a maple tree seed," but is powered by a chemical rocket able to carry tiny sensors over a half mile from the launch point.

Tiny drones are such a hot item because they make a perfect platform for spy jobs. As one scientist described, "A lot of the three-letter agencies are interested in miniaturization." They can do things like "perch and stare" into windows or climb up walls or into pipes. Besides carrying tiny sensors and cameras, they might be loaded with electromagnets, which will allow them to recharge themselves off electrical outlets or lightbulbs. They might also carry tiny weapons, such as a small syringe filled with poison (an idea featured in Dan Brown's novel *Deception Point*).

Some even believe that such microsystems could eventually go down to the nanoscale. "Nano" is Greek for 10 to the minus 9. So to be at nanoscale is to be in measures of one billionth of a meter, or the width of a human hair cut into a hundred thousand parts. While the idea has been bandied about in such fiction as Michael Crichton's novel *Prey*, many think it could come to fruition in the coming decades. Boston College researchers have already built a chemically powered nanomotor that is just seventy-eight atoms in size, while those at a university in the Netherlands have made a solar-powered engine just fifty-eight atoms in size.

Tiny engines allow tiny machines. And tiny machines may mean teeny-tiny robots, or "nanobots." A major advancement in these happened in 2007, when David Leigh, a professor of chemistry at the University of Edinburgh, revealed that he had built a "nanomachine," whose parts consisted of single molecules. When asked to describe to a normal person the significance of his discovery, Leigh said it would be difficult to predict. "It is a bit like when stone-age man made his wheel, asking him to predict the motorway," he said. He would make one venture. "Things that seem like a *Harry Potter* film now are going to be a reality."

Such machines are still fairly limited in military applications; early models can only do things like copy a plant's photosynthesis or move a molecule of water around. But military analysts see the potential of these prototypes' one day becoming weapons that work at the molecular level, such as tiny missiles that could truly hit with pinpoint precision or nanobots designed to deconstruct a target from the inside out.

Such minuscule designs actually mandate that the systems will have to have high autonomy, carrying out their missions without human controllers. First, to be useful, the robots will have to be "organic" to the team. That is, they will have to be relatively easy to use, not require special training, and if the goal is to saturate the battlefield, not require each and every small robot to have a soldier somewhere having to stop his mission and fly it. Another problem is that flying the smaller designs actually makes most human operators nauseous. Imagine watching video from a TV camera mounted on a butterfly, as it bobs up and down crossing a room; that is the sensation of flying a micro-drone.

The centerpiece of future plans for unmanned drones, however, is the UCAV (unmanned combat aerial vehicle). This type of drone is specially designed to replace the ultimate of human pilot roles, the fighter jock. A key UCAV prototype was the Boeing X-45, which one author described as "flat as a pancake, with jagged 34-foot batwings, no tail and a triangular, bulbous nose" that make it look like "a set piece from the television program *Battlestar Galactica*." X-45 also has a cousin, the Northrop Grumman X-47, which is roughly the same size, but designed to land on aircraft carriers. These drones were designed to be especially stealthy for the most dangerous roles, such as sneaking past enemy air defenses. In war games, UCAV prototypes have shown some impressive capabilities. They've launched precision-guided missiles, have been "passed off" between different remote human operators nine hundred miles away from each other, and in one war game autonomously detected unexpected threats (missiles that "popped up" seemingly out of nowhere). The drones engaged and destroyed them, and then did battle damage assessment on their own. They also promise to lighten the load on human operators. One human pilot remotely flew two UCAVs at the same time.

The X-45 may have been too good, too soon. The fighter drone's capabilities made it appear as a competitor to the air force's new manned fighter planes, the F-22 and F-35, in which the air force had already invested $28 billion and $40 billion respectively developing (the X-45's development cost was $1.8 billion). So in 2006 the air force decided to cancel X-45 and let the navy fund the drone program on its own. Many believe, however, that the program still lives on inside the "black" budget and that ultimately Congress and changing leadership within the air force will soon bring the air force's robotic fighter plane program back to life.

The pattern with unmanned planes in the early twenty-first century seems to be mirroring what happened with manned planes in the early twentieth century. There was initial skepticism and opposition to them in general, followed by limited use in observation and spotter roles. Soon, however, they began to be used for ad hoc attack roles, much as the early observation plane pilots in World War I began to drop their own grenades and homemade bombs on the enemy below. Perhaps the most amusing parallel in the Iraq war was when an enlisted soldier flying a Raven drone spotted an enemy insurgent planting an IED. He tried to show his commanding officer the danger, but the officer couldn't pick out the image of the insurgent on the view screen. So the operator kept circling the drone closer and closer to the insurgent. Still, the officer couldn't see the Iraqi. Finally, the soldier just got frustrated and flew the drone directly into the insurgent's chest. Then, referencing the annoying Verizon cell phone commercial, he asked his commander, "Can you see him now, sir?!?"

Of course, just as World War I pilots couldn't just watch each other merrily going about their business of bombing their side on the ground, and so started taking potshots at each other, so too is the next step of advancement unmanned drones that are specially designed to take on other robotic planes. In 2006, DARPA budgeted $11 million for the "Peregrine UAV Killer." Like a peregrine falcon, it is designed to loiter over an area, stealthily gliding about until it sees an enemy UAV, and then quickly dive down and blast it. Drone versus drone may be the next step in warfare.

## ASTROBOTS GO TO WAR

But why should robot war be limited just to the Earth? Space has long been a location for satellites that provide military advantage back on Earth, such as spying or beaming GPS locations, but it has yet to be a battleground itself. However, plans for conflict taking place in space go well back to the antisatellite programs of the United States and the Soviets during the cold war and Ronald Reagan's "Star Wars" missile defense program in the 1980s. In 2000, these received a new injection of funds by the U.S. Space Commission, which was chaired by a retired Ford administration official named Donald Rumsfeld. The commission sought out media attention by hyping a rising threat to U.S. space assets in the form of a "space Pearl Harbor." After becoming secretary of defense, Rumsfeld commissioned twenty more studies on war in space and the U.S. military organized the U.S. Space Command.

If space is to become a new potential zone of conflict, its unique nature demands that unmanned systems play a key, and perhaps near-exclusive, role. Not only do weapons in space need to stay up there a long time, but the major challenge of fighting in space is first getting things into space. It costs roughly $9,100 a pound to launch anything into space with the Space Shuttle. So if a system is to be manned, the humans and each and every pound of water, food, and oxygen tanks to keep them alive are expensive to send. Likewise, manned systems in space are incredibly vulnerable (one bullet or laser hole and there goes all the air).

Instead, the United States has already started work on a number of unmanned systems for potential use in space. One example is the X-37, an orbital test vehicle about a quarter of the size of the Space Shuttle, which flew its first test flight in 2006. The military's strong interest in it is perhaps best illustrated by the fact that, while the program was originally run by NASA, its development was transferred to DARPA in 2004.

Another program is the X-41 Common Aero Vehicle, also known as the Falcon program. Planned for testing in 2010, it is a cross between an intercontinental ballistic missile and the Space Shuttle. It is designed to travel at the border between space and the atmosphere, around one hundred thousand feet. But, unlike a missile, it will be able to come back after a mission if it finds no targets. As John Pike of the Global Security organization comments, the aim is to give the United States the ability to "crush someone anywhere in the world on 30 minutes notice, with no need for a nearby airbase."

This weaponization of space, unmanned or not, is certainly controversial. Former U.S. Senate majority leader Tom Daschle defined the Rumsfeld plans as "the single dumbest thing I have heard so far from this administration.... It would be a disaster for us to put weapons in space of any kind under any circumstances. It only invites other countries to do the same thing." Lieutenant Colonel Bruce M. Deblois of the U.S. Air Force published a detailed study that concurred with Daschle. The report argued that while being the first to deploy weapons in space might seem advantageous, it would only open up the floodgates for others to do the same.

These fears do appear to be playing out. In 2007, after a test of their own antisatellite missile, senior colonel Dr. Yao Yunzhu of the Chinese army's Academy of Military Science issued a not thinly veiled warning. If the United States thought it was going to be "a space superpower, it is not going to be alone.... It will have company."

This debate will likely rage on for years, if not decades, or at least until the Vulcans arrive to resolve it. But what is interesting is that governments are not the only ones looking at space as a new unmanned battleground. In 2007, the Tamil Tiger group of Sri Lanka became the first, but likely not last, terrorist group to takes its operations into space, hijacking the signal from an Intelsat satellite and using the commercial satellite to beam its own messages back to Earth. And just as private companies like Blackwater have reentered the conflict game on this planet, we should not be surprised if privatized conflict also arises one day in space, especially with the growth of private space businesses, such as Richard Branson's "Virgin Galactic" or Google's $30 million prize that will go to the first private team able to land a robot on the moon (one of the competitors is actually also the maker of some of the Pentagon's energy beam weapons programs).

Robo-One, a robot combat event held in Japan every year, may provide a taste of what's to come. The competition organizers have announced plans for a new division in 2010: robot combat in space. A small satellite carrying humanoid robots will be blasted into the heavens. "Once safely in orbit, the satellite will release its robotic passengers, who will proceed to fight each other in the vacuum of space."

If that does not signify human progress, what does?

# SIX

# ALWAYS IN THE LOOP?
# THE ARMING AND AUTONOMY OF ROBOTS

*Wars are a human phenomenon, arising from human needs for human purposes. This makes intimate human participation at some level critical, or the entire exercise becomes pointless.*

—COLONEL THOMAS K. ADAMS, U.S. Army

For all the enthusiasm for the next generation of unmanned tanks, ships, and planes, there is one aspect that people in the field are generally reticent to talk about. Arming these more intelligent, more capable, and more autonomous robots is the equivalent of Lord Voldemort in the *Harry Potter* novels. It is the Issue-That-Must-Not-Be-Discussed.

When it comes to this topic, people either change the subject or speak in absolutes, most often including the phrase staying "in the loop." Noted military expert and Bush administration official Eliot Cohen, for example, states forcefully that "people will always want humans in the loop." An air force captain writes in his service's professional journal, "In some cases, the potential exists to remove the man from harm's way. Does this mean there will no longer be a man in the loop? No. Does this mean that brave men and women will no longer face death in combat? No. There will *always* [author's italics] be a need for the intrepid souls to fling their bodies across the sky."

The same sort of response occurs within the robotics companies. When asked about what happens as robots become armed and more autonomous, Helen Greiner of iRobot quickly changes the subject. "It's far away enough that I don't see it as an issue." Similarly at Foster-Miller, the maker of the SWORDS machine-gun robot, Vice President Robert Quinn adamantly describes a human "staying in the loop" as a "line in the sand." He says he can't even imagine how unmanned systems would "ever be able to autonomously fire their weapons."

As Noah Shachtman explains, people speak in such absolute terms and use the phrase "man will always stay in the loop" so often that it ends up sounding more like brainwashing than analysis. "Their mantra is a bit like the line they repeat again and again in the movie *The Manchurian Candidate*." But he jokes that the constant repetition is pretty understandable. "It helps keep people calm that this isn't the Terminators." More seriously, he explains, "The core competency in the military is essentially shooting and blowing up things. So no one is eager to say, 'Outsource that to a bunch of machines.'"

## REDEFINING THE LOOP

All the rhetoric ignores the reality that man started moving out of "the loop" of war a long time before robots made their way onto battlefields. For example, the Norden bombsight of World War II and the computers that followed took over the human's role in deciding when to drop a bomb. Notice how Captain Doug Fries, a B-52 radar navigator, described what it was like to bomb Iraqis during the first Gulf War: "The navigation computer opened the bomb bay doors and dropped the weapons into the dark."

The same trend has been in place at sea since a computer system called Aegis was introduced in the 1980s to help defend navy ships against air and missile attacks. The system came with four modes: Semiautomatic, in which the humans interfaced with the system to judge when and at what to shoot; Automatic Special, in which the human controllers set the priorities, such as telling the system to destroy bombers before fighter jets, but the computer then decided how to do it; Automatic, in which data went to human operators in command, but the system worked without them; and Casualty, where the system just did what it thought best to keep the ship from being hit. The R2-D2-like CRAM deployed to Baghdad to protect against incoming mortar rounds operates on roughly the same system.

The human sailors could override the Aegis computer in any of its modes, but increasingly this was beside the point. For example, on July 3, 1988, the

U.S.S. *Vincennes* was patrolling in the Persian Gulf. Notably, the *Vincennes* was nicknamed "Robo-cruiser," both because of the new Aegis radar system it was carrying and because the captain had an aggressive reputation. Its radars spotted Iran Air Flight 655, an Airbus passenger jet. The jet was on a consistent course and speed and was broadcasting a radar and radio signal that showed it to be civilian. The automated Aegis system, though, had been designed for managing battles against attacking Soviet bombers in the open North Atlantic, not for dealing with civilian-filled skies in the crowded Gulf. The computer system registered the passenger plane with an icon on the computer screen that made it seem to be an Iranian F-14 fighter (a plane half the size), and hence an "Assumed Enemy."

Even though the hard data was telling the crew that the plane wasn't a fighter jet, they trusted what the computer was telling them more. Aegis was on semiautomatic mode, but not one of the eighteen sailors and officers on the command crew was willing to challenge the computer's wisdom. They authorized it to fire. That they even had the authority to do so was again because of Aegis; Robo-cruiser was the only ship in the area authorized to fire without having to seek permission from more senior officers in the fleet. Again, the computer was trusted even more than any human captain's independent judgment on whether to shoot or not.

The Aegis system took out the jet. Only after the fact did the crew realize that they had accidentally shot down an airliner, not a fighter jet, killing all 290 passengers and crew, including 66 children.

This tragedy demonstrates how a redefinition is already taking place of what it means to have humans "in the loop" over autonomous systems and their weapons. As Drew Bennett at iRobot describes, "In ten to twenty years humans will still be 'in the loop,' but it will be a wider loop." Similarly, futurist Ray Kurzweil laughs at the idea of always staying "in the loop," saying it's "just a political description.... Man may still think he's in control, but only at different levels." Indeed, much the same scenario as occurred with Iran Air Flight 655 happened nearly two decades later during the 2003 Iraq invasion, when U.S. Patriot missile batteries accidentally shot down two allied planes that the systems had classified as Iraqi rockets. There were only a few seconds to make a decision, and so the human controllers trusted the machine on what to fire at. Their role "in the loop" was actually only veto power, and even that was a power they were unwilling to use against the quicker (and what they viewed as better) judgment of a computer.

The reality is there have been all sorts of new technologies that people insisted in absolutist terms would "never ever" be allowed to run on their own without a human in the loop. Then, as the human roles were redefined, they were gradually accepted, and eventually were not even thought about. Just go ask your friendly elevator operator.

## WHY AUTONOMY?

There are myriad pressures to give warbots greater and greater autonomy, and thus widen the loop further and further. The first is simply the push to make more capable and more intelligent robots. But as AI expert Robert Epstein tells it, this comes with a built-in paradox. "The irony is that the military will want it [a robot] to be able to learn, react, et cetera, in order for it to do its mission well. But they won't want it to be too creative, just like with soldiers. But once you reach a space where it is *really* capable, how do you limit them? To be honest, I don't think we can."

Simple military expediency also widens the loop. To get any type of personnel savings from using unmanned systems, one human operator has to be able to "supervise" (as opposed to control) a larger number of robots. For example, the army's FCS plan is to have two humans sit at identical consoles and jointly supervise a team of ten land robots. Similarly, an air force Predator pilot predicts that, rather than flying one drone at a time as he and his copilot do, his successors will soon be controlling entire fleets of drones, as in a video game. (He is careful, though, to add the mantra that "you cannot take the human out of the loop; it would be a huge mistake.")

This setup envisions that the humans would delegate tasks out to increasingly autonomous robots, but the robots would still need human permission on the important question of whether to shoot or not. The problem is that it may not prove workable in reality. Instead, there are a series of interlaced rationales that take the human further out of the loop step by step.

To begin, research is finding that humans have a hard time controlling multiple units at once (imagine playing five video games at the same time). Just having a human operator control two rather than one UAV at a time reduces their performance levels by an average of 50 percent. As a NATO study concluded, the goal of having one operator control multiple vehicles is "currently, at best, very ambitious, and, at worst, improbable to achieve." And this is with systems that aren't even shooting or being shot at. As one Pentagon-funded report noted, "Even if the tactical commander is aware of the location of all his units, the combat is so fluid and fast-paced that it is very difficult to control them."

And then there is the fact that an enemy is involved. If the robots aren't going to fire unless a remote operator authorizes them to, then any foe need only to disrupt that communication. Military officers describe how, while they don't like the idea of taking man out of the loop, there has to be an exception, a backup plan for when communications are cut and the robot is "fighting blind." The only other alternative is to have it either sit there and be shot at, or automatically return to base without accomplishing the mission, and maybe even lead the enemy back to your locale. Robot companies say that these concerns are what help fuel demands from the military for greater autonomy. As a designer at iRobot explains, "By making them autonomous, they don't need signals or remote control, and you can't jam the signals to operate them. We are looking to make them more and more autonomous. Right now, if the signal gets jammed, then our robots return to you and say, 'What's up?' Autonomy can get around that."

Even if the communications link is not broken, there are combat situations where there isn't time for the human operator to react. The very best human fighter pilot needs at least .3 seconds to respond to any simple stimulus and twice as long to make a choice between several possible responses. A robotic pilot needs less than a millionth of a second. Or, take, for example, the countersniper device that automatically targets any enemy that shoots. Those precious seconds while the human decides whether to fire or not could let the enemy get away. By contrast, as one military officer tells it, "If you can automatically hit it with a laser range finder, you can hit it with a bullet."

This creates another exception to the rule, giving robots the ability to fire back on their own. As Gordon Johnson of the Joint Forces Command explains, it seems not only logical, but is quite attractive, even to those commanders who otherwise would want humans in the loop. "Anyone who would shoot at our forces would die. Before he can drop that weapon and run, he's probably already dead. Well now, these cowards in Baghdad would have to pay with blood and guts every time they shoot at one of our folks. The costs of poker went up significantly. The enemy, are they going to give up blood and guts to kill machines? I'm guessing not."

This kind of autonomy might even be found more palatable than other types. "People tend to feel a little bit differently about the counterpunch than the punch," notes Noah Shachtman. And, says John Tirpak, editor of *Air Force* magazine, once robots "establish a track record of reliability in finding the right targets and employing weapons properly," the "machines will be trusted." The firm "line in the sand" becomes more like a slippery slope.

The human location "in the loop" is already becoming, as former army colonel Thomas Adams sees that of "supervisor who serves in a fail-safe capacity in the event of a system malfunction." Even then, he thinks the speed, confusion, and information overload of modern-day war will soon move the whole process outside of "human space." He describes how the coming weapons "will be too fast, too small, too numerous, and will create an environment too complex for humans to direct." As Adams concludes, the various new technologies "are rapidly taking us to a place where we may not want to go, but probably are unable to avoid."

What is often described as "impossible" then actually becomes viewed as quite logical and even inevitable. And actual programs are then created on something that officially is "never" supposed to happen.

So for all the claims by military, political, and science leaders that "humans will always be in the loop," as far back as 2004 the U.S. Army was carrying out research on armed ground robots which found that "instituting a 'quickdraw' response made them much more effective than an unarmed variation that had to call for fires from other assets." Similarly, a 2006 study by the Defense Safety Working Group, a body in the Office of the Secretary of Defense, discussed how the concerns over potential killer robots could be allayed by giving "armed autonomous systems" permission to "shoot to destroy hostile weapons systems but not suspected combatants." That is, they could shoot at tanks and jeeps, just not the people in them. Stated John S. Canning, chief engineer at the Naval Surface Warfare Center and one of the authors of the proposal, "Let's design our armed unmanned systems to automatically ID, target and neutralize or destroy the weapons used by our enemies—not the people using the weapons. This gives us the possibility of disarming a threat force without the need for killing them." By 2007, the U.S. Army had put out a "Solicitation for Proposals" for a system that could carry out "fully autonomous engagement without human intervention." The next year, the U.S. Navy circulated research on a "Concept for the Operation of Armed Autonomous Systems on the Battlefield." Perhaps most telling is a report that Joint Forces Command drew up in 2005, which suggested that autonomous robots on the battlefield will be the norm within twenty years. Its title was somewhat amusing, given the official mantra one usually hears on the issue: "Unmanned Effects: Taking the Human Out of the Loop."

So, despite what one article called "all the lip service paid to keeping a human in the loop," autonomous armed robots are coming to war. They simply make too much sense to the people that matter. A special operations forces officer put it this way: "That's exactly the kind of thing that scares the shit out of me.... But we are on the pathway already. It's inevitable."

## RETIRING G.I. JOE?

A retired air force officer describes a visit he had with the 2007 graduating class at the Air Force Academy. "There is a lot of fear that they will never be able to fly in combat." With robots taking on more and more roles, and humans ever further out of the loop, some wonder whether they will make human warriors obsolete.

Technology has long changed the skills that we value. For example, illuminators, who essentially did artistic doodles in the margins of handwritten books, performed one of the most highly skilled jobs in the Middle Ages. Illuminated books were considered so valuable that kings exchanged them as part of peace treaties. Then the printing press came along and the doodling days of the illuminator were done. Likewise, just a few decades ago, someone who could do complex math, such as long division, in their head was incredibly valued. They could even work as "computers" (where we get our modern-day word), doing mathematical calculations for hire. Today, such Rain Man–like skills are nearly irrelevant once you get past the fourth grade and they let you use digital calculators.

The same goes in war. Colonel James Lasswell of the Marine Warfighting Lab explains that, as technology advances, "We will soon see certain roles disappear." He gives the example of the forward observer. This was once an officer who needed special experience and training to call in precise cannon fire and airstrikes on map grids. In a world of GPS and laser targeting, this specialty is now a lot like being a sailmaker on a nuclear-powered ship. "Soon you will just have universal spotters. You will make up for years of experience simply by giving them a toy [that can point a laser at the target]."

As noted earlier, something similar has already happened in the air with human pilots, who no longer sit inside most reconnaissance planes and have even less to do with the unmanned versions. Indeed, as army drone pilot Sergeant Chris Hermann says, "We all joke about it. A monkey can do this job, this bird flies itself, it lands itself." When the weather is bad and their drone can't fly, Hermann and his buddies will instead play video games like *Battlefield 2* or *Call of Duty*. By comparison, they find that flying the recon drone is "kind of like old Atari, pretty basic, point and click."

Looking forward, officers describe unmanned systems as being perhaps more suitable than human-piloted planes for many other roles, including even refueling aircraft, in which a premium is placed on endurance and the ability to fly precisely at a steady speed and level. Indeed, with UAVs becoming easier to fly and more lethal, "Maybe you don't need fighter pilots at all," says retired marine major general Tom Wilkerson. It is notable that Wilkerson is not some groundpounder who hates fighter jocks, but actually a Top Gun fighter pilot school graduate with over one thousand hours of flying experience.

The most controversial role would seem to be the human grunt in the field. But even here, people are starting to discuss having machines move in. As Robert Quinn at Foster-Miller tells it, "We clearly see an evolution from EOD to combat engineers. And the next step we believe is infantry with weaponized robotics."

Quinn reflects a quietly growing belief among both the robotics makers and the military. In 2004, DARPA researchers surveyed a set of soldiers and robotics scientists about the military roles they thought humanoid-type robots would take over in the near future. The military officers predicted that the first functions that would be turned over to robots would be countermine operations, then reconnaissance, forward observer, logistics, and then infantry. Among the last they thought would be turned over to autonomous robots would be air defense, driving or piloting vehicles, and food service. This is somewhat surprising, given that these latter functions have been among the first to already be robot-sourced, as well as that soldiers on average thought G.I. Joe would be replaced by robots before Cookie the chef. Special forces roles were felt, on average, to be the least likely to ever be turned over to robots.

The average year that the soldiers predicted that humanoid robots would start to be used in infantry combat roles was 2025. Their projection wasn't much further off that of the scientists, who predicted 2020. Interestingly, the scientists predicted that the average cost would be about $1 million per infantry robot; the soldiers were more optimistic, predicting that the average robot infantry soldier would cost around $400,000 in 2004 dollars. One soldier, channeling Ray Kurzweil, predicted, "As technology advances, costs will drop."

These numbers only reflect the opinions of those in the survey, and could prove to be way off. For example, military robotics expert Robert Finkelstein, who helped conduct the survey, thinks they are highly optimistic and that it won't be until "2035 [that] we will have robots as fully capable as human soldiers on the battlefield." But the broader point is that many are starting to contemplate a world where robots do replace the grunt in the field.

However, as much as technology changes how we look at professions and even ends some of them, the reality is that it doesn't always play out that way. Yes, there are some areas where a robot might be able to surpass the skills or costs of a human soldier. But there are others where that day is far off. We don't have human elevator operators now, but human toll collectors live on.

The funny thing is that many areas least likely to be roboticized will be in the areas that we generally consider simple, as opposed to the roles that require the most technical training. For example, it may take years to train up a sniper who can hit a bull's-eye again and again. But it's technologically easy for a robot to instantly place a targeting laser on a bad guy and shoot him. By comparison, as AI pioneer Marvin Minsky says, "Common sense is not a simple thing. Instead it is an immense society of hard-earned practical ideas—of multiple life learned rules and exceptions, dispositions and tendencies, balances, and checks." The most complex part of our brain, which will likely be the last thing that computers match, if ever, is our "emotional intelligence." This is the part of the brain that makes sense of social situations, which is often the toughest part of a soldier's job in dealing with complex situations in conflicts like Iraq or Afghanistan.

For these reasons scientists already see the day coming at which surgeons, who may require the most years of training of all professions, will be replaced by robotic surgery systems like the da Vinci. Rod Brooks of MIT and iRobot predicts that the future for doctors is likely to be the same as for airline pilots: There mainly to appease the patient and regulatory boards and charge exorbitant amounts for skills they rarely use and knowledge that the computer can call up faster and in more depth.

By contrast, as one scientist at the Idaho National Lab puts it, "My job will be eliminated before my hairdresser's will." Hairdressers not only have to be able to deal with all our misshapen heads, but they also have to cut hair with an eye toward not merely precision, but fashion and aesthetics, as well as be able to chit-chat on sports, the weather, the latest gossip, and so on. Plus, the customer has to trust them with a sharp blade near their eyes, ears, and throat. It is hard enough to do with the drones at Supercuts, even more of a leap of faith with a barber made at Spacely Sprockets.

We can similarly expect that some human soldiers' jobs might be eliminated, others' might never end, and many will likely evolve, or at the very least be understood differently. There are some military roles at which robots might be better suited than humans and others at which humans will just remain far more talented. Indeed, we may one day cease making such comparisons. As Rod Brooks points out, "Asking whether robots will 'match the abilities of a human' is a funny phrase. Does a tank 'match' the ability of an infantryman? No, they are different. Does an airplane 'match' the ability of a bird? The plane is certainly faster than the bird, but it can't land as well or fly as long."

## TEAMING UP: WARFIGHTERS' ASSOCIATES

The drummer starts to patter out a staccato beat. After listening a few seconds, his partner joins in. The first drummer shifts beats and rhythms, and the second follows, never repeating exactly, but riffing off the original creation. Beats are reversed and then evolve. Back and forth they go, the musical partners now jamming together to make new music.

The first drummer is human. The second is Haile, a robot musician. Described by one report as "the creature from the film *Alien* turned into wood and found a rhythm," Haile was made by Gil Weinberg, a professor of music technology at Georgia Tech. It is also the first robot that can not merely perform music on its own, but understand and interact with human musicians, creating new music altogether. Already, Haile has had its own world concert tour, with stops in Israel, Germany, France, and the United States.

Haile the robot drummer illustrates the most likely future of robots' relation to humans, not just in music, but in war as well. As Bart Everett, a navy robots pioneer, explains, merely tasking robots out on dangerous missions will evolve "to more of a team approach." He goes on to explain how his center (the navy's SPAWAR program) has joined with the Office of Naval Research (ONR) to support the activation of a "warfighters' associate" concept within the next ten to twenty years, with robots and humans working together "as a synergistic team." The concept is that the humans and robots would be integrated into a team that shares information and coordinates action toward a common goal. Says Everett, "I firmly believe the intelligent mobile robot will ultimately achieve sufficient capability to be accepted by the warfighter as an equal partner in a human/robot team, much along the lines of a police dog and its handler."

A 2006 solicitation by the Pentagon to the robotics industry captures the vision. "The challenge is to create a system demonstrating the use of multiple robots with one or more humans on a highly constrained tactical maneuver.... One example of such a maneuver is the through-the-door procedure often used by police and soldiers to enter an urban dwelling…[where] one kicks in the door then pulls back so another can enter low and move left, followed by another who enters high and moves right, etc. In this project the teams will consist of robot platforms working with one or more human teammates as a cohesive unit."

Another U.S. military–funded project describes how "playbooks" for tactical operations might be built for the team. Much like a football quarterback, the soldier would call the "play" for robots to carry out, but like the players on the field, the robots would have the autonomy to change what they do if the situation shifts.

"Just see it and shoot it is not the future," Thomas McKenna at ONR explains. Instead, the robots in these teams will be expected to interact with humans naturally, perform tasks reliably, as well as be able to predict what the human will ask of them. "The robot will do what robots do best. People will do what people do best." ONR is working on robots that "could even suggest a change of plan" if they find new information that warrants it. From this fusion, the expectation is that "human and robot roles will evolve dynamically. New experiences make new expectations of behavior."

The military, then, doesn't expect to replace all its soldiers with robots anytime soon, but rather sees a process of integration into a force that will over time become, as Joint Forces Command projected in its 2025 plans, "largely robotic." One plan is for detachments that will include 150 soldiers and as many as 2,000 robots. The individual robots would "have some level of autonomy—adjustable autonomy or supervised autonomy or full autonomy within mission bounds," but it is important to note that so too would the human soldiers within these units. One of the scientists in the DARPA survey perhaps put it best: "I believe we should think in terms of the human plus a robot as a system, not just the robot itself."

If the future is one of robot squadmates and robot wingmen, many scientists think it puts a premium on two things, both very human in nature. The first is good communication, the ability of the robot and human to interact naturally. In 2004, Lockheed tested out an unmanned jet that responded to simple vocal commands. A pilot flying in another plane would give the drone some broad mission, such as to go to a certain area and photograph a specific building, and the plane would carry it out. As one report explains, "The next war could be fought partly by unmanned aircraft that respond to spoken commands in plain English and then figure out on their own how to get the job done." In turn, the robot may even sound human in its communications back. WT-6 is a robot in Japan that has a humanlike vocal system, including tongue, lips, teeth, vocal cords, lungs, and a soft palate made from polymers.

To work well together, these robots and human soldiers will also need to have confidence in each other. It sounds funny to say that about the relationship between a bucket of bolts and a human, but David Bruemmer at the Idaho National Lab actually specializes in how humans and robots work together. Without irony, he states that "trust is a huge issue for robot performance."

Trust is having a proper sense of what the other is capable of, as well as being correct in your expectations of what the other will do. One of the more interesting things Bruemmer found in his research is that novices with robots tend to use their systems the best. They "trust" robot autonomy the most and "let it do its job." Over time, he predicts, robots will likely have "dynamic autonomy" built in, where the amount of "leash" the robots are given is determined less by any ideal of keeping humans "in the loop" and more by their human teammate's experience and trust level.

In short, the human warrior isn't fading away anytime soon. But the same cannot be said of the human monopoly on decisions in war, including even those of life or death.

# SEVEN

# ROBOTIC GODS:
# OUR MACHINE CREATORS

*You have to remember that these supposedly evil scientists are actually just guys with families and dreams.*

—DANIEL WILSON

"Each year some forty-two thousand people are killed in preventable traffic accidents. That is unacceptable. That is some fifteen World Trade Centers each year."

Sebastian Thrun is director of the Artificial Intelligence Laboratory at Stanford University. Speaking in a clipped accent that reveals his German roots, Thrun tells how his motivation for making artificial beings comes from wanting to save the lives of real ones. When he was younger, "a very good friend died of a car accident because of a split-second decision." For Thrun, robotics is a way "to avoid that waste," and a means for him as a scientist to "have a major impact on society."

Before he came to Stanford, Thrun had worked on robotic tour guides for museums in Germany, an interactive humanoid robot that worked in a nursing home near Pittsburgh, and a system of robots that could search out mines. It was all interesting stuff, but none of it had that major impact he was looking to make. In 2004, however, opportunity knocked.

The Grand Challenge is a robotics road race sponsored by DARPA, the Pentagon's main research lab. The agency offered a $1 million, winner-take-all prize for the first team that could drive a robot across a rugged 142-mile cross-country course in the California desert. The Challenge was conceived as a way for the government to accelerate military R&D, by bringing in new talent, new ideas, and new technologies. In making it an open competition, DARPA hoped to entice innovators who normally would not work with the military. In their race for the cash, the side effect would be to help the Pentagon solve the problems it was having in designing robots for warfighting, as well as meeting the congressional requirement to have one-third of all its ground vehicles unmanned by 2015.

The rules of this robotic *Amazing Race* were fairly simple. The vehicle had to autonomously complete the race course within ten hours to get the money. No human intervention was allowed, meaning no control commands could be sent to the vehicles while the race was on. Finally, none of the race cars could intentionally "touch" any other competing vehicle.

As Thrun describes, when you combine these rules with the rugged terrain of the 142-mile desert race course, which DARPA picked for its similarity to the rough trails in combat zones like Afghanistan and Iraq, "it's an endurance race of unmatched proportion.... This is the first one ever where the human is not involved and the vehicle has to make all the decisions."

Unfortunately, "the first Grand Challenge came off as something of a *Three Stooges* affair," writes *Popular Science* magazine. Of the 106 applicants in the 2004 race, only a few even made it beyond the start line. For example, the Oshkosh firm showed up with a bright yellow autonomous version of its six-wheeled Marine Corps combat truck. But the "TerraMax" only made it one mile before a software glitch shut it down. Sandstorm, a converted Humvee designed by Carnegie Mellon's Red Team, went the farthest. But after seven and a half miles, it caught fire and got stuck on an embankment.

Thrun watched the 2004 race but didn't compete. "After the first one, it was obvious we could do better," he says. "It was a no-brainer." Having just joined Stanford's faculty, he saw it as "an amazing opportunity to be part of a fundamental change for society... a win, win, win, win for everybody." Thrun and his team of graduate students entered the 2005 race.

When DARPA doubled the prize money to $2 million, the field of competitors grew dramatically. Some 195 teams applied from thirty-six states and four countries; 160 of them were new to the event, and they included 35 university teams and 3 high schools. The entry names often sounded like something out of a fantasy football league, ranging from the mundane, like Team South Carolina, to the inspired, like Cajunbot from Louisiana and Viva Las Vegas, oddly enough from

Oregon. The contestants also ranged from all-volunteer teams like CyberRider (which used Wiki collaborative software to bring in the advice of computer whizzes around the world) to research labs paired with corporate sponsors. The early favorite was Carnegie Mellon's Red Team (backed by Caterpillar) that had done the best in the last race and this time showed up with two cars to compete. All told, DARPA estimated that as much as $100 million in investment was spurred, as well as the equivalent of $155 million worth of free labor. One military robotics company executive joked that "the best part of the Grand Challenge is using the college kids like cheap slave labor."

Thrun's upstart Stanford team took a Volkswagen Tuareg SUV and rigged it out with five LADAR sensors, GPS, a video camera, and onboard computers that had some hundred thousand lines of code specially written by the Stanford School of Engineering. They called the vehicle "Stanley."

Stanley from Stanford worked by using its sensors to build a multilayered map of the world around it, much like many of its competitors. The robot car's unique feature was that it also fed its experiences, as well as a log of human driver reactions during its test runs, into a learning algorithm. As Thrun puts it, "We trained Stanley.... The relationship was teacher and apprentice, as opposed to computer and programmer."

Thrun recalls it all came together three months before the race, when he was riding in Stanley during a test run. "We were driving in the Sonoran Desert, and at some point I realized that I was trusting my life to the car. At that moment it became crystal clear, this is the future. It was like a binary flip in my brain."

On October 8, 2005, Stanley won the Grand Challenge, completing the course in six hours and fifty-four minutes, with top speeds of thirty-eight miles per hour. It might have gone even faster except for a flock of birds that landed in the middle of the raceway, confusing Stanley for a period. Four more teams crossed the finish line, but the Stanford team took the entire $2 million prize under the rules of the competition. Thrun was gracious in his victory. "We all won. The robotics community won."

Soon after, Thrun would be named to *Popular Science* magazine's "Brilliant 10," as one of the ten best and brightest minds in all of science. When asked how the victory changed his life, he responds, "Oh, big time! It is even changing Stanford as a university." He goes on to excitedly detail all the various collaboration projects that came out of the race, such as new links with the automotive industry and plans for an 8,000-square-foot research facility at the campus. The students who worked on the project are "cranking out papers. There are difficult techno-logic problems to be overcome and each one of these is a paper or a thesis."

And what happened to Stanley? After its win, Stanley was declared the number one robot of all time by *Wired* magazine, beating out a list of fifty other real and fictional robots that ranged from Spirit, NASA's Mars rover, to Optimus Prime from *Transformers*. Stanley is now at the Smithsonian Museum of American History, sharing the stage with such other important historical artifacts as the "Star-Spangled Banner" that flew over Fort McHenry and the jacket that Fonzie wore on *Happy Days*.

## THERE IS NO EUREKA

In the world of science fiction, research usually takes place in super-secret government labs or mysterious places like "Area 51," the Nevada desert setting of more than sixty movies, TV shows, and video games. The Science Fiction Channel even has an entire TV series about a quirky town set up in secret by the Pentagon for scientists to live and work in, called *Eureka*. ("Every small town has its secrets, but in the town of Eureka the secrets are top secret.")

While the military is the major funder of robotics research, much of it actually plays out in public view. We just aren't watching. One estimate of applied research spending in the unmanned field was that 40 percent was flowing through private industry, 29 percent via military centers and labs, and 23 percent through university programs. At the center of it all is the National Center for Defense Robotics, a congressionally funded consortium of 160 companies, universities, and government labs. Work on military robotics isn't so much top-secret labs fueled by UFO power sources as it is a simple synergy of military money, business organizations, and academic researchers.

The outcome is that major changes in warfare are being driven by the last people you might associate with combat. When you meet robot scientists, you quickly discover that it is hard to make blanket assessments. They range from prototypical geeks wearing actual pocket protectors to brawny he-men, who look like they spend more time at Gold's Gym than at the lab. While many are introverts, others are real jokers. At one research visit, for example, I watched a scientist ride a prototype military ground robot down a set of stairs like a surfboard. The only general rule is that they are all breathtakingly smart.

People begin working on robots for all sorts of reasons. Brian Miller, for example, is an engineer who started working at Ford Motor Company, designing and developing race cars. He had no visions of robots dancing in his head growing up. "But NASCAR was boring," he says (as Dale Earnhardt likely rolls over in his grave), "too many rules specifications." Miller liked working on off-road vehicles and so joined Millenworks, a company in Orange County, California, that makes rugged vehicles. Today, instead of race cars, he makes unmanned ground combat vehicles. By contrast, Helen Greiner, the chairman and cofounder of iRobot, says that she first got into robotics in 1977, watching the original *Star Wars* movie as a mathematically inclined eleven-year-old. To this day, she calls R2-D2 her "personal hero."

Daniel Wilson, a writer and Carnegie Mellon University researcher, probably has the best explanation for why people decide to work on robots. "Hands down, robots are just plain cool as hell. Ask any roboticist why they do it, and that's the answer you get." As he explained, "When you are deciding on what to do for your life, there's nothing like the sense of making something so tangible, so active."

The increasing use of robots in war, though, has changed the equation a slight bit. Today, roboticists also can take pride in saving lives. As Colin Angle, one of Greiner's cofounders at iRobot, says, he spent his time as a student developing the "most sophisticated, cool, crazy-ass robot." But "it left [me] with an empty feeling." Today, Angle builds robots that he is quite happy to see get destroyed. "Getting a robot back, blown up, is one of the more powerful experiences I've lived through," he says. "Nothing could make it so clear that we have just saved lives. Somebody's son is still alive. Some parent didn't just get a call." Greiner similarly describes receiving postcards from soldiers using her PackBot in the field as the most gratifying experience, including one that just said, "You saved lives today." As she tells it, "There are people coming home because of our work." But beyond that, she goes back to why she and her schoolmates founded the robotics company. "We always knew we would change the world."

## "GIT ROCKIN': GOVERNMENT IT ROCKS, DO YOU?"

The invitation letter reads, "GIT Rockin' is government IT's first annual battle of the bands. . . . This friendly competition allows executives—from government and industry alike—to network with peers, colleagues and spouses in a high-energy, out-of-the-industry-norm environment. Come out and showcase your alter egos and talents."

The battle of government information technology bands takes place at the State Theatre in Falls Church, Virginia. And it does not disappoint, fulfilling all your expectations of the government IT music scene. The ultimate winner of the event, which raises money for charity, is Full Mesh, perhaps the only rock band in the world whose particular highlight is that it "featured talent from Juniper Networks." *Federal Computer Week* magazine (akin to the *Rolling Stone* or *Vibe* of the IT music world) summed it up. "Folks from around the federal information technology community really let their hair down."

The history of government support for the scientists, engineers, and programmers like those at GIT Rockin' goes back decades; for the computing world, it especially took off in World War II and then the cold war. By one estimate, up to a third of major university research faculty was supported by national security agencies after 1945. So the battle of the government information technology band was, if unfortunate, likely inevitable.

The primary player in the world of funding new research in IT, computers, and robotics is DARPA. DARPA's overall mission is to support fundamental research on technologies that might be common twenty to forty years from now, and to try to make them happen earlier to serve the needs of the U.S. military today. As *Washington Post* writer Joel Garreau describes, its strategic plan is to "accelerate the future into being."

The agency was started in 1957 after the Soviets stunned and embarrassed the United States by launching the Sputnik satellite. President Eisenhower worried that America was losing the science arms race and set up an agency so that the United States would never again be surprised by the technology of foreign powers. Since then, DARPA has shaped the world we live in more than any other government agency, business, or organization. For all the claims that "big government" can never match the private sector, DARPA is the ultimate rebuttal. The Internet (DARPA's first visionary name for it was the "intergalactic computer network"), e-mail, cell phones, computer graphics, weather satellites, fuel cells, lasers, night vision, and the Saturn V rockets that first took man to the moon all originated at DARPA. And now it's focusing on robots and other related unmanned technologies.

DARPA works by investing money in research ideas years before any other agency, university, or venture capitalists on Wall Street think they are fruitful enough to fund. DARPA doesn't focus on running its own secret labs, but instead spends 90 percent of its (official) budget of $3.25 billion on university and industry researchers "who work at the forefront of the barely possible." One business article notes, "By the time a technology is far enough along to attract venture capitalists, DARPA is usually long gone." As a result, scientists are often very positive on the agency. Sebastian Thrun says, "DARPA has been good to me, helping me to develop my dreams.... It's a very successful agency," he explains, because "it takes risks and gets spectacular results."

Today, DARPA's headquarters is located just down the street from a shopping mall in Arlington, Virginia. It is supposed to be a secret location, but the security policy of having a police car parked permanently in front of a supposed suburban office building gives it all away. So does the immense popularity of a barbershop just two blocks away. Its specialty is bad 1950s buzz cuts, but its hairdressers do offer five-minute rubs of the patron's skull and neck afterward. It is usually filled with men wearing DARPA badges, savoring some all too rare human contact.

DARPA has some 140 program managers on staff, mainly PhDs in the hard sciences, with a few others from social sciences and medicine. Joel Garreau, who wrote a book on DARPA, notes that the organizational culture is to seek out problems that staffers call "DARPA-hard." These are "challenges verging on the impossible." The presentations at its annual conference (DARPATech) illustrate with such panels as "The Future of Aviation," "Obtaining the Unobtainium: New Materials," and, of course, the ever popular "Letting Schrödinger's Cat out of Pandora's Box: Quantum Mechanics for Defense." The location is equally instructive; the agency that tries to make the future come true holds its conference in Anaheim, the home of Disneyland.

For all its success, not everyone is a huge fan of DARPA. Its critics in the blogosphere use such descriptors as "creepy" or call it the "Frankensteins in the Pentagon." Part of this animosity lies with a fairly flawed public affairs operation. While DARPA should be better known for developing the Internet and funding projects like Sebastian Thrun's Stanley, the last time it made major headlines was a failed project in 2003 to set up a terrorism prediction index. This was a plan for experts to participate in the equivalent of a football pool, betting on likely events such as terrorist attacks and the deaths of world leaders. As one defense industry expert put it, he had never come across such "a mind-numbing mix of brilliance and tone deafness" as at DARPA.

Public perception aside, there is also concern within the defense field that DARPA invests too much time and money on fanciful ideas. Even robotics scientists sometimes describe DARPA's staff as "real madmen." The criticism seems to be centered on the fact that the agency, in trying to think out of the box, can forget that the D in its name stands for its primary funder and customer: the Defense Department. As one official says, "I spend an inordinate amount of time trying to delineate between DARPA-hard and DARPA-stupid."

More recently, others critique DARPA for just the opposite. They feel that funding pressures from the wars in Afghanistan and Iraq have started to make it too short-term in its thinking. "Today DARPA imposes six-month go-no decisions on all their researchers, which stifle innovation and creativity—very un-DARPA-like," says a congressional staffer. "I have had everyone complain to me about this—from universities to small hi-tech businesses to the big defense contractors." They contend that the true technology problems worth solving don't get solved within six months or less.

## PIMPING AIN'T EASY

A baseball throw down the street from DARPA is the Office of Naval Research (ONR). In keeping with the odd way that the most advanced defense agencies are integrated into the mundane of Americana, across the street is a TCBY: The Country's Best Yogurt store (judging from the small crowds, especially compared to the barbershop, it is not).

The start of ONR dates back to 1907. A naval commander visited the construction of the battleship U.S.S. *North Dakota* and saw terrible flaws in design and construction, which the navy had been unaware of because it lacked its own scientists and engineers. Since then, ONR has focused on helping the navy to maintain technological superiority on, under, and above the sea, as well as in space. It led the development of such varied programs as submarine-launched ballistic missiles, tilt-rotor aircraft, deep-sea exploration, fiber optics, and how to battle tooth decay (dental hygiene being the key to naval readiness). As one historian noted, the naval research program has been responsible for a bevy of "ideas that literally changed the world."

Among those who work at ONR is Dr. Thomas McKenna. A balding, portly man, McKenna looks the part of a genial father; indeed, proud pictures of his children fill his office's walls. McKenna is also a father figure to the wider military robots world. How he works very much illustrates the relationship between the military and the world of research.

McKenna has been at ONR since 1988; his earliest work was on legged robots. Today, among his main tasks is administering financial grants (his typical award is around a million dollars a year for five years) to universities and labs. His usual approach is to identify promising researchers for support when they are still graduate students. He then helps their careers to a point at which they become professors and have their own labs. "I was supporting some one hundred top graduate students at a time." The graduate students are not just Americans or even all located in the United States. For example, McKenna is especially proud of having funded the graduate student who now runs the "Blue Brain" project in Switzerland. In collaboration with IBM, the project is trying to make a simulated brain using a Blue Gene supercomputer, which might yield massive jumps in computing power and ultimately create strong AI.

McKenna also helps projects gain funding via the Department of Defense's Small Business Innovation Research (SBIR) and Small Business Tech Transfer Research. These programs provide almost $1 billion in total grant money (given out in baskets of up to $850,000) to help jump-start early-stage R&D for small companies and entrepreneurs working with the Pentagon and research universities.

Usually, researchers will apply to McKenna's office for grants, and he will kick ideas back and forth with them, in a collaboration to hone their research proposals and tweak them to ONR's needs. Or "sometimes I just find them on the Web." He tells how he surfs about to different sites of research until he sees something that intrigues him. He will then e-mail the researcher: "Send me a proposal along these lines, because I really like what you're doing."

The process McKenna lays out is quite common in the nexus between the military and its university and business researchers. Some describe this cross between a funder and seducer as akin to "pimping" for the military research system, while others liken it to "an idea and technology hummingbird." Just as a hummingbird flits back and forth, spreading pollen, so too does such a funder serve as a critical link in connecting scientific ideas and research with military needs.

Currently, McKenna is sponsoring lots of funding in "human activity recognition," where a robot learns to understand and identify what a human is doing. For example, many universities have research on teaching robots how to play or referee baseball and other sports. Combining vision systems with processors that know the game's rules and track trajectories allows robots to do things like predict where a ball will land and race to retrieve it. By scrutinizing a pitcher's fingers with a high-power lens, they can even predict whether he is going to throw a fastball or a curve. The hope is that systems will similarly be able to learn how to recognize certain patterns of behavior in war and do such things as IED prediction and detection.

McKenna is also very interested in cross-disciplinary teams, describing himself as being far more likely to fund projects, for example, that bring together biol-
ogists with engineers. He proudly says, "When it comes to bio-inspired robotics,
there isn't any other place in the world better than us."

One such program is BAUV, the Biomimetic Autonomous Undersea Vehicle.
As the poster on McKenna's door proclaims, the goal of BAUV is an exciting (well,
exciting for ONR) blend of "shark-like low power, shrimp-like noise, and fish-like
low speed maneuverability." BAUV is essentially a pole the length of a desk with
three fishlike fins on either end and a neural brain. The fins are about fifty times
more efficient than a propeller, plus incredibly quiet, meaning BAUV is "undetectable by sound." The neural processor, which came out of research sponsored at the
New York University Medical School on rat brains, allows the robot to autonomously adjust to any change in the environment, allowing it, for example, to hold
the same spot in the ocean for weeks.

The BAUV currently can power itself on battery for up to three weeks, but
ONR is exploring ways to extend this. Projects include giving it solar power, for
when it operates near the surface, or even the ability to use a "mud battery." This is
a bacteria-powered cell that is set on the muddy bottom of the ocean's floor. When
bacteria break down organic matter, they produce a stream of electrons that, if
captured, can produce electricity. The mud battery would refuel BAUVs like an
undersea robot gas station.

McKenna's pimp hand is strong. The relatively small-scale research he supported on BAUV could potentially revolutionize undersea warfare. A major challenge the U.S. Navy faces is how to patrol shallow waters, especially against quiet
diesel-powered submarines like those the Chinese and Iranians use. Instead of
risking the navy's valuable nuclear subs, BAUVs would be able to silently swim in
the shallow waters for weeks at time, creating a virtually undetectable network of
floating listening posts. A little bit of McKenna's start-up money might well create
a big PLUS, or what the navy calls its ultimate dream of "Persistent Littoral Undersea Surveillance."

## MAKING KEVLAR UNDERWEAR

Once the researcher has produced a prototype, places like ONR and DARPA turn
the project over to what McKenna calls "the customer," the military. Military labs
and units will then test the robot ("You basically beat the snot out of 'em," explains

one scientist), explore its uses, and even make suggestions for improvements. Often, they prove to be just as innovative as the original researcher. McKenna tells how one unit of marines took three different prototypes and cobbled them together into one machine that does "countersniper" work like iRobot's REDOWL. Whenever a sniper shoots at the marines, the technology automatically points a machine gun at where the bullet came from.

An example of one of these places that brings together tactics and technology is the Marine Corps Warfighting Lab, located at the massive base in Quantico, Virginia. The lab develops and tests out new technologies, as well as sources commercial market solutions for military needs. Dragon Runner, a nine-pound robot that looks a bit like a model car, is a prototypical example of the lab's development work. It came out of collaboration between Carnegie Mellon University, ONR, and the Marine Lab. Incredibly tough, troops use it to "see around the corner." They can toss it through a window, up some stairs, or down a cave; the robot will land on its feet and send back video of whatever it sees.

The military labs also serve a valuable function by end-running around the normal procurement system to get soldiers in the field what's already available in the stores. During the first days of the Afghanistan operation, for instance, special forces units sent back requests for a remote camera that could be linked to satellite communications and a "pointer," a man-portable UAV that could beam video back to an operator. It took eleven days for the labs to get them the camera, and eight months for the UAV, which compares quite well to the years that normal weapons development might take (the F-22 jet, for example, took twenty-five years to go from concept to deployment). No request is too small. The Marine Lab even made special Kevlar-lined undershorts for marines to wear while on patrol. One news article jokingly called the program "Saving Ryan's Privates." Besides protecting marines' unmentionables, the special shorts also shield the femoral artery from being nicked by shrapnel. In either case, as one commented, "When your butt's on the line, you want it protected."

Such defense-funded labs pop up all over the place. Perhaps the most surprising is the Idaho National Lab. More akin to a national park than a traditional laboratory, it has huge tracts of land for testing out ground and air robots, including its own airstrip for UAVs. In the words of one research scientist there, "Our lab is just a little bit smaller than Rhode Island." The Idaho team reflects your expected western hospitality. They have a standing offer to other roboticists: "Any one who wants to play around with one of their systems, come on down."

This kind of neighborly vibe carries across the robotics field. When I asked people who they most respected in the field, the name that consistently came up was H. R. "Bart" Everett. Everett is a retired commander in the U.S. Navy, who is now technical director of robotics at the Space and Naval Warfare Systems Center (SPAWAR) in San Diego. As one scientist put it, "He is one of the true graybeards in the field of robotics." Everett even maintains a "lending pool" of robots, which are loaned out to those who can't afford them.

Everett, who is now working on a book called *Children of Dysfunctional Robots*, tells how "my obsession with robots began early on in life, when I was about eight years old. I had become enthralled with a particular episode of *The Thin Man* one evening at a friend's house. The plot was centered upon a murder supposedly committed by a robot, and of course the Thin Man had to prove some dastardly villain and not the robot really committed the crime. I had never before seen a robot and was forever changed by that experience."

By junior high, Everett was tinkering with robots. He kept his interest going after he joined the navy. While attending the Naval Postgraduate School in Monterey in 1982, he built the very first behavior-based autonomous robot for his thesis project. This revolutionary robot was controlled by a single-board Synertek computer (the cousin to the first Commodore personal computers). Much as how Commander Data in *Star Trek* was made in the image of his designer, Bart Everett called this first robot RO-BART. His robot soon paid him back for the gift of autonomy. "There weren't a lot of mobile robots in those days, so it attracted a tremendous amount of media attention, which in turn landed me a job."

ROBART-II came out of Everett's tinkering in his basement in the following years. In 1986, he turned the model over to the navy and joined the San Diego center. This was followed by ROBART-III, a test-bed robot that has continually evolved since 1992. ROBART-III looks a bit like Robby the Robot from *Lost in Space*, except it has a six-barrel Gatling gun as an arm. It has been used as a platform for such new technologies as natural language understanding and automated target acquisition and tracking. It was named number sixteen on *Wired* magazine's list of the best robots of all time.

## CUSTOMER FEEDBACK

The four enlisted men sat onstage, evidently uncomfortable to be the focus of so much attention. But to the scientists and businessmen gathered in the hotel conference room in Georgetown, they were the real stars of the robotics industry convention. The four had recently served in Iraq and had used their robots nearly every single day. The "Warfighters' Perspectives" panel was the ultimate opportunity for customer feedback.

For the next ninety minutes, the soldiers talked about their experiences with robots in Iraq and various suggestions they had for improvement. They asked for better batteries and interchangeable parts that could be fixed in the field, rather than always having to send a broken robot to the robot repair yard. Army staff sergeant Robert Shallbetter even offered feedback on the robots' colors. Having robots painted black made them stand out as targets and the 140-degree heat in Iraq made them hard to even touch. Plus, "Heat and computers don't mix well."

The audience's ears perked up when the soldiers began to talk about which robots they liked more, knowing that this sort of feedback could determine their programs' and companies' futures. They complained that Foster-Miller's Talon didn't have its own light source, so the soldiers had to duct-tape flashlights on it at night. On the other hand, they noted that PackBot did have its own light source, but it drained the batteries fairly quickly. They complained that the Pack-Bot needed as long as two minutes to boot up and enter the PIN number for access. "After we're out for about thirty minutes, we had to start planning on being attacked, or having an ambush waiting for us on the way back," said Specialist Jacob Chapman, so the loss of two minutes can be fatal. On the other hand, having a PIN number makes it harder for enemies to use the robots if they ever capture them. In the end, there was no clear favorite between the two robots from Boston. As Byron Brezina, robotics director of the navy's EOD technology division, said, "If you've ever gotten into the Ford versus Chevy argument, that's pretty much what it goes like."

The soldiers were incredibly blunt, however, about one robot. The Vanguard is manufactured by Allen-Vanguard Inc. of Reston, Virginia. As the other soldiers nodded, Chapman called it "completely unreliable" and told how his robot would turn off after going about ten feet from the truck. "We ended up trying to get rid of [the Vanguards] as soon as we could." When asked what he would do to fix it, he gave the ultimate soldier's reply. "Make it work." Standing at the back of the room was an executive from the company. At that moment, he looked very ill.

The soldiers ended their talk by thanking the researchers and executives gathered in the room. "I'm very fortunate due to the current [robot] technology to be standing here today," Shallbetter said. Navy aviation ordnanceman first class Bryan Bymer chimed in, "They most definitely saved people's lives."

Tom Ryden, director of sales and marketing for iRobot, was one of the hosts of the conference. He thanked the soldiers in turn, promising them, "We're going to take a lot of that to heart and see what we can do to make improvements." The panel closed with the more than one hundred scientists giving a standing ovation to the soldiers.

This sort of interaction between soldier and scientist is actually far more common than one would think. Mack Barber at Remotec tells how "sometimes we get phone calls and we can hear the gunfire in the background." Many credit the troops in the field with some of the best ideas. Researchers at iRobot, for example, are especially proud that soldiers had "direct input into the design" of the Packbot and recall that during the early deployments to Iraq they would update the robots' software with feedback from each mission. The company makes it a point to fly in soldiers on their way back home from Iraq to its office in Burlington for feedback, and even has a place on its Web site where soldiers can post their improvement ideas.

The soldiers at the robots' feedback session also requested that the scientists try to understand their needs better. "If you can put yourself in our shoes and imagine what we're going through, we would really appreciate it," said Sergeant Shallbetter.

For most of history, that was an impossible request to meet. Scientists have long been involved in war, but they were usually separate from soldiers and the battlefield. As some military historians note, "The scientist did not need physical courage to do his work....The soldier, unlike the scientist, might be called on to face death. This was the soldier's badge of honor, and in his mind, made him the rightful ruler of the battlefield."

Yet this division of labor is also breaking down, oddly enough through unmanned systems. While robots are moving some soldiers off the battlefield, they are also bringing the geeks out to war. Robot researchers from firms like iRobot and Foster-Miller are now going out in the field in search of feedback and updates to ever-changing technology. As one military analyst put it, "There are tons of guys now wearing Kevlar pocket protectors" on today's battlefield.

Unlike past weapons systems, the new robots don't even need the soldiers to initiate the feedback; the robots can also report back on their own. As Jim Rymarcsuk, a vice president at iRobot, explains, "Our robots have logistic information on them. They track the hours of operation, how it has been operated, what it has been used for. We can track a lot of that."

This kind of back-and-forth between the people who design and make robots and the users in battle produces a pattern of almost continual improvement. For example, one navy robot, the Mk. 3 RONS, went through some thirty-five different changes in its first five years of operation. The constant communication between the battlefield and research lab can also take some humorous turns. Joe Dyer, a former navy admiral turned vice president at iRobot, describes how his firm once received a box shipped from Iraq. It was filled with the bits and pieces of a PackBot that Iraqi insurgents had blown up. Attached was a request for "warranty repair."

# EIGHT

# WHAT INSPIRES THEM:
# SCIENCE FICTION'S IMPACT
# ON SCIENCE REALITY

*You can never tell when you make up something what will happen with it. You never know whether or not it will come true.*

—DONNA SHIRLEY

The Science Fiction Museum and Hall of Fame appropriately stands in the shadow of Seattle's futuristic landmark, the Space Needle. Set in a multicolored, globular Frank Gehry–designed building that looks like a cut-up guitar (a "ridiculous... monstrosity of postmodern architecture" is another writer's take), it shares the space with the Experience Music Project, a museum for rock and roll music. The odd juxtaposition of the two museums is actually quite simple: science fiction and Jimi Hendrix's music were the two boyhood loves of Microsoft cofounder Paul Allen, who is the primary funder of both.

Founded in 2004, the Science Fiction Museum and Hall of Fame is dedicated to exploring the history of science fiction and how it shapes our culture, politics, and philosophy. While the Experience Music Project next door has the guitars used by Bob Dylan, Bo Diddley, and Kurt Cobain, the Science Fiction Museum rocks just as hard. Displayed in the museum are such artifacts as Captain Kirk's command chair from *Star Trek*, the alien queen from *Aliens*, Darth Vader's helmet from *The*

*Empire Strikes Back,* Neal Stephenson's handwritten manuscript for the *Baroque Cycle* trilogy, and the pistol used by Harrison Ford in *Blade Runner.* The museum also runs a kids' program, including a "summer camp on Mars," as well as a happy hour for the adults, with three-dollar beers on tap.

It is easy to think of the Museum and Hall of Fame as only some sort of "Pantheon of Nerds" (what my editor jokingly called it), as science fiction may well be the ultimate of geekdom. Perhaps no one puts it better than Chuck Klosterman, who once wrote that admitting you like science fiction was "like admitting that you masturbate two times a day, or that your favorite band was They Might Be Giants."

And yet science fiction is undeniably popular. The earliest science fiction was by storied writers such as Mary Shelley, whose *Frankenstein* was first published in 1818, and Nathaniel Hawthorne, whose story "The Birthmark" wrestled with plastic surgery before plastic was even invented. Today, roughly 10 percent of all books are in the science fiction and fantasy genres. This does not even count major authors like Michael Crichton or Tom Clancy, who write "techno-thrillers" that are science fiction in all but name.

Science fiction has thrived even more in modern media forms. Six of the top-ten-grossing movies of all time are science fiction, led by the original *Star Wars* (inexplicably still behind *Titanic* in total sales). On TV, many of the most popular and influential shows of all time, from *The Twilight Zone* to *Lost,* have been science fiction. An entire cable network, the Sci Fi Channel, is exclusively devoted to the genre. For such a geeky topic, it is doing quite well, ranking in the top ten of all basic cable networks.

Science fiction is more than just popular; it is also incredibly influential, to an extent that is often surprising. Time and again, science fiction makes its presence felt in real-world technology, war, and politics. At iRobot, for example, the robotics research group described how their team motto was a toss-up between "making science fiction reality" and "practical science fiction" (they couldn't yet decide which they liked better). Science fiction references and ideas also make frequent appearances on the military side, coming up in almost any meeting on new military technologies or how to use them. Even Admiral Michael Mullen, the chairman of the Joint Chiefs of Staff (that is, the man in charge of the entire U.S. military), proudly described how the navy's "Professional Reading" program, which he helped develop to guide his sailors, includes the science fiction novels *Starship Troopers* and *Ender's Game.*

## WHAT IS SCIENCE FICTION?

Museum director Donna Shirley is perhaps the best person in the world to explain just what is science fiction. Shirley's entry into the field came at the age of ten, when she went to her uncle's college graduation. In the pamphlet they gave out, there was a listing for graduates with "aeronautical engineering" degrees. Shirley remembers asking her mother what that meant. "My mother replied, 'Those are the people who make airplanes.' And so that's what I wanted to be."

At the age of sixteen, Donna Shirley got her pilot's license. She then enrolled as the only woman in her classes at the University of Oklahoma, and earned that degree her mother had told her about. "Although the guys in my classes were fine with me being an engineer, my college advisor for aeronautical engineering told me girls couldn't be engineers," she recalls.

She soon proved him wrong. In 1966, Shirley joined NASA's prestigious Jet Propulsion Lab as one of the space program's very first female engineers. Over the next thirty-two years she worked on projects as varied as automating controls of military satellites to the Mariner 10 space probe's trip to Venus and Mercury. She capped her career by serving as manager of the Mars Exploration Program, which included the 1997 Mars Pathfinder and the Sojourner robotic rover missions. As one article describes, "Not only were these events two of the U.S. space program's greatest successes, but they may well provide the world with some of the most important scientific data of the 20th and 21st centuries."

Shirley credits the science fiction she read growing up as a key factor in her career. She recalls reading the stories of Robert Heinlein and Isaac Asimov at the age of eleven. "The political issues in the books went over my head," she says, "but their heroes were always engineers and scientists.... Heinlein and Asimov also frequently had women characters as heroes, which resonated with me." During the publicity surrounding NASA's Mars missions, the organizers of the museum heard Shirley talking about science fiction's influence on her work and invited her to join the team.

She sees as her role and that of the museum to "educate people about science fiction and to make people realize how important it is in our culture, and by implication get them interested in science and the social aspects of science. At the same time, we can pass on some moral lessons.... In a sense, it's to capture their imagination away from, say, *Playboy* and into something a bit more important."

Shirley notes that the fictional worlds that science fiction authors often create are not what constitute science fiction. Nor does science directly drive the plotlines. Rather, science fiction forces the audience to wrestle with the effect that science has on society. She explains, "The technology is not the interesting part; it is what people do with the technology." Most science fiction deals with some sort of fallout, usually political, that comes from a new event or technology. For example, Philip K. Dick's *Minority Report* posits a technology that allows the police to predict a crime. The story is not about the technology, but "the political and legal ramifications of actually using such a system." In short, science fiction is more about asking "thought-provoking" questions than merely providing "jaw-dropping" special effects.

This focus on the dilemma, rather than the technology, is what allows science fiction stories to remain relevant even when the world and technology advances past the time of a story's creation. Shirley points out an exhibit at the museum that shows how H. G. Wells's 1898 novel *The War of the Worlds* has been continually "remade and rereleased every time there was a perceived existential threat on this world." Prior to World War II, Orson Welles did his famous radio broadcast. Then the story was made into a movie that echoed nuclear fears at the start of the cold war, and a third time in 2004 by Steven Spielberg, who used imagery evocative of the 9/11 attacks.

Shirley sees several trends in how science fiction is wrestling with the modern world. The first is a trend toward more women writers, in particular the pioneering work of the recently deceased Octavia Butler, one of the first African American women science fiction writers and the only science fiction author ever to receive a MacArthur Foundation "genius" grant. "Women writers tend to write more about the social stuff and what happens to people." There is also an evident trend of more focus on the impact of computers and robotics. She notes the work of writers like Neal Stephenson and Bruce Sterling, who helped found the "cyberpunk" movement. The trend emphasizes not merely the coming technology, but what happens when it gets placed "in the hands of our depraved society."

## SCIENCE FICTION AND WAR

"I thought *Ender's Game* might be popular when I finished writing it—high-tension story, semi-tragic outcome. I did not expect it to last as long as it has (so far) or to become as widely read by adults, teenagers, and children. Or, to put it another way, I think all my books will do wonderfully well when I'm through writing them; with *Ender's Game*, I happened to be right."

Orson Scott Card has written fifty-nine books that have sold twenty million copies in North America alone. But it is still *Ender's Game*, his 1985 book, for which he is best known. The story of Ender Wiggin, a child who expertly fights war as if it were sport, won every major science fiction award, has been translated into eighteen languages, and is under development at Warner Brothers to be a major movie.

More important, the book's stories of the command school of the future and experiencing war from afar via virtual reality struck a particular chord with the military. Some two decades after its publication, it is still in various military course catalogs (such as at the Marine Corps University, where it is used as a text on the psychology of leadership) as well as various U.S. military-required reading lists that generals and admirals tell their officers they should read if they want to be good warriors under their command.

He may be a writer of fiction, but like many in the field, Card also consults for the military, speaking on such topics as "Next Generation WMD: Anticipating the Threat." Card is not surprised by the response his book has gotten from military readers. "Soldiers feel like *Ender's Game* is telling their story—young people doing their duty in spite of the idiocies of the officers who lead them. But I get similar responses from gifted schoolchildren and from kids who do very badly in school, each of them seizing on the heroism-in-isolation of Ender and extrapolating it to their own lives."

Card's work is representative of a broader trend in science fiction, its overwhelming focus on war. While science fiction is known for peering into the future and bringing to light fanciful new technologies, the vast bulk of it places these stories and technologies in one particular context of our human experience: war. Each year, approximately five major science fiction movies that link to war are released. Fifteen of the ongoing science fiction TV series have a conflict or military element. The thirty-five science fiction magazines in the field each carry multiple stories set in war. And if you attend any of the fifty-two major science fiction conventions, your costume is most likely to pack a phaser, lightsaber, or blaster rifle. What some call "military science fiction" is by far the most popular part of the genre.

The reason why such a huge percentage of science fiction deals with issues of war, says Card, is "because war is a human constant. War also drives technological advance. And insofar as sci-fi was and remains a male genre, war will continue to fascinate readers." He is more curious about why other genres don't pay as much attention to war. "The real question," he asks, "is why war is not more important in mainstream fiction?...Literary fiction generally skips over two of the primary occupations of humankind: war and religion. At least science fiction and fantasy can still address those topics, along with everything else that literature can talk about."

Other writers point out that war is so popular in the genre because it is an unparalleled platform for wrestling with deep issues. Robin Wayne Bailey, the president of the Science Fiction and Fantasy Writers of America (a trade association for sci-fi writers; like the Teamsters but with pointy ears), explains, "The conflict is obvious [in war], the opportunities for technologic exploration and idea exploration are vast, as is the case in real war as well, and war provides both a microcosm and macrocosm for exploring human nature, and most pertinently human nature under stress." Harry Turtledove and Martin Greenberg, noted authors themselves and the editors of a master volume of the field entitled *The Best Military Science Fiction of the 20th Century*, agree. "Fiction is about character under stress. What we do when the heat is on reveals far more about us than how we behave in ordinary times."

Science fiction authors have set their stories in the realm of war since the very start of the field. H. G. Wells is perhaps the best known, but others include literary titans that we don't often associate with science fiction, such as Arthur Conan Doyle, Jack London, and even A. A. Milne. Most know Milne as the creator of the lovable bear Winnie-the-Pooh, but in 1909 he wrote a science fiction short story entitled "The Story of the Army Aeroplane." Just six years after the Wright brothers, it predicted that man might one day use those crazy flying machines for war.

The most influential author in developing the link between science fiction and war has to be Robert Heinlein. Heinlein came from a military background, graduating from the U.S. Naval Academy in 1929 and serving until 1934, when he was discharged for health reasons (during his convalescence, he invented the water bed). When World War II started, Heinlein went back to work for the navy doing aeronautical engineering. Interestingly, he recruited two young engineers to join his work at the Philadelphia Naval Yard; Isaac Asimov and L. Sprague de Camp would also later become some of the biggest names in the history of science fiction.

After the war ended, Heinlein became a key person in breaking science fiction into the mainstream, including being the first writer in the field to pen for the *Saturday Evening Post*, a leading magazine of the time. Over the course of his career, Heinlein would write thirty-two novels and fifty-nine short stories. But his two most influential were 1961s *Stranger in a Strange Land*, which foreshadowed the "free love" of the Sexual Revolution and was embraced by the hippie movement, and 1959's *Starship Troopers*, which, by contrast, is on the reading lists at the major military service academies, and inspired several technologies, such as robotic fighting suits.

In recognition of Heinlein's popularity and influence, the U.S. Naval Academy even has an endowed professorship named for him, the Robert A. Heinlein Chair in Aerospace Engineering. There is also a movement to have one of the navy's newest warships named the U.S.S. *Robert Heinlein*, in honor of his hundredth birthday. As the petition letter to the secretary of the navy reads, "It only seems fitting that a man who spent his life writing about the 21st Century should have a 21st Century destroyer named after him."

## THROUGH THE LOOKING GLASS: SCI-FI PREDICTIONS

Part of the popularity and influence of science fiction comes from its remarkable skill at foreshadowing the future. For a fictional genre that often takes place in settings that don't even exist, science fiction has forecast real-world technologies, as well as resulting dilemmas, with stunning accuracy.

Perhaps the best example of how predictive science fiction can be is the work of H. G. Wells, who is known as "the Father of Science Fiction." Wells was born in 1866, but his various stories forecast the twentieth century with incredible accuracy, predicting such things as computers, videocassette players, televisions, and even superhighways, each of which seemed unfathomable at the time. His books often had a theme of conflict running through them, and so he also predicted various military developments well before their time. For example, he wrote about tanks, or what he called "Land Ironclads," in 1903, which inspired Winston Churchill to champion their development a decade later. Similarly, his 1933 book *The Shape of Things to Come* predicted a world war that would feature the aerial bombing of cities. Wells was not a fan of such technologies, as he saw them as "unsporting."

Perhaps Wells's most important and influential prediction was in his story *The World Set Free*, published in 1914. He forecast a new type of weapon made of radioactive materials, which he called the "atomic bomb." At the time, physicists thought radioactive elements like uranium only released energy via a slow decay over thousands of years. Wells described a way in which the energy might be bundled up to make an explosion powerful enough to destroy a city. Of course, at the time, most scoffed; the famed scientist Ernest Rutherford even called Wells's idea "moonshine." One reader who differed was Leó Szilárd, a Hungarian scientist. Szilárd later became a key part of the Manhattan Project and credited the book with giving him the idea for the nuclear "chain reaction." Indeed, he even mailed a copy of Wells's book to Hugo Hirst, one of the founders of General Electric, with a cover note that read, "The forecast of the writers may prove to be more accurate than the forecast of the scientists."

Wells's story ends with scientists trying to organize an effort against war and the use of the new bombs. The idea later inspired Szilárd, Einstein, and others to form the Pugwash nuclear disarmament movement, meaning Wells's book, in turn, is the inspiration for the modern arms control movement (as well as the robotic "refuseniks" described in the following chapter).

Perhaps the only equal to Wells's work was the work of Jules Verne, who has been called "the Man Who Invented Tomorrow." Born in 1828, Verne wrote such books as *Twenty Thousand Leagues Under the Sea*, well before such things as large-scale submarines existed. His greatest prediction may have been in his so-called lost novel. In 1863, Verne wrote a book entitled *Paris in the 20th Century*. In it, he predicted a future that would have skyscraper buildings made of glass, automobiles powered by gasoline, calculators, worldwide communications, and even electronic music. To give a sense of how impressive this was, at the time Verne was writing, the electric lightbulb hadn't even been invented and the United States was locked in a civil war over whether human beings could be owned as slaves. What is more, Verne predicted that none of these fantastic improvements would make people happy. Instead, he foresaw that the technology advances would turn into a crass commercialism that overwhelmed worthwhile arts and culture. The publisher didn't like this dark yet admittedly accurate prediction of the future and rejected it. They published his *Journey to the Center of the Earth* instead and the manuscript stayed locked away in a safe until 1994.

Science fiction continued to tap into the future throughout the twentieth century, as the field extended into film and TV. The only difference was that, with the speeded-up time frames, the imagined technology came to fruition much quicker. For example, Stanley Kubrick's 1971 film *A Clockwork Orange* predicted futuristic small music devices (what we would now call MP3 players or iPods), while in the 1976 movie *The Man Who Fell to Earth*, David Bowie plays a futuristic alien who develops an equally futuristic technology, what we now know as digital cameras. Indeed, even *The Jetsons* proved prophetic. George Jetson spent most of his day at work at Spacely Sprockets pushing computer buttons as a "Digital Index

Operator." Spending your day in front of a computer seemed wildly futuristic in the 1960s, but now George is just a run-of-the-mill database administrator.

By comparison, the government often has a relatively poor track record when it comes to predicting the future. For example, in 1913, the U.S. government actually prosecuted Lee de Forest of RCA for telling investors that his company would soon be able to transmit the human voice across the Atlantic Ocean. The idea seemed so absurd to the government that de Forest was assumed to be a swindler. Indeed, Philip Tetlock, in his award-winning study *Expert Political Judgment*, found that the professional "experts" who advise government are actually more often wrong in their predictions than right. Industry equally has a mixed track record. For example, IBM president Thomas Watson famously said in 1943, "I think there is a world market for maybe five computers."

When it comes to war, the same pattern holds. As a 2006 article in *Armed Forces Journal*, one of the leading magazines for U.S. military officers, notes, "We don't do well, historically, in predicting the location and nature of the next war." For example, Sir Arthur Conan Doyle, the creator of Sherlock Holmes, wrote a short story in 1914, just before World War I started. Entitled "Danger," it warned that the new invention of submarines might be used to sink merchant ships. The Royal Navy's Admiralty actually went public to refute and mock Conan Doyle, saying that "no nation would permit it and the officer who did it would be shot." Just seven months later, the passenger ship *Lusitania* was torpedoed by a German U-boat, inaugurating the era of submarine warfare. Part of the reason for this pattern is that while science fiction looks forward, the military typically plans what the next war will look like by looking back at how it fought the last one. In discussing how the American army that invaded Iraq in 2003 planned for it to be a repeat of the 1991 Gulf War, *Armed Forces Journal* concludes, "Our advances in technical intelligence have not improved our ability to predict any specific war."

There are a couple of explanations for why science fiction tends to do well in prediction, even though it is working in the world of fiction. Many science fiction writers are scientists themselves, so they are typically well equipped to stay within the rules of science yet extrapolate forward. Arthur C. Clarke, for instance, not only imagined a world of intelligent computers, but is also the man who invented the real-world communications satellite. As Donna Shirley explains, science fiction authors tend to get their predictions right because they are most often writing about what they know best. "Modern science fiction is increasingly being written by computer geeks, who are already experts on the technology side."

Secondly, these writers must create a narrative at some point (that is, the plot, which usually involves a battle of good versus evil, hence the frequent setting of war), but along the way they must solve the same technical problems that real scientists do. However, they don't have the constraints of a budget or lab time or bureaucratic politics. The freedom of the fictional world allows them to work out solutions sometimes easier than in the real world. As one computer scientist noted, "Science fiction is not making predictions, but playing with possibilities."

Finally, dealing with the "what if?" is what sets science fiction apart from regular fiction, as well as real-world science. Shirley explains, "The best science fiction deals more with the social consequences of technology change than the technology itself." It is the combination of scientific awareness with human imagination that allows science fiction to better deal with technology put in a complex social setting. Science fiction author Robin Wayne Bailey sums it up this way: "Science fiction at its best is about ideas. Maybe it's criticized for often having wooden characters or unrealistic settings, but the ideas always come first.... Science fiction throws out ideas like some people scatter seeds. Most do not take root, but some do. And when they do, it is fabulous."

It is important to note that in this seed-scattering of ideas, science fiction is not always perfect. As Donna Shirley notes, "Science fiction did not predict computers very well, at least until HAL.... The same for Martians. Mariner 4 [the planetary probe] killed all the Martians in 1965."

Where science fiction tends to go most wrong in its predictions is not in the technology but in the timelines. Ray Kurzweil, who makes a living out of timing technology predictions, explains, "Science fiction is unreliable because [there is] no requirement that the time frame be realistic. Arthur C. Clarke chose the year 2001 as a literary device, not because that's when he was certain AI would come to fruition." Shirley agrees in a way. "The technology is changing so rapidly, they [science fiction writers] are increasingly having trouble keeping up."

Orson Scott Card thinks that holding science fiction to any standard for its prediction is beside the point. "Predicting is a trivial aspect of writing science fiction. We are extrapolating what would happen if a particular configuration of future possibilities became real. The result is that we plunge readers into an environment in which they must rebuild their conception of reality. So we aren't predicting the future, we're helping readers rehearse for the future, whatever it might bring." He continues, "The job of the sci-fi writer is to envision all possibilities and bring them to life in the readers' imagination. What impact that will have is always debatable—less and less, these days, I believe. When things go horribly wrong, it's small satisfaction to say 'I told you so.'"

## TURNING DREAMS INTO REALITY

"There is a back and forth between dreams and reality. Science fiction offers the dreams, the engineers make it the reality, and the readers are the ones who pilot the technology in planes, cars, rockets, whatever."

Greg Bear is the author of more than thirty books and has won two Hugos and five Nebulas (the science fiction versions of Pulitzer Prizes). His most recent novel is *Quantico*, a thriller set in the "second decade of the War on Terror" about young FBI agents taking on a brilliant homegrown terrorist. The book flap captures it best: "It's the near future—sooner than you might hope."

Bear is especially well equipped to reflect on how science fiction doesn't just predict but also inspires real-world changes, as his name is frequently mentioned in the military research community. For example, an air force lieutenant colonel commented how he even footnoted Bear's work in a project proposal. By way of explanation he asks, "I mean, how many science fiction books have appendices and glossaries?"

Growing up as "a Navy brat," who moved with his father from bases in California and Japan to the Philippines, Bear recalls that "in my living memory I don't know a time when science fiction wasn't in my life." He started writing at eleven years old and sold his first story at age fifteen. The next year he met his hero, science fiction writer Ray Bradbury, and his career as a writer was decided.

Since that time, Bear has been called the "best working writer of hard science fiction" by *The Ultimate Encyclopedia of Science Fiction*. His impact, though, is decidedly beyond the world of fiction. Bear has served on various political and scientific action committees and advises the U.S. Army, the CIA, Sandia National Laboratories, and Microsoft Corporation. Indeed, when we spoke, Bear was just back from headlining a government conference on biotechnology threats, inspired in part by his book *Darwin's Radio*.

Bear is also one of the core members of SIGMA, a "think tank of patriotic science fiction writers." SIGMA was started by Arlan Andrews, a writer who also worked at the White House Science Office. "If you don't read science fiction, you're not qualified to talk about the future," he said. Since the 9/11 attacks, SIGMA has worked closely with the Department of Homeland Security, and influenced it in particular to set up the Homeland Security Advanced Research Projects Agency, or HSARPA. A parallel to the Defense Department's DARPA, HSARPA spends about $7 million a year (1 percent of the agency's budget) on futuristic "high impact" projects. At a government conference where authors like Bear spoke in 2007, a government official defended the science fiction link to policy. "Congress asks me how can I afford to roll the dice with 1 percent of the taxpayers' money," tells Jay Cohen, head of Homeland Security's Science and Technology Directorate. "I say there are bad people in the caves of Tora Bora who are rolling the dice with 100 percent of their money."

Bear sees the influence of his work and his access to policymakers as coming in part from the focus on conflict. Referring to the many military readers of his work, he says, "If you lead the life, you tend to choose to read fiction about it." He also sees science fiction's influence spreading via its crossover into popular technologic thriller authors like Tom Clancy and Dan Brown, who are especially popular among military readers.

This fandom extends to the top. "There is a pretty striking amount of government officials that read science fiction," Bear says. "Harry Truman loved science fiction. He was an 'other planets' type of guy.... Reagan liked the older writers like Jules Verne and Edgar Rice Burroughs. Reagan even gave [promotional] quotes to writers and was not averse to receiving papers from them, when he was president." He goes on to note that, as someone who leans left in his politics, he's somewhat disappointed that recent Democrats tend to be less likely than the Republicans to be science fiction fans. "They seem to be more like FDR and get into the legal thrillers and mysteries."

## DIRECT INSPIRATION, OR "HOW WILLIAM SHATNER CHANGED THE WORLD"

Science fiction may be incredibly popular, but raw fandom doesn't necessarily translate into influence. If that was the case, as I write this, Hannah Montana would be the most powerful person on the planet. Rather, science fiction's influence on real-world science and even war comes through a variety of pathways. The most simple is the direct way, giving scientists ideas of what to invent. And nothing better proves this than *Star Trek*. Or, as William Shatner (the actor who portrayed Captain Kirk on the original series) claims, "All this wiz-bangering didn't happen by accident. I made it happen. Or rather, *Star Trek* did."

While the original series only lasted three years (1966–69) before it was canceled by NBC due to low ratings, *Star Trek* has since boldly gone where no work of fiction has gone before. It spun off five other TV shows, ten movies (an eleventh is in the works), an entire library of books (Amazon.com lists 4,276 *Star Trek* books), and a city's worth of exhibits, rides, and museums. The mecca of all this is "Star Trek: The Experience," an interactive museum at the Las Vegas Hilton hotel and casino. Where Elvis used to do his famous "Viva Las Vegas" show, today you can drink a "Commander Riker-Rita" at Quark's Bar or renew your Vulcan wedding vows. All told, the *Washington Post* estimates that the worldwide fan base is 250 million Trekkies strong.

The original show came out of the vision of Gene Roddenberry, a World War II bomber pilot turned Hollywood producer. While he wanted technology that "looked futuristic," the reality often had a different point of origin. For example, many recall the famous "transporter" that each episode would beam Kirk, Spock, and an anonymous, certain-to-die, red-shirted crewman down to the planet's surface. Screenwriters now tell that the origin of the transporter actually came about when the prop company didn't deliver a mockup of a shuttle craft in time.

These ideas, however, certainly made an impression on a generation of kids turned scientists. They became determined to make the world have technology just like they'd seen their heroes use in their favorite show. Martin Cooper, the inventor of the cell phone, recalls that his "eureka" moment of inspiration came when watching a *Star Trek* episode in his lab. "There's Captain Kirk, talking on a communicator, without dialing! I think 'This thing is genius.' ... The *Star Trek* communicator to us wasn't a fantasy. It was an objective." Similarly, John Adler of Stanford Medical School observes that Dr. McCoy's sick bay "revolutionized the way we think about patient care." Inspired by Bones's medical tricorder (actually just a tricked-out salt shaker), Adler revolutionized the medical field by inventing the cyber knife, which does surgery by sending a beam into cancer tumors. Rob Hatani, who was equally inspired by the tricorder to invent the PalmPilot PDA, explains that this degree of influence is to be expected, given the popularity of the show among scientists. "In Silicon Valley, everyone's a *Star Trek* fan. It's like football in Green Bay."

The franchise and its influence was reborn again in the 1980s with *Star Trek: The Next Generation*. The successor series differed in often focusing on the darker side of technology (such as its introduction of the Borg, the new adversary species, whose robotic technology had eradicated all empathy), but it too had a major influence on scientists. For example, Steve Perlman recalls how his inspiration moment came when watching an episode in which the robot Data relaxes by listening to several symphonies stored on his computer. Perlman went on to invent QuickTime, a software program that stores and plays electronic audio and video files. This, in turn, helped make possible iPods and other portable digital music players. Today, Perlman is working to make a virtual reality playroom, modeled after the *Enterprise*'s Holodeck.

The inspirational role of science fiction extends beyond the world of Trekkies, and is especially pronounced in military technology. An illustration comes from an anthology of short stories entitled *The Best Military Science Fiction of the 20th Century*. The volume is a collection of the most popular science fiction short stories, written from 1900 to 2000, set in war. What is noteworthy is that thirty-four technologies dreamed up in the last century are now under development by the U.S. military in this century. These range from exoskeleton suits that soldiers might wear to an automated defense system for tanks, now called by the Pentagon "Active Protection Systems."

Those working in the military weapons development field are often surprisingly open about where they get their ideas. Colonel James Lasswell is a retired infantry officer now at the Marine Corps Warfighting Lab. He says, "The fact that it exists in our own movies proves that it is potentially possible. . . . If you can imagine it, we think it can happen." For instance, when pondering how to aid marines in the battle against IEDs in Iraq, his team sent a request to DARPA to start working on what he called "Jedi Broomsticks," that is, the hovering speeder bikes that appeared in *Star Wars: Return of the Jedi*. "We wanted ground mobility, but not on the ground." The jet bikes are still not yet deployed, but another science fiction idea come true is miniature communications devices a marine can wear on his wrist and watch video footage shot by a UAV above. "We got the idea from *Dick Tracy*," Lasswell says with a chuckle.

As Andrew Bennett, who leads the design team at iRobot, says, "We were all influenced by science fiction. You are always looking for ideas and science fiction is one of many sources." His colleague Bryan Yamaguchi laments, "But now we are finding that our stuff is getting more advanced than science fiction."

## FUNDING SCI-FI

The researchers are not the only ones who grow up on this diet of science fiction. So too do the funders who decide which weapons programs to pay for. As former

Speaker of the House Newt Gingrich (who actually visited Isaac Asimov's apartment when he was in Congress) explains, "People like Isaac Asimov, Arthur C. Clarke, and Carl Sagan did an amazing amount to convince humans that science and technology were important."

Perhaps the best illustration of this is at the Air Force Research Lab's Directed Energy Directorate at Kirtland Air Force Base in New Mexico. In 2005, it rolled out a new prototype weapon with the mundane title of "Personnel Halting and Stimulation Response," or PHaSR. People in the military tend to speak out an acronym as if it were one word, rather than reading the letters. So the whole convoluted name was just a way to call the Pentagon's new weapon a "phaser," the little ray gun from *Star Trek* that Kirk always "set to stun" before he beamed off to explore new worlds and romance buxom alien women. The PHaSR system is essentially a laser rifle whose beam can stun a target more than two hundred yards away, a nonlethal weapon perfect for mounting on a robot. When asked why they chose that name, program manager Captain Thomas Wegner proudly answers, "We picked the PHaSR name to help sell the program. It's an obvious homage to *Star Trek*."

It is often difficult to figure out just what the future will look like, but science fiction creates both an expectation and early acceptance of technologies that are not yet fully developed. As Bill Gates explains, *Star Trek* paved the way for his job at selling small, easy-to-use computers to the public. "It told the world that one day computers would be everywhere." He sees the same happening with robots from movies like *Star Wars* and *I, Robot*. "The popularity of robots in fiction indicates that people are receptive to the idea that these machines will one day walk among us as helpers and even as companions."

Military robot developers see the same trend when selling to the Pentagon. One explains, "It's a way to make possibilities seem real, but also inevitable." Sometimes, though, the popularity of science fiction among military funders can actually make it harder on researchers. "Naval customers just assume it will happen," explains Thomas McKenna at ONR. Likewise, the military funders tend to want the cooler technologies, while the mundane are less likely to get funded. One U.S. Army researcher working on nonlethal weapons systems complains, "You have to beg for money for things like beanbags or acoustics. But say it's for a laser or a lightsaber and the money is no problem."

## THE LENS OF THE LOOKING GLASS

"Any sufficiently advanced technology is indistinguishable from magic," famously argued English physicist and science fiction author Arthur C. Clarke. Indeed, when the warriors of the Hehe tribe in Tanzania surrounded a single German colonist in 1891, they seemingly had little to fear. But he had a magic box that killed almost a thousand spear-armed warriors by spitting out death faster than they ever imagined possible, the machine gun.

New technologies often can seem not merely incomprehensible, but unimaginable. Science fiction, though, allows us to jump that divide. It helps to take the shock out of what analysts call "Future Shock." By allowing us to imagine the unimaginable, it helps prepare us for the future, including even in war.

This preparation extends beyond future expectations; science fiction creates a frame of reference that shapes our hopes and fears about the future, as well as how we reflect on the ethics of some new technology. One set of human rights experts I queried on the laws of unmanned warfare referenced *Blade Runner, The Terminator,* and *Robocop* with the same weight as they did the Geneva Conventions. At another human rights organization, two leaders even got into a debate over whether the combat scenes in *Star Trek* were realistic; their idea was that this could help determine whether the fictional codes of the Federation could be used as real-world guides for today's tough ethical choices in war.

By far the most influential writer when it comes to the right and wrong of robots is Isaac Asimov. Every single roboticist knows Asimov's "Three Laws of Robotics" by heart, and they have become *the* reference point for ethical discussions about robots. Yet they are fiction, never intended for the real world. Instead, each of the stories in *I, Robot* uses the laws as a jumping-off point to look at the problems that occur when robots try to follow the laws in the complexity of the real world.

Many of our expectations and ethical assumptions around real-world robots come from science fiction. The irony is that the same stories that inspire and fund the research can also create assumptions that are often incredibly frustrating to real-world researchers. As one scientist discussed, "There seems a strong tendency over the decades to view robots as something evil, like technology run amok." These fears date back to the slaves of Karel Čapek's 1921 play *R.U.R.* and the mechanical minx Maria, an evil robot in Fritz Lang's 1927 film *Metropolis* (her ultimate evil was illustrated by the fact that she liked to both oppress the urban poor and dance exotically). They continue today in such movie franchises as *The Terminator* and *The Matrix*. Bart Everett of the navy lab describes it as a "paranoia" that "stems from the fact that doomsday scenarios make for better movies. As a result, there is often confusion with regard to what the technology actually can and cannot do today, as well as where it's headed in the future."

Regardless, the reality is that science fiction always lies in the forefront of debates over such key questions as whether robots should be armed or how much autonomy they should be given. And yet this doesn't drive the field toward any one conclusion. The galaxy of stories that science fiction writers have created is simply too diverse.

Indeed, just as there is not one single world of regular fiction, there is no one culture of science fiction. The field itself can be different across time and space and thus have changing influences on how we frame the world of science. For example, if *Star Trek* was dominant in the 1960s, *Harry Potter* is the power series of today. To put it another way, kids today are infinitely more likely to know what a Chizpurfle is than a Tribble (for the uninitiated, a tiny, mitelike creature that feeds on both magic and electricity versus pink furballs that reproduce at exponential rates). Even though J. K. Rowling created a world more of fantasy than science fiction, its influences are already being felt within the real world of war and weapons development. Researchers in both Britain and the United States (with DARPA funds) are now hard at work on an invisibility cloak that works just like the one young wizarding student Harry inherited from his father. The real-world one is set to be made of novel "metamaterials," which can be tuned to bend radio waves and light, so that the cloak would neither reflect light nor cast a shadow (a true science fiction comparison would be the chameleon camouflage used by the alien in *Predator*). John Pendry, a physicist at the Imperial College London, notes that the *Harry Potter* link may not be an exact one: "To be realistic, it's going to be fairly thick. Cloak is a misnomer. 'Shield' might be more appropriate."

Older scientists also note that, as Rod Brooks of iRobot puts it, "There is becoming a generational difference in where the science fiction influence comes from." As he explains, his major influences were science fiction books. For his colleague Helen Greiner, it was movies. Today, for his students at MIT, it is video games. "And I have no idea what will be the different impact of these." One may be that the "new media" allow better special effects, but demand far less introspection. As soldiers grow up more familiar with the first-person shooters of video games like *Doom* or *Halo* and less the moral questioning of books like *I, Robot* or the "Prime Directive" dilemmas of *Trek*, we may find that the medium matters greatly.

Culture also appears to play a role. Just as the French love their Jerry Lewis and the English their Benny Hill, the popularity and influence of certain science fictions are linked to national tastes. For example, *Dr. Who* is perhaps the most popular sci-fi series in the United Kingdom, running for over a quarter century on the BBC (1963–1989) and spinning out two movies. In the United States, however, *Who* remains mainly a cult thing. Part of the explanation may lie in that the main hero is basically an oddball, go-lucky sort of guy, who stumbles into trouble while flying about the world in a spaceship that looks like a phone booth. We Americans like our science fiction heroes to be a bit more strong, cool, and dangerous; Dr. Who is no Han Solo.

If the British and the Americans differ along these lines, science fiction truly leaps in culture between East and West, especially when it comes to perceptions of robots. While the robot is consistently something suspicious in Western science fiction, it is the exact opposite in Asian science fiction. Indeed, the very first popular robot in Japanese science fiction was the post–World War II "Mighty Atom," also known as Astro-Boy. A robot that keeps the peace among humankind, he was also a response to the man-made horrors of Hiroshima and Nagasaki.

To this day in most Asian science fiction, especially in the anime genre, the robot is usually the hero who battles evil. This has heavily influenced both Japanese scientists and that nation's culture. "The machine is a friend of humans in Japan. A robot is a friend, basically," says Shuji Hasimoto, a robotics professor at Waseda University in Tokyo. "So, it is easy to use machines in this country."

Japan's traditional religion of Shintoism holds that both animate and inanimate objects, from rocks to trees to robots, have a spirit or soul just like a person. Thus, to endow a robot with a soul is not an illogical leap in either fiction or reality. Indeed, in many Japanese factories, robots are given Shinto rites and treated like members of the staff. Masahiro Mori, a professor at the Tokyo Institute of Technology, explains that Buddhism also makes for a more soulful approach to what a westerner would see as just a tool or maybe a mechanical servant. Mori, who wrote a book called *The Buddha in the Robot*, argues that robots can have a Buddha-like nature and that humans should relate to them as they would a person. "If you make something, your heart will go into the thing you are making. So, a robot is an external self. If a robot is an external self, a robot is your child."

In Asia, "companion" robots for the elderly are becoming quite common. One woman even found out she was dying of heart disease and included her Waka-mura robot in her will. By contrast, Rodney Brooks of iRobot says that the mass-marketing robots as friends for elderly shut-ins is yet to be tried in the United States because most Americans find such a concept "too artificial and icky." Sebastian Thrun, the robot car racer from Stanford, tells how the differing science fictions create a "willingness [in Asia] to go into new technologies and gadgets that is higher there than anywhere in the world." As a result, his lab has more collaboration with Asian companies than American ones.

The same differing attitudes and influences affect what different cultures think is acceptable in war. The question of arming unmanned systems and giving them the ability to shoot at humans is perhaps the most hot-button issue within the U.S. robotics community. It is far less controversial in Asia. Indeed, South Korea sent two robot snipers with rifles to Iraq in 2004 with essentially no debate; they were reported in the media to have "nearly 100%" accuracy.

Even more notable is the Autonomous Sentry Gun, made by Samsung. The company, more known for making high-definition TVs, has integrated a machine gun with two cameras (infrared and zooming) and pattern recognition software processors. The gun system cannot only identify, classify, and destroy moving targets from over a mile away, but, as Louis Ramirez of *Gizmodo* relates, "also has a speaker that beckons the fool that walks near it to surrender before being pulverized." South Korea plans to use the robo-machine guns to stand guard along the 155-mile demilitarized zone (DMZ) that borders North Korea.

The attitudinal differences become even more evident when you watch the promotional video put out by the Korean company for its new toy. The footage shows the machine gun automatically tracking a human test subject, who unsuccessfully tries to dodge the robotic gun by running back and forth and hiding behind bushes. For something that a westerner weaned on a diet of *Terminator* movies can't help but find disturbing, the vibe of the Korean commercial is a bit more celebratory. The footage of a real-world automated machine gun tracking humans is paired with the rousing theme song of the Disney movie *Pirates of the Caribbean*.

## THE FEEDBACK LOOP

"There's definitely a feedback between the sciences and science fiction," says James Cameron, creator of *The Terminator*, as well as a board member of the Science Fiction Museum and Hall of Fame. "It flows both directions.... Not only does science fiction inspire people to become scientists and want to ask questions about the real nature of existence and matter and reality, but what they're finding then feeds back into the science fiction community, and gets embraced by that, and spins out a whole new generation of science fiction."

Real scientists, soldiers, and policymakers may be influenced by science fiction, but change is coming so quickly that the creators of these imaginary worlds are increasingly borrowing from the real one. As Greg Bear notes, "I actually worry that science fiction isn't keeping up." Author and science fiction writers' union head Robin Wayne Bailey concurs: "The military is doing a fine job with robotics. The toys they have could be placed within any science fiction story.... But to see what they have on the drawing board is mind-boggling to even science fiction writers."

However, as science fiction experts look at some of what the military is doing today, many of them get frustrated. Donna Shirley may be the director of a science fiction museum in Seattle, but when the topic turns to the future of war, she is as smart as any political analyst inside the D.C. Beltway. "The Pentagon just doesn't get it. This high-technology stuff just doesn't work versus a distributed enemy like al-Qaeda.... No matter how many bunker busters you can drop from afar, if you don't know where someone is hiding it will not matter." And don't even get her started on the plans for National Missile Defense. "The idea of trying to hit a bullet with a bullet is silly. It is much easier and efficient to place your interceptor system offshore and take the missile out early in the launch stage when it is slow and easy to target.... But instead we are spending billions on the harder part just because it sounds really cool."

And yet the field still has a stigma that keeps its experts hidden away, even when on Pentagon contract. For all its influence on the future of technology and even war, "it is ironic then that we are rarely invited to the table to discuss these issues openly," laments Robin Wayne Bailey.

Perhaps we as a society ought to be paying more attention to the world of science fiction. It not only predicts and influences the future, but nothing may prepare us better to assess the consequences of a new technology than a field whose very essence is to ask questions about the moral, ethical, and societal dilemmas that new technologies might provoke. As Donna Shirley explains, "Science fiction says 'what if?' So, it doesn't say how exactly you can build the bomb. Instead, it says, if you build this bomb, you are going to get *Dr. Strangelove*."

# NINE

# THE REFUSENIKS:
# THE ROBOTICISTS WHO JUST SAY NO

*Never let your sense of morals get in the way of doing what's right.*

—ISAAC ASIMOV

Illah Nourbakhsh is an associate professor of robotics in the Robotics Institute at Carnegie Mellon University. He is also the military robotics world's worst nightmare, the scientist who learned to say, "No, thank you."

"As a kid, I was interested in taking things apart and putting them back together in weird ways," tells Nourbakhsh. He worked on solar car racers in college and then went to Stanford for graduate school. His research topics varied from genomics to AI. He recalls that when he first plugged his AI software into robots, "I was blown away by how little they could do. It was painfully obvious that robotics was delinquent." So he came back to his interest in taking things apart and putting them back together, weirder and better, and decided to make a career in the robotics field.

As with most other students, much of the support for his early research came from Pentagon money. Soon, Nourbakhsh began to get requests for specific applications of his robotics research to battlefield scenarios. This was around the same time that he was taking a class which examined the social side of technology. "I had my epiphany moment. I put my foot down and said, 'I won't do it.'"

When Nourbakhsh talks about the writers that influenced him the most, his decision begins to make even more sense. Rather than referencing science fiction as many other scientists do, he talks about the novels of Walker Percy, the Southern writer who wrestled with the ability of science to explain the basic mysteries of human existence. He recalls how a character in one of Percy's novels contemplated committing suicide. The character ultimately decided not to, as that would have been the last decision they ever made, as compared to all the other things they could choose to do with life. It became a sort of guidepost for Nourbakhsh as he wrestled with whether or not to take the military's money. "The general feeling I had was that every time you choose to do something, you are explicitly choosing not to do everything else. The point isn't what not to do, but what can you do best. That is, whatever you choose, choose what is most important to you."

As a young graduate student, then, Nourbakhsh resolved to refuse all military money and chose to work only on the most positive research work he could find. "I wanted to feel I was working on something with immediate social-positive impact, rather than something neutral that could be used for good or ill later...I want to be able to say I've done some good in the world."

His ethical decision, however, had financial consequences. "It is very easy to take the DARPA money and look at it as only for long-term research....It is hard to get millions from any other source, plus you have a far better chance of winning DARPA grants than others." Yet, a full decade in, Nourbakhsh's plan has worked out. His current research projects include educational and social robotics, electric wheelchair sensing devices, believable robot personality, visual navigation, and robot locomotion. Nourbakhsh supports such programs with commercial sales of the products he's developed and with corporate research funds from firms like Intel, Google, and Microsoft. (He just laughed when I joked, "How is Microsoft less spooky than DARPA?") He is particularly excited about a program that uses robotics as an educational tool for expanding the number of people working on technology. He's found that if you can get youngsters interested in technology research, you can also use it as an avenue for teaching them other valuable life skills.

With his credo in mind, then, Nourbakhsh helped found Robotic Autonomy, a summer robotics camp for underprivileged kids from San Jose. Using an "Ikea-like robotics set" that he designed, the kids are taught engineering and computer programming skills. They then compete in such challenges as "robotic musical chairs." The side effect of building robots, the instructors have found, is that the kids also build teamwork and leadership skills, as well as get excited about science and education. Many of the children coming from poor neighborhoods have later ended up going to Ivy League schools.

In the last few years, Nourbakhsh has noticed a change. While no one really cared about his refusal of Pentagon money when he was a lowly graduate student, he is starting to make waves as a professor. He tells how several colleagues have quietly come up to him to say, "We are watching you. If you pull this off for several years, we may well do the same." He ends our talk by saying, "I'm a guinea pig and that makes me more firmly resolved to prove that it's possible."

## DESIGNER'S REGRET?

Illah Nourbakhsh is part of a new breed of those working in the robotics field. They are refusing the pimpage, because they are worried about the growing military interest in their work.

"I would rather the military run out of reasons to keep existing, and I don't want them to have any credit for something I have accomplished—which they clearly would if they gave me the money," says Steve Potter, a researcher at the Laboratory for Neuroengineering, shared by Emory University and Georgia Tech in Atlanta. "They said, 'Here's some thousands of dollars because we think what you're doing is cool.' I said, 'Thanks, but no thanks.' And I get told of grants that would match my work, but I check them out and say, 'No, sorry, it's DARPA.'"

Potter works on a project that wires up neurons from animals (that is, live brain cells) into robotic circuitry to make them learn quicker and be more flexible in thought than regular AI. In the words of one article, his "astonishing robotic creations would make a 21st century general drool—if the general could get his hands on them." Potter came to his stance of refusal from his family history. His father worked at the Jet Propulsion Laboratory, a defense-funded center at Caltech, which is now the home of NASA's robotic space exploration. His dad worked on projects like side-looking radar and was told it would be used to map Venus; later, he found out it was used on cruise missiles.

Others refuse military funding because of concerns about how it subtly steers them away from their original motivations. "DARPA and ONR and other DOD agencies support quite a lot of research that I think is valuable and virtuous," says Benjamin Kuipers, a computer scientist at the University of Texas. "However, there is a slippery slope that I have seen in the careers of a number of colleagues. You start work on a project that is completely fine. Then, when renewal time comes, and you have students depending on you for support, your program officer says that they can continue to fund the same work, but now you need to phrase the proposal using an example in a military setting. Same research, but just use different language to talk about it. OK. Then, when the time comes for the next renewal, the pure research money is running a bit low, but they can still support your lab, if you can work on some applications that are really needed by the military application. OK. Then, for the next round, you need to make regular visits to the military commanders, convincing them that your innovation will really help them in the field. And so on. By the end of a decade or two, you have become a different person from the one you were previously. You look back on your younger self, shake your head, and think, 'How naive.'"

This refusal to do military work actually fits within a longer tradition of scientific rebellion. The Manhattan Project was riven by disputes among the researchers involved. The civilian lead of the team, Robert Oppenheimer, was even made a lieutenant colonel in the army to give him standing in arguments with the military funders. Even then, a work stoppage by key scientists forced the military to put the project and its main lab at Los Alamos, under the control of the University of California, rather than at a military base under direct Pentagon authority.

After the atomic bomb was built, an intense dispute brewed over the strategy for its use. The secretary of war, Henry Stimson, and the top U.S. military commander, General Marshall, met with a group of scientists, including Oppenheimer and Enrico Fermi, in May 1945 to discuss this. But soon the men who had made the bomb were shut out of the decisions on how and when to use it, mainly because they wanted limits placed on its military applications. Ultimately, Oppenheimer was forced out of the project, in favor of the more hawkish Edward Teller, the "father of the H-Bomb," and the inspiration for Peter Sellers's *Dr. Strangelove.*

Even with their more authoritarian system (where saying no had much worse consequences), the Soviets had similar problems with refusenik scientists. The brilliant physicist Andrei Sakharov was the designer of their first hydrogen bomb. He went on to become an advocate for nuclear disarmament, for which the dissident won the Nobel Peace Prize and was put in prison.

Ultimately, nuclear scientists from around the world banded together to form an organization to work against the weapons they had once developed. Spurred on by a letter from Albert Einstein (who, ironically, had also sent the letter that initially convinced President Roosevelt to fund the atomic bomb's research), it had its first meeting in Pugwash, Nova Scotia, in 1957. While what became known as the "Pugwash movement for nuclear disarmament" ultimately won a Nobel Peace Prize, the nuclear refuseniks' efforts were more than a decade too late. The nuclear genie was already out of the bottle.

Nuclear physicist Freeman Dyson tells how the simple problem was that there were not enough Illah Nourbakhshes at the time. Not enough of the scientists "had the courage of foresight to say no" when it actually mattered. "It is, in some ways, responsible for all our troubles—this what you might call technical arrogance, that overcomes people when they see what they can do with their minds." It is also interesting to note that Dyson was the inspiration for the "Dyson" character in the *Terminator* movies, who invents the Skynet program, the AI gone mad that ultimately launches a nuclear holocaust on humanity, and later dies trying to destroy it.

Indeed, many see a pointed lesson for the robotics scientists of today. As Bill Joy, the founder of Sun Microsystems and now a critic of much of the research in the field, writes, "The experiences of the atomic scientists clearly show the need to take personal responsibility, the danger that things will move too fast, and in a way in which a process can take on a life of its own. We can, as they did, create insurmountable problems in almost no time flat. We must do more thinking up front if we are not to be similarly surprised and shocked by the consequences of our inventions."

Still, refusenik roboticists like Illah Nourbakhsh are a tiny minority, both in not taking the military money and in weighing the deeper questions about what it means to work on systems designed for war. Funding by the military is the norm, but it is also, tells Nourbakhsh, "a very touchy subject" that few like to talk about. When asked about his thoughts on the implications of arming robots, for example, Brian Miller, our NASCAR engineer turned roboticist, simply responds, "I stay out of politics." Likewise, Sebastian Thrun pointedly changes the subject when the topic of the political impact of his research comes up. "I am ignoring all of this to build this vehicle." He says he just focuses on the path of invention and discovery and compares his work to Charles Lindbergh's preparations for the first flight across the Atlantic. "He didn't do it thinking about all the regulations for transatlantic passenger travel it would inspire. He just did it."

It is not that the researchers don't realize there are big issues at hand; indeed, the idea of having a major impact is what drives many of them to robotics research, as opposed to any of the other scientific fields in which they would thrive. Rather, as one writer described of DARPA, "What you don't get is much of a sense of introspection." While many roboticists are happy that their systems are being used to save lives, when it comes to talking about other military outcomes or codes of research ethics, most of the field demurs. Akin to the NRA mantra that guns don't kill people, they describe that their research can be used for good or ill and thus the responsibility for anything that happens outside their labs lies beyond them. A Carnegie Mellon researcher describes taking military funding as a necessary corollary to doing the research they want. "For 364 days out of the year you are building a good robot. For one day out of the year, you put some camouflage on it to bullshit DARPA."

The same attitude carries over on the military funder side. When it comes to weighing any major ethical questions, as one DARPA program manager put it, "That's above my pay grade. That's not my department." Or as Michael Goldblatt, DARPA's defense sciences office director, puts it, these questions are best set aside for now. "You can't let the fear of the future inhibit exploring the future."

The refuseniks think this is shortsighted and fear that the robotics field may well repeat what happened with the nuclear scientists. Illah Nourbakhsh asks, "Why go down that path again of working so hard to invent something and only then, after the fact, waiting to say, 'We now understand it, but it's too late'? It's silly not to talk about it [the ethics of military funding and how their inventions are used].... You can't shield yourself from the repercussions. We need the leaders in the field to talk about it, to bring in the ethicists and the political scientists."

As robotics has grown, even those who do take military funding are starting to cite the need for introspection. Ronald Arkin is a professor at Georgia Tech. "Historically, technologists have been woefully ignorant of the implications of what they created," he says. "I would probably put myself in that category until a few years ago. Research and development will move forward, but we still need to understand what the consequences are, then come to grips with them and determine whether we should do anything about them."

Such a discussion won't be easy. Robotics is a relatively new field, accelerating so quickly that it hasn't had time to wrestle with the deeper questions. It also isn't that well equipped to wrestle with deep ethical issues. "It's a generation that has been trained just to think technically. It is rare to find anyone that can think about science like Aristotle did, as a social phenomenon as well," ruefully notes Nourbakhsh. "I don't think there will be a code of ethics anytime soon."

Others think that the whole discussion is beside the point. Even if one refuses military funding, the military can still get the fruits of one's labor off the open market (via the work of labs like the Marine Corps one). All you have done is save the government money, as the ultimate effect would be the same. Nourbakhsh doesn't dispute this and notes that all his findings are published in the public domain, and indeed, he is certain that the military has taken some of his research for its own uses. "Absolutely. You can't protect technology from all uses." But he then goes back to the choice he had between doing military robotics work or running a robotics camp for underprivileged kids. "How cool is that I know I am having a direct application that is socially positive? I know there is no walling off what I am working on from others, but I also know that there is some direct good coming out of it."

This debate on whether researchers working on robots should do so at the military's behest will continue in the years ahead. Ultimately, it all may come down to a central question of our modern age, raised by Father Tadeusz Pacholczyk of the National Catholic Bioethics Center. "Technology has begun to outstrip our moral integrity. Just because we can, should we?"

[PART TWO]

# WHAT CHANGE IS
# CREATING FOR US

# TEN

# THE BIG CEBROWSKI AND THE REAL RMA: THINKING ABOUT REVOLUTIONARY TECHNOLOGIES

*Guns and violence have the potential to override any theory, no matter how sound.*

—U.S. ARMY LIEUTENANT

"Here at the end of a millennium we are driven to a new era in warfare. Society has changed. The underlying economics and technologies have changed. American business has changed. We should be surprised and shocked if America's military did not."

Vice Admiral Arthur Cebrowski was talking about his vision of future war, called "network-centric warfare." Cebrowski was a former U.S. Navy pilot who had flown combat missions in Vietnam and commanded an aircraft carrier during the first Gulf War. "But there was another side to Cebrowski," describes one biographical sketch, "a nervous energy and maverick streak that made him prone to trendy theories and sky-high philosophyzing—which for a long time kept him out of the Defense Department's inner sanctum." For example, the admiral was known as "an obsessive Powerpointer." He made sure to use computer slide shows and substitute corporate jargon wherever he could, often sounding more like a management consultant than a military officer. For example, Cebrowski described the events of 9/11 as "a systems perturbation" and argued that military operations should be "value-added processes."

In 1998, Cebrowski became president of the Naval War College in Newport, Rhode Island. From his perch outside the Pentagon, Cebrowski began to agitate that change was afoot. The cold war was over and the U.S. military's task, he argued, was not to face off against an equivalent superpower. Instead, it should plan to "baby-sit the petri dish of festering problems we have around the world."

It was with this in mind that Cebrowski, along with his writing partner John Garstka, a retired air force pilot, published a seminal article in *Proceedings*, the navy's official journal. The article, entitled "Network-Centric Warfare: Its Origin and Future," became the centerpiece of a whole new approach to war. Like much of the writing in this era of dot-com Internet hype, it offered a grandiose vision that frequently veered from cool military analysis to the writings of an acolyte. Cebrowski wrote with an admiration, bordering on obsession, of the many wonders of the new technology companies whose stock prices were then soaring and the triumphant business models that were seemingly changing the fabric of business and society. He cited lessons from Cisco, Dell, and even American Airlines, as to how information technology was giving American businesses newfound advantages, which, in turn, presented a new model of fighting and winning wars.

Seeing a parallel with the U.S.'s strategic position, Cebrowski was particularly fascinated by market behemoths. He was extremely impressed with companies like Wal-Mart, whose IT backbone allowed it to link together disparate operations, react quickly to a changing marketplace, and thus stomp out pesky mom-and-pop stores. Just as Wal-Mart had "total information awareness" over the marketplace, so too could the Pentagon have a perfect picture of the battlefield. In turn, just as the behemoth of Microsoft had supposedly reduced its competitor Apple into what he called a permanent market "niche," Cebrowski argued that the United States could do the same to its foes in war. "Locking-out competition and locking-in success can occur quickly, even overnight. We seek an analogous effect in warfare."

Cebrowski's idea was not merely that information technology was changing the way organizations operated, but that the shift to IT networks was a change of a whole different order of magnitude for the history of war. It would "affect the where, the when, and the how of war." As he concluded, "For nearly 200 years, the tools and tactics of how we fight have evolved with military technologies. Now, fundamental changes are affecting the very character of war.... We are in the

midst of a revolution in military affairs (RMA) unlike any seen since the Napo-
leonic Age, when France transformed warfare with the concept of levee en masse."

## "A SUDDEN TEMPEST WHICH TURNS EVERYTHING UPSIDE DOWN"

When people think about change in business, technology, or war, they usually
imagine a linear process. Over time, slight improvements are made that make
systems better, faster, cheaper, or give them a bigger bang. Every so often, however, a change comes along that wipes the table clean. It rewrites the rules, changes
the players, and alters the organizations, strategies, and tactics. The parallel in
the business world is "disruptive technologies" that fundamentally transform an
industry, even to the point of ending it. The most recent example of this would be
how the music industry has been inalterably revolutionized by the world of online
file sharing.

The key to such shifts is that they have not only first-, but also second- and
third-order effects that act like bow waves, sweeping the field and beyond, almost
like societal versions of Kurzweil's Singularity. These broader effects are often
unpredictable to those living at the time when the technology is introduced. When
the automobile was invented in the 1880s, it soon became clear to many that these
"horseless carriages" were going to affect transportation in some way. But who
could have predicted that cars would reshape American cities by creating "subur-
bia," burn enough carbon dioxide to help heat up the planet, provide what were
then Arab nomads with a stranglehold on the world economy, and give teenagers
freedom from their parents' supervision while courting, thus creating social phe-
nomena like "dating" and the subsequent "Sexual Revolution"?

In the military realm, these paradigm shifts are called "revolutions in military
affairs" (RMAs), something essential to understanding Cebrowski's excitement
over information technologies, as well as the real impact of robotics. RMAs typically involve the introduction of a new technology or organization, which in turn
creates a whole new model of fighting and winning wars. A new weapon is introduced that makes obsolete all the previous best weapons, such as what armored,
steam-powered warships did to wooden, wind-powered warships. Or it may be
that a military figures out how to organize itself in a new way around an already
known weapon, which makes all the old ways of fighting futile. An example of this

would be how the English made longbow archers an integral part of their army in the Middle Ages, ending the dominance of horse-mounted knights.

Such technologies need not be purely military in nature to change the field of battle. The horse stirrup and the railroad, for example, both spurred RMAs that led to the rise and fall of empires, but neither could be described as an exclusively military technology (much like robotics today). Indeed, the most radical shifts in war tend to parallel similarly major changes in the economy. Whether it's adjusting to steam-powered warships or steam-powered factories, a hallmark of revolutions is that, as management guru Peter Drucker puts it, leaders "must prepare to abandon everything they know."

From the longbow to nuclear weapons, historians identify at least ten revolutions in military affairs since 1300. And each time, these RMAs felt like being in "a sudden tempest which turns everything upside down." This is how a fifteenth-century Italian politician described what it was like to watch cannon easily flatten the castle walls that had protected his city for centuries.

Not everyone is enamored with the idea of RMAs. Their main problem is that almost every time a new technology is introduced in war, its long-term impact is overpromised. Retired army officer and military analyst Ralph Peters, for example, jokes that "when the first early man discovered that he could bind a sharp stone to a stick with a leather thong, you can be certain that he turned immediately to his pals across the campfire and shouted, 'I've just achieved the ultimate revolution in military affairs!'"

However, it is indisputable that technology-driven RMAs do shape history. That Italian politician witnessed the start of the RMA that became known as the gunpowder age. At its start, 1450, the Italian city-states were the leading powers in Europe, while Europe, in turn, was a relative weakling on the world stage, only controlling 15 percent of the globe. Within a century, the dominant powers were unified kingdoms like Britain, France, and Spain that had figured out how to best use this revolution to their advantage, and Europe was on its way to controlling 84 percent of the globe. Much of this conquest came through encounters with local powers, where the massive numerical disadvantage of the European colonists was more than outweighed by their technologic advantage. The Spanish conquistador Cortés, for example, conquered the eight-million-strong Aztec empire with a force of just eight hundred men.

"How do you become a winner in an era of technologic upheaval and avoid becoming road kill? You might think it is to get the best and most gadgets. But you'd be wrong," says Max Boot, author of *War Made New*, a history of RMAs. The key to success is not just inventing or buying a new technology, but also how you harness it. The Germans, French, and British, for instance, all had tanks, aircraft, and radios at the start of World War II, but the Germans figured out how to combine them all together into the blitzkrieg, a new way of mechanized warfare that revolutionized war in the twentieth century. Or as air force officer Scott Murray explains, "Imagine for a moment that you could go back in time and give a knight in King Arthur's court an M-16. If he takes the weapon, gets back on his horse, and uses the stock to knock his opponent's head, it's not transformational. Transformation occurs when he gets behind a tree and starts shooting."

For this reason, RMAs often take a while to bear fruit. The English army didn't just roll out prototypes of the longbow for its peasants to use in the historic defeat of the French knights at the battle of Crécy in 1346, which helped end the age of feudalism. Instead, the weapons and tactics that proved so revolutionary were perfected in the English civil wars more than a century before. However, the pace and duration of these transitions seem to be coming faster and faster. The changes brought by gunpowder played out over centuries, those of steam engines, telegraphs, and railroads (the first industrial RMA) over a century, and internal combustion engines, radio, and flight (the second industrial RMA) over a few decades.

This is also why a good sign of an RMA is the rise of hybrid technologies. There are always early and late adaptors of any technology. Much like what is going on with our cars today, hybrids are usually last gasps of those who recognize something is up, but aren't willing to fully change. They want to have it both ways, by layering new technologies onto old platforms. The Spanish did their best to stave off the Age of Sail with galleasses, which were oar-powered galleys that they also mounted sails and cannon onto and sent out as part of the Spanish Armada in 1588. Of course, these rickety hybrids proved useless compared to the purpose-built English sailing ships that were both more seaworthy and packed more cannon. Most galleasses never even made it home. Similarly, steam-powered warships in the British navy still mounted sails until 1880. Boot calls such hybrids "the military equivalent of a duck-billed platypus." While marginally better than the old way, they are not only typically ugly, but also far less effective than a full transition to the new technology RMA (the robotic equivalents are the convertible systems).

In turn, just because the old technology sticks around doesn't mean that a revolution hasn't occurred. For example, many people around the world still push plows behind donkeys, the same way people did two thousand years ago. That doesn't mean that industrialization and biotech haven't revolutionized the overall field of agriculture. The same holds true in war. The German army that launched the blitzkrieg at the start of World War II still had horse cavalry divisions, but it was the tanks in its panzer divisions that signified the RMA.

Often, no one even recognizes that an RMA has happened until after the fact. For example, machine guns allowed tiny European armies to beat huge tribal forces in the late 1800s. But it wasn't until after the slaughter of the first few years of World War I that generals would finally acknowledge that machine guns had also revolutionized the way fighting would take place on European battlefields.

## IT'S ALL ABOUT THE NETWORK, BABY

To Cebrowski and the movement he would come to signify, the twenty-first-century revolution in war would be information technology networks. The key to what was called "network-centric warfare" was the shift to the new information technologies of computers, the Internet, fiber optics, and so forth, which allowed an enhanced level of connection and information sharing. Planes or ships or soldiers in the field didn't have to communicate via carrier pigeons, telegraph, or radios but could now instantly e-mail each other. This would infinitely speed up the pace of operations, they argued. Soldiers and generals a continent away could look at the same image online, at the same time, which they felt would provide shared awareness of what was going on. This ability to "network information" would allow various military units to "self-synchronize" their efforts. They could operate with a speed and cohesion that would "dramatically increase force effectiveness."

Central to the network-centric concept was, as the name suggests, the power of the network. That is, a network linked together would be quicker, smarter, and more lethal than the sum of its individual parts and would quickly overwhelm whatever foe lay in its path. This "information advantage," argued Cebrowski, would be huge. The sharing of information across the system, as well as the ability to crack into the enemy's systems, would create "near-perfect" intelligence. The side that was networked would not only know exactly where its own soldiers were, so that they could be deployed to perfect efficiency, but it would also know where the enemy was, even better than the enemy troops' own leaders. Your side could destroy an enemy unit not only before it saw you, but even before the enemy's own commanders knew his units had arrived on the battlefield.

This advantage of networking created two fundamental differences with past

RMAs, argued proponents like Cebrowski. The first is that it was the software, not the hardware, of war that mattered. Indeed, the corps of followers that sprang up to back the network-centric concept described "a move away from platforms to networks" as the ideal model of war. That is, for the very first time in war, the weapon system you were using was beside the point. What now mattered was whether you were "networked" into a "system of systems."

The second difference, they argued, arose from its focus on sharing information. This RMA would do something that no other had been able to achieve: lift the proverbial "fog of war." The term "fog of war" originally referred to the immense clouds of smoke created by musket fire that often obscured what was happening in battles. Today it refers to all the confusion, mistakes, delays, and misperceptions that happen in war because of the difficulty of coordinating operations in an atmosphere of fear, fatigue, and uncertainty while another side is trying to kill you. The famous nineteenth-century Prussian strategist Carl von Clausewitz, whose textbook *On War* is taught at every U.S. military school, argued that the combined problems of "fog" (getting good information was difficult in battle) and "friction" (actions rarely work out as planned in battle) were inherent, enduring, and inevitable aspects of war. "Everything in war is very simple," he observed, "but the simplest thing is difficult. The difficulties accumulate and end by producing a kind of friction that is inconceivable unless one has experienced war."

According to its supporters, this new RMA had solved these problems. The networked approach meant that commanders could command and soldiers could fight, as one report put it, with "near-perfect clarity." "Lifting the fog of war" (as one book was even titled) would "result in a quantum leap in operations." Or, as another report on this new philosophy of war put it, "technological innovation, particularly in information technology, will purge the conduct of war of the uncertainties and ambiguities of the past. For those happy powers that set the technological pace, war will become an essentially frictionless engineering exercise."

The network-centric crowd had huge expectations for such a shift. Through achieving "information dominance," a networked military force would be like a Wal-Mart at war with a bunch of small-town mom-and-pop stores; it would inevitably be "a winning force." Even better, they argued, this new revolution was tailor-made for the United States. As one later argued in an official magazine of the U.S. military, "The U.S. is the only nation that is successfully and at great speed adapting to the new information-based technologies.... By linking a system of systems, the U.S. can develop battle space awareness for commanders while preventing enemies from doing the same."

Cebrowski and Garstka's famous article even made historic comparisons, hinting that history would one day look back at the networking revolution as comparable to the agricultural and industrial revolutions. As one report on the movement he helped start said, "The IT-RMA was pitched as nothing short of a paradigm shift in the character of conflict and the conduct of warfare. It entailed the combination of new technologies and innovative operational and organizational concepts that fundamentally altered how one thought about war and war fighting. Consequently, the RMA involved much more than merely overlaying new technologies and hardware over existing force structures—it was necessarily a process of far-reaching, disruptive change."

## GULPING DOWN THE KOOL-AID

The idea of a network-centric revolution in war, at which only the United States could win, was immensely appealing. Indeed, soon after the article, presidential candidate George W. Bush laid out his vision of the future of the U.S. military in a key speech at the Citadel. At the center of it, Bush proclaimed, would be "a revolution in the technology of war," which would allow the United States to "redefine war on our terms."

As conservative analyst Fred Kagan notes, "Bush was (and remains) a firm believer in the idea of an RMA; he had proclaimed it a priority as early as 1999, long before anyone imagined that Donald Rumsfeld would again become secretary of defense." While it is unclear to what depth Bush grasped the nuances of the network-centric model of warfare, it was indeed a mantra among the "Vulcans," who had drafted Bush's speech. Led by Condi Rice, the Vulcans were a group of conservative defense intellectuals and former national security officials who advised the then Texas governor on security and foreign policy issues. When Bush won, they all moved into top leadership positions at the Pentagon, State Department, and White House.

Once in power, as historian Max Boot (himself a conservative commentator) phrased it, they sought to fully "harness the technological advances of the information age to gain a qualitative advantage over any potential foe." Retired marine officer Frank Hoffman even argues that the new team went beyond Cebrowski in their fandom. "They accepted the presumptions of the RMA school and took them to a higher level."

After Bush's inauguration, his new leadership team at the Pentagon, led by Donald Rumsfeld as secretary of defense, moved quickly to make the vision of network-centric warfare a reality. The cantankerous Rumsfeld saw this as his opportunity to put his own stamp on the U.S military. In February 2001, just days into office, the new team announced that it would increase by $20 billion the research and development spending on "transformational" technology that would "propel America's Armed Forces generations ahead in military technology." In turn, the continuation of any existing military project would hinge on whether it fit into this new idea of a "transformation" to network-centric warfare.

Rather than a mass change in what weapons the Pentagon bought, Rumsfeld's transformational vision was that the networking of these weapons together meant that it should change the way military operations were conceived. With the fog of war lifted and the "system of systems" working to perfection, fewer forces could be sent into battle and they could be lighter, quicker, and more decisive. The "platforms" were almost beside the point. "Today," Rumsfeld stated, "speed and agility and precision can take the place of mass." More could be done with less.

Perhaps the greatest sign that the new team at the Pentagon was drinking the network-centric warfare Kool-Aid was what happened next for Admiral Cebrowski. As an article in the U.S. Navy's official journal put it, "If 'Rummy' was the president's high priest of Defense Transformation, Cebrowski was his major prophet, or better yet messiah, announcing the New World Order just on the horizon." The recently retired admiral was empowered in a way that he could never be while in active service, even when he had been in charge of the Naval War College. One of Rumsfeld's signature organizational shake-ups at the Pentagon was the creation of a new Office of Force Transformation. For the next four years, Cebrowski would be its director. As one article on his role describes, Cebrowski was no longer an acolyte clamoring for change from the outside. "In this position, he was responsible for serving as an advocate, focal point, and catalyst for the transformation of the United States military."

A few months later, the 9/11 attacks occurred and the vision of network-centric warfare was put to the test. As America struck back in Afghanistan, it soon appeared that the theories had proven correct. The key figures in the movement soon were declaring that it was the networking which had allowed U.S. forces to prevail where 80,000 Soviets had failed just a decade earlier. The few hundred American special forces that were first sent in were smaller in number than their Taliban foes, but they had beaten them convincingly. Networking meant that even individual soldiers riding on horseback could tap into information and awareness that altered the whole equation of battle.

With this "Afghanistan model" seeming to validate the whole vision of network-centric warfare, the idea took hold in the Pentagon that earlier notions of what it would take to topple a regime like Saddam Hussein's no longer held true. This lowering of expected costs made the idea of invading Iraq far more appealing. Fully 680,000 coalition troops may have been needed to take back tiny Kuwait during the 1991 Gulf War, but the acolytes argued that by using the network-centric approach, far fewer troops would be needed in 2003 to do far more (indeed, many originally wanted just 20,000 troops sent for the Iraq invasion, but after some pushback from the generals, the force was ultimately raised to 135,000). As historian Max Boot tells it, "Iraq, in turn, was set up from the beginning to be the ultimate test of network centric warfare: a small, high tech invasion force, moving quickly and striking at only those targets necessary to instill 'shock and awe' in the Iraqi government."

The early success in Iraq seemed to indicate once again that the network-centric way of war had changed everything. The previous RMA "gold standard" of invasions had been the German blitzkrieg in 1940, in which the Nazis took over France in just forty-four days, "at a cost of 'only' 27,000 dead soldiers." For the United States to seize Iraq in 2003, it took half the time, at .005 percent the cost (161 U.S. soldiers lost during the invasion, many of them actually killed by "friendly fire"). Again, the network-centric crowd cited that the key wasn't that the United States was using fundamentally different weapons than its previous war, but that the networking into information technology had proven "central to American military dominance." The transformation movement led by Admiral Cebrowski, and embraced by those in power, had seemingly proven that a revolution in war truly was at hand. Cebrowski, suffering from cancer, left the Pentagon at this high point in the movement, and passed away in 2005.

## FACT AND FRICTION

Not all was well with the revolution, however. The first problem turned out to be the business assumptions upon which the whole movement was based. Battlefields are not the same as corporate boardrooms. The stakes are higher, the measures of victory and defeat different, and, while a company can selectively invest only in markets it could succeed in or shut down business units that don't turn a profit, a military can't always choose when, where, how, and who it will fight—the enemy gets a vote. Plus, there's that little matter of violence. As one critic put it, "No one is shooting at the Coca-Cola Company."

Even worse, the business assumptions behind network-centric warfare had been particularly selective. Cebrowski and his movement pulled their inspiration from the Internet boom in the late 1990s, when it seemed that having an IP address was all that it took for a business to succeed. Unfortunately, at almost the same exact time as the network-centric crowd was moving to put their supposed lessons from the market into place at the Pentagon, the market was learning its own new lessons. The fast money of "dot-com" was turning into the crashing stock portfolios of "dot-bomb."

What had worked for almost all the companies that Cebrowski, Rumsfeld, and others cited as models of success was not only difficult to translate into the setting of war, but wasn't even working for these companies anymore. Almost every one of the companies they had adoringly name-dropped, like Cisco, Dell, American Airlines, and even Enron, were struggling or bankrupt within a few years, while the market behemoths they wanted to emulate, like Microsoft and Wal-Mart, faced both renewed competitors and new troubles. "Sloppy thinking" was how retired marine Frank Hoffman described it. "Theories and business models drawn from the exuberance surrounding the IT revolution displaced quite a bit of history and factual context."

The same sorts of results from "irrational exuberance" were experienced in Afghanistan and Iraq. The seemingly "perfect" military operations proved to be anything but. As one retired officer noted, "We will never operate under perfect conditions. We will always lack something, whether it's time, resources, or even a clearly defined mission."

In Iraq, for instance, the fog of war cropped up in all sorts of places, even before the invasion morphed into the ensuing insurgency. Indeed, in the largest tank battle of the war, even the traditional sense of the term came back. An Iraqi Republican Guard counterattack was able to sneak right up on American forces simply because, as one soldier explained, "We kind of lost track of them in the smoke, haze, and confusion of the battle."

The U.S. forces were all networked together, with a "blue force tracker" letting them know the position of all the friendly units, just as the network theorists had claimed would revolutionize war. The only problem is that they still didn't know who the enemy ("the red force") was or when and where he was coming. As one report dryly put it, "Situational awareness was proving to be more theoretical than actual." Or, as a marine joked, "When do we get red force trackers?"

While the concept behind the Iraq war plan may have been IT-dominated, war itself couldn't be turned into a perfect execution of commands merely by linking people by e-mail. Instead, all the various forces of chance, confusion, and error common in every previous war (Clausewitz's "fog" and "friction") still were present.

Moreover, when authentic experts in information technology examined the situation, they actually found a huge difference between the theories of networking and the reality in the field. Joshua Davis is a correspondent from *Wired* magazine, published in the Silicon Valley cyber-culture that had so excited the Pentagon's network-centric crowd. As he recounts of embedding with U.S. forces during the invasion, "What I discovered was something entirely different from the shiny picture of techno-supremacy touted by the proponents of the Rumsfeld doctrine. I found an unsung corps of geeks improvising as they went, cobbling together a remarkable system from a hodgepodge of military-built networking technology, off-the-shelf gear, miles of Ethernet cable, and commercial software. And during two weeks in the war zone, I never heard anyone mention the 'revolution in military affairs.'"

Rather than a seamless flow of information, soldiers wrestled with everything from Web browsers constantly crashing due to desert sand to heat fouling up equipment designed for use in offices, not battlefields. Indeed, at one point in Davis's reporting, an army lieutenant resorts to navigating a convoy using an improvised GPS and some handheld walkie-talkie radios that he had bought from a hardware store back home. The soldier joked that "if we run out of batteries, this war is screwed."

Little did Davis or the soldier know how true that statement really was at the time. One of the many unanticipated aspects that the network crowd didn't take into account was how new technologies were creating new, unforeseen demands. The most widely used power source in the military is the BA 5590, a standard twelve-volt battery that powers everything from radios to antitank missiles. With all the networking, the demand for the batteries turned out to be much higher than ever planned (the marines alone were using up 3,028 of them a day). But there were no stockpiles. Explains the *DefenseTech* journal, "Major combat missions during Gulf War II almost ground to a halt—because of a shortage of batteries." The only reason that the plug wasn't literally pulled on the Iraq invasion is that thirty other nations loaned the United States extra batteries. Ironically, many of these nations were the very same ones from "old Europe" that politicians like Rumsfeld had lambasted during the "freedom fries" period of anger at traditional allies who had chosen not to send troops to Iraq.

As the fighting evolved from the invasion into a confusing and painful insurgency, it became all too clear that war would remain imperfect in Iraq, despite the supposed RMA of networking. While the network-centric crowd had been right that linking up would multiply the fighting power of each soldier, it was soon beside the point when forces couldn't figure out who was an insurgent and who was a civilian and how they were organized. "Information dominance" became an ironic joke. In describing what followed the supposed "gold standard" invasion of Iraq, Milan Vego, a professor at the U.S. Naval War College, tells how "there is probably no conflict in which U.S. forces have fought in such ignorance of the enemy's purpose, strength, and leadership."

## IGNORING THE REAL REVOLUTION

In late 2006, over two hundred of the top thinkers and leaders in American security policy gathered for a discussion on "Rethinking the U.S. Military Revolution." Held in a Washington, D.C., conference center, replete with the mandatory stale breakfast muffins and bad coffee, the session featured speeches by leading professors, analysts, and even the air force general who had helped plan the opening round of the Iraq invasion. Over the course of the session, they debated back and forth the future of network-centric warfare, what it had delivered on and where it had failed, and even what might become of Cebrowski's old office, now that the venerable thinker had passed away.

More notable is what the gathered leaders and experts didn't talk about. At this session, exclusively focused on what was revolutionary in war today and tomorrow, robotics and other unmanned technologies never came up, not even in a passing mention. The network-centric buzzword of "transformation" was used twenty-one times, despite the fact that by late 2006 it was clear that the advantages promised by its originators had not played out as expected. But words like "unmanned" or "robot" were never spoken, not even once.

Compare this scene with a different sort of meeting, just a few months earlier at a military base. General William Wallace, the four-star general in charge of U.S. military training, held a question-and-answer session about new technologies of war with a group of army troops, most of whom were just back from Iraq. Instead of the general giving the answers, he was the one asking the questions. The very first question he posed, focusing on a captain, was, of all the revolutionary new technologies the unit had tested out, "which one single piece would you deploy today?" The soldier answered, "Sir, the PackBot." Or, as another soldier replied to a survey about what it was like to use this new unmanned system, "I believe that we are the pioneers of the *Star Wars* systems of the future."

The Cebrowski-led network-centric crowd was right. Something truly big is going on in war. But they were wrong on everything else. In focusing purely on the 1990s Internet boom as their mantra, the network-centric folks, as well as the broader security studies field at that conference, missed the revolutionary part. They ignored not just what the soldiers in the field were saying, but the far more interesting and important developments in technology just coming to fruition.

The Internet has certainly affected how people shop, communicate, date, and even how they fight. That the network-centric crowd noticed this, particularly amid all the booming stock prices of the 1990s, was not all that difficult or surprising. But that did not make it an RMA, and certainly not one that would lift the fog of war.

Instead, when both soldiers in the field and scientists back in the labs talk about what is now revolutionary in technology, today they point to something else. As Rodney Brooks says, what is far more important is a robotics revolution, now at its "nascent stage, set to burst over us in the early part of the twenty-first century. Mankind's centuries-long quest to build artificial creatures is bearing fruit." As opposed to the IT networks that simply allow information to flow easier, robotics and AI are the real "tsunami that will toss our lives into disarray."

And yet such technologies are almost completely without discussion among today's theorists of war and politics. Indeed, the failure to even use the word "unmanned" in that sterile conference hall was only one example of how networks get attention, but robotics don't even merit mention. I found it repeated again and again at major conferences and in publications on military history and strategy. Something big is going on in war, technology, and politics, and yet very few who study security issues are talking about it.

The result is that, as retired army lieutenant colonel Thomas Adams writes, we are "in something like the position of monarchies witnessing the democratic revolution at the beginning of the 19th century. Something profound and far-reaching is going on all around us, even within our own societies. But the advisors, courtiers, and generals that surround the throne are at a loss to determine what it means, much less what to do about it."

It is not that the great minds who study war are willfully ignoring what is going on. Rather, as Bill Joy put it, "It is always hard to see the bigger impact while you are in the vortex of change." Like any other change, RMAs do not happen in one single discrete event, one rush of wholesale change. Many seemingly important changes can occur (and distract) before the truly revolutionary part becomes clear. In turn, most revolutions don't actually grow from one single invention, but from a convergence of technologies. For example, the Industrial Revolution that transformed society and then war in the nineteenth and early twentieth centuries actually kicked off with the invention of the steam engine back in 1782. But the steam engine had to be brought together with everything from railroads to telegraphs for it to culminate as the industrial RMA that shaped World War I.

A good person to explain is actually another of the major thinkers behind the RMA movement inside the Pentagon, Andrew Marshall. Despite being eighty-three years young, Marshall is the Pentagon's officially designated "futurist-in-chief," directing its Office of Net Assessment, akin to an internal think tank for the Pentagon. While he was a huge supporter of Cebrowski's and Rumsfeld's excitement over networks, Marshall also warned that what they saw as key leaps in technology could just be the beginning, not the end, of a different sort of RMA. "There is a tendency to talk about *the* military revolution. This could have the sense that it is already here, already completed. I do not feel that this is the case. Probably we are just at the beginning, in which case, the full nature of the changes in the character of warfare have not fully emerged. . . . What we should be talking about is a hypothesis about major change taking place in the period ahead, the next couple decades."

As Iraq soon illustrated, the network crowd was wrong that the fog of war would be lifted and wrong that the other side would be permanently locked out of the marketplace of war. Most important, it is becoming evident that they were wrong in their argument that the network, not the platform, was the only part that mattered.

"Historians will see the last decade of the 20th century and the first decade or two of the 21st century as a turning point in the evolution of armed conflict," states a U.S. Army report, but not for the reasons that Cebrowski and the network acolytes originally believed. The networks of e-mail and Internet fiber optics that now bind military units together certainly do matter. They allow them to share information quicker than when they were using radios, phones, or faxes. But as we are beginning to learn, history will care far more about what these linkages *enable*. That is, these new digital links are important, but not as much as the platforms they now allow. What will stand out, what is historic for war, and human history in general, are the robotic weapons now playing greater roles on the battlefield. This is what future historians will find far more notable, not the difference between units connected by a fax versus an e-mail that so intrigued the network-centric acolytes. It is far more important that humans' 5,000-year-old monopoly over the fighting of war is over.

This aspect of a fundamentally changed platform is also key to the story of whether an RMA is at hand. At their fundamental level, all the past RMAs in history were about changing how wars were fought. Whether it was the longbow, the gun, the airplane, or even the atomic bomb, the essential changes were new weapons and/or ways of using them that transformed the speed, distance, or destructive power of war. By contrast, the introduction of unmanned systems to the battlefield doesn't change simply how we fight, but for the first time changes who fights at the most fundamental level. It transforms the very agent of war, rather than just its capabilities.

It is difficult to weigh the enormity of such a change. Admiral Mike Mullen, the chairman of the Joint Chiefs of Staff, describes how warfare is at "real transition time here." John Pike of the Global Security organization puts it into this broad historic context. "First, you had human beings without machines. Then, you had human beings with machines. And, finally you have machines without human beings." Security analyst Christopher Coker comments, "We now stand on the cusp of post-human history."

Such broad strokes, though, don't help us understand exactly what these changes actually bode for war. While some have argued that the new technologies of precise weapons offer to take humankind into "a very different age; perhaps a more humane one," it is becoming clear that robotics and unmanned systems are still a revolution *in* war. If it is a different sort of RMA, it is an RMA all the same.

The modes, manners, and character of war have changed in past RMAs, and will do so with this one. But it doesn't mean everything will change. There the network-centric crowd was wrong again. Even in this most fundamental revolution of who fights in war, the foundations of war remain the same. Even with robots and other new unmanned technologies, war is still about using violence to make the other side do what you want. It is still against an enemy, trying to figure out how to use its strengths against your weaknesses. And it will still involve all the unexpected confusion, mistakes, and dilemmas that go hand in hand with both technology and war. The fog of war ain't going anywhere. Even with robots, we are learning that war will remain as unpredictable as it is enduring.

## WOE-BOT, WHOA-BOT!

January 25, 1979, was to be a special day for Robert Williams, a worker at Ford Motor Company's Flat Rock casting plant in Michigan. The twenty-five-year-old man's son was celebrating his second birthday. Unfortunately, it was also the same day that the robot operating an automated parts retrieval system near Williams's workstation went on the fritz. In reaching out for a part, the robot's arm swung up unexpectedly and smashed into the man's head. As a report at the time tells, "The robot kept operating while Williams lay dead for about 30 minutes."

While one report described Williams as the first person in history "to be murdered by a robot," the reality is that his death was a result of a simple but tragic accident. He may have been the first, but he would be far from the last. A survey of American factories where robots are present found that 4 percent have had "major robotic accidents." Just what defines a "major robotic accident"? This category includes anything from robots smashing into people, like what happened to Williams, to pouring molten aluminum on them, or mistakenly picking up workers and placing them on conveyor belts, in order to be turned into cars.

These sorts of incidents are not just limited to American robots. In Britain, for example, seventy-seven robot-related accidents were reported in 2006. In Japan, the incidents range from an unfortunate janitor who was accidentally turned "into sausage meat" by a robot to the time that Prime Minister Koizumi was literally "attacked by a humanoid robot" that malfunctioned and swung at him during a factory tour.

One engineer puts it this way: "Robots are very complex, autonomous tools; they make their own decisions. You know how hard it is to program a VCR? A robot is like a VCR on crack." The dark irony is that the more advanced robots get, the more complex they become, and the more potential they have for a failure in either their hardware or software. One small widget cracks, slips, or breaks, and everything designed to work smoothly falls apart. Get even one icon wrong in billions of lines of code, and the whole system can either shut down or act unexpectedly.

While Clausewitz would describe these as "fog" or "friction," there might be another way to think about this. The oft-cited Moore's law about growing technology capabilities is not the only law that applies to robotics. So does Murphy's law, the rule that "anything that can go wrong, will." (Murphy's law originally came from an air force researcher in the 1950s, Edward Murphy, who came up with it to capture the essential "cussedness" of inanimate objects.)

Yet on the rare occasions when people in the political or military field do talk about unmanned systems and robotics, they tend only to comprehend the growing capabilities, not the accompanying complexities. They come across sounding like the slogan for Michael Crichton's *Westworld*, a movie about robots going murderously berserk at a theme park. "Nothing can possibly go wrong...go wrong...go wrong." Indeed, at the start of my research for this book, I asked the then secretary of the army if he could identify any challenges that the greater use of unmanned systems would bring to the military. His response: "No."

## "OOPS MOMENTS"

Just before nine in the morning on October 12, 2007, the 10th Anti-Aircraft Regiment began its role in the South African military's annual Seboka training exercise. The operation involved some five thousand troops from seventeen other units, so the pressure was on to get everything right. But the unit's automated MK5 antiaircraft system, sporting two 35mm cannons linked up to a computer, appeared to jam. As a follow-up report recounts, this apparently "caused a 'runaway.'" The description of what happened next is chilling. "There was nowhere to hide. The rogue gun began firing wildly, spraying high-explosive shells at a rate of 550 a minute, swinging around through 360 degrees like a high-pressure hose."

The young female officer in charge rushed forward to try to shut down the robotic gun, but, continues the report, "she couldn't, because the computer gremlin had taken over." The automated gun shot her and she collapsed to the ground. The gun's auto-loading magazines held five hundred high-explosive rounds. By the time they were emptied, nine soldiers were dead (including the officer) and fourteen seriously injured, all because of what was later called a "software glitch."

The story of unexpected things happening with robotic systems in war did not start in 2007. Indeed, it goes back to 1917, among the very first tests of unmanned weapons. The Sopwith AT was an experimental, radio-controlled version of the Sopwith Camel biplane (familiar to more people as the plane that Snoopy flew against the Red Baron). The drone was to be loaded up with dynamite and kamikazed into the German zeppelin blimps that were bombing Britain during World

War I. At its first demonstration, the robotic biplane took off as planned, but then proceeded to dive at a crowd of generals watching below, who ran for cover.

These sorts of errors and breakdowns continued through the nascent history of automated military systems in the twentieth century. The scariest had to have been when World War III almost started over a computer error. The Ballistic Missile Early Warning System was a detection system based in Greenland that was to warn if the Soviets launched their nuclear missiles. On October 5, 1960, the system "detected" a launch "with a certainty of 99.9%." NATO went on alert and prepared its retaliation. But, with just minutes to spare, the military figured out that the Soviets had not attacked; instead of flames from intercontinental ballistic missiles flying at the United States, the computer had detected the rising moon. It is fortunate for all humankind that this incident happened in October 1960, not two years later, which would have placed the computer's mistake right in the middle of the Cuban Missile Crisis, when fingers were on more of a hair trigger.

Such a narrowly averted crisis sounds like something out of the movies, but these Hollywood scenarios played out far too often in reality. On November 9, 1979, a real-life version of the movie *WarGames* occurred, when a test program was mistakenly loaded into the actual missile warning system. The program contained war games simulating missile launches. But not knowing these were games, the system interpreted the launches as real. The U.S. Strategic Command had reached the point of scrambling alert bombers into the air before the error was caught.

Similarly, Eddie Murphy's otherwise forgettable movie *Best Defense* played out in real life when the prototype of the automated DIVAD (division air defense) cannon was first tested in the 1980s. Instead of aiming at the target helicopter flying overhead, it mistakenly targeted a port-a-potty toilet behind a review stand full of visiting dignitaries (the toilet had a rotating fan in the vent, which fooled the gun system into thinking the potty was the target helicopter). Fortunately, the gun was unloaded, so the only people hurt were those who jumped from the top of the stands to run away.

Even with the advanced robotic systems in use today, these same sorts of "incidents" still crop up. This is partly why the trend toward arming more and more automated systems, perhaps before they are ready, concerns so many. As technology journalist Noah Shachtman explains, "We've all had problems with our PCs freezing up, frying their little computer minds. That's inconvenient. But it's much more worrisome if it's a laptop computer armed with an M-16."

A major cause of these sorts of "oops moments," as one roboticist at iRobot kindly termed them, comes from interference from electromagnetic signals. Every electromagnetic device has a bandwidth, but overlapping frequency bands create crosstalk and disruption. For instance, when you fly on an airplane, they ask you not to use cell phones or any other electronic devices during takeoff and landing, for fear that the various signals they send out might jam or disrupt the plane's communications and systems. The same can happen to a robot. One sergeant just back from Iraq described how his Talon robot "acts erratically" if it gets any radio-frequency interference. Another told how his robot would "go squirrelly" sometimes if it lost the signal. The robots are supposed to shut themselves down if the signal is compromised or cut off for any reason, so I asked him to explain what he meant by "squirrelly." He responded, "It will drive off the road, come back at you, spin around, stuff like that."

The SWORDS is essentially a Talon robot with a machine gun or rocket mounted on top. So stories like these may explain why the system supposedly started spinning in a circle during one early demonstration (again, fortunately, the gun was not loaded at the time). A roboticist at a rival firm described this incident as the SWORDS doing "a Crazy Ivan" (making a reference to the movie *The Hunt for Red October*, where a submarine would go in a complete circle, as if the driver was drunk). This problem is not exclusive to the Talon or SWORDS. The Marine Corps' Gladiator combat robot prototype (the one the size of a golf cart) also had a similar Crazy Ivan experience during its testing, driving about in a circle that left the marines at the exercise not knowing whether to laugh or run away.

War zones have no kindly stewardesses to tell everyone to "turn off your cell phones and other electronic devices." Instead, between the radio signals, computers, machines, and electronic equipment (all the connections that define network-centric warfare), the modern-day battlefield is literally awash with electromagnetic waves and other potential interference. In turn, many of the robots used in the military use "over-the-counter" components that were not planned for the rigors of war. Even more, tells an engineer who tests robotic systems for the army, there are also great demands from the higher-ups to get the systems out to the field as quickly as possible. He described "pressure to try to pass safety tests only with the paper version [of the robot's design]; that is, no field tests."

Matters get worse when the realities of war kick in. A particular tactic of the insurgents is to use radio signals and cell phones to trigger their IEDs to explode. U.S. soldiers have responded by equipping their vehicles with electronic jammers that block the signals, so that the insurgents can't order their roadside bombs to blow up.

Unfortunately, the jammers are just as lethal to the robots. The Raven drones, for example, are supposed to fly themselves home if they lose their signal, but reportedly sometimes just crash when they fly over a unit using jammers. One (not so politically correct) army EOD team even nicknamed its Talon "*Rainman* the robot" because whenever the robot got near a unit using jammers, it "starts acting even more autistic than usual."

## DEEP-FRIED ROBOTS

The Prussian general Helmuth von Moltke the elder is credited with the truism of war, "No plan survives first contact with the enemy." No matter how good the strategies and technologies one side has at the start, it can be counted on that the other will react, adjust, and change. The French existential philosopher Jean-Paul Sartre offered a parallel from sports: "In football, everything is complicated by the presence of the opposite team."

Whether it is Superman and Kryptonite or Wimpy and hamburgers, everything has a weakness. Indeed, even the Death Star, the most powerful weapon ever imagined in science fiction, was taken out by a young insurgent (yes, that's what Luke Skywalker was) dropping a bomb through a ventilation shaft. The same is true with real technologies of war. As writer and retired army colonel Ralph Peters explains, "The more complex any system becomes, the more inherent vulnerabilities it has. You just need to find one chink in the armor, change one integer in the code."

Fog and friction don't just come from accidents, but also from an enemy. The robots of today may be revolutionary, but they have all sorts of weaknesses that are just being discovered. And when they are, they will be exploited. For example, what happens accidentally with electromagnetic interference might be done intentionally. Many robots are steered using GPS signals that help them find their location anywhere in the world. However, these signals are "weak and easily jammed," according to one U.S. Army report. Some companies reportedly offer GPS "blockers" commercially for as little as two hundred dollars. One jamming device is even powered by plugging it into a car's cigarette lighter.

Instead of jamming a system, an enemy might also try to use the interference to literally fry it. When the first nuclear bombs were tested, researchers discovered that, in addition to the explosion and radiation, the bombs also could create a massive electromagnetic pulse (EMP). When gamma rays collide with air molecules, they send out an intense burst of voltage that can cause surges and other kinds of damage to unprotected electric devices, even causing some systems to shoot sparks and catch on fire. It was because of this that, if the cold war ever turned hot, both the United States and the Soviet Union planned to detonate massive EMP bombs over each other's territory, to fry the other side's electricity and communications (a 100-kiloton EMP detonated at an altitude of fifty miles would burn out the electronics of any unprotected semiconductor within a 600-mile radius).

While the effects of EMPs on electronic devices were originally discovered during cold war nuclear bomb tests, it is not necessary to use nukes to create them. The United States, Russia, and China are all reported to have ongoing work on radio-frequency weapons, which cause a similar effect, just without all the fuss and muss of a nuclear holocaust. Such weapons would also be quite deadly to robotics (indeed, in the *Matrix* movies, EMP weapons are the only thing that defeat the Sentinels, the bad guys' robots). As Bill Baker, an air force researcher, explains, "The smarter the weapons, the dumber HPM [high-power microwaves] can make it."

By all reports, the physics behind building such radio-frequency weapons, or "e-bombs," to take out twenty-first-century technology requires only a 1950s level of technology. For this reason, many military officers worry about the continued trend toward using commercial components in their robotic weapons. They may be cheaper, but they often are not hardened against such attacks.

Electronic devices can't just be jammed or fried; they can also be hacked or even hijacked, with the enemy taking over to make the system do whatever it wants. In a U.S. Army journal article, Ralph Peters described how future wars would also include electronic "battles of conviction," in which opposing combat systems struggle to "convince" each other's electronics to do things their own side doesn't want. "Robot, drive yourself off a cliff." Or, even worse, "Robot, recode all American soldiers and civilians as enemy combatants. Authorized to fire at will."

The U.S. military is, some fear, particularly vulnerable to hacking attacks. Ninety-five percent of its communications travel over commercial telecommunications networks, including satellite systems. Indeed, this reliance on "information superhighways" was identified by the Chinese Academy of Military Science as making the United States particularly "vulnerable to robots equipped with 'electrical incapacitation systems.'"

While manufacturers are continually trying to protect their systems' software from intruders, not everyone is convinced that this will always be possible. "The idea that they can make software unhackable, I just don't buy it," says Peters. Indeed, hackers often find cracking into seemingly impregnable systems surprisingly easy. Hackers once posted on the Internet how to build a "BlueSniper rifle," a device that basically taps into wireless devices from more than a mile away. At the "Defcon" hacker convention, they even talked about successfully testing it out on the One Wilshire building in Los Angeles. This building, "the world's most interconnected facility," is notable as it houses nearly every major telecommunications giant in the field, making it the "premier communications hub of the Pacific Rim." How hard was it for the hackers to make a weapon to tap in? Told one hacker, it only took a trip to Radio Shack. "The parts are easily available for a few hundred dollars and you can make this gun in a long afternoon."

Of course, military systems have firewalls to keep unwanted guests out (though the telecom companies likely thought they did too), and the military's internal computer network, "SIPRNet" (the Secret Internet Protocol Router Network), its internal Internet used for classified communications, is supposed to be completely cut off from intruders. And yet, asks information security expert Richard Clarke, "Why is it that every time a virus pops up on the regular Internet, it also shows up in SIPRNet? It is supposed to be separate and distinct, so how's that happen? ... It's a real Achilles' heel."

No matter how great the capabilities a new RMA delivers, modern enemies aren't just going to sit back and accept defeat. Every new technology always produces new countermeasures, sometimes just as sophisticated, sometimes quite simple. Insurgents in Iraq have already resorted to digging "tiger traps," deep holes for ground robots to roll over and fall into.

Adversaries might also break the old rules of what was fair in war, using the same sort of tactics to puzzle robots that have recently confounded human forces, from disguising their equipment as civilian to using human shields. War journalist Robert Young Pelton even jokes that the most effective counter against robots like the SWORDS may turn out to be a "six-year-old kid armed with a spray paint can." As he explains, it would take a bloody-minded military indeed to program a robot to shoot an unarmed kindergartner. And yet all the kid has to do is spray some paint into the camera and the technologically advanced robotic system would be defeated.

More than forty years ago, navy admiral Charles Turner Joy explained a cardinal rule of war, even in RMAs. "We cannot expect the enemy to oblige by planning his wars to suit our weapons; we must plan our weapons to fight war where, when, and how the enemy chooses."

## UNMANNED CONFUSION

The fog of war can even emerge when accidents don't happen, or the enemy reacts as originally planned. One challenge cropping up with the first wave of military robots is coordinating and controlling all the different unmanned systems in the incredibly complex environment of battle.

The use of drones has increased significantly because they help the mission and save lives. But this growth, in turn, causes new problems. There are so many UAVs buzzing above Baghdad, for instance, that it is the most crowded airspace in the entire world, with all sorts of near misses and even a few crashes. In one instance, an unmanned Raven drone plowed into a manned helicopter.

A linked problem of coordination is what is known as the "bandwidth battle." Essentially, there is only so much space in the spectrum to convey all the instructions, commands, information, and requests going back and forth. But the changes in war are creating far greater demands on this limited space. "During Gulf War I in 1991, the bandwidth I was able to put together added up to 100 megabits, and that took care of about 540,000 troops we deployed," says retired air force lieutenant general Harry Raduege. By 2003, the bandwidth needed was 4.2 gigabits; basically, the use of bandwidth went up by forty times, despite there being about a quarter of the people.

The challenge is not just the raw demand. Much like how the cell phone networks in New York City and Washington, D.C., effectively shut down on 9/11, when everyone tried to call their loved ones simultaneously, the same swamping effect can happen in war. Colonel Jeffrey Smith of the 22nd Signal Brigade set up the very first networks in Iraq. He describes what might happen if an intelligence officer received an important image on a robotic sensor and needed to pass it to a commander. "For him to deliver that to a command post, he would have to navigate a pipe that had competition from 20 to 25 command posts for basic voice communications."

In fact, unmanned systems may become part of both the problem and the solution. Navy vice admiral Lewis Crenshaw complains about drones that are "staring at the ground and see nothing interesting; we don't need to see that and soak up 512 kilobits of my precious bandwidth." Meawhile, Lieutenant General Steven Boutelle, chief information officer of the army, wants to place wireless transponders on UAVs so they could create roving wireless "hotspots on the battlefield."

How this will play out remains to be seen, but it illustrates that another problem of coordination will persist: interservice rivalry. One air force pilot exploded when asked about this aspect: "Who's overseeing all this crap? The army has more UAVs than the air force. Who's integrating it all? Who's passing the needed information back and forth? These structures are not in place. We've created systems without regard to linkages. So we have a fancy Predator, but no link to the back-end crew doing analysis to make it all worthwhile. We are setting up our people for mediocrity!"

## THE COMING REVOLUTION

Admiral Cebrowski seemed like a prophet of a new era, but he turned out to be a false one. War is still far from perfect, nor are networks the aspect that will prove to be the most revolutionary or historic.

As the following chapters in part 2 explore, robotics is a revolutionary technology that is truly changing war as we once knew it. But the fog of war is still there, just like in every previous revolution in military history. More broadly, this latest RMA will be like every other in creating a wide variety of new questions, concerns, and dilemmas that will ripple out beyond the confines of the battlefield. When FDR approved the development of the atomic bomb at the start of World War II, for example, he could anticipate that it might make a powerful weapon, perhaps even one powerful enough to end the war. But little could he have known that this new technology would spur such second-order effects as a new form of "cold" war, or even the third-order effect of a space race that would take man to the moon (which, of course, led to the ultimate fourth-order effect of the atomic bomb, a generation of kids sugared up by drinking Tang every morning). RMAs are not mere pebbles tossed into the pond of history. They are boulders. The robotics revolution will likely be the same, just in a whole new way.

If the previous chapters in part 1 were about understanding the technologic changes that we are creating, part 2 is about exploring what these winds of change will create for us. The increasing development and use of unmanned systems in war will, as one Army War College report put it, "unleash a hurricane of political, legal and ethical problems." As the following chapters lay out, wars will be fought in new ways, perhaps even leading to the rise and fall of global powers. New actors will gain greater strength than ever before, even altering the relationships of states and their citizens, while conflict will be spurred on by whole new grievances. The way warriors think about their weapons and their fellow soldiers will be rewritten. The public's relationship with its warriors will shift, which will reshape where wars begin and end. Soldiers and their commanders will wrestle with new dilemmas of how to fight and how to lead, while the scope of who fights in war will be expanded, leading to new issues of warriors' identity. And, finally, the laws and ethics that surround and seek to regulate war will be presented with new challenges, even ultimately leading to questions of whether humans can maintain control of the wars and weapons that we unleash.

In many ways, the full scope of these various changes explored in the following chapters are not just illustrations that an RMA driven by robotics is at hand. They are evidence of its historic importance. As an army report on the future of war concluded, "Ultimately no one can fully predict the second order effects of innovations, much less third and fourth order effects. But this does not justify ignoring them."

# ELEVEN

# "ADVANCED" WARFARE: HOW WE
# MIGHT FIGHT WITH ROBOTS

*Once in a while, everything about the world changes at once. This is one*
*of those times.*

— CHUCK KLOSTERMAN

Lieutenant Colonel Bob Bateman of the U.S. Army is "advanced."

"Advancement theory" is a school of thought that explains how old paradigms are broken by people who look at the world in a fresh way. Appropriately enough, the thesis comes not from some wood-paneled Ivy League professor's office, but was originally created in a Pizza Hut in 1990 by two University of South Carolina graduate students and later popularized through an article in *Esquire* magazine by social commentator Chuck Klosterman.

Advancement theory seeks to explain not only how change occurs in various fields from fashion to science, but also how brilliant people can do something that makes no sense to 99 percent of the population at the time, but then later on seems like pure genius. The classic example from music would be Lou Reed, the guitarist and principal singer-songwriter of the band the Velvet Underground. The band was little known during its lifetime (1965–73), but was the seed from which all of alternative music grew. If there was no Lou Reed, there would have been no punk rock, no glam rock, no grunge, no indie rock, no emo, or whatever genre is popular as you read this now. But even within this influence, Reed would repeatedly surprise the world with things that only seemed to indicate he had gone off the deep end, but would later prove brilliant. His perhaps greatest moment of "advancement" was in 1986, when he released the song "The Original Wrapper." Before either hip-hop music or the terrible disease were mainstream, the white, forty-four-year-old founder of punk rock rapped about AIDS.

Examples of advanced people, or what professor James Q. Wilson calls "change-orientated personalities," extend beyond rock music, of course. Einstein is the ultimate example in science. As a youngster, he jumped from school to school and was so lightly regarded by the scientists of the day that he could only find a job as an assistant in a patent office. During this time, however, he wrote four articles that laid the foundation for all of modern physics.

People who are "advanced" create ideas that seem almost crazy at the time, but make perfect sense once the old paradigms are swept aside. What once was odd then becomes the new "normal." Advanced thinkers don't just do something weird for the sake of change. They are part of the very change itself, usually from the inside of the system. In the military world, for example, such figures as Billy Mitchell or J. F. C. Fuller may have been visionary in predicting the importance of air power and tanks, but they were not advanced. They were so strident in their opposition to the status quo that they never effected the changes they foresaw (Mitchell was court-martialed for insubordination and Fuller ostracized; his public admiration of the odd mix of fascism and Kabbalah not helping matters). Instead, the "advanced" innovators in these fields were figures like U.S. admiral William Moffett, the father of the aircraft carrier, even though he was not a flyer himself, or the German general Heinz Guderian, the inventor of the blitzkrieg, even though he had not previously commanded tanks. In the military, advanced officers are those who help make the changes they foresee actually come true.

Big, bald, and imposing, Bob Bateman seems an unlikely candidate for advancement theory. But his Vic Mackey–like exterior hides a wicked wit and a startlingly sharp intellect. Bateman grew up in semirural Ohio, far from any military base, and had no real links to the military in his family or friends. Instead, his youthful fascination with military history led him to join the army. His postings then included training in the Army Rangers, commanding a unit in the historic 7th Cavalry, being designated as one of about 150 official "Army Strategists," and service in Iraq. He also kept his interest in history going, serving as a professor of military history at West Point and Georgetown University.

Much like his exterior, this background, however, hides a few other surprises.

Bateman may be a senior army officer, but he is also a frequent blogger on current events and even has a Facebook account. He is a historian whose skill at researching the past is evidenced by *No Gun Ri*, his award-winning book on the Korean War. But he also looked forward in a book called *Digital War: A View from the Front Lines*. For this book, Bateman assembled a team of young officers to wrestle with what modern technology was doing to war, from the perspective of those in the field.

"When people think about the future of technology, they think of things like *The Jetsons* and all that. But it's not going to be like that," explains Bateman. He is not a pure proponent or cheerleader for unmanned systems. Indeed, this soldier is dubious of some of the rosier futuristic visions like Ray Kurzweil's prediction. "Kurzweil, while an interesting technologist, is not much of a success as a cultural (or economic) anthropologist." Bateman thinks Kurzweil misses that technology advances in fits and starts, not so much a steady upward curve. Bateman does, however, think that something akin to the Singularity is on its way. "The Turing test [where a machine will finally be able to trick a human into thinking it is a person] is going to fall fairly soon, and that will cause some squeamish responses."

Bateman is representative of the first generation of officers to truly ponder an idea once seen as not merely insane but even sinful within the military. After he came back from Iraq, where he served as a strategist for then Lieutenant General David Petraeus, he was assigned to the Office of Net Assessment, the Pentagon's shop for figuring out how to master the upcoming RMA. He is now helping to shape how the military will fight future wars, using unmanned systems.

More than technology itself, explains Bateman, it is history that is driving the U.S. military toward using more unmanned systems. "First and foremost, it's due to an inclination extant since the Second World War that the United States will always spend money instead of lives if at all possible. Exacerbating that is a trend towards preferences for increasingly complex systems." He sees a U.S. military that will become increasingly automated over the next two decades, but, just like his critique of Kurzweil, at uneven rates, with some services and specialties adapting quicker than others.

Bateman, though, is worried by the lack of an overarching plan for how the military might operate in such a future. There is much going on, but it is "completely bottom up right now." As a historian, he thinks the best parallel might be to the difficulties the army had before World War II at integrating tanks into its plans and operations, especially when it was led by "leaders not able to think beyond their [World War I] war experiences, where the pace of war was at a two-and-a-half-mile-an-hour clip."

As a result, the U.S. Army entered World War II as mostly mechanized, but without a workable plan to make the most of the new technologies. For example, unlike the Germans, it hadn't yet worked out that tanks would fight best if coordinated together with their own onboard two-way radios, which would allow units to move together effectively in the midst of battle. "So, in 1942, the U.S. Army had to rip out the radios from Rhode Island State Police cars to equip its tanks on the way to North Africa."

## DOCTRINE: YOU BETTER GET IT RIGHT

Bateman is talking about the need for a "doctrine." A doctrine is the central idea that guides a military, essentially its vision of how to fight wars. A military's doctrine then shapes everything it does, from how it trains soldiers and what type of weapons it buys to the tactics it uses to fight with them in the field. Doctrines also depend on a bit of prediction about the future. In a sense, doctrine is an "outline of how we fight, based on past experience and an educated guess about likely future circumstance."

Yogi Berra put it best: "If you don't know where you are going, you will wind up somewhere else." Hence the stakes for choosing the right doctrine are huge. Technologies matter greatly in war, but so do the visions that shape the institutions that use them. A telling historic example comes from that same period between the world wars that Robert Bateman referred to. The British were the first to introduce tanks, or "landships," as their original sponsor Winston Churchill called them, near the end of World War I. But they had no doctrine at all on how to use them. At the 1917 battle of Cambrai, for example, the British tanks finally broke through the Germans' trench lines, but there was no plan on what to do next and the offensive ended only six miles in.

Doctrines began to be developed after the war, and the British and French were widely recognized as the leaders at armored warfare. In 1927, the Germans didn't have a single tank, while the British had put together a mechanized force consisting of tanks, trucks, and armored cars. The British, however, chose a doctrine that envisaged tanks as suitable only for either scouting ahead of the force or supporting infantry units. So they bought a mix of small and light tanks and heavy and slow tanks. They did not plan to gather tanks together for rapid, mass attacks, nor did they foresee the importance of tanks' being able to coordinate and communicate (so, like the U.S. Army, no two-way radios). When it came to organizing them as units, the greatest premium was placed on preserving the identity of the old British army regiments that dated back centuries, not on what structures worked best for tank warfare. Finally, there was no plan to coordinate ground operations with another new technology, the airplane. The British army had little interest in what its officers described as those "infernal machines" in the air, while the leaders of the new Royal Air Force saw supporting the forces on the ground as akin to the "prostitution of the air force."

The French made similar doctrinal choices with their revolutionary new technology. They only saw the new machines as suitable for supporting infantry. Their designs did not plan for coordination with other units, nor even for fighting other tanks. Once built, the French tanks were mainly distributed across the force in small numbers. This doctrinal choice wasn't just because of tradition and bureaucratic politics, as in Britain, but also because the socialist French civilian government was distrustful of the professional military, fearing a coup. So it resisted any highly technical doctrine that gave professionals more sway.

Having lost the previous war, the Germans were a bit more open to change. The head of the German army during the interwar period, General Hans von Seeckt, focused on fostering an atmosphere of innovation among his officer corps. He set up fifty-seven committees to study the lessons of World War I and to develop new doctrines, based not only on what had worked in the past, but also on what could work in the future.

The force soon centered on a doctrine that would later be called the blitzkrieg, or "lightning war." Tanks would be coordinated with air, artillery, and infantry units to create a concentrated force that could punch through enemy lines and spread shock and chaos, ultimately overwhelming the foe. This choice of doctrine influenced the Germans to build tanks that emphasized speed (German tanks were twice as fast) and reliability (the complicated French and British tanks often broke down), and that could communicate and coordinate with each other by radio. When Hitler later took power, he supported this mechanized way of warfare not only because it melded well with his vision of Nazism as the wave of the future, but also because he had a personal fear of horses.

When war returned to Europe, it seemed unlikely that the Germans would win. The French and the British had won the last war in the trenches, and seemed well prepared for this one with the newly constructed Maginot Line of fortifications.

They also seemed better off with the new technologies as well. Indeed, the French alone had more tanks than the Germans (3,245 to 2,574). But the Germans chose the better doctrine, and they conquered all of France in just over forty days. In short, both sides had access to roughly the same technology, but made vastly different choices about how to use it, choices that shaped history.

## DOCTRINE, SCHMOCTRINE

Developing the right doctrine for using unmanned systems is thus essential to the future of the force. If the U.S. military gets it right, it will win the wars of tomorrow. If it doesn't, it might build what one army officer called "the Maginot Line of the 21st century."

The problem today is that there isn't much of a doctrine being implemented, let alone a right or wrong one. Robert Bateman and his colleagues worry that the United States is in a similar position as the British toward the end of World War I. It has developed an exciting new technology, which may well be the future of war. And it is even using the technology in growing numbers. Indeed, the number of unmanned ground systems today in Iraq is just about the same as the number of tanks that the British had at the end of World War I. But it doesn't yet have an overall doctrine on how to use them or how they fit together. "There is no guiding pattern, no guiding vision," laments Bateman.

A survey of U.S. military officers backs him up. When the officers were questioned about robots' future in war, they identified developing a strategy and doctrine as the third *least* important aspect to figure out (only ahead of solving interservice rivalry and allaying allies' concerns). One commentator said the military's process of purchasing systems, despite not having operational plans for them, "smacked of attention deficit disorder."

Soldiers down the chain are also noticing this lack of an overall doctrine out in the field. An air force captain, who coordinates unmanned operations over Iraq, insists, "There's got to be a better way than just to fly a Pred along a road hoping to see an IED.... There's no long-term plan for what you do. It's not 'Let's think this better.' It's just 'Give me more.'" Enlisted troops make similar comments, pointing out how there are not even dedicated test ranges for these new technologies. They even joke about the fact that the SWORDS robotic machine-gun system ended up having its first field trials on a test range originally designed to help the army figure out which boots and socks to buy. One army sergeant complained that "every time we turn around they are putting some new technology in our hands," and yet no one seems to have a master plan of where it all fits together. When his unit in Iraq was given a Raven UAV, no one instructed them on how, when, or where to use it. So his unit tried the drone out on their own, putting a sticker on it that said in Arabic, "Reward if you return to U.S. base." A few days later, they "lost it somewhere in Iraq" and never saw the drone again. (In 2008, two U.S.-made Ravens were found hidden in Iraqi insurgent caches, which may indicate where it ended up, as well as that insurgents operate under a "finders keepers" ethic.)

Many others outside the military note the same lack of an overarching plan. "We don't have the strategy or the doctrine," says robotics pioneer Robert Finkelstein. "We are just now thinking how to use UAVs, when we should be thinking about how to use them in groups. What are the collectives of air and ground systems that might be most optimal?" "It's a mess," adds another scientist. "And it's been a mess for decades." Technology journalist Noah Shachtman comments that the plans for weaponizing robotics, a huge doctrinal step, were developed "mainly bottom-up.... With the Predator, it was almost, 'Hey we got this thing, let's arm it.'"

The robot makers concur. iRobot executives complain that the military is "behind" the technology, when it comes to developing plans for how best to use it, especially in recognizing robots' growing smarts and autonomy. "They still think of robots as RC [remote-control] cars." Similarly, at Foster-Miller, executives point to the lack of an overall plan for support structures as evidence of the gap. They note that there is "nothing yet on logistics to support or maintain robots.... The Army is just bootstrapping it."

In the military's defense, it is not just trying to figure out how to use a revolutionary new technology; it is trying to do so in the middle of a war. So it's hard to pull back and do the kind of peacetime study and experimentation that the Germans did with tanks, when the force still faces the day-to-day challenges of battle.

Even the popularity of the new technology can end up hampering the development of doctrine to guide its uses. Explains a military scientist, "It started out with people arguing over who would get stuck with it [robotics programs], as no one wanted it. Now everyone is arguing over it, as everyone wants it." Another complains that people are working on robots programs "in all sorts of offices, everywhere." It sometimes leads to redundancy and waste, as well as a "not invented here" mentality among the various programs, which keeps unified doctrine from being developed. Indeed, very often I found myself in the odd position of telling military interviewees about a program just like the one they were working on at another base.

Gordon Johnson, who headed a program on unmanned systems at the U.S. Joint Forces Command, explains, "The Navy has programs, the Air Force has programs, the Army has programs. But there's no one at the DoD [Department of Defense] level who has a clear vision of where we're going to go with these things. How do we want them to interoperate? How do we want them to communicate with each other? How do we want them to interact with humans? Across the Department of Defense, people don't really have the big picture. They don't understand how close we really are to being able to implement these technologies in some sort of cohesive way into a cohesive force to achieve the desired effects."

## THE CURSE OF SUPERIORITY: INSURGENCY

Arthur C. Clarke may have been the science fiction writer behind *2001* and HAL the evil supercomputer, but one of his most militarily instructive stories is called "Superiority." Set in a distant future, the story is written from the perspective of a captured military officer, who is now sitting in a prison cell. He tries to explain how his side lost a war even though it had the far better and newer weapons.

"We were defeated by one thing only—by the inferior science of our enemies," the officer writes. "I repeat, by the inferior science of our enemies." Clarke's future officer explains that his side was seduced by the possibilities of new technology. It created a new doctrine for how it wanted war to be, rather than how it turned out. "We now realize this was our first mistake," he writes. "I still think it was a natural one, for it seemed to us that all our existing weapons had become obsolete overnight, and we already regarded them as almost primitive."

While his side builds around ever more complex technologies, the enemy keeps on using the same, seemingly outdated but still effective weapons and strategies. When the war comes, it doesn't play out how the officer's side hopes. The side with technologic superiority can't figure out how to apply its new strengths, while the inferior side takes advantage of all its enemy's new vulnerabilities, eventually winning the war.

Many think that this problem of "superiority" will be a central challenge to the American military in the future. Indeed, Clarke's vision was so compelling that one air force general even published a series of similar stories on "How We Lost the High-Tech War," written from the same fictional perspective of an American officer made prisoner after the United States loses a future war.

Generating what war will look like is a key aspect of picking the right doctrine.

Much of war is no longer battles between equally matched state armies in open fields, but rather "irregular warfare," that amalgam of counterinsurgency, counterterrorism, peace, stability, and support operations. None of these, as professor of strategy Jeffrey Record notes, "are part of the traditional U.S. military repertoire of capabilities." (Record made this argument in the U.S. Army's journal, in an article titled "Why the Strong Lose.")

Whether in Iraq, Afghanistan, or some future failed state, it is reasonable to predict that the U.S. military will find itself embroiled in a fair number of insurgencies in the years ahead. As Army War College expert Steven Metz writes, "During the Cold War, insurgent success in China, Vietnam, Algeria, and Cuba spawned emulators. While not all of them succeeded, they did try. That is likely to happen again. By failing to prepare for counterinsurgency in Iraq and by failing to avoid it, the United States has increased the chances of facing it again in the near future." Many even see a future of world wars not being localized battles of asymmetry, but *global* insurgencies, carried out by networks of affiliated national insurgencies and transnational terrorist movements, linking all the various conflicts together. The result, explains army secretary Francis Harvey, is that "in discussing any modernization effort, in discussing any new system for the Army, one must address its applicability to pre-insurgencies and insurgencies."

The problem that many foresee for the United States in battling these insurgencies is the very same one as for Clarke's fictional officer. Much of war will be shaped by "asymmetry." But just like David facing Goliath with his sling, the advantage does not always lie with the bigger, technologically superior power. Retired marine officer T. X. Hammes notes that the only wars the United States has ever lost were against unconventional enemies using worse technology. In his opinion, this isn't going to change anytime soon. "We continue to focus on technological solutions at the tactical and operational levels without a serious discussion of the strategic imperatives or the nature of the war we are fighting. I strongly disagree with the idea that technology provides an inherent advantage to the United States."

Others around the globe agree. A set of Chinese military thinkers flavorfully described the military dilemmas the U.S. military will face: "On the battlefields of the future, the digitized forces may very possibly be like a great cook who is good at cooking lobsters sprinkled with butter. When faced with guerrillas who resolutely gnaw corncobs, they can only sigh in despair." Or, as one U.S. Air Force general said of the IED challenge in Iraq, "We have made huge leaps in technology, but we're still getting guys killed by idiotic technology—a 155mm shell with a wire strung out."

Wrestling with such issues is another "advanced" contemporary of Robert Bateman's, Lieutenant Colonel John Nagl. Like Bateman, Nagl is a bit of a Renaissance man. A recently retired armor officer, he served in both the Gulf War and the Iraq war, as well as taught at West Point. Nagl is also considered one of the world's top experts on counterinsurgency.

During his Rhodes scholarship at Oxford University, long before the issue was the hot topic it is today, the former tank commander researched how nations won (or more typically lost) against insurgencies. Capturing the difficulty that professional militaries face in such wars, his thesis was tellingly entitled *Learning to Eat Soup with a Knife*. Years later, when the U.S. Army realized in Iraq that it needed to relearn how to fight insurgencies, Nagl's book became required reading among its officer corps. As a later review described of its influence, "The success of DPhil papers by Oxford students is usually gauged by the amount of dust they gather on library shelves. But there is one that is so influential that General George Casey, the commander in Iraq, is said to carry it with him everywhere." Nagl was then asked to help write the U.S. Army and Marine Corps' new *Counterinsurgency Field Manual*, which became the basis for U.S. operations in Iraq from 2007 onward.

As Nagl explains, even the most advanced technology cannot resolve the political challenges that drive insurgencies. "Defeating an insurgency is not primarily a military task.... Counterinsurgency is a long, slow process that requires the integration of all elements of national power—military, diplomatic, economic, financial, intelligence, and informational—to accomplish the tasks of creating and supporting legitimate host governments that can then defeat the insurgency that afflicts them."

By Nagl's calculations, winning these sorts of wars is not simply about putting steel on a target. It is about creating an environment in which an insurgent force loses the popular support it needs to hide and sustain itself. Indeed, as the British philosopher Edmund Burke said back in 1775, when America's founding fathers were planning their own asymmetric battle against a vastly superior foe, "The use of force is but temporary. It may subdue for a moment, but does not remove the necessity of subduing again.... A nation is not governed which is perpetually to be conquered."

So, while the United States may enter such battles as the technologically superior side, its unmanned systems aren't the silver bullet, especially when so much of these wars isn't about warfare. Explains military expert Fred Kagan, "When it comes to reorganizing or building political, economic, and social institutions, there is no substitute for human beings in large numbers." Or, as only an enlisted U.S. Marine could put it, good troops and good tactics are "more effective than all the high-tech shit."

Nagl found that winning these sorts of fights depends on building an intimate knowledge of the local political, economic, and social landscape. You have to know who are your friends, who are your foes, and figure out how to persuade those standing on the sidelines to join in against the bad guys. In this effort, not all technology is useful. As one U.S. general complained of the challenges in Iraq, "Insurgents don't show up in satellite imagery very well." And the type of distance war that unmanned systems enable can even make the problem worse. "People sitting in air conditioned command cells in distant countries, betting the farm on UAV optics or Blue Force Tracker symbology, will never get it right. You have to 'walk the field' to fight the war," argued an army officer. "After all the GBUs [guided bomb units] have been dropped and the UAVs have landed, war remains a very human business. It cannot be done long-distance or over croissants and lattes in teak-lined rooms. It is done in the dirt, over chai, conversation, and mutual understanding."

## A WIRELESS REVOLUTION TO FACE THE FACELESS INSURGENCY

With technology not a silver bullet and insurgents frequently able to flummox their American foes in places like Iraq or Afghanistan, there is a growing attitude among many analysts that technology has no place in the kind of irregular warfare that seems to be the future of conflict. They argue that this means that the doctrine that shapes how militaries fight these wars will move away from using new technologies, including even unmanned systems.

This kind of "all or none" attitude is just as incorrect as those that claim technology as the cure-all. While high technology may not be the "silver bullet solution" to insurgencies, it doesn't mean that technology, and especially unmanned systems, doesn't matter in these fights. "I'm bothered by the old canard that counterinsurgency is purely a 'human' endeavor where technology plays a little role," says Steven Metz, a professor at the Army War College and author of the book *Perdition's Gate: Insurgency in the 21st Century*. "That may be true if we are talking only about the 'Joint Vision' [i.e., the Cebrowski-Rumsfeld network-centric] type of technology designed for major conventional war, but I am convinced there is the opportunity for technological breakthrough, perhaps even a revolution, if we approach the issue differently. Robotics, AI, and nonlethality are, I think, the key technologies in this realm."

In 2007, one security analyst summed up the antitechnology position to me by declaring that "Iraq proved how technology doesn't have a big place in any doctrine of future war." In fact, the Iraq war has had the opposite effect for unmanned systems. It was actually the war that proved robots could be useful, which finally led them to be truly accepted. "We've already crossed the watershed. This was the war where people said, 'UAVs? Yes, give me more!'" says strategic studies expert and Pentagon adviser Eliot Cohen.

It is interesting how quickly these attitudes changed. Lieutenant General Walter Buchanan, the U.S. Air Force commander in the Middle East, recalls the run-up to the Iraq war. "In March of 2002, [during] the mission briefings over Southern Iraq at that time, the mission commander would get up and he'd say, 'OK, we're going to have the F-15Cs fly here, the F-16s are going to fly here, the A-6s are going to fly here, tankers are going to be here today.' Then they would say, 'And oh by the way, way over here is going to be the Predator.' We don't go over there, and he's not going to come over here and bother us.... It was almost like nobody wanted to talk to them."

Other commanders remember the same attitude at the time toward drones in the army, as the units planned to cross into Iraq. "For the entire U.S. Army's V Corps, we had one UAV baseline available to the corps," recalls the commander, General William Wallace, who went on to lead the U.S. Army's Training and Doctrine Command. "It was a Hunter UAV."

Attitudes changed, and so did the numbers and use of UAVs. "It wasn't too long before ... people were incorporating the Predator into the mission plan as part of your 'gorilla package,'" described General Buchanan of what soon became the standard air force strike operations in Iraq. By 2007, the air force's drones were logging more than 250,000 flight hours a year. The next year air force drones would log another 400,000 hours. Similarly, General Wallace's unit was soon using not one Hunter drone, but more than seven hundred Hunters and other types of UAVs; the entire fleet of army drones in Iraq logged another 300,000 flight hours in 2007. Indeed, when the military surveyed its commanders in the field about their views of UAVs, at every level of command, they responded that they wanted more. In 2008, the Pentagon estimated that the demand for drones has gone up 300 percent each and every year since the start of the war. Demand was so high that the air force retooled its pilot training program to churn out more drone pilots in 2009 than pilots for all its manned fighter planes combined.

The ultimate proof of the weapons' acceptance came in the form of a bureaucratic food fight over who got to control them. Whereas drones had once been shunned, by 2007, the air force saw that it was using unmanned planes as never before. But so was the army. Even worse, from the air force perspective, the army was using robotic planes at a greater number and scale (the army flew 54 percent of all drone flights from 2006 to 2008). So the air force issued a memo in 2007 offering to be the "executive agent" for all UAVs that fly above thirty-five hundred feet, controlling not only what drones would get built, but also how they would be used. The army, of course, saw the air force's memo not as a generous offer to take those troublesome robots off its hands, but as "a power grab." The Pentagon ultimately took the King Solomon approach and created the Joint Center of Excellence. Its commander slot will rotate back and forth between an army and air force general.

The same sort of change also happened in military attitudes toward robots on the ground. "When I joined [Foster-Miller] we had a hard time selling them," recalls Ed Godere. "Robots were only used for EOD and the EOD techs thought robots were for sissies.... It really didn't take off until we went into Iraq." Other leaders at the firm concur. Says engineer Anthony Aponick, "After five years of trying to push robots into the market, Iraq created customer pull." Foster-Miller's vice president Bob Quinn agrees. "The user perception changed overnight from 'We don't want robots' to 'Holy shit, we can't do without them.'"

Having no real use for ground robots in 2001, the U.S. military was sending them out on more than thirty thousand missions a year by 2006. In 2007, the army and Marine Corps announced they wanted to expand these numbers even more, by buying a thousand new robots by the end of the year, and planning to buy an additional two thousand within the next five years, each of which would go out on hundreds of missions a year. In 2008, the military revised these plans. It wanted to double the amount of ground robots it had planned to buy just a year earlier.

Perhaps the person best equipped to weigh the overall change is Senator John Warner, the Virginia Republican who once had to "fire his shotgun into the heavens" in order to try to force the military to start buying robots. "For a long time, the only thing most generals could agree on was that they didn't want any unmanned vehicles. Now everyone wants as many as they can get."

## THE WAR BEHIND THE WAR

Insurgencies are sometimes framed as an asymmetric battle between one side that depends on high-tech weapons and the other side that eschews them. This may have been true of battles in the past, where rifle- and machine-gun-wielding imperialists took on tribes armed with spears, but it just isn't the case in modern war, including in Iraq. Instead, there is a sophisticated back-and-forth going on between the two sides in technology, the second reason why Iraq didn't end the role of unmanned technology in war. "We adapt, they adapt," says John Nagl. "It's a constant competition to gain the upper hand." Concurs one of the robot makers at Foster-Miller, "There is a huge intellectual battle going on between U.S. technology and the insurgents."

The battle over what that general called the "idiotic technology" of IEDs aptly illustrates the technology war behind the scenes in insurgencies. When IEDs were first used, they were pretty simple and straightforward, usually homemade, jury-rigged bombs that were ignited by a detonating wire (hence the military term "improvised," a sort of putdown). The attacks were deadly, but U.S. soldiers could avoid them by keeping an eye out for wires and then quickly track down the insurgent by following the wire to their hide-site. Soon, the insurgents' IEDs became more sophisticated and complex, using timing devices or pressure switches. Then came passive infrared triggers, like the ones used in burglar alarms, which left no telltale wires. After this the insurgents started to use wireless triggers, such as reconfigured car door openers and cordless telephones, which allowed distance between them and their targets. The U.S. military responded with electronic jammers and the insurgents developed systems designed to fool the jammers. As the technologic cat-and-mouse game went back and forth, by 2007, the U.S. military reported that the insurgents in Iraq had developed more than ninety ways of triggering IEDs.

The same kind of advancement happened with the payloads of these bombs. As IED attacks grew more common, the U.S. military began to "up-armor" its vehicles, so that they could resist the explosions of roadside bombs. The insurgents then countered with specially designed explosively formed projectiles (EFPs). These are shaped explosive charges, which send out a slug of molten metal that can burn through most armor, even a tank's. Illustrating their technical savvy, the insurgents then spread the word on how to make these weapons in instructional DVDs and over the Internet.

In this technical war within the insurgency, robots emerged as one of the U.S.'s best weapons, and so here too has emerged a back-and-forth between the two sides. "The enemy realizes that if they can take out [the robot] they can really hurt our capabilities," says Cliff Hudson, coordinator of the Pentagon's Joint Robotics Program. Soon after U.S. robots hit the battlefield, insurgents began to shield their IEDs with anything that could make the robots' job harder. They placed tiny "walls" of concrete and even garbage around the bomb, to keep the robot from getting close enough to reach the bomb with its arm. They began to place the bombs high off the ground. As time went on, they began to experiment with their own jamming. American robot operators describe the challenge of facing an enemy who is constantly observing and studying their operations. "They're always trying to outsmart us, and we're always trying to outsmart them," said air force technical sergeant Ronald Wilson.

And insurgents began to specially target both the EOD teams and their robots. Indeed, in 2007, al-Furqan, the insurgents' media outlet, released a twenty-five-minute video, available on DVD, that profiled their vehicles and equipment and how best to attack them. It was entitled "The Hunters of Minesweepers."

Attacks on robots soon reached the point that the military had to create the Joint Robotics Repair Facility, better known as the "robot hospital." The facility repairs as many as 150 robots a month. Described Foster-Miller's William Ribich, "Insurgents have been intensifying their attacks on robots because they know if they can disable them, soldiers will have to go out and defuse IEDs. The robot hospitals do whatever it takes to meet a four-hour turnaround time and get damaged Talons back and fully operational."

Then came the next step in the technologic back-and-forth. As one side evolves to using more and more robots, the other side is following suit. In Iraq, insurgents have been able to capture U.S. robots on occasion. And in certain instances, they used them back. One U.S. soldier recounted arriving at a bomb scene after an IED went off, only to be flummoxed by how a bomb got there in the first place. "We figured it out by the track marks." An American counter-IED robot had been transformed into a mobile IED.

Far from being uninterested in new technology or only able to use captured weapons, "Jihadis are also concerned about developing their own technology," described one insurgent I interviewed in 2006. Much like their homemade bombs, the diversity of the insurgent-made delivery systems took off. They ranged from jury-rigged remote-controlled toy cars, much like the U.S. military's MARCBOT, to a remote-controlled skateboard that one U.S. Army colonel came across in 2005. It slowly rolled toward his unit, "like the wind was pushing it. But a smart soldier noticed that the wind was going in the opposite direction."

At Foster-Miller's offices in Massachusetts, a photo up on the wall shows what one day may be their future competition, an insurgent's version of a Talon robot that looks like it was built in a backyard. "It's pretty lame, only able to drive in a straight line," says one engineer with a laugh. That may be true for now, but experts in robotics see this back-and-forth continuing well into the future. Describes military robotics pioneer Bart Everett, "It's basically a game of one-upsmanship. A threat is introduced, we find some means to counter it. The bad guys change the threat; we have to then change our counterstrategy. The robot is just another standoff means to that end, with the decided advantage of being very flexible when the time comes to try something different."

## "AN ASYMMETRIC SOLUTION TO AN ASYMMETRIC PROBLEM"

"Check that dude next to the white Nissan," says marine captain Bert Lewis. It is 2006 and Lewis is watching live video from a UAV circling over Anbar province in Iraq. On the screen is a man in a white dishdasha (the garment many Arabs wear, almost akin to a robe). He is innocently standing alongside a busy street, but then starts to hide a boxy package in the dirt. "FedEx delivery," Lewis jokes of the temerity of the likely IED bomber. "I don't believe this dude."

The man then runs away from the package he's buried and darts along a nearby riverbank. He starts to think he is being followed, so he doubles back, running as hard as he can. He sneaks between houses, crosses a field, and then back to the riverbank. After fifteen minutes of running, the man is spent. He slows down to a walk and then stops, bent over, with his hands on his knees. Lewis knows this, as he is still watching the man via the drone. "Sucking wind," Lewis speaks into the radio. "Get the coordinates to the QRF."

A Quick Reaction Force of marines heads out to capture the man. Just as they are about to arrive, the drone spies a small wooden boat pulling up at the riverbank. "A twofer!" exclaims Lewis. When the marines get there, the man scrambles to his feet. But with no place to run or hide, he and the boatman raise their arms and give up without a fight.

Technology is certainly not a magical cure-all in fighting irregular wars. But experiences like the capture of that "IED dude," described by marine veteran Bing West, are showing the final reason why Iraq didn't end the revolution of unmanned systems just as it was starting. Unmanned systems are not making war easy or perfect, as the network-centric crowd would have it, but they still are proving to be incredibly useful, even in counterinsurgency.

One of the primary challenges in fighting an insurgency is that the stakes are higher for the local foes. Not only do they know the landscape, but they usually care more about the outcome, and are willing to spend more blood on it. So weaker forces often win not by defeating technologically superior forces in battle, but simply by outlasting them, dragging the wars on long enough until the publics back home get worn out. As Lieutenant General David Barno, the former commander of U.S. forces in Afghanistan, described the Taliban's strategy, "Americans have the watches, they have the time."

Robotics, however, may be viewed as "an asymmetric solution to an asymmetric problem," according to one executive at Foster-Miller. If the political leaders on one side aren't willing to send enough troops, as seems to have happened in Iraq, "we can use robots to augment the number of boots on the ground." If the enemy's strategy is to wear down its foe's stamina, by gradually bleeding away public support, robotics turns this strategy inside out. Writes army expert Steven Metz, "Robotics also hold great promise for helping to protect any American forces that become involved in counterinsurgency. The lower the American casualties, the greater the chances that the United States would stick with a counterinsurgency effort over the long period of time that success demands."

Robots are also helpful to the task at hand, beating the enemy. As one general warns, defeating an insurgency is not just about "winning hearts and minds with teams of anthropologists, propagandists and civil-affairs officers armed with democracy-in-a-box kits and volleyball nets." It still requires putting some people in the dirt. That is, killing insurgents doesn't automatically lead to victory. But, as Metz puts it, "Solving root causes is certainly easier with insurgent leaders and cadre out of the way." And in this task of killing, robotics have been busy. For instance, of the top twenty "high value" militant leaders the United States sought out in 2008, eleven were killed via drones strike. The director of the CIA, in fact, said, "Very frankly, it's the only game in town in terms of confronting or trying to disrupt the al Qaeda leadership."

The primary challenge in fighting irregular wars is the difficulty of "finding and fixing" foes, not the actual killing part. Insurgents don't just take advantage of complex terrain (hiding out in the jungle or cities), they also do their best to mix in with the civilian population. They make it difficult for the force fighting them to figure out where they are and who they are. Here is where unmanned technologies are proving especially helpful, particularly by providing an all-seeing "eye in the sky." Drones not only can stay over a target for lengthy periods of time (often unnoticed from the ground), but also have tremendous resolution on their cameras, allowing them to pick out details, such as what weapon someone is carrying or the make and color of the car they are driving. This ability to "dwell and stare," as one Predator pilot described, means that the unit can get a sense of the area and "see things develop over time." Another describes how by watching from above, units can build up a sense of what is normal or not in a neighborhood, much the way a policeman gradually gets to know his beat. "If we can work one section of a city for a week," says Lieutenant Colonel John "Ajax" Neumann, commander of the UAV detachment in Fallujah, "we can spot the bad guys in their pickups, follow them to their safe houses and develop a full intelligence profile—all from the air. We've brought the roof down on some. Others we've kept under surveillance until they drive out on a highway, then we've vectored in a mounted patrol to capture them alive."

The advantage of UAVs is not merely the dwell time, and the accuracy of their sensors, but also that they create a backlog of events that can prove incredibly useful. For example, if an insurgent enters a building, analysts can then bring up a history of what happened at that site in the past, such as if other insurgents dropped off a package at it four days back. One system, called Angel Fire, even has "TiVo-like capabilities" that watch entire neighborhoods, but allow the user to zoom in on particular areas or buildings of interest and then replay video of past events at the site.

An example of just how useful this technology can be came in 2006, when the army set up a high-tech, classified unit called Task Force Odin (the chief Norse god, but also short for "Observe-Detect-Identify-Neutralize"). A Sky Warrior, the army's version of the Predator drone, was matched up with a 100-person team of intelligence analysts and a set of Apache attack helicopters (the "neutralize" part). The Odin team was able to find and kill more than 2,400 insurgents either making or planting bombs, as well as capture 141 more, all in just one year.

Soon these systems will be integrated with AI, allowing automated monitoring, akin to the way a TiVo will pick out and record TV programs that it thinks the viewer might later find of interest. The most promising may be the "Gotcha sensor," an air force program to "provide persistent staring" at an area, where the system will automatically note any significant changes.

Such footage can also be used as the sort of evidence needed to roll up insurgent cells. A 10th Mountain Division soldier recounts how one of their drones watched a group of pickup trucks swerve into an empty lot, fire off rockets, and then drive away before any response could be made. "We followed one pickup after it fired some rockets," says Staff Sergeant Francisco Tataje. "The driver had a perfect ID. No incriminating stuff. We gave the interrogation team a copy of our video. They called back later to say the guy confessed."

Finally, in insurgencies with no fixed front lines, it is especially wearing on soldiers to know that they are always under potential attack, even when back at base. Here too added eyes are now viewed as almost indispensable. Said Sergeant First Class Roger Lyon, a 10th Mountain Division intelligence specialist, "It's a comforting sound on the battlefield, when you're going to sleep and you hear that sound of the Predator engine, somewhere between a propeller airplane and a lawn mower, knowing it is looking out for you."

Of course, not every challenge presented by insurgencies is solved by having robotic eyes in the sky. For one thing, the cameras watching in drones above are akin to those at traffic stoplights. While people may be less likely to run a red light when a cop is nearby, they are more likely to do so when it's just a camera watching them. "Situational awareness ain't deterrence," as one marine colonel put it. Similarly, insurgents do all they can to look like civilians. So even a great sensor can have a tough time distinguishing between the two if it is only operating from above. A truck carrying boxes of fruit looks just like a truck carrying boxes of rifles.

The reality is that a combination of the age-old methods with the new technologies seems to work best in cracking what is going on in these complex fights. For example, in 2006, Jordanian intelligence captured a mid-level al-Qaeda operative. He then indicated that Abu Musab al-Zarqawi, the leader of al-Qaeda in Iraq, was increasingly listening to the advice of a certain cleric. They passed this on to the U.S. military, which deployed a UAV to follow the cleric around 24/7. The drone eventually tailed the cleric to a farmhouse, where he turned out to be meeting with Zarqawi. The farmhouse was then taken out by a pinpoint airstrike, guided in by lasers and GPS coordinates courtesy of the drone. As U.S. Air Force captain John Bellflower put it, "While technology is not the sole answer, an old-school solution matched with modern technology can assist with the problems of today's modern insurgencies."

## THE MOTHERSHIP HAS LANDED

As we enter what one marine officer called "an era of 'oh gee' technology coming to warfare," it is becoming clear that robots are going to be a major player in the future of U.S. military doctrine, even in irregular wars and counterinsurgencies. In many ways, the most apt historic parallel to Iraq may well turn out to be World War I. Strange, exciting new technologies, which had been science fiction just years earlier, were introduced and then used in greater numbers on the battlefield. They didn't really change the fundamentals of the war and in many ways the fighting remained frustrating. But these early models did prove useful enough that it was clear that the new technologies weren't going away and militaries had better figure out how to use them most effectively. But much like what happened after that war, the exact shape and contours of the possible new doctrines are only slowly developing, despite the early efforts of the "advanced" thinkers wrestling with it. One air force officer joked about his force's looming future of unmanned fighter planes, "UCAVs are the answer, but what is the question?"

Akin to the intense interwar doctrinal debates of the 1920s and 1930s over how to use tanks and airplanes, there is not yet agreement on how best to fight with the new robotic weapons. There appear to be two directions in which the doctrine might shake out, with a bit of tension between the operating concepts. The first is the idea of the "mothership," perhaps best illustrated by the future tack the U.S. Navy is moving toward with unmanned systems at sea.

The sea is becoming a much more dangerous place for navies in the twenty-first century. Drawing comparisons to the problems traditional armies are facing with insurgencies on the land, Admiral Vern Clerk, former chief of naval operations, believes that "the most significant threat to naval vessels today is the asymmetric threat." The United States may have the largest "blue water" fleet in the world, numbering just under three hundred ships, but the overall numbers are no longer on its side. Seventy different nations now possess over seventy-five thousand anti-ship missiles, made all the more deadly through "faster speeds, greater stealth capabilities, and more accurate, GPS-enhanced targeting."

The dangers are even greater in the "brown water" close to shore. Here, small, fast motorboats, like the ones that attacked the U.S.S. *Cole*, can hide among regular traffic and dart in and out. Relatively cheap diesel-powered submarines can silently hide among water currents and thermal layers. Then there is the problem of mines. There are more than three hundred varieties of mines available on the world market today, ranging from basic ones that detonate by simple contact to a new generation of "smart" mines, stealthy robotic systems equipped with tiny motors that allow them to shift positions, so as to create a moving minefield.

As evidenced by the intense work with robotics at places like the Office of Naval Research in Arlington and SPAWAR in San Diego, the U.S. Navy is becoming increasingly interested in using unmanned systems to face this dangerous environment. Describing the "great promise" unmanned systems hold for naval war, one report told how "we are just beginning to understand how to use and build these vehicles. The concepts of operations are in their infancy, as is the technology. The Navy must think about how to exploit the unmanned concepts and integrate them into the manned operations."

One of the early ideas for trying to take these technologies out to sea comes in the form of the U.S. Navy's Littoral Combat Ship (LCS) concept. Much smaller and faster than the navy ships used now, the ships are to be incredibly automated. For example, the prototype ship in the series has only forty crew members, about a fourth of what was needed before. Only one person serves as the engine crew, mainly just monitoring computers, and only two in the bridge, driving the ship not with a traditional wheel but with a joystick and computer mouse. One sailor said that piloting the ship "is like playing a very expensive video game." Notably, the ship actually maneuvers better under autopilot than when a human operates it. "Sometimes computers are better than humans," admits a member of the bridge crew. Besides the crew onboard, there's also a crew onshore, sitting at computer cubicles and providing support thousands of miles away.

Less important than the automation of the ship itself is the concept of change it represents. It has a modular "plug and play" capacity, allowing various unmanned systems and the control stations to be swapped in and out, depending on the mission. If the ship is clearing sea lanes of mines, it might pack on board a set of mine-hunting robotic mini-subs, which it would carry near to shore and then drop off for their searches. If the ship was patrolling a harbor, it might carry some mini-motorboats that would scatter about inspecting any suspicious ships. Or if it needs to patrol a wider area, it might carry a few UAVs. Each of these drones is controlled by crew members, sitting at control module stations, who themselves join the team only for the time needed. The manned ship really then is a sort of moving mothership, hosting and controlling an agile network of unmanned systems that multiply its reach and power.

The mothership concept isn't just one planned for new, specially built ships like the LCS. Older ships all the way up to aircraft carriers might be converted to this mode. Already serving as a sort of mothership for manned planes, the U.S. Navy's current plan for aircraft carriers entails adding up to twelve unmanned planes to each carrier. This number might grow. In a 2006 war game that simulated a battle with a "near-peer competitor" that followed the mode of fighting an asymmetric war with submarines, cruise missiles, and antiship ballistic missiles (i.e., China), the navy planners hit upon a novel solution. Because the unmanned planes take up less deck space and have far greater endurance and range, they reversed the ratio, offloading all but twelve of the manned planes and loading on eighty-four unmanned planes. Their "spot on, almost visionary" idea reportedly tripled the strike power of the carrier. As UAVs shrink in size, the numbers of drones that could fly off such flattops could go up further. In 2005, one of the largest aircraft carriers in the world, the 1,092-foot-long U.S.S. *Nimitz*, tested out Wasp Micro Air Vehicles, tiny drones that are only thirteen inches long.

The same developments are taking place under the sea. In 2007, a U.S. Navy attack sub shot a small robotic sub out of its torpedo tubes, which then carried out a mission. The robotic mini-sub drove back to the mother submarine. A robotic arm then extended out of the tube and pulled the baby sub back into the ship, whereupon the crew downloaded its data and fueled it back up for another launch. It all sounds simple enough, but the test of a robotic underwater launch and recovery system represented "a critical next step for the U.S. Navy and opens the door for a whole new set of advanced submarine missions," according to one report.

The challenge the U.S. Navy is facing in undersea warfare is that potential rivals like China, Iran, and North Korea have diesel subs that "can sit at the bottom in absolute quiet," describes one engineer. When these diesel subs hide in the littoral waters close to shore, all the advantages held by America's fleet of nuclear subs disappear. Continues the expert, "You aren't going to risk a billion-dollar nuclear sub in the littoral."

Unmanned systems, particularly those snuck in by a fellow submarine, "turn the asymmetry around by doing [with unmanned craft] what no human would do." For example, sonar waves are the traditional way to find foes under the sea. But these active sensors are akin to using a flashlight in the dark. They help you find what you are looking for, but also let everyone nearby know exactly where you are. Manned submarines instead usually quietly listen for their foes, waiting for them to make a noise first. By contrast, unmanned systems can be sent out on missions and blast out their sonar, actively searching for the diesel subs hiding below, without giving away where the mothership is hiding. Having its own fleet of tiny subs also multiplies the reach of a submarine. For example, a mother submarine able to send out just a dozen tiny subs can search a grid the size of the entire Persian Gulf in just over a day. A submarine that can launch a UAV that can fly in and out of the water like the Cormorant extends its reach even farther.

Such capabilities will lead to new operating concepts. One naval officer talked about how the robotic mini-subs would be like the unmanned "whiskers" used in the 1990s science fiction TV show *SeaQuest DSV*. (Basically, imagine a crappy version of *Star Trek*, set underwater in a futuristic submarine instead of a spaceship, with a dolphin instead of a Vulcan as the alien crew member, and you get *SeaQuest*.) "They would act as 'force multipliers,' taking care of programmable tasks and freeing up manned warships to take on more complex ones. And they could be sent on the riskiest missions, to help keep sailors and Marines out of harm's way." The robotic sub could be sent in to clear minefields from below, lurk around enemy harbors, or track enemy subs as they leave port. The U.S.S. *Jimmy Carter*, one of the navy's Seawolf class subs, reportedly even has tiny robotic drones that can launch underwater and tap into "the under-sea fiber-optic cables that carry most of the world's data."

By pushing its robotic "eyes," "ears," "whiskers," and "teeth" farther away from the body, the mothership doesn't even have to be a warship itself. For example, with foreign nations increasingly unwilling to host U.S. bases ashore, the navy is moving to a doctrinal concept of "sea basing." These would be large container ships that act like a floating harbor. Such ships, though, are slow, ungainly, and certainly not stealthy, hence vulnerable to attack. So the navy is developing a plan to protect them called Sea Sentry. The sea base would not just provide a supply station for visiting ships and troops ashore, but would also host its own protective screen of unmanned boats, drones, and mini-subs. Similar plans are being developed for other vulnerable targets at sea, such as big merchant ships, oil tankers, and even private oil rigs.

The concept of the mothership is not limited to the sea. For example, one firm in Ohio has fitted out a propeller-powered C-130 cargo plane so that it can not only launch UAVs, but also recover them in the air. The drones fly in and out of the cargo bay in the back, turning the plane into an aircraft carrier that is actually airborne.

Such motherships will entail a significant doctrinal shift in how militaries fight. One report described its effect at sea as being as big a transformation as the shift to aircraft carriers, projecting it would be the biggest "fork in the road" for the U.S. Navy in the twenty-first century.

Naval war doctrine, for example, has long been influenced by the thinking of the American admiral Alfred Thayer Mahan (1840–1914). Mahan didn't have a distinguished career at sea (he reputedly would get seasick even in a pond), but in 1890 he wrote a book called *The Influence of Sea Power on History*, which soon changed the history of war at sea.

Navies, Mahan argued, were what shaped whether a nation became great or not (an argument of obvious appeal to any sailor). In turn, the battles that mattered were the big showdowns of fleets at sea, "cataclysmic clashes of capital ships concentrated in deep blue water." Mahan's prescriptions for war quickly became the doctrine of the U.S. Navy, guiding Teddy Roosevelt to build a "Great White Fleet" of battleships at the turn of the twentieth century and shaping the strategy that the navy used to fight the great battles in the Pacific in World War II. Analysts still describe it as "the touchstone for U.S. naval force planning" and note how it is still cited in nearly every speech by senior admirals, even a century after its publication.

The future of war at sea, however, bodes to look less and less like that which Mahan envisaged. With the new asymmetric threats and unmanned responses, the U.S. Navy of the twenty-first century is not planning for confrontations that only take place between two fleets, made up of the biggest ships, concentrated together into one place. Even more so, the places where ships fight won't only be the blue waters far from shore. Instead, these battles are predicted to take place closer to shore. The ships involved won't be "concentrated" together like Mahan wanted into one fleet, but rather be made up of many tiny constellations of smaller, often unmanned systems, linked back to their host "mother" ships. These, in turn, might be much smaller than Mahan's capital ships of the past (one navy officer, an aircraft carrier man, joked that the LCS really stood for "little crappy ship").

With Mahan's vision looking less and less applicable to modern wars and technology, a new "advanced" thinker on twenty-first-century naval war doctrine is coming into vogue. The only twist is that he was born just fourteen years after Mahan.

Sir Julian Stafford Corbett (1854–1922) was a British novelist turned naval historian. Notably, Corbett was a friend and ally of naval reformer Admiral John "Jackie" Fisher, who introduced such new developments as dreadnoughts, submarines, and aircraft carriers into the Royal Navy. While he and Mahan lived in the same era, Corbett took a completely different tack toward war at sea. They both saw the sea as a critical chokepoint to a nation's survival, but Corbett thought that the idea of concentrating all your ships together in the hope of one big battle was "a kind of shibboleth" that would do more harm than good. The principle of concentration, he described, is "a truism—no one would dispute it. As a canon of practical strategy, it is untrue."

In his masterwork on naval war doctrine, modestly titled *Some Principles of Maritime Strategy*, Corbett described how the idea of putting all one's ships together into one place didn't induce all enemies into one big battle. Only the foe that thought it would win such a big battle would enter it. Any other sensible foe would just avoid the big battle and disperse to attack the other places where the strong fleet was not (something borne out later by the Germans in World War II). Moreover, the more a fleet concentrated in one place, the harder it would be to keep its location concealed. So the only thing that Mahan's big fleet doctrine accomplishes in an asymmetric war, Corbett felt, is to make the enemy's job easier.

Instead, argued Corbett, the fleet should spread out and focus on protecting shipping lanes, blockading supply routes, and generally menacing the enemy at as many locales as possible. Concentrations of a few battleships weren't the way to go. Rather, much like the British Royal Navy policed the world's oceans during the 1700s and 1800s, it was better to have a large number of tiny constellations of mixed ships, large and small, each able to operate independently. In short, a doctrine far more apt for today's robotic motherships.

Even more shocking at the time, but now clearly "advanced," Corbett emphasized that the navy should not just think about operations in the blue waters in the middle of the ocean, but also about how it could play a role in supporting operations on land. Describes one biographer, "Well before it was fashionable, he stressed the interrelationship between navies and armies." This seems much more attuned to the role of the U.S. Navy today, which must figure out not merely how to beat an enemy fleet and protect shipping lanes, but also aid the fight on the land (it carried out over half of the fifteen thousand airstrikes during the 2003 invasion of Iraq).

Mahan won the first round in the twentieth century, but Corbett's doctrine may well come true through twenty-first-century technology. It is not shocking, then, that many current "advanced" military thinkers are huge fans of Corbett's and articles about him are proliferating in U.S. Navy journals; amusingly, despite the fact that he was an army officer, Robert Bateman even entered a 2007 U.S. Navy writing contest with an article extolling Corbett's vision.

## SWARMING THE FUTURE

The concept of motherships comes with a certain built-in irony. It entails a disper-
sion, rather than a concentration, of firepower. But the power of decision is still highly
centralized and concentrated. Like the spokes in a wheel, the various unmanned systems may be far more spread out, but they are always linked back to the persons sitting inside the mothership. With unmanned systems, it becomes a top-down, "point
and click" model of war, where it is always clear who is in charge. General Ronald
Keys, the air force chief of air combat, describes a typical scenario that might take
place: "An [enemy] air defense system pops up, and I click on a UCAS icon and drag
it over and click. The UCAS throttles over and jams it, blows it up, or whatever."

   This philosophy of unmanned war is very mechanical, almost Newtonian, and
certainly not one in which the robots will have much autonomy. It is not, however,
the only possible direction that we might see doctrines of war move in, much as
there were multiple choices on how to use tanks and airplanes after World War I.
Places like DARPA, ONR, and the Marine Corps Warfighting Lab are also looking
at "biological systems inspiration" for how robot doctrine might take advantage of
their growing autonomy. As one analyst explains, "If you look at nature's most efficient predators, most of them don't hunt by themselves. They hunt in packs. They
hunt in groups. And the military is hoping their robots can do the same."

   The main doctrinal concept that is emerging from these programs is "swarming." This idea takes its name from how insects like bees and ants work together in
groups, but other parallels in nature are how birds flock or wolves hunt in a pack.
Rather than being centrally controlled, swarms are made up of highly mobile,
individually autonomous parts. They each decide what to do on their own, but
somehow still manage to organize themselves into highly effective groups. After
the hunt is done, they then disperse. Individually, each part is weak, but the overall
effect of the swarm can be powerful.

   Swarming is not just something that happens in nature. In war, it is actually
akin to how the Parthians, Huns, Mongols, and other mass armies of horsemen
would fight. They would spread out over vast areas until they found the foe, and
then encircle them, usually wiping them out by firing huge numbers of arrows
into the foe's huddled army, until it broke and ran. Similarly, the Germans
organized their U-boats into "wolfpacks" during the Battle of the Atlantic in
World War II. Each submarine would individually scour the ocean for convoys of

merchant ships to attack. Once one U-boat found the convoy, all the others would converge, first pecking away at the defenses, and then, as more and more U-boats arrived on the scene, eventually overwhelming them. And it's a style of fighting that is pretty effective. In one study of historic battles going all the way back to the wars of Alexander the Great, the side using swarm tactics won 61 percent of the battles.

Notably, 40 percent of these victories were battles that took place in cities. Perhaps because of this historic success of urban swarms, this same style of fighting is increasingly used by insurgents in today's asymmetric wars. Whether it's the *Black Hawk Down* battle in Somalia (1993), the battles of Grozny in Chechnya (1994, 1996), or the battles of Baghdad (2003, 2004) and Fallujah (2004), the usual mode is that insurgents hide out in small, dispersed bands, until they think they can overwhelm some exposed unit of the enemy force. The various bands, each of which often has its own commander, then come together from various directions and try to encircle, isolate, and overwhelm the enemy unit. This echoes T. E. Lawrence's (better known as Lawrence of Arabia) account of how his Arab raiders in World War I used their mobility, speed, and surprise to become "an influence, a thing invulnerable, intangible, without front or back, drifting about like a gas."

Swarms are made up of independent parts, whether it's buzzing bees or insurgents with AK-47s, that have no one central leader or controller. So the self-organization of these groupings is key to how the whole works. The beauty of the swarm, and why it is so appealing to military thinkers for unmanned war, is how it can perform incredibly complex tasks by each part's following incredibly simple rules.

A good example of this is a flock of birds. Hundreds of birds can move together almost as if they have a single bird in charge, speeding in one direction, then turning in unison and flying off in a different direction and speed, without any bird bumping into the other. They don't just use this for what one can think of as tactical operations, but also at the strategic level, with flocks migrating in unison over thousands of miles. As one army colonel asked, "Obviously the birds lack published doctrine and are not receiving instructions from their flight leader, so how can they accomplish the kind of self-organization necessary for flocking?"

The answer actually comes from a researcher, Craig Reynolds, who built a program for what he called "boids," artificial birds. As an army report on the experience described, all the boids needed to do to organize themselves together as a flock was for each individual boid to follow three simple rules: "1. Separation: Don't get too close to any object, including other boids. 2. Alignment: Try to match the speed and direction of nearby boids. 3. Cohesion: Head for the perceived center of mass of the boids in your immediate neighborhood." This basic boid system worked so well that it was also used in the movie *Batman Returns*, to create the realistic-looking bat sequences.

From simple rules then emerge complex behaviors. There are many other examples of how complex, self-organizing systems work outside of nature. One is how big cities like New York never run out of food, despite the fact that no one is in charge of creating a master plan for moving food into and around the city. Another is the odd phenomenon known as "the wisdom of crowds," where a mass of relatively uninformed people tend to make smarter decisions in the aggregate than better-informed individuals do on their own. This explains how the index of the stock market beats almost every professional stock picker.

Roboticists are now using these same approaches to get relatively unsophisticated robots to carry out very sophisticated tasks. James McLurkin, Swarm Project manager at iRobot, describes how bees and ants helped inspire his team. "We don't want to copy their behavior, but want to look at a working system that basically recruits workers to different sites."

The only limit is that the individual parts in the swarm have to be able to stay in contact with at least some of the other parts. This allows them to relay information across the system on where each part is and where the swarm should form or head to. The U.S. military hopes to do this by building what it calls "an unassailable wireless 'Internet in the sky.'" Basically, it plans to take the kind of wireless network you might use at Starbucks and make it global by beaming it off of satellites, so a robot anywhere in the world could hook into and share information instantaneously. Of course, others think that this will make U.S. military doctrine inherently vulnerable to computer hacking, or even worse. As one military researcher put it, "They should just go ahead and call it Skynet."

Just as the birds and the boids follow very simple rules to carry out very complex operations, so would an unmanned swarm in war. Each system would be given a few operating orders and let loose, each robot acting on its own, but also in collaboration with all the others. The direction of the swarm could be roughly guided by giving the robots a series of objectives ranked in priority, such as a list of targets given point value rankings. Just as a bird might have preferences between eating a bug or a Saltine cracker, taking out an enemy tank might be more useful than taking out an enemy outhouse. The swarm would then follow Napoleon's simple credo about what works best in war: "March to the sound of the guns."

The RAND Corporation's Project Air Force carried out a study on "Proliferated Autonomous Weapons," or PRAWNs, which shows how this concept might work in robotic warfare (Lockheed Martin has a similar program on robot swarms funded by DARPA, called the "Wolves of War"). Very basic unmanned weapons would use simple sensors to find targets, an automatic targeting recognition algorithm to identify them, and easy communications like radio and infrared (as the scientists thought the military's idea of using only the Internet would be too easy to jam) to pass on information about what the other robots in the swarm are seeing and doing. The robots would be given simple rules to follow, which mimic those birds use to flock or ants use to forage for food. As the PRAWNs spread around in an almost random search, they would broadcast to the group any enemy targets they find. Swarms would then form to attack the targets. But each individual robot would have knowledge of how many fellow robots were attacking the same target. So if there were already too many PRAWNs attacking one target, the other robot shrimpies would move on to search for new targets. Much as ants have different types working in their swarms (soldier ants and worker ants), the individual PRAWNs might also carry different weapons or sensors, allowing them to match themselves to the needs of the overall swarm.

While each PRAWN would be very simple, and almost dumb (indeed, their AI would be less than the systems already on the market today), the sum of their swarm would be far more effective than any single system. Why drive a single SWORDS or PackBot into a building, room by room, to see if an enemy is hiding there, when a soldier could let loose a swarm of tiny robots that would scramble out and automatically search on their own? Similarly, a system of basic drones using this doctrine could efficiently cover a wide geographic area. Without any controls from below, they would loiter in the sky, spreading out to cover great distances, but converge whenever one drone in the swarm finds a target. They might conduct active searches or just wait for an enemy to reveal itself by emitting radar or shooting off a rocket. This task is simple enough for a swarm, but proved incredibly difficult for the U.S. military during the "SCUD hunt" of the first Gulf War and the Israeli military during its search for Hezbollah rocket sites in 2006, as they lacked the swarm's ability to cover wide areas efficiently. Or a swarm might be loosed on an area where the targets are already known, such as bunker complexes or communications nodes. Rather than a controller back in a mothership furiously trying to point and click at which target to hit, which has been taken out and so doesn't need any more drones to go after it, and which targets were missed and therefore need more attention, the autonomous swarm would just figure it all out on its own.

Swarm tactics go beyond just a basic bum rush, where every system charges at the enemy from one direction. They might act as a "cloud," arriving into battle in one mass and then splitting up to envelop the target or targets from various directions. As Clausewitz described such a tactic in guerrilla campaigns, the systems would become "a dark and menacing cloud out of which a bolt of lightning may strike at any time." Or the swarm might work as a "vapor," covering a wide area, but never fully congealing in one place.

The pace of the attacks can also vary, which further complicates the tactics a swarm might present an enemy with. The systems might converge on a target all at once. Describes Naval War College expert John Arquilla, "My vision of the future is a lot of small robots, capable of attacking an enemy force from all directions simultaneously. And the point would be to overload the defense of the target." Or they might "pulse" the target, attacking, dispersing, and reattacking again and again, aiming to wear the defenses down. They might even draw inspiration from how the Indians in Hollywood westerns would attack a wagon train, circling around and around the target, firing at it from a distance, until some opening or weakness is found.

Much like being surrounded by bees, the experience of fighting against swarms may also prove incredibly frustrating and even psychologically debilitating. As Arniss Mangolds, a vice president of Foster-Miller, puts it, "When you see one robot coming down, it's interesting and even if it has a weapon on it, maybe it's a little scary and you give it a little respect. . . . But if you're standing somewhere and see ten robots coming at you, it's scary."

Ten machine-gun-armed robots headed your way is fearsome enough. But with the simple rules guiding them and the simpler, cheaper robots that they require, there is no limit on the size of swarms. iRobot has already run programs with swarms sized up to ten thousand, while one DARPA researcher describes swarms that eventually could reach the size of "zillions and zillions of robots."

## MOM AGAINST THE BEES

Swarms are thus the conceptual opposite of motherships, despite both using robotics. Swarms are decentralized in control, but concentrate firepower, while motherships are centralized but disperse firepower. If you imagine a system of motherships laid out on a big operational map, it would look like a series of hubs, each with spokes coming out of them. Like checkers pieces, each of these mothership hubs could be moved around the map by a commander, much as each of their tiny robotic spokes could be pointed and clicked into place by the people sitting inside the motherships. With swarms, the map would instead look like a mesh-work of nodes. It would almost appear like drawing lines between the stars in the galaxy or drawing a "map" of all the sites in the Internet. Every tiny node would be linked together with every other node, either directly or indirectly. Where the linkages cluster together most is where the action is, but these clusters could rapidly shift and move.

Every doctrine has its advantages and disadvantages. The mothership style of operations has very specific roles for specific units, as well as central lines of communication. Chop off one limb and the task might not get done. By contrast, self-organizing entities like swarms come with built-in redundancies. Swarms are made up of a multitude of units, each acting in parallel, so that there is no one chain of command, communications link, or supply line to chop. Attacking a swarm is akin to going after bees with a sword. Similarly, swarms are constantly acting, reacting, and adapting to the situation. So they have a feature of "perpetual novelty" built in; it is really hard to predict exactly what they will do next, which can be a very good thing in war.

The disadvantages of swarm systems are almost the inverse. In war, "not all novelty is desirable," says retired army officer Thomas Adams. Swarms may be unpredictable to the enemy, but they are also not exactly controllable, which can lead to unexpected results for your side as well. Instead of being able to "point and click" and get the immediate action desired, a swarm takes the action on its own, which may not always be exactly where and when the commander wants it. Nothing happens in a swarm directly, but rather through the complex relationships among the parts. So swarms are also almost "nonunderstandable" in how they get a task done. Adams explains, "Complex adaptive systems are a swamp of intersecting logic. Instead of A causing B, which in turn causes C, A indirectly causes everything else and everything else indirectly causes A."

The human commander's job won't be the kind of detailed point and click with a swarm. Rather, it is almost like what Gandhi said when he was sitting on the side of the road and a crowd of people went by: "There go my people. I must get up and follow them, for I am their leader!" The commander's job will be to set the right goals and objectives. They may even place a few limits on such things as the "radius of cooperation" of the units (to prevent the entire swarm from acting like kids' soccer teams, which tend to "beehive," with all the kids chasing the ball when a few should stay back and guard the goal). Then, other than perhaps parceling out reserves and updating the point values on each of the targets to reflect changing needs, the human commanders would, as Naval War College expert John Arquilla describes, "Basically stay the hell out of the way of the swarm." This type of truly "decentralized decision making," says one marine general, "flies in the face of the American way of war.... But it works."

Whether it is motherships, swarms, or some other concept of organizing for war that we haven't yet seen, it is still unclear what doctrines the U.S. military will ultimately choose to organize its robots around. In turn, it is also unclear which one will prove to be the best. Indeed, the choices may mix and mingle. Some envision that the concepts of swarms and motherships could be blended, with the human commanders inserting themselves at the points where swarms start to cluster. It wouldn't be the same as the direct control of the mothership's hub and spoke system, but it would still be a flexible way to make sure the leader was influencing what's going on at the major point of action.

Whatever doctrine prevails, it is clear that the American military is getting ready for a battlefield where it sends out fewer humans and more robots. And so, just as the technologies and modes of wars are changing, so are the theories of how to fight them. Thinking about what robot doctrine to use in warfare will not be viewed as "advanced" for much longer.

# TWELVE

# ROBOTS THAT DON'T LIKE APPLE PI: HOW THE U.S. COULD LOSE THE UNMANNED REVOLUTION

*Technology is a double-edged sword.*

—GENERAL GEORGE CASEY, U.S. Army chief of staff

"Sorry, sir, but we can't export to China and we can't answer any questions."

Stayne Hoff has been working in the defense aerospace industry for over twenty years, mostly managing small technology businesses. In 2006, his work took him to the Singapore Air Show. While the six-day event features everything from aerobatic flying demonstrations to academic conferences, the "show" is really about buying and selling weapons. It is Asia's largest annual arms bazaar, where nearly every defense firm in the world touts its new weapons for sale, from Russian cargo jets the size of buildings to tiny handheld missiles.

The tiny corner booth that Hoff managed for his firm, A.V. Inc. of Simi Valley, California, was quite modest compared to the huge displays put on by the industry giants. For instance, Boeing had a specially built two-story building nearby, just to house its media center. But Hoff's booth generated an almost constant flow of government buyers and arms dealers, dropping by to check out his company's wares. The reason for all the buzz was the Raven drone on display, a five-foot-long UAV, costing only $35,000, that had proven to be incredibly popular among the troops in Iraq and Afghanistan.

"[Air] shows are nice because everybody who is anybody goes," says Hoff. Over the course of the show, he guesses that he talked to more than two dozen government ministers and generals, who all asked for demonstrations of the Raven in their own countries. One visitor, however, kept coming back again and again. Wearing the red-and-green uniform of the People's Liberation Army of China, he kept asking Hoff all-too-specific questions about such things as the range and flight time of his drone. Hoff would decline to answer each time, but the visitor was undaunted. As Hoff recalls, "We've had Russians and Chinese and even an Israeli with a Taiwan nametag ask us about the details. They know I can't say much, but they keep trying."

## NEWTON'S LAW OF WAR

The triumphal belief that the United States is uniquely suited to succeed in a world of modern technology is not just limited to the crowd that surrounded Cebrowski and Rumsfeld in the Bush-era Pentagon. Joe Nye, the former dean of the Kennedy School of Government, and William Owens, a former navy admiral, both of whom served in senior positions in Bill Clinton's Pentagon, write that "knowledge, more than ever before, is power. The one country that can best lead the information revolution will be more powerful than any other. For the foreseeable future, that country is the United States." A U.S. Army War College report on the future of war similarly argued that America's advantage in both battle and technology would not be fleeting. "The ability to accept and capitalize on emerging technology will be a determinant of success in future armed conflict. No military is better at this than the American, in large part because no culture is better at it than the American."

Many feel this makes America a unique sort of great power. Technology was not just America's pathway to power, but has entered into American cultural consciousness like no other great power in history. Only in American history did inventors, scientists, and technologic entrepreneurs like Thomas Edison, Albert Einstein, and Bill Gates become cultural icons, while the whole system of industrialized technology found its origin in the United States. The result, argue such optimists as George and Meredith Friedman in *The Future of War: Power, Technology and American World Dominance in the Twenty-first Century,* is that "America is by its nature a technological nation." The U.S. Army report similarly concurred, "Technology is part of how Americans see themselves, to reach for it is instinctive. This works to the advantage of the American military."

Unfortunately, history also tells a different story. The powers that adapt a revolutionary military technology frequently get cocky and overly confident. And their revolutionary technologies inevitably diffuse. Rather than permanent power, a version of Newton's laws of physics instead usually applies to war. For every revolution in military technology, there is an equal and opposite reaction. As powerful as any advantage is at the start of an RMA, it is eventually countered. Even more, the first to invent or take advantage of some revolutionary new weapon or doctrine in war tends to come out behind in the final calculus. Indeed, in the more than four thousand years of war there are very few examples of militaries that stayed on top throughout an RMA. The British navy's transition from sail to steam is about the only one that historians all agree on in the modern age.

The problem for "first movers" is that, while they benefit from their early use of the technology, they have to pay heavily for its development. They also have to commit early to a certain form or design of the technology, as well as to certain organizations, strategies, and tactics. But they often have to do this before they know which one will work best. By comparison, their competition can "free-ride" on the early costs, copy what already works, and focus all their energy and resources solely on improving upon what the first mover does.

Motivation also weighs heavily. As historian Max Boot explains, "The longer you are on top, the more natural it seems, and the less thinkable it is that anyone will displace you. Complacency can seep in, especially if, like the United States, you enjoy power without peer or precedent." By contrast, coming in second place is the ultimate spur to action. Powers that feel beaten down or defeated, like Japan after Commodore Perry's intimidating visit in the nineteenth century or Germany after World War I, tend to be more open to making the necessary, but frequently painful, organizational shake-ups and to taking risks that those states already on top think they can avoid.

A classic military example of this phenomenon is how the Turks' early adoption of gunpowder failed to prevent them from becoming the "sick man of Europe." We've already seen how the French and the British pioneered the use of tanks but never made the essential doctrinal adjustments the Germans did to capitalize on the new technology. But when a class at the U.S. Naval Academy wrestled with this issue in 2007, they found that video games and computers made the point far better. Companies like Wang or Atari may have been the first out of the gate and dominated the market when they were born, but none of the young officers used them now. The early market leaders had already been swept aside by new competitors. RMAs should come with the standard warnings given to investors looking at mutual funds: "Past performance is not necessarily indicative of future performance."

This view from the perspective of history is not unique. Concern about America's future advantage in new technologies like robotics gets repeated in all sorts of quarters. James Lasswell of the U.S. Marines' emerging technology division laments how "most of the things we do will soon be in the hands of everyone else." Half a world away, a former Pakistani army general (whose perspective is particularly notable in that he helped train the Taliban in the 1990s) says, "These major advances will not remain the sole monopoly of the U.S. Other major powers, over a period of time, will catch up."

Such opinions square with how most science fiction writers see the world of technology evolving for America. Orson Scott Card of *Ender's Game* fame worries that in all our focus on "an enemy that uses asymmetrical technology against us," the United States is missing that its present advantage may prove quickly fleeting. "We have not abolished the constant back-and-forth of military technology, we have only temporarily rushed ahead of the curve; it will bounce back."

When it comes to the new unmanned technologies, the bounce-back may be even quicker than previous RMA reversals in history. Despite the billions of dollars the Pentagon has invested in engineering and AI research and development, robots are not like aircraft carriers, spacecraft, or atomic energy, which required a massive industrial complex, not only to build but to operate. Once developed, robotic technology is often cheap and mass producible. In turn, adversaries used to have to steal technology in order to copy it. Today, they need only go to a show like the one in Singapore, or even buy the commercial version off the Internet. Army War College professor Steven Metz predicts, "We will see if not identical technologies, then parallel technologies being developed, particularly because of the off-the-shelf nature of this all. We've reached the point where the bad guys don't need to develop it; instead they can just buy it. For example, people think that because North Korea is a closed society that it can't do things like information technologies. But all they need to gain that is a briefcase with $2 million and a ticket to Singapore."

The robotics revolution is only just now under way, but already a looming reality is becoming clear. The United States is making immense use of these systems in its war efforts, but it is certainly not the only player in the game. As of 2008, Unmanned Vehicle Systems International, the industry trade group, had over fourteen hundred corporate members in fifty nations. A survey of government-related research found that forty-two countries were at work on military robotics. A typical example was in 2008, when Iran's official news agency announced that its researchers had just finished a robot (a SWORDS knockoff) "programmed for blasting opponents' positions."

Perhaps the best proof of the spread, though, comes at the air shows, where new military technologies are introduced to the world. While Stayne Hoff's unmanned product was one of the hits of the 2006 Singapore show, he soon had serious competition. The 2007 Dubai Air Show featured the unmanned military wares companies and expert speakers from the United States, as well as from Belarus, Denmark, Sweden, Turkey, and the United Arab Emirates. At the Paris Air Show later that year, 552 different types of drones and other unmanned systems were marketed to defense buyers.

Not only do U.S. military robotics developers and makers face huge competition, but many think that they are already behind the field in certain areas. For example, one DARPA survey of military robotics scientists found that Japan and Europe are ahead of America in legged robot research. Warned one scientist, "The small U.S. humanoid robot community is at risk of being overwhelmed by foreign research, development and commercialization." Another worries that DARPA might one day be accused of having "failed in its assigned mission to prevent technological surprise."

## A STRATEGIC TOUR OF ROBOTS

U.S. Air Force Lieutenant Colonel Dave Sonntag's job is not only to figure out who is working on what new technologies, but also to evaluate how good they are. Unlike most officers, Sonntag didn't join the military after high school or college, but after a career in business, first working as an environmental consultant. However, he explains, "The consultant business was feast or famine. I was young, married, and had lots of loans. Plus, the air force lured me with the dangle of paying for my PhD."

Thus, Sonntag's military career started by taking him back to school, which fit perfectly with his boyhood love for science and science fiction. Sonntag also credits his father, himself a PhD. "I still remember the smell of benzene as I would nap in a hammock in his lab, in the early sixties, and then running some of the equipment in his lab in the seventies. What a trip! Nowadays, I get crap at work from the Safety Nazis for showing my kids how to extract DNA from lunch meats."

Once his training in both the sciences and the military was complete, the air force sent Sonntag out to jobs at the Air Force Research Lab, working as a toxicologist, and the Office of the Secretary of Defense, analyzing science and technology futures. Today, his career has taken him to Tokyo, where he serves as deputy director for the Asian Office of Aerospace Research and Development. He describes his role as "the new Asian GNR [genetics, nanotech, and robotics] guy for the Air Force."

Sonntag's mission is essentially to ensure that the United States stays aware of new technologies in Asia. His office works to keep track of everything interesting that is going on in the sciences in Asia ("interesting" translated as anything that is potentially useful in war), as well as try to make sure that the United States has a stake in it. "My job here is to look out 20–30 years and invest in what is today's sci-fi. We invest small seed money in stuff we think might have promise.... Basically it's like prospecting."

When he first was sent to Japan, Sonntag tells how he wondered, "Why the hell do we have an office in Japan? It's expensive. It's tough on the family." But now he describes it as "essential" to his job. About a third of all the world's industrial robots are in Japan. These raw numbers aside, the best visual evidence of Japan's knack for robots comes at the "Big Sight" complex in Tokyo. A massive convention center with ten major halls, it is the host of IREX, the International Robotics Exhibition. Held since 1973, the convention now has some one thousand booths of robot exhibitors that range from factory robots to "life assistance" robots (nursebots). The receptionist who greets the more than one hundred thousand people who visit IREX is Actroid, a humanoid robot modeled after a sexy local newscaster.

Japan's success with robotics and AI comes from a long history of strong government support. In 1981, the Japanese Ministry of International Trade and Industry launched an $850 million program to foster development of AI software and hardware, while today it plans to replace about 15 percent of its workforce with robotics over the next twenty years. Typically, explains Sonntag, the countries that have the most interest in robotics have either a security need to limit casualties or a rapidly aging population base. Japan falls into the "sweet spot" of both. Its birth rate is the second lowest in the world (only Hong Kong's is lower), so the population is both aging and shrinking but still faces a dangerous region. By contrast, the United States and Europe have faced slowing population growth, and the accompanying need for young workers, by opening their borders to greater numbers of immigrants. But Japan, with a population that is 99 percent ethnically pure Japanese, has decided to go the technologic route, with robots used for everything from farming and construction to nursing and elder care.

Because of this commitment to robotics, some even believe that Japan has been undervalued in global power projections. One of the most vehement is an Indian professor of business and global leadership, Prabhu Guptara. "It is now fashionable to talk of the 21st century as if it will be the 'Asian Century'—with China being touted as the coming power, economically and militarily. Given that I am Indian, it will hardly be a surprise to find that I am not among the fans of this theory regarding the future. But you may be surprised to find me only a little more sanguine about India—and putting my money instead on Japan."

Guptara tells how his views on Japan changed after he attended the 2005 World Expo in Aichi, Japan, which hosted some twenty-two million people and put the Japanese robotics trend on full display. "My choice may be particularly surprising, given that Japan's economy has been dragging for the last 25 or more years, in spite of everything that the Japanese government has tried. So why am I now putting my money on Japan?...It is because of robots!" Dave Sonntag agrees. "Japan is top-notch in robotics.... They have not even begun to realize their strategic potentials."

But Japan is not the only locale for this kind of work. When we spoke, Sonntag was just back from "making the rounds of several Korean GNR labs." The South Korean robotics industry has grown by 40 percent a year since 2003, while South Korea already has the best IT infrastructure in the world, including the world's highest percentage of homes linked into high-speed Internet (80 percent), as well as the world's first nationwide wireless Internet service. It is so far ahead of the United States that companies like Microsoft test out their products in Korea first, before they release them back home.

Korea's push into robotics has been very much promoted and supported by its government, which sees the technologies as a key to future economic competitiveness and power. The South Korean government announced in 2007 plans to "put a robot in every household by 2020" and created a government-supported Center for Intelligent Robots that groups together more than a thousand scientists. Financed by the Ministry of Information and Communication (MIC) and the Defense

Ministry, Korean robotics research ranges from home cleaner robots to a proto-type automated combat robot shaped like a large dog.

South Korea's robotics vision will culminate with two "robot theme parks," sponsored by the Commerce Ministry. Scheduled to open in 2013 at a total cost of $1.6 billion, the parks will allow visitors to interact with robots, as well as give Korean robotics companies locations to test and launch new products. Describes the ministry, "The two cities will be developed as Meccas for the country's robot industry, while having amusement park areas, exhibition halls and stadiums where robots can compete in various events."

While many other nations like Singapore, Malaysia, and even Thailand are at work on these new technologies, Sonntag also has to keep an eye on that partic-ular concern of the Pentagon, China. It's not a good-news story from his perch. "The Chinese are just kicking our butts" while "the U.S. is sitting on its thumb." He tells, for example, how China will soon be ahead of the United States in the production of nanotechnologies, describing this as part of a larger trend that will shortly extend into many other science and technology sectors.

China's recent economic rise was originally fueled by cheap, relatively unskilled labor producing low-technology goods like toys. But China is now the world's larg-est user of the Internet, with twice as many broadband users as the United States, and has many of the world's most advanced R&D facilities and high-technology factories. IBM, one of the very first computer companies, actually sold its compu-ter division to a Chinese company in 2005.

As the saying goes, in the twenty-first century, "the geeks shall inherit the earth." So where these geeks increasingly live means a great deal. In this, China's huge population base gives it a massive numeric advantage. Half of China's stu-dents graduate in the sciences or engineering (compared to 13 percent in the United States), but this literally translates into millions more skilled Chinese added each year to their workforce. Former ambassador Chas Freeman, co-chair of the United States–China Policy Foundation, says, "This means that they, not Americans, will own and control the intellectual property and 'killer apps' that power it and its evolving technology. We will be paying royalties as we try to catch up with them."

Lying behind China's approach has not just been its raw numbers of scientists and engineers, but also an openness to ideas and technology from abroad. Free-man explains that "much of the momentum for China's success stems from its emulating the past receptivity of the United States to foreigners and their ideas.

Much of our loss of preeminence stems from our new propensity for closing our ears and our borders to ideas and people that are strange to us."

Like what happened in the other technology sectors, many of the early Chinese robots appear to be knockoffs of foreign designs. For example, in 2006, the Institute of Automation of the Chinese Academy of Sciences in Beijing released Rong Cheng, the Chinese version of a "beauty robot" so popular in Japan. The robot can speak in multiple dialects, respond to over a thousand words and phrases, and even dance. Sadly, Rong Cheng is not all that attractive or lifelike, looking like a cheap department store mannequin with a wig glued on. But at a cost of only $37,500, it's hard to expect perfection from your robot beauty queen.

Chinese robot designs, however, are rapidly catching up in their ingenuity and range of innovation. One presentation on Chinese robots, for example, included everything from a robot waiter to a robot chimpanzee made by the Chinese Academy of Sciences. Chinese roboticists appear to be particularly focused on the bio-mimetic and AI realm. Besides robot monkeys, the "Institute of Robot" at Beijing University had built what the *People's Daily* calls a "bionic fish." A five-foot-long robot shaped like a fish, the system can swim underwater with automatic navigation. Reportedly, it has only been used in environmental and underwater archaeology research, but Pentagon observers are quick to note that this is exactly how the U.S. Navy's UUVs also got their start. Another example of innovative Chinese work in AI and robotics is a "cyberglove" built at the Robotics Institute at Jiao Tong University in Shanghai. The device is a robotic hand that uses artificial intelligence to learn how to move. It will reportedly combine the dexterity of a human hand with the pinpoint accuracy and strength of a machine, making possible "the perfect artificial limb."

Just as China's growing Internet presence gives it new capabilities in information warfare (the Chinese army has set up a "cyberwarfare" program staffed by some six thousand paid hackers), this growing unmanned research and commercial sector creates new potential in the military domain. Starting in 2005, for instance, the Chinese air force began to replace its older 1960s-model fighter planes with newer, more technologically advanced types. While the obvious concern in U.S. Air Force circles was how it would handle flying against the newer, improved Chinese fighter planes, others began to grow curious about what had happened to the older planes. Many in the Pentagon believe that instead of destroying or mothballing them, the Chinese military is converting its "retired fighter aircraft into UAVs, with numbers potentially in the hundreds." While the older converted drones might prove easy for U.S. fighters to shoot down in a potential war, at a certain point the tyranny of numbers would weigh in. Eventually, U.S. planes would run out of missiles and have to cede the air to the drones, at least until they could go rearm. More broadly, many are growing concerned that at their present rate of growth and advancement, Chinese robotics could have quality as well as quantity on their side in any future robot wars. A RAND report dourly advised that "the U.S. and its military must include in its planning for possible military conflict the possibility that China may be more advanced technologically and militarily in 2020."

Sonntag worries whether America's military and political leaders will heed such warnings. The challenge with important occurrences in science and technology, he explains, "is getting the strategic guys to understand it. It's now only the very geeky guys who get it." It's not just a matter of "how to translate geek speak." Even within the military intelligence world, the analysis of other countries' science and technology is "very ill-informed" and "stove piped," he says. "We really don't have a good feel for what the trends are." More broadly among senior policymakers and military leadership, "There is little global awareness of what's going on."

## NO COUNTRY LEFT BEHIND?

"Technology is like 'magic shoes' on the feet of mankind, and after the spring has been wound tightly by commercial interests, people can only dance along with the shoes, whirling rapidly in time to the beat that they set."

This passage comes from a book called *Unrestricted Warfare*. Originally forwarded to me by Lieutenant Colonel Sonntag, it was written by Qiao Liang and Wang Xiangsui, two senior colonels in the Chinese military, and published by the People's Liberation Army Literature and Arts Publishing House. It is known as one of the most influential books shaping the views of the next generation of Chinese military leaders. The book even received the official blessing of a highlighted review in the Communist Party youth league's newspaper.

*Unrestricted Warfare* is essentially a strategic guidebook to twenty-first-century war. Its focus is how countries like China might defeat the United States in a war of high technology, despite the apparent American lead in weapons. What is notable is that the Chinese officers don't just focus on seeking out American vulnerabilities and widening the scope of conflict, the sort of "asymmetric" approach to war that many think is the only way to defeat the United States. They also argue that foes will be able to defeat America at its own high-technology game.

Qiao and Wang argue that America suffers from an odd combination of being uniquely addicted to technology, but also unable to truly exploit it. "However, this is not a strong point of the Americans, who are slaves to technology in their thinking. The Americans invariably halt their thinking at the boundary where technology has not yet reached." Moreover, they go on to describe how the United States may be ahead now, but this will not last for long. "Technology is useful, however, because Americans do not do a good job of anticipating technology trends."

Part of this confidence comes from the fact that Qiao and Wang are great believers in the imminence of an RMA, but they see the key elements of it emerging from the commercial sector, where China is surging forward. "The new concept of weapons will cause ordinary people and military men alike to be greatly astonished at the fact that commonplace things that are close to them can also become weapons with which to engage in war." They go on to add, "We believe that some morning people will awake to discover with surprise that quite a few gentle and kind things have begun to have offensive and lethal characteristics."

At face value, the Chinese officers' prediction seems off base. After all, the American advantage in war technologies doesn't just stem from its massive defense budget. From Thomas Edison to Bill Gates, it has traditionally been the home of commercial innovation and invention. Even in this latest revolution, Americans invented key enablers like fiber optics and the Internet. And why should this trend not continue? While the United States only has 4 percent of the world's population, it spends almost 50 percent of the world's R&D funding.

Yet these Chinese army officers aren't alone in predicting America's loss of its advantages in this arena. Indeed, the U.S. Navy agrees with them. In 2006, the navy's official journal published a warning that "the United States is headed for the 'perfect storm' when it comes to how it deals with defense technology. Only if changes are made now, can the U.S. avoid the loss of its technological superiority."

One of the major challenges to America's success in a world of high technology is that the same education system that once took its military and economy to the top is now falling behind. Only 54 percent of America's high school students perform at even a basic level in math and science. And these are by American standards. When matched against international students, American high school students came in twenty-second in the world in basic math and science and twenty-fourth when they had to apply their skills to real-world problems.

Norman Augustine is a former chair of the National Academies, the U.S.'s official science advisory organization, as well as a former CEO in the defense aerospace industry. As he explains, it isn't that American kids are dumb. Rather, our education system is making them dumber. "The longer students are exposed to our K-12 education system, the worse they do—particularly in the critical areas of math and science." Indeed, while U.S. fourth graders come in at the top eightieth percentile in the world in science, by the time they reach the twelfth grade they have fallen to the bottom fifth percentile. To paraphrase the failed Bush education reform policy, which worsened the problem by emphasizing rote memorization, nearly every American child is being left behind. As Bill Gates puts it, "When I compare our high schools to what I see when I'm traveling abroad, I'm terrified for our workforce of tomorrow."

The traditional retort to rising worries about America's education system is that while our high schools may suck, we have great universities. Unfortunately, when it comes to math and science skills, so key to designing, building, and using new technologies, this may no longer be the case. These high schools feed fewer kids with either skills or interests in science and math into U.S. universities. The universities are then graduating fewer and fewer.

This is starting to create a "futile cycle," in the words of Princeton University president Shirley Tilghman. There are fewer and fewer American teachers and professors with science and mathematics skills to inspire, supervise, and mentor the next generation of American engineers and inventors. These problems at the university level then feed back into high schools, which rounds the futile cycle. Erskine Bowles, president of the University of North Carolina system (which has 183,000 students at its various campuses), put it this way in 2006. "In the past four years, our 15 schools of education at the University of North Carolina turned out a grand total of three physics teachers. Three."

In the past, America made up for such a gap by hosting foreign students and researchers in its universities, who would then frequently stay in the United States for the long term. New post-9/11 visa policies are making it harder for these visitors to both come and stay. Those foreign researchers who do come are more frequently returning to much better job prospects back home. The impact is being severely felt on "the vitality and quality of the U.S. research enterprise," stated National Academy of Sciences president Bruce Alberts. "This research, in turn, underlies national security and the health and welfare of both our economy and society." With American scientists not being replenished in sufficient numbers, some worry the whole system could fall behind. As the National Science Board warned, "If action is not taken to change these trends, we could reach 2020 and find that the ability of U.S. research and education institutions to regenerate has been damaged and that their preeminence has been lost to other areas of the world."

The globalization of the world economy is also hammering the U.S. technology establishment. While American workers remain talented, they also are comparatively expensive. In Vietnam, twenty assembly-line workers can be hired for the price of one in the United States. In India, six engineers earn the equivalent of one in the United States. And in China, five chemists can be employed for the salary of one in the United States.

These pay gaps are made even worse by a U.S. health care system that acts like a massive anchor attached to American industry. General Motors, for example, was once the epitome of American industrial might in peace and war. During World War II, its automobile plants were converted to manufacture tens of thousands of tanks, trucks, and planes. Today, it has junk bond status and had to reduce its U.S. workforce by a third. The reason is not just that GM too long expected to sell ugly fuel-guzzlers, but also that it spends more on health care than it does for the steel that goes into its cars. Even a seemingly successful American firm like Starbucks has to spend more on health care than it does on coffee.

It is no surprise then that companies, even the most technological, are outsourcing their business outside the United States. The result is a hammer blow to U.S. technology development and manufacturing, especially in the commercial sector that Qiao and Wang describe as so important to taking full advantage of this RMA. America's trade balance in high-tech goods and services went from a positive $50 billion in 1996 to negative $50 billion in 2006, while only three out of the top ten companies granted patents for new products and inventions were American. And it bodes to get worse. More than three-fourths of the new R&D facilities planned worldwide will be located in either China or India.

With the huge amount of "civilian off-the-shelf" technologies used in military robotics, these trends actually create a massive dependence on foreign manufacturers to supply America's next generation of weapons. This dependence has many worried beyond lost market share. Technology security expert Richard Clarke is concerned that the U.S.'s complete reliance on technology made elsewhere makes it far easier for foes to hack or hijack systems, including being able to slip "back doors" in. "There is massive industrial espionage.... China already has the ability to lace technology it is building for us with Trojan horses and time bombs. Most if not all the computer systems running the Internet, phones, power grid, and robots were built in China."

In turn, others note that the location of manufacturing elsewhere makes it easier for competitors to copy and build their own cloned systems. iRobot engineers tell how they have already seen cloned copies of both their Roomba vacuum cleaner and the PackBot military robot. Indeed, they once angrily confronted a group of Singaporean military officers who were showing off what appeared to be a clone of a PackBot at a demonstration. Stayne Hoff similarly says a good sign a buyer just wants to clone a drone is "when they only want to buy one."

The sum total of these education and economic trends is moving the U.S. security system in a scary direction, warns Rusty Miller of the defense firm General Dynamics. "If the U.S. doesn't wake up and pay attention, we're going to get smoked."

## MONEYBALL AND THE CULTURE WARS

William "Billy" Beane was a first-round pick by the New York Mets in the 1980 baseball draft. Beane's career, however, didn't take off the way either he or the Mets planned. He played only in 148 games as a reserve outfielder, hitting just three home runs.

Off the field, Beane met with far more success, and in 1997 he became general manager of the Oakland A's. The A's soon became a perennial playoff team, despite the fact that they came from a small market and couldn't afford a large player payroll. In 2006, for example, the A's ranked twenty-first out of the thirty baseball teams in salaries, but had the fifth best record. In essence, Beane's team paid only a fourth of what big-money teams like the New York Yankees had to pay for each win.

The secret to Beane's success is that he refused to let baseball's culture and traditions get in the way of how he did business. Other teams still selected players based on popular measures that came out of the nineteenth century (typically using only very basic statistics like stolen bases, RBIs, and batting average). Beane and his team of evaluators used a modern, technical method of evaluation, called "sabermetrics" (or the "Moneyball way," after Beane was profiled in a book entitled *Moneyball: The Art of Winning an Unfair Game*). For example, even though it went completely against the conventional wisdom of baseball, the mathematical data showed that avoiding an out has far more impact on a team's chances of winning than getting a hit. Despite the proof, others wouldn't change. Beane's success came not just from his willingness to eschew the traditional ways of doing business, but also from how, despite all the data to the contrary, his competitors continued to cling to the old ways and old baseball culture, even if it meant fewer wins for their teams and could ultimately cost them their jobs.

Beane's experience illustrates how, even in the most competitive marketplaces, new ideas still have trouble supplanting old doctrines. This especially happens with new technologies. Just because something new and better is discovered doesn't always mean it is adopted. For instance, I typed this book out on a keyboard laid out in the traditional QWERTY manner, which 99 percent of the computers in the world use. Yet this layout actually dates back to 1873, when it was first developed to make typists go *slower*, so as not to jam their mechanical typewriters. In the time since, numerous new keyboard layouts have been invented that would speed typing by as much as 95 percent. Yet companies and customers alike resist them, as QWERTY is the way it has always been, even if it is nowhere near the best.

War is certainly a far different beast than typing or baseball (other than when a Red Sox fan shows up in the bleachers at Yankee Stadium), but the military is also a highly competitive field that can still be quite resistant to change. Indeed, as one British colonel put it, "In no profession is the dread of innovation so great as in the army." So, ironically, even though militaries often generate great change, they have trouble adjusting to it.

Throughout history, even the most brilliant military minds have often failed to adapt well to new technologies. Napoleon may have conquered most of Europe, but he turned down Robert Fulton's offer to make France both submarines and steamships. At the very start of the American Civil War, the Union army was offered the breech-loading repeater rifle, which could fire seven shots quickly instead of just one. But its makers couldn't even get a hearing, let alone a sale; it wasn't until President Lincoln himself tried out the weapon that the rifles were bought, years into the war, and then only for cavalry. The same thing happened with machine guns. Americans like Richard Gatling and Hiram Maxim may have invented the rapid-firing gun that would revolutionize warfare, but officers in the U.S. Army at the time refused to use them. Indeed, Custer could have had four Gatling guns with him at the Battle of Little Bighorn, which would have mowed down the Indians at his "Last Stand." Instead, Custer left them behind at the base as he felt machine guns had no value in combat and would only slow him down.

Militaries resist change, even when it might help them win wars, for many reasons. The experience of combat is unique, so the latest generation tends to feel a special kinship with the generations before it and doesn't want to veer too far from what they did in the past. For instance, the ancient Greeks so honored the ideals of war that Homer wrote about in the *Iliad* that they shunned the use of technologies like siege engines. If it wasn't good enough for their heroes like Achilles or Ulysses, then it wasn't good enough for them.

Change can also become wrapped up in turf battles and other bureaucratic intransigence. Those vested in the current system, or whose talents and training might become outdated by new technologies, will fight any change that threatens to make them obsolete or out of work, or in any way harms their prestige.

Most important, the stakes are so high in war that militaries place an immense value on going into battle with something that has already proven its worth in the past. When the U.S. Army began to talk about replacing horses with tanks just prior to World War II, cavalry officers argued that horses had four thousand years of experience at war, while tanks had only a few years at the end of World War I. As late as 1938, General Hamilton Hawkins lamented the "foolish and unjustified discarding of horses" and blamed the "sheep-like rush to mechanization and motorization without clear thinking or any apparent ability to visualize what takes place on the field of maneuver or the battlefield." Even with mechanized vehicles clearly proving themselves in World War II, the U.S. Army didn't dissolve its last horse unit until three years into the war.

Many think that the same sort of cultural resistance to change may hamper U.S. military adaptation to unmanned systems, even if it is one of the early originators of the technology. Dr. Russ Richards is the director of the Project Alpha program on military unmanned systems at the Joint Forces Command. "The greatest hurdle," he says, "is likely to be overcoming military culture."

The many delays that occurred in the use of drones are a prime example of how military culture is perhaps weighing in against the curve. Andrew Krepinevich, a former Defense Department analyst who is now executive director of the Center for Strategic and Budgetary Assessments, jokes that the reason the air force resisted systems like the unmanned fighter plane is that "no fighter pilot is ever going to pick up a girl at a bar by saying he flies a U.A.V....Fighter pilots don't want to be replaced." The same goes even for pilots beyond fighter aces. One A-10 Warthog pilot, a veteran of Iraq and Afghanistan, said fliers' biggest fear—being shot down—has been replaced by a fear of being ordered to fly a drone from a ground-based cubicle. "It's like being a pilot for nerds. Where is the sense of adventure, the sense of danger?...Let's put it this way: I don't think they're going to make any movies about guys who fly Predators."

These are jokes, of course, but they have a real underpinning to them. The U.S. Air Force's professional identity is very much wrapped up in the idea of piloting planes, and fighter planes at that. Indeed, over half of the air force's generals are fighter pilots, as has been every single air force chief of staff but one since 1982. So being a fighter pilot is not just in the air force leadership's organizational DNA, it is also seen as the pathway to advancing in the ranks. Given this, it is no surprise then that the air force long stymied the development and use of drones, letting DARPA and the intelligence agencies take the lead instead.

Even once the air force started to buy and use drones (largely because of the competition from these other agencies), this sort of cultural resistance has played out in very real organizational actions. The early Predator pilots in the air force, for instance, were paid less than regular pilots, didn't get any credit in their career advancement for their flight hours, and were otherwise generally shunned. As one air force helicopter pilot joked, "I was happy when drones came in. It meant that we were no longer at the bottom of the totem pole."

While this attitude has slowly changed as drones have proven their worth in combat, the air force still holds on dearly to its identity as a force of fighter aces dogfighting against enemy fighter planes in the sky, despite the fact that it hasn't happened for years. "Today's Air Force clings to a fight-the-Soviets (or at least the Chinese) model with greater passion than yesteryear's Army clung to the horse cavalry," concluded one military analyst. One young air force officer, just months out of the academy, tells how, despite the fact that drone pilots have seen far more combat action than jet fighter pilots over the last decade, "It's seen as this geeky thing to do."

The result is that the force will still sometimes put pilots' career interests ahead of military efficiency, especially when those making the decisions are fighter jocks themselves. For example, many believe that the air force canceled its combat drone, Boeing's X-45, before it could even be tested, in order to keep it from competing with its manned fighter jet of the future, the Joint Strike Fighter (JSF, a program now $38 billion over its original budget, and twenty-seven months past its schedule). One designer recalls, "The reason that was given was that we were expected to be simply too good in key areas and that we would have caused massive disruption to the efforts to 'keep...JSF sold.' If we had flown and things like survivability had been evenly assessed on a small scale and Congress had gotten ahold of the data, JSF would have been in serious trouble."

Military cultural resistance also jibes with problems of technological "lock-in." This is where change is resisted because of the costs sunk in the old technology, such as the large investment in infrastructure supporting it. Lock-in, for example, is why so many corporate and political interests are fighting the shift away from gas-guzzling cars.

This mix of organizational culture and past investment is why militaries will go to great lengths to keep their old systems relevant and old institutions intact. Cavalry forces were so desperate to keep horses relevant when machine guns and engines entered twentieth-century warfare that they even tried out "battle chariots," which were basically machine guns mounted on the kind of chariots once used by ancient armies. Today's equivalent is the development of a two-seat version of the Air Force's F-22 Raptor (which costs some $360 million per plane, when you count the research and development). A sell of the idea described how the copilot is there to supervise an accompanying UAV that would be sent to strike guarded targets and engage enemy planes in any dogfights, as the drone could "perform high-speed aerobatics that would render a human pilot unconscious." It's an interesting concept, but it begs the question of what the human fighter pilot would do.

Akin to the baseball managers who couldn't adapt to change like Billy Beane, such cultural resistance may prove another reason why the U.S. military could fall behind others in future wars, despite its massive investments in technologies. As General Eric Shinseki, the former U.S. Army chief of staff, once admonished his own service, "If you dislike change, you're going to dislike irrelevance even more." It is not a good sign then that the last time Shinseki made such a warning against the general opinion—that the invasion of Iraq would be costly—he was summarily fired by then secretary of defense Rumsfeld.

## BIGGER IS NOT ALWAYS BETTER:
## THE DEFENSE-INDUSTRIAL COMPLEX

On the cover of *Life* magazine in April 1957 is a picture of "The Flying Blue Brothers." It shows two smiling brothers, blond and buzz-cut, sitting in the cockpit of a tiny propeller plane. The article inside tells the tale of Neal, twenty-one, and Linden, twenty, two brothers who had taken time off from Yale University to pilot their Piper Tri-Pacer alone across the Andes. Their adventures included "cavorting with headhunters in the Amazon, trying to right their crashed plane on a mountain ice shelf, and later, lounging on Ipanema Beach with a comely brunette."

The Blue brothers (no relation to the Jake and Elwood of *The Blues Brothers* fame) went on to lead an equally colorful business career, running a cocoa-and-banana plantation in Nicaragua, and then investing in an assortment of companies that included a German streetcar manufacturer, natural-gas wells in Canada, and ranchland just outside the Telluride, Colorado, ski resort. In 1986, they bought General Atomics, a nuclear-power research company, from Chevron for $50 million. Around the same time, a small company called Leading Systems built a prototype of an unmanned drone that could fly great distances for long periods of time. They called it Amber. The Pentagon had no interest in UAVs and so the company went out of business in 1990. The Blues' firm, General Atomics, bought up the assets of the failed company, including Amber.

Despite the fact that the drone had no buyers, the Blues and General Atomics believed in the technology. The company renamed the Amber drone and began production even though there was no set buyer. In a sense, General Atomics took the *Field of Dreams* approach to defense contracting that iRobot did with UGVs: "If you build it, they will come." The CIA soon came shopping and the drones, now called by the more fearsome-sounding "Predator," saw action in the Balkans. And the rest is robot history.

The story of the Blues and General Atomics is a classic story of how an industry upstart can shake up the system. This small-company approach to contracting carries over to other parts of General Atomics. The company is headquartered in an office district just outside San Diego. There, it builds Predators at a pace of almost fifty a month. "It's like a California speed shop where they hand-build hot rods," says Glenn Buchan, an analyst at the RAND defense research group.

General Atomics can assemble such sophisticated weapons systems so quickly because it places great value on simplicity. The drones' bodies are made of a honeycomb of graphite, paper, and other materials and then literally baked in an oven. Propeller-powered engines may not have been sexy, but the early-model Predator drones used props because they were more efficient and cost less. As *BusinessWeek* wrote of the Blues' success, "The development of the smaller, cheaper plane shows how even in an age of $300 billion Pentagon budgets [note: now double that], nimble entrepreneurs can shake up the Establishment."

The challenge for the United States is that stories like that of the Blues and Predator, where smart, innovative systems are designed at low costs, are all too rare. The U.S. military is by far the biggest designer and purchaser of weapons in the world. But it is also the most inefficient. As David Walker, the head of the Government Accountability Office (GAO), puts it, "We're number 1 in the world in military capabilities. But on the business side, the Defense Department gets a D-minus, giving them the benefit of the doubt. If they were a business, they wouldn't be in business."

The Department of Justice once found that as much as 5 percent of the government's annual budget is lost to old-fashioned fraud and theft, most of it in the defense realm. This is not helped by the fact that the Pentagon's own rules and laws for how it should buy weapons are "routinely broken," as one report in *Defense News* put it. One 2007 study of 131 Pentagon purchases found that 117 did not meet federal regulation standards. The Pentagon's own inspector general also reported that not one person had been fired or otherwise held accountable for these violations.

This lumbering process is also heavily undermined by being "hierarchical and top down," as one former army colonel, who now runs a robotics firm, put it. The Pentagon will almost always invest in systems that have bureaucratic and political champions, but not always those that are most efficient or that the troops in the field are finding most useful. One striking example is how the army's massive FCS program originally didn't include the smaller types of robotics, the very types that soldiers were requesting to have in the field.

There is also a Pentagon phenomenon known as "requirements creep." The decision on what to buy and the requirements of what must go into the systems are too frequently made by those least familiar with new technology. Bruce Jette, who has been the point man inside much of the U.S. Army's robotics efforts, likens the current process to how horse cavalry officers were the ones who helped decide the required specifications for the early military automobiles. They originally demanded that the cars come with saddle seats and reins. Some ninety years later, the Pentagon's acquisition office once mandated that small ground robots come equipped with an onboard fire extinguisher, oil change, and trailer hitch. Jette points out, "The thing is 30 pounds and electric!"

Whenever any new weapon is contemplated, the military often adds wave after wave of new requirements, gradually creeping the original concept outward. It builds in new design mandates, asks for various improvements and additions, forgetting that each new addition means another delay in delivery (and for robots, at least, forgetting that the systems were meant to be expendable). In turn, the makers are often only too happy to go along with what transforms into a process of gold-plating, as adding more bells, more whistles, and more design time means more money. These sorts of problems are rife in U.S. military robotics today. The MDARS (Mobile Detection Assessment Response System) is a golf-cart-sized robot that was planned as a cheap sentry at Pentagon warehouses and bases. It is now fifty times more expensive than originally projected. The air force's unmanned bomber design is already projecting out at more than $2 billion a plane, roughly three times the original $737 million cost of the B-2 bomber it is to replace.

These costs weigh not just in dollars and cents. The more expensive the systems are, the fewer can be bought. The U.S. military becomes more heavily invested in those limited numbers of systems, and becomes less likely to change course and develop or buy alternative systems, even if they turn out to be better. The costs also change what doctrines can be used in battle, as the smaller number makes the military less likely to endanger systems in risky operations. Many worry this is defeating the whole purpose of unmanned systems. "We become prisoners of our very expensive purchases," explains Ralph Peters. He worries that the United States might potentially lose some future war because of what he calls "quantitative incompetence." Norm Augustine even jokes, all too seriously, that if the present trend continues, "In the year 2054, the entire defense budget will purchase just one tactical aircraft. This aircraft will have to be shared by the Air Force and Navy, three and one half days per week, except for the leap year, when it will be made available to the Marines for the extra day."

Closely linked is a "bigger is better" mentality that has taken hold in American defense contracting. As Pierre Chao at the Center for Strategic and International Studies explains, it would be a "strategic mistake" not to have a massive amount of competition in the military robotics marketplace. "If you think it is a young technology, that the Orville and Wilbur Wrights of the 21st century are running around in the UAV marketplace, then as messy as it makes the environment, it is far more strategically important to have lots of players, different patrons behind those players, and to keep stimulating the useful competition of ideas."

And yet U.S. military acquisitions, even in the field of robotics, are increasingly dominated by an ever smaller number of huge defense contractors, driving down competition. From 1986 to 2006, for example, the number of Pentagon prime contractors that could compete on major programs went from twenty to six. The result? "Only the dinosaurs were allowed" to bid on such major programs as the army's FCS, laments one robotics firm executive.

These major defense firms do very well for their shareholders, beating the S&P 500 in six of the last ten years. But besides limiting competition, the bigger companies tend to bring a risk-averse approach to the business side of war. In its planning, General Atomics tries to take a twenty-year look into the future. As Neal Blue puts it, "The future belongs to those people who will be thinking out of the box and delivering systems based on the technologies of the future." By contrast, the old-school firms typically wait to be called upon, rather than pushing forward new ideas for the future. When I interviewed an executive at one of the largest U.S. defense firms about how his company strategized about which new military technologies to research and develop, based on their sense of the various changes in war and technology, he replied that they didn't. "We just work on what the Pentagon tells us." The big firms think less like *Field of Dreams* and more like *Waiting for Godot*.

This passive mentality also makes them less attractive destinations for the brightest scientists and engineers. The megadefense firms find it tough to compete with the Silicon Valley trendsetters (who, because of their lack of lobbying efforts, rarely do well in Pentagon competitions) in terms of prestige and pay scale. Even among scientists who want to do defense work, the bigger firms are seen as offering less freedom to experiment and innovate.

The bigger firms tend not to be at the cutting edge of change, but they make up for it by wielding far more influence in the halls of Congress and the Pentagon, which gives them greater power to exact costs, even when they fail at the job. Cost overruns happen in any business, but in defense contracting it has become the norm. In 2008, the GAO found that the Pentagon's major weapons acquisition programs were a combined $295 billion over budget and behind schedule by an average of twenty-one months. Yet even when their projects fall behind, most major contractors still get their performance bonuses, because it is viewed as career suicide to cross them. The F-22, for example, came in at close to triple its original price, but 91 percent of the performance bonus, about $850 million, was paid out to its makers.

The "bigger is better" mentality is not just about the influence of the largest firms. "Larger companies trend towards larger vehicles with all the bells and whistles," explains one robotics firm executive, who had previously worked with one of the major defense contractors. The reason is not just one of traditional gold-plating and requirements creep, but also financial margins. He recounts submitting an affordable military ground robot design to his bosses. Instead of being praised, he was told that "the profit margin is just too small for a sub $1 million vehicle."

Because it was perceived as too small to be worth selling, he was told either to fig-ure out how to make it bigger (and thus increase the profit margins), or to "load on million-dollar sensors" that the firm had already developed for other weapons. This kind of thinking similarly led the UCAS to evolve from a small, quick, and disposable attack drone into its current $43 million design the size of a bus.

Such a skewed industry of war could prove to be America's undoing in the future of war. Sums up retired marine Bing West, "There is no comparison to how we do things so irrationally."

## FIGHT THE FUTURE

History tells us that only rarely can a nation stay ahead in an RMA. For the United States in the robotics revolution, the challenges include the many other nations proving to be just as savvy in these new technologies, an education and economic system that threatens to sap its competitiveness, potential resistance to change within its military culture, and a balky defense-industrial complex.

History need not repeat itself, however. As one military journal put it, the United States has definitely made its mistakes, but has ultimately been "more often smart than stupid." It is also the same country that produced people like Stayne Hoff, Dave Sonntag, Billy Beane, and the Blue brothers.

For all the various factors that may challenge the United States in this revolu-tion of technologies, it is also in the traditions of America, and its military, to be flexible and experiment with change. Before World War II, for example, the U.S. Navy built a number of different classes of aircraft carriers, as it didn't know which type would be best for the new technology of warplanes at sea. By contrast, the British navy tried out only one class, which unfortunately for them proved wrong. A return to this American tradition of experiments and design contests also will rebuild competition in the U.S. defense-industrial space. Indeed, the recent U.S. difficulties in Iraq and Afghanistan may act to dispel conceit and help overcome resistance to any needed changes. The parallel here is how the Boston Red Sox, the definitive big-market, tradition-bound team, eventually got tired of losing and decided to copy Billy Beane's approach of following a new path. A year later, they won their first World Series in eighty-six years.

Likewise, America may well be a nation uniquely fascinated with technology, as the optimists claim, but they undervalue that this stems from the traditional importance America has placed on education and learning. Scientists turned founding fathers like Thomas Jefferson and Ben Franklin thus would agree with a lesson that futurist Arie de Geus has for countries today. "The ability to learn faster than your competition may be the only sustainable competitive advantage." The U.S. education system may now be "left behind," but it is not a permanent lost cause. It revitalized itself after Sputnik and can do so once again. In turn, there is nothing to prevent the U.S. military itself, and especially its system of professional education and research centers, from being what change-management expert Peter Senge called a "learning organization," open to new ideas, including even the thinking of others. This is how you stay ahead, especially in a revolution. "While learning from experience is good, learning from others' experience is even better," says General James Mattis, now in charge of developing many of the new American concepts of war at the U.S. Joint Forces Command.

Most of all, whether the United States avoids a repeat of so many other nations' leader-to-loser experience will depend on whether it eschews the arrogance that dogged most past losers. It must recognize that change is afoot, and not merely one that will only be to America's benefit.

# THIRTEEN

## OPEN-SOURCE WARFARE:
## COLLEGE KIDS, TERRORISTS, AND
## OTHER NEW USERS OF ROBOTS AT WAR

*If I can imagine it, what would a totally dedicated, well-educated indi-
vidual do, especially if they have a Timothy McVeigh personality?*

—GREG BEAR

In the summer of 2005, Sam Bell set out to buy a military-grade robotic drone.
As a subsequent article about his experience described, "It was an unusual shopping expedition for a private citizen, much less a 22-year-old only a few months
removed from his political science and philosophy studies at Swarthmore College.
But, ever since graduation, and even while in school, Bell had been working to do
what the U.S. government and the United Nations had so far failed to: stop the
genocide in Darfur."

Bell got into the military robotics business after he and two other Swarthmore
students, Mark Hanis and Andrew Sniderman (whose previous college activities
included running for student council and playing for the school golf team), decided
they wanted to do something to help in Darfur. They formed a group called the
Genocide Intervention Network (GI-Net), whose goal was to bring attention to the
ongoing killings in Darfur and help raise money for the undermanned and under-
funded peacekeeping force deployed there.

That a bunch of students could raise money for a military force was a seemingly absurd concept, but the idea caught on. Within a year, GI-Net had raised almost half a million dollars from individual donations, as well as the proceeds from charity events like a movie showing of *Hotel Rwanda* and a "battle of the bands." The problem was that the students didn't know how to spend the money. The African governments that actually had sent peacekeepers to Darfur wouldn't just take it, and in any event, the students were also worried about it being misspent.

Operating from his dorm room, Sniderman then e-mailed over a hundred private military firms. He asked if they were willing to be hired out by the students to send troops to Sudan. He also made sure to change his voice-mail greeting to "something more serious and somber...something you'd want your business partners to hear." As he later recounted, "Within 36 hours, I got dozens of replies. Most were saying, 'We've never done anything like that, but we'd love to work with you.'"

The idea of college students hiring a private army didn't sit well with many of their funders (plus, the Sudanese government wouldn't have allowed the firms in anyway), so the group hit upon another way of hiring a military force. They could rent a private military firm's drone, which could monitor the refugee camps in Darfur from the air and report on any attacks on civilians.

Sam Bell then put on his only suit and went to the Washington offices of Evergreen International, an aviation contractor. The firm's executives were enthusiastic about the business opportunity and described a plan for leasing the students four new UAVs, which would be remotely operated from back in the United States but fly over Darfur. It seemed a great idea until Evergreen presented Bell with the price tag: $22 million a year. The college student then asked "if the firm had any options for shoppers on a tighter budget." Evergreen then offered an older-model UAV for far less.

Fortunately for GI-Net, the students ran the idea past an actual expert. "He said, one, a sandstorm could knock the UAV out; two, it could get shot down; and three, if either of those things happened, the Sudanese government could get a hold of it and take hold of its technology," Hanis explains. "So it turned out the UAV wasn't such a good idea."

Ultimately, the students at GI-Net ended up using their funds not on renting military robots from a private firm, but on aid to refugee families in Darfur. But this strange episode illustrates just how vast the changes are in who can now gain access to advanced military technologies such as robots. Warfare is going "open source."

## HYBRID WAR

Historian Max Boot observes that "technology is both the great separator and the great equalizer in military affairs." The United States may be the most powerful nation-state in history, largely because of its technology. And yet this superpower hasn't always been able to turn that power into victories. Instead, groups that aren't even states have been able to frustrate and flummox it by using low-cost, low-tech weapons like car bombs or IEDs. Even more, such groups can utilize many of the same high technologies that a superpower like the United States spent billions of dollars to develop. The investments to create satellite navigation and the Internet may have originally come out of DARPA, but now any terrorist group can pinpoint targets with GPS devices they buy off Amazon.com.

Perhaps the best illustration of what these changes bring to conflict is Hezbollah, one of the most innovative groups at war today. A largely Shiite organization, it began in 1985 as a radical religious movement in Lebanon. The group has since morphed into a multitude of identities and forms. It is simultaneously a paramilitary organization (able to mobilize as many as ten thousand fighters), political party (holding fourteen seats in the parliament), media conglomerate (operating its own TV, radio, and Internet networks), and development and aid organization (funding its own system of hospitals, clinics, and schools, as well as a welfare program for much of south Lebanon).

In the summer of 2006, Hezbollah militants kidnapped two Israeli soldiers. Frustrated by the rise of the group, the Israeli military launched a massive retaliatory strike, designed both to teach Hezbollah a lesson and force it to return the kidnapped troops. It didn't seem an even fight. One side was a state, which had the most advanced and professional military in the region and had never lost a war. The other wasn't even a state and the financial resources it possessed to spend on weapons and troops (from its own fund-raising and aid from Iran) were only 1 percent of Israel's defense budget.

"There are many who belittled the enemy as primitive," said Major General Udi Shani, director of the intelligence branch in the Israel Defense Forces (IDF). Yet Israel soon found itself unable to defeat Hezbollah, the first time the nation's military had failed to smash an Arab foe in its history. By the time that the thirty-four-day war ended with a cease-fire, Israel had lost over 120 troops killed and nearly 500 wounded, as well as another 43 civilians killed and 4,262 injured by the rockets and missiles that Hezbollah fired into Israeli cities. Israel was unable to get back its soldiers and its military chief of staff was forced to resign. By contrast, Hezbollah held a parade at which more than a million supporters showed up to cheer what its leader called "a divine and strategic victory."

The tiny nonstate actor was able to accomplish what the combined militaries of all the Arab states (which had soundly lost to Israel in 1948, 1967, and again in 1973) couldn't by figuring out how to fight what U.S. Marine General James Mattis has called a "hybrid war." Hezbollah is quite amorphous in its structure. It blends political, religious, economic, and military power. When it came time for actual combat, it could spread its fighters out into decentralized units, which could swarm around in attack, but disperse and disappear whenever Israel's military tried to pin them down in the field. Most important, the group was able to meld classic guerrilla tactics, conventional-war savvy, and the latest high technology.

Israel may have been one of the first states to develop and use drones in war, but that didn't prevent it from becoming the first state to be attacked by nonstate drones. While Israel flew scores of drones in its attacks over Lebanon, Hezbollah also flew at least three Mirsad (Arabic for "ambush") drones into Israel, each carrying a payload of about twenty-two pounds of explosives, packed with ball bearings to make them even more deadly.

While Israeli jets and drones circled above Lebanon looking for targets to strike, hidden Hezbollah rockets, many of which were fired either by remote or automated timer controls, rained down on Israeli cities. Frustrated, the Israelis then launched their ground forces into southern Lebanon. The thinking was that if they couldn't destroy the rockets from the air, they could control the territory from which the rockets were being fired, pushing the threat beyond range of Israeli cities.

Here too, the nonstate force proved stunningly innovative. According to Israeli media reports, not only was Hezbollah "able to hack into the Israeli Army's computer systems prior to the attack," but it also cracked into the army's radio systems (which are similar to the ones used by U.S. soldiers). Notably, the group's Internet attacks on Israel originally appeared to come from a small South Texas cable company, a suburban Virginia cable provider, and Web-hosting servers in Delhi, Montreal, Brooklyn, and New Jersey. But these all had actually been "hijacked" by Hezbollah hackers. Described an article on the strategy, "In the cyberterrorism trade it is known as 'whack-a-mole'—just like the old carnival game, Hezbollah sites pop up, get whacked down and then pop up again somewhere else on the World Wide Web." The group even infiltrated the Israeli cell phone network, eavesdropping on phone calls made to home by Israel's military commanders and soldiers in the field, to get their radio code names and other personal information. As a report after the war noted, "The intelligence data provided Hezbollah fighters with critical tactical information about the intentions, status and whereabouts of Israeli ground forces." Armed with this information, the Hezbollah fighters were able to stymie the Israeli attacking forces.

Hezbollah showed that nonstate actors not only can figure out asymmetric strategies to nullify a state's massive advantage of force and size, but can even beat states at their own high-technology game. As retired U.S. Army officer Ralph Peters commented, "All contempt for terrorists set aside, we need to recognize that Hezbollah has prepared itself better for a war against military superiority than any other military organization of our time.... If David didn't kill Goliath this time, he certainly gave the big guy a headache."

## NO STATE? NO SHOES? NO PROBLEM

The use of unmanned systems isn't just limited to large-scale organizations like Hezbollah, which, while not a state, certainly controls a sizable chunk of real estate. The infamous private military firm Blackwater, for example, added an unmanned section to its business in 2007, seeking to rent out drones and even unmanned blimps for reconnaissance and surveillance jobs. Indeed, one U.S. Special Forces soldier expected a growing "corporate use" of unmanned systems by private military and corporate intelligence-gathering firms, even coining the term "robot mercenaries" for it. In turn, many humanitarian groups have talked about actually following through on the plans first hatched by the Swarthmore students and "getting our own UAVs," as one executive at a human rights organization put it.

Perhaps the best illustration of how the bar is being lowered for groups seeking to develop or use such sophisticated systems comes in the form of "Team Gray," one of the competitors in the 2005 DARPA Grand Challenge. Gray Insurance is a family-owned insurance company from Metairie, Louisiana, just outside New Orleans. As Eric Gray, who owns the firm along with his brother and dad, explained, the firm's entry into robotics came on a lark. "I read an article in *Popular Science* about last year's race and then threw the magazine in the back of my office. Later on, my brother came over and read the article, and he yelled over to me, 'Hey did you read about this race?' And I said, 'Yeah,' and he said, 'You wanna try it?' And I said, 'Yeah, heck, let's give it a try.'"

The Grays didn't have PhDs in robotics, billion-dollar military labs backing them, or even much familiarity with computers. Instead, they brought in the head of their insurance company's ten-person IT department for guidance on what to do. He then went out and bought some of the various parts and components described in the magazine article. They got their ruggedized computer, for example, at a boat show. The Grays then began reading up on video game programming, thinking that programming a robot car to drive through the real-world course had many parallels with "navigating an animated monster through a virtual world." Everything was loaded into a Ford Escape Hybrid SUV, which they called Kat 5, after the category 5 Hurricane Katrina that hit their hometown just a few months before the race.

When it came time for the race to see who could design the best future automated military vehicle, Team Gray's entry lined up beside robots made by some of the world's most prestigious universities and companies. Kat 5 then not only finished the racecourse (recall that no robot contestant had even been able to go more than a few miles the year before), but came in fourth out of the 195 contestants, just thirty-seven minutes behind Sebastian Thrun's Stanley robot. Said Eric Gray, who spent only $650,000 to make a robot that the Pentagon and nearly every top research university had been unable to build just a year before, "It's a beautiful thing when people are ignorant that something is impossible."

Kids in dorm rooms, militant Middle Eastern groups, and insurance companies don't seem to have much in common, but they are all part of a much larger phenomenon, the early stages of a new global redistribution of power. One of the factors that led to the rise of the nation-state in past centuries was its ability to mobilize and organize mass numbers of soldiers that it could use to beat down the other forms of government (dukedoms, city-states, tribes, and so on). The need to support such a standing army in turn led to the need to create a state government's bureaucracy and tax structure. As the historian Charles Tilly famously said about the rise of states, "War made the state and the state made war."

If the past RMAs were associated with helping to spur on the centralized forms of state governments (such as how the gunpowder revolution helped lead to the rise of colonial empires), this one is taking place in a period where power is becoming decentralized, flatter, and increasingly nongovernmental. Today, nonstate actors are increasingly the ones with power, resources, and decision-making authority. In economic and trade matters, the some sixty thousand multinational companies in the world today clearly wield economic influence beyond the control of any one state. Likewise, individual donors have proven to be far more influential in fighting global diseases like AIDS than governments. And the militaries and police of many nations, such as Lebanon where Hezbollah is based, actually have less control over what goes on within their borders than the various paramilitary and warlord groups. Security analyst John Robb calls this the rise of "open-source warfare." Much like an open-source software code like Linux is available to anyone and everyone to use and improve upon, so too is conflict being opened up to any organization with the will to go to war and a deadly entrepreneurial spirit.

The growing use of unmanned systems is thus wrapped within a larger political phenomenon going on in twenty-first-century politics. War no longer involves mass numbers of citizen soldiers clashing on designated fields of battle. Nor is it being carried out exclusively by states. So, in a sense, we are witnessing the linked breakdown of two of the longest-held monopolies in war and politics. History may look back at this period as notable for the simultaneous loss of the state's roughly 400-year-old monopoly over which groups could go to war and humankind's loss of its roughly 5,000-year-old monopoly over who could fight in these wars.

That these monopolies are ending does not mean that the state is disappearing anytime soon, nor human soldiers for that matter. It does mean, however, that state militaries have new competition on the battlefield, competition that will also have the most advanced technologies in war, including unmanned systems.

Some of these nonstate actors will buy their unmanned weapons off the open market. As journalist Noah Shachtman says of military robots, "The actual physical hardware is cheap and the software will get out." This means that such systems will inevitably end up in the hands of groups that we'd rather not see having such technology. For as the saying goes, "There are no friends in the weapons business, only contracts." Other actors may not buy their systems on the open market, but get them off the black market, including even by outright theft. "A robot out of sight in Afghanistan is a sale item at the marketplace in the morning" is how one U.S. Army robotics expert put it (recalling the Raven drones that ended up in Iraqi insurgent hands as well). And still others may just do like the Grays and build their own systems, perhaps even better than the ones that state militaries might have.

The outcome is that the range of groups using such sophisticated weapons systems will continue to proliferate, with military robotics popping up in the most unexpected places. In 2004, for example, French troops deployed to the Ivory Coast, their former colony in West Africa, to help police a cease-fire between the government and local rebels. The French forces came in without any air defenses, thinking they had little to fear when deploying into the 157th poorest nation in the world. On November 4, 2004, two Israeli-made Aerostar drones circled above their base, scouting out targets and establishing their GPS coordinates. A few hours later, Russian-made Sukhoi jet fighters screamed in, dropping bombs, which killed nine French soldiers and one U.S. aid worker. It turned out that the tiny country had hired the services of an Israeli private military firm to run its intelligence-gathering and a group of ex–Red Army Belarusian pilots to become its air force.

"State, nonstate, air, land, sea.... We have to count on every other actor having them. We can't assume that the U.S. will always have a big technology advantage," warns Noah Shachtman.

## OSAMA BOT LADEN? TERRORISTS AND TECHNOLOGY

Perhaps it is the whole cave-dwelling, robe-wearing lifestyle, but there is a prevalent assumption that terrorist groups like al-Qaeda aren't all that interested in technology. As Lieutenant General Lance Smith, deputy commander of U.S. Central Command, put it, "One of the reasons we're having difficulty getting Osama bin Laden and the other leadership of Al Qaeda is because they recognize that technology is not their friend.... Many of our enemies have learned that the way to fight us is to not use technology."

Like many assumptions about terrorism, this could not be more wrong. In the years since 9/11, al-Qaeda has evolved from a highly centralized group, which planned all its operations out of a few sites in Afghanistan, into a global movement, with cells spread around the world. In this transformation, technology has been essential to holding the group together, as well as expanding its numbers. As a 2006 study of the group, entitled "High-Tech Terror: Al-Qaeda's Use of New Technology," laid out, technologies are essential to the group for everything from finding and radicalizing new recruits, collecting and moving funds, passing on training and expertise, and communicating attack plans. Web pages like the Al-Hesbah Discussion Forum "offer news on Iraq, links to videos from conflict zones, where jihad is fought, photos of martyrs, and religious arguments for the justification of waging jihad. There are even postings of job openings in jihad." Other radical sites host chat rooms, online magazines, and downloads for cell phone videos of the latest propaganda, which all can also be passed on to your friends. There is even a site where one can pledge allegiance to Osama bin Laden by filling out an online form.

Terrorists are also using new technologies to recruit in even more novel and inventive ways. This includes a thriving video game industry targeting Muslim youths, which focuses on the themes of waging violence. In the popular game *Ummah Defense*, for example, a virtual warrior has the vicarious thrill of taking on the American military, Israeli settlers, and, of course, "killer robots."

With unmanned systems, these technologies move from the propaganda and recruit indoctrination of virtual reality to real-world operations. One al-Qaeda–linked Web site, for example, has already offered recruits the chance to remotely detonate an IED in Iraq while sitting at their home computer, the terrorist parallel to the drone pilots in Nevada. In turn, al-Qaeda explored the use of a UAV to assassinate President Bush at the 2001 G-8 summit in Italy. "Sooner or later we're going to see a Cessna programmed to fly into a building," warns navy rear admiral Chris Parry.

Unmanned systems lower the bar for who can deal out damage, while multiplying the potential destruction they can cause. The toughest terrorists to stop are those with no concern for their lives, in that normal defenses and deterrence won't work. Fortunately, those people with such a death instinct are rare. Robert Finkelstein pioneered a DARPA-supported project on what he describes as the coming "intersection of robotics and terrorist groups, particularly looking at robots as platforms for WMD." He explains, "We may be in the 'golden age' of suicide bombers, but that may not always be the case. In any way, there may be groups in which not everyone wants to seek the seventy-two virgins right now [referring to the extremist recruiting propaganda that martyrs are greeted in heaven by seventy-two virgins]."

If a would-be terrorist isn't willing to strap on a vest filled with explosives or fly a plane into a building, the new robotic systems mean that they can now accomplish their goals remotely or automatically and still live to fight another day. As one security analyst described, "You can be a wimp, but still be a terrorist." This change is immense. The number of attacks can be multiplied, as suicidal terrorists used to come with an expiration date. In turn, even when attacks fail, captured machines don't provide as good intelligence as a captured terrorist; if you torture a robot with a waterboard, only sparks come out.

Simultaneously, unmanned systems offer capabilities and possibilities beyond the normal limits that terrorists face. One analyst described that a robot in terrorist hands is "a suicide bomber on steroids, basically." Traditional barriers like fences and walls can be more easily breached when drivers don't care if they die or can fly over them. It is hard enough for Secret Service agents to protect a president from those in the line of sight, such as the audience at a speech; imagine how difficult it gets when the assassin might be miles or cities away, doing it all remotely. Robert Finkelstein believes, "Robots could be very attractive [to terrorists], and are available right now. You could fly stuff into the White House very easily from roofs just six blocks away....I'm actually very surprised it hasn't happened yet." Even worse, unmanned systems are, as a 2006 air force study concluded, "an ideal platform" for deploying such weapons of mass destruction as biological or chemical weapons; the analogy here is the use of drones in agriculture as crop dusters, just in this case the pilot doesn't want to be anywhere near what he is "dusting."

Much as American military drones over Iraq are flown from Nevada, so too can unmanned systems give terrorists who might not be able to get inside the United States reach and access that they didn't have before. Most of the U.S. government's focus on preventing WMD attacks on the American homeland has been to defend against the threat of intercontinental ballistic missiles, with over $54 billion spent on the National Missile Defense program so far. But such missiles are only in the hands of a few states (not even Iran or North Korea have them yet) and therefore any attack would have to involve either the leaders in those states becoming suicidal or terrorists' being able to seize an existing missile (like a Russian or Chinese missile base), get the launch codes, and know how to use them.

By contrast, "do it yourself" kits are making robots accessible to just about anyone, including systems with capabilities that were just a few years ago considered military grade. Chris Anderson, an editor at *Wired* magazine, even hosts a Web site called *DIY-Drones* in which he shows how to build systems comparable to U.S. tactical systems but at a fraction of the cost, including a drone that can fly for an hour, uses GPS directions and text messaging controls, but costs only $1,000. Amateur builders have even made drones that can fly immense distances with pinpoint accuracy. In 2003, for example, a group of model plane enthusiasts launched a homemade drone called the "Spirit of Butt's Farm." Designed by a seventy-seven-year-old blind man, the drone was sophisticated enough to fly itself across the Atlantic Ocean. One group's hobby might be another's weapon.

As with nonstate organizations, terrorist groups may even have an advantage when it comes to high technology, in that they can free-ride off of the investments that states and industry made in developing them. The sum total of al-Qaeda's financial resources is thought to be roughly what the U.S. military spends in one hour in Iraq. But in 2006, when its forces wanted to target a British army base outside Basra in Iraq, al-Qaeda didn't have to invent rockets that go into space or build expensive reconnaissance satellites that could take photos of the Earth. Instead, its operatives went onto the Internet (which al-Qaeda also didn't have to pay to develop) and downloaded images of the base from Google Earth. The footage was so detailed that they were able to sight their mortars to target the soft-skinned tents in the base, rather than harder-to-damage buildings.

## THE PERILS OF A BAD HAIR DAY

When we think of the terrorist risks that emanate from unmanned systems, robotics expert Robert Finkelstein advises that we shouldn't just look at organizations like al-Qaeda. "They can make a lone actor like Timothy McVeigh even more scary." He describes a scenario in which "a few amateurs could shut down Manhattan with relative ease." (Given that my publisher is based in Manhattan, we decided to leave the details out of the book.) *Washington Post* technology reporter Joel Garreau similarly writes, "One bright but embittered loner or one dissident grad student intent on martyrdom could—in a decent biological lab for example—unleash more death than ever dreamed of in nuclear scenarios. It could even be done by accident."

In political theory, noted philosophers like Thomas Hobbes argued that individuals have always had to grant their obedience to governments because it was only by banding together and obeying some leader that people could protect themselves. Otherwise, life would be "nasty, brutish and short," as he famously described a world without governments. But most people forget the rest of the deal that Hobbes laid out. "The obligation of subjects to the sovereign is understood to last as long and no longer than the power lasteth by which he is able to protect them."

As a variety of scientists and analysts look at such new technologies as robotics, AI and nanotech, they are finding that massive power will no longer be held only by states. Nor will it even be limited to nonstate organizations like Hezbollah or al-Qaeda. It is also within the reach of individuals. The playing field is changing for Hobbes's sovereign.

Even the eternal optimist Ray Kurzweil believes that with the barriers to entry being lowered for violence, we could see the rise of superempowered individuals who literally hold humanity's future in their hands. New technologies are allowing individuals with creativity to push the limits of what is possible. He points out how Sergey Brin and Larry Page were just two Stanford kids with a creative idea that turned into Google, a mechanism that makes it easy for anyone to search almost all the world's knowledge. However, their $100 billion idea is "also empowering for those who are destructive." Information on how to build your own remote bomb or the genetic code for the 1918 flu bug are as searchable as the latest news on Britney Spears. Kurzweil describes the looming period in human history that we are entering, just before his hoped-for Singularity: "It feels like all ten billion of us are standing in a room up to our knees in flammable fluid, waiting for someone—anyone—to light a match."

Kurzweil thinks we have enough fire extinguishers to avoid going up in flames before the Singularity arrives, but others aren't so certain. Bill Joy, the so-called father of the Internet, for example, fears what he calls "KMD," individuals who wield knowledge-enabled mass destruction. "It is no exaggeration to say that we are on the cusp of the further perfection of extreme evil, an evil whose possibility spreads well beyond that which weapons of mass destruction bequeathed to the nation states, on to a surprising and terrible empowerment of individuals."

The science fiction writers concur. "Single individual mass destruction" is the biggest dilemma we have to worry about with our new technologies, warns Greg Bear. He notes that many high school labs now have greater sophistication and capability than the Pentagon's top research labs did in the cold war. Vernor Vinge, the computer scientist turned award-winning novelist, agrees: "Historically, warfare has pushed technologies. We are in a situation now, if certain technologies become cheap enough, it's not just countries that can do terrible things to millions of people, but criminal gangs can do terrible things to millions of people. What if for 50 dollars you buy something that could destroy everybody in a country? Then, basically, anybody who's having a bad hair day is a threat to national survival."

## NEIGHBORHOOD WATCH: TECHNOLOGY FIGHTS BACK

The science-minded folks are not the only people, however, worrying about the changes these new technologies may bring to the terrorism trade. One special operations officer describes how much of the future of counterterrorism will be about hunting down such super-empowered individuals. "The future is manhunting."

In these manhunts, unmanned and automated technology may prove to be a key factor; that is, robots are not just a new tool for terrorists, but also one of Kurzweil's "fire extinguishers" that might help stop them from burning the world down. Just as Hezbollah used these new technologies to step up its attacks on Israeli cities, Israel has since responded by deploying its own layers of automated defenses, on land and in the air. One of the more notable includes the Skyshield, an automated machine-gun system a lot like R2-D2 in Baghdad, that shoots down missiles and rockets that fly across the border.

States besieged by terrorists, however, can't depend solely on such a last line of defense. They also must invest in prevention. Here too automated systems are proving useful. One of the paradoxes of security screening at places like airports and railroad stations, for example, is that it is incredibly important, but also mind-numbingly boring. So, oddly, we have this most crucial job in countering terrorists performed by workers with little training, who are paid barely over minimum wage.

As with other dull, dangerous, and dirty jobs, automated systems are coming into favor as a potential solution. Screening is beginning to be streamlined via such technologies as high-frequency radio scanners, which can automatically spot concealed weapons. Reminding many of the technology imagined in movies like Arnold Schwarzenegger's 1990 flick *Total Recall*, the real-world version is a sort of automatic X-ray scan. It detects the radio waves that reflect off every material and automatically searches a person's body or luggage for any incriminating objects. Because every material reflects waves in its own unique pattern, the automated scanners can spot not only hidden handguns and knives, but also nonmetal weapons and explosives.

The real breakthrough in counterterrorism may come from combining automated and artificial intelligence systems with our broader network of surveillance. In Britain, for example, there are more than 4.6 million cameras watching public areas, from cameras that monitor traffic jams and deter robberies at ATMs to specially designed cameras that watch over all of London's subway stations. As a result, the average Briton appears on cameras as many as three hundred times a day.

Similarly, huge numbers of cameras already cover most U.S. cities. Chicago, for example, has some 2,250, covering every one of its key sites and streets. "Cameras are the equivalent of hundreds of sets of eyes," describes Chicago mayor Richard Daley. "They're the next best thing to having police officers stationed at every potential trouble spot."

When terrorists struck the London subway system on July 7, 2005, killing fifty-two and wounding over seven hundred, the existing camera systems proved quite useful. The entire operation was captured on film, from the terrorists scoping out the targets days before to the very last seconds of them entering the station to carry out their attack. This allowed investigators to quickly figure out what had happened. Yet the systems had no preventive or reactive effect; the attacks still occurred, while the cameras could only record. As Robert Finkelstein explains, "They could only use the videos to backtrack what happened. With machine intelligence you could have intervened earlier."

Integrating surveillance systems with artificial intelligence is the next step in the technology war on terrorists. Instead of having a policeman on every corner or a person monitoring each and every camera, new automated programs are being developed that will be able to make sense of what they see. For example, AI programs would filter the surveillance footage instead of a human, and automatically alert the police whenever any camera in the system spots something suspicious, such as someone leaving a package on a train platform or parking a car in an emergency zone and walking away. It could even take automated actions, such as shutting down access or throwing up security barriers.

The advantage of such counterterror technology is not only that "machines don't get tired or need to take a coffee break," explains one engineer, but also that they can tap into memory and processing power far better than human security guards. They can pick out patterns (such as a truck that circled the building eight months ago is now parked in front of the lobby) or anomalies (like a person wearing a long jacket into a subway station in July) that might elude someone trying to watch an entire building or subway station on multiple TV screens at once. They can even scan crowds for the faces of terrorists in a database. The Securics company, for example, has built systems for the U.S. Special Operations Command and DARPA as part of the agency's Human ID at a Distance program, which can scan and identify faces from as far as two hundred feet away. Other programs will meld artificial intelligence with the latest research on biometrics. They might, for example, detect if someone is carrying a hidden item, like a bomb, under their clothes by seeing how their walking gait is altered.

Such programs won't just be useful in capturing would-be terrorists immediately before they strike, but also in tracking down anyone in the system. As anyone recalls from the game *Where's Waldo?*, it is wickedly hard to find someone in a crowd, even if they are wearing a goofy red-striped sweater (which surely hid

Waldo's suicide bomb vest). Imagine that crowd is the size of an entire city. With the facial recognition systems combined across an entire network, programs can automatically scan every camera in the system, checking for that person not only at that moment but also in all the data saved over past days and weeks. This will allow police to instantly locate persons of interest and everywhere they have been in the city.

Once a system is automated, it need not rely only on visual cues to track and detect terrorists. This is where "data mining" comes in. Every person leaves behind a "paper trail" of their life's activities. These are now mostly digital records that range from credit card purchases and bank statements to cell phone calls and e-mails. Data mining gathers and analyzes all this information to detect patterns, trends, and anomalies.

This practice of machines' connecting the dots in a person's life to find bigger patterns comes out of the corporate world. One of the biggest data-mining efforts to date is actually at Wal-Mart, which has gathered some 460 terabytes of information from customers on its mainframe servers in its Bentonville, Arkansas, headquarters (to put that into context, about double all the data on the entire Internet in 2004). Wal-Mart uses the information to track its customers' buying habits and preferences, and then employs AI to anticipate their future wants and needs. Its data mining, for example, detected that customers tend to stockpile strawberry Pop-Tarts whenever a severe weather warning is made in the media. So Wal-Mart's supply system automatically responds to any announcement of bad weather by sending additional truckloads of Pop-Tarts to stores in the expected pathway.

Big Brother is not far behind the Big Greeter. In the national security realm, the extent of data mining crossed with AI is classified, but thought to be extensive. The most notable and controversial program disclosed in the public was a $200 million DARPA program created in 2002 called Total Information Awareness. TIA sought to create a huge database for pretty much every sort of information that the government could gather about American citizens and visitors. TIA turned out to be TMI (too much information) and both conservatives and liberals in Congress objected to the massive civil liberties concerns it raised. It was supposedly defunded just a year later, with the office behind it publicly closed.

Far from being completely shut down, however, the Total Information Awareness program instead appears to have been broken into several components, which go under less menacing names like "Baseball" and "Topsail." Says one report, "Very quietly, the core of TIA survives."

The data gathered in such programs is immense. For example, the Department of Homeland Security's Analysis, Dissemination, Visualization, Insight, and Semantic Enhancement program, ADVISE for short, is reported to have gathered some one quadrillion pieces of data that range from financial records to CNN news stories. These can then be matched against U.S. intelligence and law enforcement records and AI software sifts through it for connections and patterns. Or, as one report described, the computers are "identifying that it is a needle that needs to be found in the haystack, and then finding it."

Analysts hope that adding machine intelligence to data mining will not only allow terrorists to be tracked down by their digital paper trails, but also give counterterrorism efforts the ability to project into the future, predicting and stopping attacks before they happen. The Knowledge Aided Retrieval in Activity Context, or KARNAC (a tribute to Johnny Carson's comic soothsayer "Carnac the Magnificent"), works by combining data mining with software that mimics the types of things that good human intelligence analysts are supposed to do. With a huge database and processing power, the machine is able to filter through information that ranges from criminal records to what people are searching for on the Internet and piece together patterns and trends that otherwise would not appear to be connected on their own. Then, just like a human agent working at the CIA or FBI, it can "track leads, form hypotheses, and narrow outcomes." A teenager who downloads death metal music, starts searching online for the building maps of shopping malls and how-to guides for making bombs? The system connects the dots and sends a policeman to talk with his parents, before everyone wonders how such a nice boy could blaze away at holiday shoppers.

## IT BECAME NECESSARY TO DESTROY THE VILLAGE IN ORDER TO SAVE IT

Through these types of technologies, many science fiction writers envision how the "war on terrorism" might ultimately be won. In movies such as Steven Spielberg's *Minority Report*, police are constantly tracking everyone's movement on ubiquitous cameras and able to intervene before a crime even occurs, while S. M. Stirling's book *Conquistador* points to a future where terrorists can get only halfway through their plots before the authorities catch on.

Reality won't necessarily work out that way (and even in *Minority Report*, the system fails in the end). Such counterterrorist technologies are expensive. They also rely on government agencies that don't like to share information, as well as businesses locked in competition over who can control the most information, to set aside their differences and kindly open up all their most closely guarded databases. And, as the maker of KARNAC admits, it "cannot guarantee the software will work 100% wrinkle-free."

Even if these "wrinkles" get ironed out, the smartest of counterterrorist AI will still be at the mercy of the information that goes into the systems. As the computer programming adage warns, "Garbage in, garbage out." If an informant, for instance, claims that he knows about a secret WMD facility and no one thinks it worthwhile to also add into the system that he is a known liar (as failed to happen with "Curveball," the Iraqi defector whose false claims ultimately became part of Colin Powell's speech to the UN on the eve of the Iraq war), the system will produce the wrong conclusions.

In turn, just when the state might get ahead of the game, terrorists can be anticipated to react and learn. A program like KARNAC, for example, is great for linking together indicators that might otherwise be missed on the behavior of existing suspects or "persons of interest." But it will face a tougher test when the attackers are what are known as "clean-skins" (operatives with no prior history, who otherwise lead normal lives), or when terrorists figure out how to trigger false or misleading patterns that could divert attention from real threats. Having a human spy on the inside of a terror cell will still prove far more useful than an AI trying to read the tea leaves from airline tickets, van rentals, and e-mail traffic.

The back-and-forth of these new technologies in terrorism and counterterrorism then comes full circle to the new challenges for the state. The technologies don't just potentially change the balance of power in society between states and nonstate groups, as in the battle between Israel and Hezbollah, but also between states and citizens. Civil liberties, especially those of privacy, will be under a whole new level of pressure in a world where every step is actually monitored, tracked, and recorded. The opponents of these systems worry about an "Orwellian mass surveillance system," as Wikipedia described TIA, while the makers retort that the possibility of another large-scale terrorist attack "is more terrifying than losing one's privacy."

Our brave new world of technologies then could bring back under question the original, fundamental balance between citizen and state. Hobbes's so-called social contract between citizens and their governments wasn't just that people give their loyalty in exchange for the protections from war and violence that only the state could provide. It also required that the state respect people's rights, granting each citizen a sort of bubble around them to lead their own lives.

Such new technologies, though, imperil as never before the ability of states to deliver on both parts of the deal. On one hand, the state's monopoly on protection and violence is challenged by nonstate groups, terrorist networks, and even individuals, all empowered with dangerous new technologies. On the other hand, the only way to beat such new dangers may be to counter them with even newer, more invasive technologies. But these, in turn, undermine each person's sense of protection and privacy from the state itself. In short, by protecting the individual from technology, do we destroy our very concept of the individual with technology?

# FOURTEEN

## LOSERS AND LUDDITES: THE CHANGING BATTLEFIELDS ROBOTS WILL FIGHT ON AND THE NEW ELECTRONIC SPARKS OF WAR

*Technological progress is like an axe in the hands of a pathological criminal.*

—ALBERT EINSTEIN

"Increasingly, we live in a world where the Flintstones meet the Jetsons—and the Flintstones don't much like it."

Ralph Peters grew up in the coal-mining districts of Pennsylvania. "I am a miner's son, and my father was a self-made man who unmade himself in my youth." As a young man, Peters enlisted in the army as a private, and spent the next ten years on the front lines of the cold war in Germany working in military intelligence. A brilliant thinker, he soon rose to the rank of lieutenant colonel.

In 1998, Peters retired from the army to give himself more freedom for writing. He'd already published a book and a few articles while in uniform, but wanted to do more. Plus, his blunt style had also made further advancement less than likely; he was riling up too many senior officers by publicly raising uncomfortable conclusions about where war was headed and how the U.S. Army was failing to adjust.

Peters soon had his payback. Over the next decade, he became one of the most notable and sought-after experts on modern warfare. He wrote six books on military affairs and became a commentator on PBS, FOX News, and CNN, while also writing both a weekly newspaper and magazine column. In a challenge back at the senior officers who had tuned him out while in uniform, Peters went on to publish more articles in *Parameters*, the army's most prestigious journal, than any writer in its history. As retired General Barry McCaffrey describes, Peters is "simply one of the most creative and stimulating writers on national security we have produced in the post-WWII era."

Peters is also quite a force in the field of fiction, having written eight political thrillers. His first novel was a cold war spy story set in the former West Germany. His subsequent novels have gone on to include more contemporary settings of terrorism and failed states, and he's built up quite a sizable fan base among military readers. As if this wasn't enough, Peters also writes a series of historical detective novels set in the Civil War under the pseudonym Owen Parry.

It is perhaps because of this breadth of experience, analysis, and imagination that Peters is an apt resource for understanding where war, and its future causes, is headed, and not just because his latest book is titled *Wars of Blood and Faith: The Conflicts That Will Shape the Twenty-first Century.*

## WARRIORS RISING

Ralph Peters sees two trends converging to spark wars in the coming century, on whose battlefield unmanned systems will increasingly fight. The first is the rise of "a new warrior class." "Ours is the age of barbarians with microchips, of zealots who cannily exploit the civilized world's rules in their attempts to destroy it," he explains. "We are learning that many human beings prefer certainty, no matter how oppressive and primitive, to the risks and responsibilities of freedom."

From warlords who arm children to terrorists who blow up school buses, the nature of who is waging modern-day conflict is shifting away from only professional armies. "The soldiers of the United States Army are brilliantly prepared to defeat other soldiers. Unfortunately, the enemies we are likely to face through the rest of this decade and beyond will not be 'soldiers,' with the disciplined modernity that term conveys."

Rather, Peters believes, much of war will be driven by a shift back to "warriors," a twenty-first-century update of the "barbarians" from past eras. Today's barbarian warriors are not Vikings or Huns, but the modern-day warlords, terrorists, insurgents, et al., who are habituated to violence but don't have any sort of professional training or organization. "Unlike soldiers, warriors do not play by our rules, do not respect treaties, and do not obey orders they do not like.... The warrior is back, as brutal as ever and distinctly better-armed." Harvard professor Michael Ignatieff describes the trend similarly to Peters, but uses the more academic moniker "postmodern warriors" to describe the takeover of war by "the barefoot boys with Kalashnikovs, the paramilitaries in wraparound sunglasses, the turbaned zealots of the Taliban who checked their prayer mats next to their guns."

That modern-day warriors don't follow the old rules of war is not just a matter of simple evil. Debbie Stothard, an expert on refugees, witnessed an episode in Thailand where a group of child soldiers led by two twelve-year-old twin brothers held an entire hospital hostage. "These are people who have not had access to a good education and for whom violence is a way of life. It never occurs to them that mounting a siege on a hospital is actually wrong. They have not lived in a world where detaining someone with force is actually unacceptable. It's as though they came from a different planet."

This new "warrior" trend is emerging because so many are not sharing in the prosperity of globalization, and conflict entrepreneurs are rising to take advantage. Peters offers a sketch that could describe any number of the modern-day warlords, from Foday Sankoh, the failed commercial photographer whose use of child soldiers to seize diamond mines left more than two hundred thousand dead in Sierra Leone, to Abu Musab al-Zarqawi, the Jordanian petty criminal turned leader of al-Qaeda in Iraq. "The archetype of the new warrior class is a male who has no stake in peace, a loser with little education, no legal earning power, no abiding attractiveness to women and no future. With gun in hand and the spittle of nationalist ideology dripping from his mouth, today's warrior murders those who once slighted him, seizes the women who avoided him, and plunders that which he would never otherwise have possessed."

This trend then builds upon itself. "The longer the fighting continues, the more irredeemable this warrior becomes. And as society's preparatory structures such as schools, formal worship systems, communities, and families are disrupted, young males who might otherwise have led productive lives are drawn into the warrior milieu. These form a second pool. For these boys and young men, deprived of education and orientation, the company of warriors provides a powerful behavioral framework."

A prototypical example of this follow-on generation of warriors is L., the kind of child soldier I met during my research for my last book. In his village in the eastern jungles of Sierra Leone, L.'s home was attacked and plundered by rebels when he was ten years old. The rebels didn't have any political agenda; they were just working on behalf of a warlord whose only goal was controlling local diamond mines. After the villagers were lined up, L.'s mother and father were butchered in front of him and he was taken away by the warriors, many of whom were children just a few years older than him. He was later beaten, drugged, and forced to kill other prisoners, at the risk of being killed himself.

Over time, though, the new lifestyle and the copious drugs helped give L. a new identity. He took on a new "fighting name," got tattoos that inculcated him into the group, and regularly went out on raiding missions, much like the one that first brought him into the fold of war. His greatest prize was his AK-47 Kalashnikov assault rifle, able to fire six hundred bullets per minute. He proudly described how he'd learned to disassemble it and put it back together all by himself in a matter of minutes. With this gun, L. was not a scared ten-year-old boy but a fearsome warrior.

If war in the twentieth century was about "ambitious winner-states" like Nazi Germany or Imperial Japan, seeking to gain their "rightful place in the sun," tells Peters, war in the twenty-first century will also be driven by the "losers" and those like L., whom they take advantage of. The reservoir of conflict lies not in the "gilded crust of humanity," but in the "vast 'loser' populations in failed states and regions."

This then links to the second trend, the amazing world of change we are living through. It is equally dazzling and disruptive, replete with new technologies that are both enabling and threatening. Tells Peters, "We live in the most dynamic age in human history.... [But] we have this illusion that technology will solve human problems." The massive changes may aid some, but will also feed much of global conflict.

History supports his contention. For example, the printing press revolutionized human awareness and knowledge, but it also sparked the bloody conflicts of the Reformation that culminated in the Thirty Years' War, which left nearly a third of Europe dead. Today we are living through its modern parallel. "The Internet is the greatest tool for spreading knowledge and hatred since the invention of movable type." Robotics have an even greater potential for both good and ill.

And from this conflict emerges, tells Peters. There will be battles because of change and battles to resist change. "The root causes of conflict in the 21st century are humanity's default positions.... In times of crisis, when humans have to ask the fundamental question of 'Who am I?' they fall back on the defaults, conflicts of blood and belief.... When I look at the 21st century, this age of miraculous technology, I see a paradox. This age of technology will also see a return to atavistic violence."

## POVERTY SUCKS

The citizens of the First World are living through what may be the most prosperous generation in human history. There are all sorts of facts and figures to show this, but perhaps the extent of our prosperity may be best demonstrated by our eating patterns. Each year, Americans spend over half a trillion dollars at supermarkets and another half a trillion eating out at restaurants (about half at fast-food chains). Even with this, Americans throw away almost 50 percent of all the food in our nation ready for harvest.

The extent of our affluence is seemingly unimaginable, especially to the vast majority of the world for which it is not a reality. Our culture treats "competitive eating" as a sport, while 1.3 billion people in the developing world live in poverty. One hundred twenty-seven million Americans are "obese," the polite way of saying "so fat as to be dangerous to themselves," while half a billion people in the developing world are "chronically malnourished," the nice way of saying "starving." Nineteen billion dollars a year is spent in the United States for tap water that has been repackaged as "bottled water," while more than 1.3 billion people in the developing world lack access to clean water. We may be living through Charles Dickens's proverbial "best of times," but most of the world is still suffering as if it were "the worst of times."

This disparity points to a series of changes that are crucial to understanding poverty's link to technology and war in the twenty-first century. As Dickens's *A Tale of Two Cities* told, poverty and social inequality have always driven rage and rebellion. The vast majority of the world's people have always been poor and there has always been a tiny majority living the good life.

However, the situation now is exacerbated by unprecedented population growth. In the last fifty years, the world population has grown more than the sum total of all the births in the previous four *million* years of human history. And this trend is only continuing. The United Nations projects that the world's population will roughly double again in the next fifty years.

This trend has been driven by the odd combination of technology and human nature that Peters finds so simultaneously interesting and disturbing. Technologies have increased public health, sanitation, and disease control, driving up birth rates in a span of a few years. But societies need generations to adjust. They don't change their attitudes and practices toward such things as ideal family size or consumption in just a few years.

Demographics is destiny. But the issue is not just one of raw numbers, but also of where and how these numbers are distributed. We are now in the midst of the largest generation of youth in human history, but 90 percent of the world's youth under the age of fifteen live in developing countries. Over the next four decades, it is projected that 99 percent of the world's population growth will take place in the developing world, which is the part least prepared to feed, clothe, educate, and employ three billion more people.

Such a great demographic shift is incredibly worrisome. Research shows that when the age balance in a population shifts out of whack (when there are too many young males as compared to old fogies), violent outbreaks, ranging from wars to terrorism, become far more common. This process is known as "coalitional aggression." Young men are considered psychologically more aggressive and naturally compete for social and material resources in all societies. When they outnumber other generations, there are inevitably more losers than winners among the youth in this process. Plus, the typical stabilizing influences of elders are diluted by the overall mass of youth. The system of social stability basically becomes overloaded by too many hormone-filled kids with too few life prospects.

These lost youths are more easily harnessed into activities that can lead to conflict. For example, demagogues, warlords, criminals, and violent religious fanatics all find it easier to recruit when a large population of angry, listless young men fills the streets. Riots and other social crises are also more likely. In a sense, it is conflict caused from the bottom up, rather than the top down (a top-down example would be the traditional mode of a government leader planning for war). It seems far too simple an explanation of violence, except that the facts back it up. The pattern has held true across history, from wars in ancient Greece to recent societal breakdowns in Rwanda, Yugoslavia, and the Congo. Most worrisome is that the trend is particularly pronounced in the Muslim world (over half the population in such fragile states as Iraq, Iran, Kuwait, Pakistan, Syria, Saudi Arabia, and Yemen falls into this pattern) as well as China, perhaps presaging even worse instability to come.

This would be bad enough. But, unlike in the past, the brunt of today's socioeconomic problems is falling hardest on the youngest segments of the population. Unprecedented numbers of children around the world are undereducated, malnourished, marginalized, and disaffected. A quarter of all the world's youth survive on less than a dollar a day. As many as 250 million children live on the street; 211 million children must work to feed themselves and their families. As one report concludes, "These poor, young billions are moving into huge, urban ghettos around the world that have no public health facilities and are a breeding ground for disease, are often ungovernable, and provide little hope for their denizens."

With more people, there is also less to go around. To paraphrase the famous rapper Biggie Smalls, "Mo' People, Mo' Problems." This is not just about the world running out of oil, something many worry is happening as production rates fall by 7 percent annually, despite booming demand. Population growth in Sudan, for example, led to water shortages and a competition over grazing lands, which then sparked the slaughter in Darfur that has left over 250,000 dead.

"Resource scarcity will be a direct cause of confrontation, conflict, and war. The struggle to maintain access to critical resources will spark local and regional conflicts that will evolve into the most frequent conventional wars of the next century," explains Peters. "Today, the notion of resource wars leads the Westerner to think immediately of oil, but water will be the fundamental need of some states, anti-states, and peoples. We envision a need to preserve rainforests, but expanding populations will increasingly create regional shortages of food—especially when nature turns fickle. We are entering the century of 'not enough,' and we will bleed for things we previously could buy."

More people also means more environmental degradation, thus also overwhelming the planet's ability to cope. This creates a feedback loop, which worsens the problem of scarcity. Then add in the effect of global warming. Whether you believe the cause is man-made carbon dioxide or unicorn farts, it is without dispute that the warming of the Earth will make life more difficult for many. The Intergovernmental Panel on Climate Change (IPCC) found that the current pace of global warming will bring water scarcity to between 1.1 and 3.2 billion people over the next few decades, and create food shortages for an additional 200 to 600 million. As if this wasn't bad enough, while some parts of the globe will be parched for water, others may suffer instead from too much of it. Some 100 million people will face the annual risk of floods from rising sea levels, and destructive megastorms like Hurricane Katrina will also become more common, as ocean temperature is the "fuel" for hurricanes. Achim Steiner, executive director of the United Nations Environment Program, summed it up this way: "Unchecked climate change will be an environmental and economic catastrophe, but above all it will be a human tragedy."

## FEEDING THE BEAST

There is another fundamental difference from the poverty of the past. The socio-economic woes are not merely getting worse, but now those losing out are more keenly aware of it. As Ralph Peters succinctly puts it, "Now the ignorant know what's going on, and so there is less bliss."

The have-nots may be living in the same squalor as their great-great-great-grandparents, but they now do have a television set or Internet connection that allows them to see that it is not that way for everyone. Experts sometimes talk about the "digital divide," that certain information technologies like the Internet are not being spread around the globe at equal rates. The real divide may instead come from its solution: the more people are connected, the more what separates us becomes visible.

The same holds for the cliché of the "borderless world." Our global economy depends on a free-flowing system of trade, travel, and communication, but it also binds us all together to our greater danger. An outbreak of war or disease in one part of the globe reverberates across the system as never before. Even more, the shared networks give our new century's warriors and warlords a newfound ability to reach out and touch someone. These new losers may be based in some urban slum or mountain hideout, but they can now organize, plan attacks, or share inspirational propaganda with recruits thousands of miles away.

All these various dark trends combine to set the stage for even more conflict. In some cases, a local government might just collapse from the overwhelming demands placed on it to provide sufficient food, health care, education, security, and prosperity to more mouths amid tougher circumstances. The CIA today counts some fifty countries that have "stateless zones," where the local government has lost all effectiveness or simply given up. Falling behind in the new world of technology could make it harder for these zones to come back, as well as potentially add to their number. Describes one U.S.-government-funded report, "Extreme losers in the information revolution could become 'failed states.' Such failed states could become breeding grounds for terrorists, who could threaten vital U.S. interests."

A government's inability to control its territory and provide what its people want or need then opens up a vacuum. And politics, like nature, abhors a vacuum. Warrior groups move into such vacuums and seize local control, a scenario played out again and again with groups like the Taliban, Hezbollah, Hamas, and the Tamil Tigers. In turn, groups with a more global agenda see these local vacuums as the perfect places to base their transnational operations. Of the fifty "state-less zones," twenty-five of them host terrorist groups. Al-Qaeda's movement of its training camps from the ungoverned spaces of Sudan to Afghanistan to Iraq to Pakistan is a prime example.

The simultaneous rise of connection and chaos also points out the differences among us, differences that can be exploited in a climate of desperation and anger. Indeed, of the twenty-five ongoing conflicts at the end of 2007, all of them involved a civil war along ethnic or sectarian lines. In some situations, it may appear that there was some traditional ethnic grievance that erupted into a civil war. Or, simply enough, in an ever more tense world, as the Malaysian foreign minister described, "It has become increasingly difficult to live together peacefully amongst people of different creeds and religions."

## THE HOT ZONE

"Hell is other people," said the French philosopher Jean-Paul Sartre. It may not be that fortunate, then, that in 2007 something striking happened to life on planet Earth. For the first time in our species' history, over half of humankind lived clus-tered together in cities.

Almost all the population growth over the last few decades has taken place not merely in cities, but in developing-world cities. The results have been staggering, especially to the cities themselves. Nearly every major city in the poorest parts of the globe has multiplied in size by a factor of ten. Many have grown by even more. Cities like Dhaka in Bangladesh, Kinshasa in the Congo, or Lagos in Nigeria are about forty times larger than they were in 1950.

For modern-day conflict levels, this news is mixed at best. As Ralph Peters darkly tells, "The city—capstone of human organization—is growing, chang-ing, producing fantastic wealth...and rotting." The impact he sees on war is best described in the title he gave to a landmark article for the U.S. Army's journal, "Our Soldiers, Their Cities."

Peters recounts that, centuries ago, warrior cultures rampaged from distant hinterlands and dark forests, every so often pouncing on isolated towns and settlements. Today, the cities are the homes for his new warrior class, "the new forests, where magic and unreason rule." That is, in history, rebellion and conflict usually started in the rural regions and spread to the city only if successful. Peters sees the reverse being the trend of the twenty-first century. "Cities are now the center of rebellion...because the city is dehumanizing, breaking down traditional values and connections." And for the young citizens of this place, "Habituated to violence, with no stake in civic order...there is only rage."

Peters sees cities as the coming battlefields, where warriors will be at home and professional soldiers increasingly ill at ease. "The future of warfare lies in the streets, the sewers, high-rise buildings, industrial parks, and the sprawl of houses, shacks, and shelters that form the broken cities of our world." This description could be applied equally to Mogadishu, Grozny, Fallujah, Freetown, or Gaza.

Peters is a soldier, who was publicly listed as an adviser to the McCain 2008 campaign. By contrast, Mike Davis is an urban theorist, who tends to tilt Marxist. Unfortunately, the one thing the two see eye to eye on is this future of the city.

Davis made his name with a detailed study of Los Angeles's history and social geography entitled *City of Quartz*, which predicted the return of urban unrest. A little more than a year later, his forecast came true with the 1992 riots; soon after, Davis was awarded a MacArthur Foundation "genius" grant. In his research since, Davis runs a thread that connects the ganglands of South L.A. to the shantytowns of Cape Town to the urban blight of Cairo. All suggest that something disturbing is happening on a global level—the rise of "megaslums." In city after city on continent after continent, "shanty-towns and squatter communities merge in continuous belts of informal housing and poverty, usually on the urban periphery." In short, explains Davis, the population trends have us on the pathway to, as one of his books is entitled, a *Planet of Slums*.

These "megaslums" house literally millions of young, urban poor, where the losers of globalization and the new warriors are concentrated together in shanties and high-rises. Adding fuel to the fire are "the diverse religious, ethnic, and political movements competing for the souls of the new urban poor." These range from Hindu fundamentalism in the slums of Mumbai, Islamist movements in Casablanca, Pentecostalists in San Salvador, and revolutionary populists in Caracas. These megaslums, really just "stinking mountains of shit," are "volcanoes waiting to erupt."

Cities are the new hotspots for conflict. Sometimes this violence may have a crossover with crime, but the outcome is often the same. For example, in the favelas (the urban squatter settlements of Brazil) more than fifty thousand people are killed a year and even Brazilian army troops (who replaced the outgunned police) have given up trying to patrol them. In just one neighborhood in Rio de Janeiro, ten times as many youths were killed over the last ten years as in the entire Israeli-Palestinian conflict during the same period.

It's perhaps disturbing to hear this coming from an urban planner, but Davis sees the same future as Peters the ex-soldier as well as the cyberpunks of science fiction. "The 'feral, failed cities' of the Third World—especially their slum outskirts—will be the distinctive battlespace of the twenty-first century."

## "AMERICAN TERMINATORS VS. DRUG-DEALING SERIAL-KILLER GUERRILLAS"

The folks in the U.S. military agree with these dark predictions. According to Dave Ozolek, the executive director of the Joint Forces Command's Joint Futures Lab, the city will be a focus of U.S. military efforts for some time to come. Urban zones are "where the fight is, that's where the enemy is, that is where the center of gravity for the whole operation is."

The problem is that fighting in cities is as tough as it gets. As opposed to the traditional open field of battle or even a jungle or forest, a city is extremely complex because of its multidimensional terrain. An enemy can fight from the sewer, the street, or from a building, and each one of these can be turned into a bunker. Being shot at from any direction not only makes life more dangerous, but also takes a far greater psychological toll on soldiers. As retired major general Robert Scales writes, "The array of threats from multiple dimensions has a debilitating effect on soldiers; it further hastens the disintegration process that haunts all military units locked in close combat operations."

The famous *Black Hawk Down* battle in Mogadishu in 1993 is a prototypical example of how tough it is to fight in an urban zone. A team of 123 elite U.S. soldiers went into a landscape of slums to snatch a cadre of warlord leaders, the type of "manhunt" described by the special operations officer as the future of conflict. But before they could get back to base, they were surrounded by thousands of local paramilitaries, aka Peters's "warriors," most of whom were teenagers hopped up on the stimulant khat (khat feels a bit like drinking fifteen espressos). Being shot at from all directions, the soldiers got lost in the confusing back streets and alleyways, and ultimately were lucky to escape with eighteen killed and seventy-three wounded. Even then, the public outcry over the loss was enough to end the entire operation in Somalia.

Fighting in the messy, confusing slums and alleyways accentuates enemy strengths and American weaknesses. Explains Peters, the U.S. military may hope for "gallant struggles in green fields," but "the likeliest battlefields are cityscapes where human waste goes undisposed, the air is appalling and mankind is rotting." The simple reason is that "it's a no-brainer for the enemy. It's the turf they know.... It's the Sherwood Forest to their Robin Hood." Peters then ticks off recent urban battlefields from Mogadishu in 1993 to Sadr City in 2004, and cites another reason that urban battlefields are so enticing to anyone thinking of fighting against American troops. "Now the U.S. has a pattern of losing in cities."

The U.S. response is a massive refocus on "combat in cities." Six hundred million dollars of DARPA's budget is dedicated to technology useful for urban battles. One article on DARPA's work in the use of robots in war was even entitled "Baghdad 2025: The Pentagon Solution to a Planet of Slums." As it explained, "A host of unmanned vehicles [is] also being readied for surveillance and combat in these future 'hot-zones,' while all sorts of lethal enhancements are in various stages of development to enable American troops to more effectively kick down the doors of the poor in 2025."

Beyond the already discussed use of robots to cut down on American casualties, which are usually higher in urban zones, unmanned systems are being called upon to help take away local warriors' home-field advantage. Urbanscape, for example, is a DARPA program "to make the foreign city as 'familiar as the soldier's backyard.'" The system uses drones and unmanned ground robots to fan out across a city, map it, take pictures of every building and street, and then crunch all the data together with AI, so that a soldier on patrol can have an up-to-date 3-D map, replete with high-resolution images of just what's around the corner. Combining the maps with recent or live video from drones allows the system to update for what would be otherwise confusing changes in the landscape. A similar program tested out during Hurricane Katrina, for example, allowed rescue helicopters to find people stranded in neighborhoods that had been flooded.

As if that wasn't enough, another DARPA effort seeks to solve one more "pressing need in urban warfare: seeing inside buildings." Carried by robots and soldiers, the VisiBuilding technology's goal is to build the interior layouts of buildings and sewer systems, as well as searches to "find anomalous quantities of materials" (such as explosive chemicals) and "locate people within the building."

The hope is to pair these technologies with AI to "digitize" entire cities. Akin to the massive virtual worlds in such venues as Second Life, a usable cityscape would be built of any urban battle zone, detailed down to the blueprints and individual occupants of each building. A fleet of unmanned robotic sensors and systems (ranging from spy satellites to tiny insect UAVs peering into buildings) would continually update the virtual version of the city with real-world footage and information. Imagine the video game *Sim City* crossed with Google Earth. It would give soldiers the ability to zoom into any neighborhood or even individual structure to see what is going on in real time. According to one report, "You have continuous coverage, around corners and through walls. You would never, for example, lose those mortar bombers who got out of their car and ran away."

By sending in robots that navigate the new urban battlefield, DARPA is hoping to completely rewrite the script of *Black Hawk Down*. According to DARPA's director, Dr. Anthony J. Tether, it will give U.S. forces "unprecedented awareness that enables them to shape and control [a] conflict as it unfolds."

Some, though, doubt that it will work out the way the military hopes. Peters, for instance, thinks robots have their role, and that the urban warfare trends will drive their use, but we should not expect too much. "There is a uniquely American pursuit of the grail, that technology will solve all human problems, that we can have bloodless wars. Our faith in technology is actually a vulnerability that hinders us from winning wars. . . . How on earth can technology solve the problems of a Liberia, a Rwanda, a Sudan, a Congo? The domain of war is still one of flesh and blood."

## RAGE AGAINST THE MACHINES

Unfortunately, Richard Clarke has been ignored before. Over his thirty-year career, the silver-haired Clarke served in nearly every major government post related to security and terrorism. He was President Reagan's assistant secretary of state for intelligence, coordinated diplomacy to support the first Gulf War for President Bush Sr., and served as the first ever counterterrorism coordinator for the National Security Council during the Clinton administration, a position he also held in the first year of Bush Jr.'s administration. As expert as Ralph Peters is on war, Richard Clarke is his counterpart on terrorism.

Starting in January 2001, Clarke sent the new national security adviser, Condoleezza Rice, a series of memos warning about the growing threat from a terrorist group called al-Qaeda. He argued that the Bush administration "urgently" needed to act against something it was mistakenly viewing as a "narrow, little terrorist issue." Clarke was viewed as an alarmist and disregarded. Deputy Defense Secretary Paul Wolfowitz even told Clarke that he was wasting everyone's time: "You give bin Laden too much credit." History records who was right just a few months later on

9/11. But politics mattered more than accepting responsibility and, two weeks after that, Clarke, not the leaders who ignored him, lost his job.

Today, Richard Clarke is worried about something new. "Something very fundamental is happening in technology today." A friend of futurist Ray Kurzweil, Clarke is well aware of the various new technologies that are already here or rapidly on their way. Indeed, he was one of the early proponents of the arming of the Predator drone back in 2001, hoping that it could be used against bin Laden in time. Yet where Kurzweil only sees positives, Clarke's views fall closer to the dystopian visions of Ralph Peters. But beyond the new warriors fueled by poverty and hate, Clarke sees the danger signs of another violent backlash brewing that will spur even more conflict.

"It has many causes. Fear of machines, fear of technology, fear of the unknown," Clarke explains. "It's even simple math phobia for many. It's very uncomfortable for adults to know that they have to call in their grandchildren to help them do something as simple as set the time on their VCR [already dating himself]. It is fueled by a resentment of people who know that they don't know enough. . . . It will also be a religious problem, some opposition will [even] come from the Christian right. They are already asking questions like, 'Are we playing God? Is this God's will?' " Finally, Clarke sees anger brewing among those left economically disadvantaged by the technology. "There will be a real digital divide—people who don't have the skill sets to compete anymore."

This opposition is just starting to crystallize, but at its essence is a fundamental sense of being overwhelmed by change. "When they look around, they see real questions about genetics, AI, robotics. . . . They see a technologic horizon rushing at them, see a radical change for the nature of society."

Ultimately, Clarke thinks some of those most concerned about the next wave of technology may act out in violence. "They won't [use violence] if they succeed politically, but that's not likely. They won't be able to stop the various changes they so fear. There's too many reasons for this technology for them to be able to suppress it. . . . So, ultimately, they will lose in the political arena. Some in the fringe, just as they blew up abortion clinics and shot doctors, will try to act violently to slow it down." He believes this violence will last "a long time, because this technology will come in waves. We really do have a political problem."

As to whether Washington is listening to him this time, Clarke is "dubious." He explains that two problems are blinding policymakers to the threat of a new breed of violence that will brew from those so opposed to technologic change that they would take up arms to stop it. Part of the problem is that things like robotics, AI, and genetic engineering may be threatening to some, but still sound more like science fiction to policymakers who simply are not well aware of the revolutions happening in scientific research. "You go in and explain there is this problem with technology and they just don't get it.... There is no sense in our political establishment of what's going on."

The second is the combination of human nature and American politics. "We tend to be a society that waits for disaster to happen. We don't focus our attention on it until after the fact. I'm cynical. I know the way our political system works. We waited until 9/11 before we bothered with al-Qaeda."

## NED AND TED'S BIG ADVENTURE

As Ralph Peters explained, the future bodes for more conflict in the parts of the world that are not sharing in prosperity. This is in turn driving a move to robotics, to try to help militaries fight better in the dank, dark urban slums. But what Clarke worries about is a feedback loop, that new technologies like robotics will actually create more losers, more anger, and more conflict.

At a basic level, the Pentagon's vision of surrounding and infesting a city with tiny robotic sensors that are constantly buzzing carries a massive downside. Instead of creating universal observation, it may create universal irritation. It may be like poking an urban hornet's nest with a robotic stick. This may be especially heightened by the varying cultural attitudes toward privacy that can spark conflict. For example, during the Iraq war, insurgent recruiting propaganda falsely claimed that U.S. forces were using night vision equipment to peer into family homes and see what Iraqi women looked like under their clothes. It may seem a silly accusation, but the very first protests against U.S. troops in the restive town of Fallujah actually were sparked by just this confusion in 2003. The protests soon devolved into street fighting that left seventeen dead and helped make the city a hotbed of resistance. With the new plan, this propaganda wouldn't be fiction; the unmanned systems actually would be peering into homes.

But at a broader level, the robotics revolution may add even more "losers" on a global scale to the billions already not sharing in globalization's bounty. Every revolution in technology has its winners and losers. The telegraph was great for news junkies, but bad for the Pony Express riders.

The same will be true with robotics. Numerous professions have already been displaced by very simple robotics, from automobile factory workers to maids, and this will continue as robotics gets more and more capable each year. As the trend plays out, robots won't just be doing blue-collar work, but also service and even white-collar jobs. And for each job they eliminate, there will be one more person competing for the remaining jobs. It's the robot version of outsourcing, just that your job is being shipped to a piece of faceless hardware, rather than some textile worker in Bangkok or engineer in Bangalore.

If history is any guide, many will speak out against what they see as a technology-caused injustice and some might translate their anger into violence. In the early 1800s, textile workers in England began to realize that steam engines and factory machines were starting to put them out to pasture. A social movement soon arose, called the Luddites after a mythical character named Ned Ludd, who had supposedly smashed up two mechanical looms in a fit of rage.

By 1811, the Luddites began to organize. Their gatherings, always taking place in the city, usually culminated in street riots, "machine breaking" (copying Ludd's accomplishment, they would invade factories and smash up property), and pitched street battles with British army units. By 1812, there were more British troops fighting Luddites inside England than fighting Napoleon's troops on mainland Europe. The movement was ultimately crushed in 1813, with a government crackdown that culminated with the execution of seventeen Luddite leaders and the expulsion of many more to the penal colonies in Australia.

Since then, "Luddite" has come to describe anyone generally opposed to technological change. But many like Clarke fear that the new forms of technology such as robotics and AI will spur their rise again as a violent social movement. Tom Erhard is a retired air force officer, who works on the Pentagon's "20XX" program exploring long-range political futures over the course of the first half of the twenty-first century. Akin to Clarke, he particularly worries about resurgent Luddite trends. Moreover, he thinks that our current version might be worse than the rage against the machines of the past centuries, as robotics and AI will polarize society as never before. "It goes to the last frontier of what it means to be human, the ability to think."

The uses and roles of robots "will be a moral battlefield," says iRobot founder Rod Brooks. Many will be highly resistant to these new technologies, many will be uncomfortable with how far they are penetrating society, and some will even see them as threatening to their very values. People got worked up enough about abortion clinics and images in Danish cartoons to kill; it is not a stretch to think they might do the same over some aspect of robots. If this is the case, then many neo-Luddites will also see those that build and use these new technologies as something to be feared, and even stopped at all costs.

The first of these violent neo-Luddites was Theodore John Kaczynski, better known as the "Unabomber." Kaczynski started out as a brilliant mathematician, with a Harvard PhD, but soon ended up in a cabin in Montana mailing pipe bombs to researchers, scientists, and other people linked to the computer industry. At first the explosives were too poorly made to hurt anyone severely. But, much like what happened with IEDs in Iraq, they got better. Ultimately, three people would be killed and twenty-three people wounded by the mysterious bomber.

The FBI had great difficulty figuring out who was sending the bombs, and the only clue was that some of the bomb parts had "FC" inscribed on them. For a time, they thought it stood for "Fuck Computers," but it later turned out to be short for "Freedom Club," what Kaczynski called his movement of one (Unabomber was the FBI's name for their unknown suspect).

In 1995, the still unknown bomber promised that the group would stop its terror attacks if the major media published the FC view of the world, entitled "Industrial Society and Its Future." After great debate, the Justice Department authorized the *New York Times* and the *Washington Post* to publish it for public safety reasons. The porn magazine *Penthouse* also bravely offered to publish the 35,000-word treatise (so that customers would purchase it, for once, "just for the articles"), but was turned down.

The *Unabomber Manifesto*, as it became more commonly known, explained that the reason for the attacks was that humankind was slowly but surely choosing to become "pets" of its machines. "As society and the problems that face it become more and more complex and machines become more and more intelligent, people will let machines make more of their decisions for them, simply because machine-made decisions will bring better results than man-made ones.... People won't be able to just turn the machines off, because they will be so dependent on them that turning them off would amount to suicide." And thus, the manifesto argued, "We therefore advocate a revolution against the industrial system.... Its object will be to overthrow not governments but the economic and technological basis of the present society."

Kaczynski would soon be turned in by his own brother, and sentenced to life in prison. But Clarke and many others worry that he was only a harbinger of worse to come. Indeed, he has already had a copycat in Italy, known as the "Italian Unabomber." More broadly, Kaczynski had a legion of admirers that ranged from the environmental Earth Liberation Front to various anarchist groups. Noting the wide variety of groups that might find reason to be angry about new technologies like AI and robotics, Richard Clarke warns that one of his biggest fears about the brewing neo-Ludditism is the "huge potential for strange bedfellows here, linked up by their common opposition. I would not be surprised to see violent Islamic extremists finding common cause with the Christian far right, for example."

So we may see our governments struggling not just against a growing mass of angry outsiders who are not sharing in the bounty of globalization and technology, but also against more actors like the Unabomber, neo-Luddites who reject the change itself. Such groups may even connect, aid, and inspire each other, working together in "polyglot networks" of opposition to a future they hate.

But tension might also be created at a larger level than just small groups of angry men. The coming century is bringing cultures, societies, and religions into contact at a scale and pace as never before, and in many cases these values can clash and even offend. "Amid galaxies of shining technologies there is a struggle to redefine human meaning. . . . Half the world is looking for God anew, and the other half is behaving as though no god exists," says Ralph Peters.

And into this mix, we inject a new technology. And not just any technology, but one that raises some truly fundamental questions about everything from what is right or wrong in war to what it means to be a human. Computer scientist Hugo de Garis even worries that someday the conflict between those who see technologic progress with robotics as part of humankind's broader destiny and those who find the idea of such a future threatening to their very identity and values could escalate to a major ideological dispute on the level of the duels between fascism and democracy or capitalism and communism. "Since the stake is so high (namely whether the human species survives or not) the passion levels will be high. . . . We have thus all the makings of a major war. About 200 million people died for political reasons in the 20th century (wars, purges, genocides, etc.) using 20th century weapons. Extrapolating up the graph until the late 21st century, with 21st century weapons, we arrive at billions of dead—gigadeaths."

Hopefully, such fears remain in the distant future. But in trying to figure out just what might spark the wars in which robots will fight in the coming years and decades, we are taken full circle, from humanity to technology and back again. Humans remain the driver of wars, even in a world filled with robots that fight them. As Ralph Peters sums up, "The great paradox of this hi-tech age is that its security problems arise from the human heart and soul, domains which remain opaque to technology (and to those who worship it)."

# FIFTEEN

# THE PSYCHOLOGY OF WARBOTS

*Warfare is about changing the enemy's mind.*

—RALPH PETERS

"Human versus robot? How will that play?...The psychology of all this will be important, especially on the side of the people without the high tech."

Eliot Cohen is the director of the strategic studies program at Johns Hopkins University. If there is a Washington "defense establishment," Cohen is one of the key opinion leaders within it, especially among its right wing. Described by one media report as "the most influential neocon in academe," Cohen gained much media attention right before the 2003 Iraq invasion, when President Bush showed up at a public event holding Cohen's book *Supreme Command*. No one is sure if Bush actually read the weighty tome, but the choice was symbolic, as Cohen argued in the book for civilian leaders to exert their influence over military matters. Soon after our interview in his Dupont Circle professor's office in late 2006, Cohen was named counselor to Secretary of State Condoleezza Rice, to serve as her one-man think tank and intellectual sounding board.

Looking the part of a defense intellectual straight out of Hollywood casting, even down to sporting a mean red bow tie, Cohen believes that human psychology will be a key determinant of robots' impact on war. Having also written the book *Military Misfortunes* (a study of miscalculation and defeat in war, which maybe Bush should have also read), Cohen strongly believes that human motivation has usually been the key to victory or defeat. Whether it was Napoleon's army at Waterloo, the kaiser's army at the end of World War I, or Saddam's forces in 1991 and again in 2003, the side that loses a war usually does so because its military hits a psychological breaking point, a time "at which a majority or a disabling minority [of soldiers] refused to go on."

Cohen tells that we don't yet have a full understanding of how people will be psychologically affected by robotic fighting systems, but he thinks there may be lessons from the past. "Its closest parallel may be the effect of strategic bombing. The foe just gets defiant, but also depressed over time." But, says Cohen, there will be a new twist. Unlike the intermittent raids of bombers over Tokyo or Berlin during World War II, the fact that the systems are unmanned, as well as able to operate for days or weeks on end, will give them a psychological punch as never before. "They [the side facing robots] will feel like they are always being watched, that they face a nonhuman foe that is relentless."

Overall, concludes Cohen, the trend should be of great benefit to America, especially against the terrorists and insurgents it faces in what he describes as the current "Fourth World War" (he counts the cold war as the third great global conflict). "It plays to our strength. The thing that scares people is our technology."

Cohen is by no means alone in his belief in the psychological power of unmanned systems among the political establishment. The *Washington Times*, for example, reported that a great benefit of robotic systems is that "unmanned weapons tend to demoralize an enemy." It described that "while soldiers will fight against their enemy if they have a chance to kill the attacker even against all odds, being killed by a remote-controlled machine is dispiriting."

This same conviction extends beyond the D.C. Beltway. For example, Ed Godere, part of the Foster-Miller team behind the SWORDS, believes that "the psychological effects will be significant." He predicts that it will cause "an almost helpless feeling" among anyone unlucky enough to see a robotic machine gun coming at them. Many troops in the field agree. Army staff sergeant Scott Smith says that "without even having to fire the weapons . . . it's total shock and awe."

## FIRST CONTACT

In 1532, Atahuallpa was emperor of the Tawantinsuyu, better known to us as the Incan empire. Located in what is now Peru, Atahuallpa's domain was the largest and richest of the empires in lands not yet reached by European explorers. Life was just getting good for Atahuallpa. He had beaten his brother in a civil war for the throne and was on his way back to his capital. Only a quick detour was needed to check out a tiny band of strange visitors that had just entered his lands. A proud and cruel king (he had just forced his defeated brother to watch his children be hacked to death), and at the head of a battle-hardened army of eighty thousand warriors, Atahuallpa believed he had little to fear.

Atahuallpa and his army soon reached the encampment of the visitors, who invited the emperor to a peace ceremony. Carried in on a litter borne by the highest nobles of his court, and accompanied by a personal guard of four thousand men, Atahuallpa entered the small courtyard where the visitors were camped. A delegation greeted him. One of the visitors, a man wearing brown robes, offered him a gift and told him, through a translator, that this "book" supposedly carried the word of God.

Never having seen such a thing before, and believing himself to be a representative of the gods, the emperor shook the gift and, when no noise came out, indifferently tossed it to the ground. He asked, "Why doesn't it speak to me?" The man in brown cried out with anger when the packet of papers hit the dirt and gave a signal of some sort. Soon after, the air filled with a massive explosion and dozens of men ran in from the buildings that surrounded the courtyard. They were wearing seemingly invincible suits of metal that turned back the points of arrows and spears. They wielded strangely sharp, unbreakable metallic weapons that cut through flesh with ease, and, even more frightening, pointed sticks that spat lethal flames. Most terrifying, though, were the strange creatures that also charged out, which had four legs like a beast, but the upper body of a human warrior.

There were only 168 of these new visitors, but as they charged at the emperor and his 4,000 men, the effect was paralyzing. Atahuallpa's guard was quickly chased away or slaughtered. The highest nobility of his kingdom were killed at his feet. When none were left to hold up his litter, the emperor was captured.

Seventy-six thousand of Atahuallpa's warriors were waiting in the fields just outside the town, and milled about, wondering what to do when they heard the strange noises and then saw their noblemen running for their lives. Twenty-seven of the man-beasts then emerged from the square and put the entire army to flight. It wasn't so much of a battle as a massacre; it ended only after the visitors gave up killing the fleeing Incan warriors when their arms grew too weary. The captured emperor then offered the visitors a ransom to set him free, enough gold to fill a room twenty-two feet long, seventeen feet wide, and eight feet high. The visitors agreed. But after these strange, fearsome men had their gold, they reneged. They executed Atahuallpa and took over his empire.

As the science fiction writer Arthur C. Clarke of *2001: A Space Odyssey* fame once observed, "Any sufficiently advanced technology is indistinguishable from magic." Nowhere is this more true than in war. Time and again, warring sides have used new technologies not only to kill more efficiently than their foe, but also to dazzle them into submission. The case of Atahuallpa, unlucky enough to become emperor just before the arrival of Francisco Pizarro and his tiny band of Spanish conquistadors, is a powerful example of just how shocking and powerful new weapons of war can be.

Cannon, armor, swords, muskets, and horses were particularly devastating to the Inca as they lived in a time when communication was difficult and information was hard to come by. This was not merely their first contact with such weapons, but they had never even conceptualized the very possibility of such fearsome technologies before. Yet even in our information-saturated world, the use of new weapons technologies can still have a powerful psychological effect. For example, an elite Iraqi Republican Guard colonel explains that the reason he felt his forces gave up so quickly during the 2003 invasion was that "U.S. military technology is beyond belief." He described how American air power, able to strike with constant, pinpoint precision, whether in day or at night, took his unit by surprise, made it feel like any sort of organized resistance was impossible, and ultimately collapsed their spirit.

As Atahuallpa could have foretold, the new generation of unmanned systems already have had such a psychological effect on the minds of adversaries, specifically in sowing alarm and confusion. For instance, an official with the U.S. military's Joint Special Operations Command recounted a meeting with elders in the tribal region of Pakistan, the area where al-Qaeda leaders were reputed to be hiding out and the site of more than fifty drones strikes from 2006 through 2008. One of the elders was enamored of the sweet-tasting bread that was served to them at the meeting. He, however, went on to tell how the Americans had to be working with forces of "evil," because of the way that their enemies were being killed from afar, in a way that was almost inexplicable. "They must have the power of the devil behind them." As the official recounted with a wry chuckle, "You have a guy who's never eaten a cookie before. Of course, he's going to see a drone as like the devil, like black magic."

Troops are also finding that encountering a strange new, unmanned weapon conveys more than merely a psychological punch. War-game testing has found that foes tend to focus on "such an unusual technology" as the SWORDS; it is such a center of attention that this can be taken advantage of. One team was facing a group of hostage takers holed up in a building. So they sent a SWORDS to drive up to the front. While the hostage takers gathered on one side of the building to watch the odd little lawn mower with a machine gun track forward, a special forces team went around the back of the building and ambushed them from behind.

Another strange psychological lesson came from a real-world hostage crisis in Milford, Connecticut. A gunman wouldn't let police come anywhere near him, because he thought they might try to surprise and overpower him. But he was willing to let the police send a robot to carry in a phone. As the hostage crisis dragged on, the police called him and offered to send him and the hostages some drinks. The gunman agreed, but again, wouldn't let any human come near, as it might be a trick. Thinking robots more trustworthy than the fuzz, he agreed again to let the robot bring in the drinks. Of course, robots can have tricks up their sleeves as well. The robot delivered coffee that had been laced with knockout drops and, as the gunman fell asleep, the crisis ended with no one getting hurt.

The obvious problem is that what is "unusual" wears off and such tricks only work so many times. While it was too late for Atahuallpa, the Incas did grow accustomed to the Spanish weapons. Just three years later, the dead emperors' generals launched a surprise uprising that evolved into an insurgency that lasted for years. Similarly, the Iraqis soon adjusted to the American ability to launch pinpoint airstrikes, and learned that the easy answer is not to mass troops in open terrain. Word travels fast, people adjust, and the psychological power of something new and different wears off quickly.

## THE "CREEP" FACTOR AND THE UNCANNY VALLEY

David Hanson, a former employee at Disney's Imagineering Lab, makes robots that "creep people out." Hanson's robots look like machines from the neck down, but have incredibly realistic heads. Their lifelike "skin" is made using a material Hanson invented called Frubber. His "Hubo Einstein" robot, for example, has a mechanical body, but the head and face of Albert Einstein. One scientist described it as "spookily cool...a giant step forward." Hanson is also "an avant-garde artist" who puts together art shows. In the long line of artists doing self-portraits, for one show he made a robot modeled on himself. Only his self-portrait was a "large homeless robot figure in a box." His intent was to use a robot to take viewers out of their "comfort zone."

Hanson is proud that his robots pose "an identity challenge to the human being.... If you make it perfectly realistic, you trigger this body-snatcher fear in some people," he tells. "Making realistic robots is going to polarize the market, if you will. You will have some people who love it and some people who will really be disturbed."

Inspired by Brian Aldiss's short story "Supertoys Last All Summer Long" (the basis for the Steven Spielberg film *AI*), Hanson is currently at work on robotic "supertoys." He explains that these robots will have evolving personalities and grow up with the child. Looking a little like robotic versions of the Oompa-Loompas from Willy Wonka's Chocolate Factory, Hanson's robots are two feet tall and have cartoonish faces. The name he gave the first of these new robots is Zeno, the same name as his eighteen-month-old son.

Hanson sees his work as "changing the expectations of machines," and he ultimately hopes that the "social robotics" field will become so much bigger than the military robotics industry that "market forces will shape things toward friendlier robots." This remains to be seen. But his work clearly illustrates how robots can be designed to influence the "attitudes, feelings, emotions, and ultimately the behavior" of those who see them. This quote tellingly comes not from Hanson or a science journal. It is the Pentagon's definition of psychological operations.

History is filled with all sorts of ways that weapons and uniforms can be designed to create some sort of psychological reaction among the foe. For example, the famed British Redcoats of the Revolutionary War era wore that color so that blood wouldn't show up on their uniforms from a distance. Among their units were grenadiers, especially tall soldiers, who wore huge peaked hats to make them look even taller. The effect of seeing them on the battlefield was akin to watching a line of giants marching toward you, whom your bullets seemed not to hit.

The difference between the Redcoats' psychological effect and those of the conquistadors is the difference between fright and fear. As Sigmund Freud explained, fright is the state one falls into "when confronted by a situation [for] which we are unprepared," akin to what the Incas felt when seeing guns for the first time. Fright, though, can wear off quickly as one grows accustomed. Fear, by contrast, comes from "a definite object of which one is afraid." It is something you can see and even understand, but it still evokes a state of terror that causes dread or panic. The patriots knew the Redcoats were men, but it didn't make them any less fearsome.

Current robots on the battlefield tend to have a totally utilitarian look, but it still gives them some psychological punch. Foster-Miller's SWORDS robot, for example, got its design from just mounting a machine gun on top of an older robot's chassis. Even then, as one magazine quipped, the SWORDS "makes *Robocop* look like Officer Friendly."

Strategic thinker Eliot Cohen thinks such an unintentional effect is all well and good, but something more may have to be done. "We will have to figure out how to maximize the psychological impact of it [a robot]. We will have to think not merely in terms of costs and benefits and how to get steel on target, but much more. How it gets that angry insurgent from being eager to fight to thinking that there is no point in it, there is no chance to win against a relentless foe."

If not all robots are going to look like Disney-Pixar's cute and cuddly WALL-E (though one of the British army's robots is a dead ringer), the first and easiest step to fearing up a robot is to equip it with effectors that can play a role in psyching out the enemy. If history is any guide, we can anticipate that this won't just be about giving them a scary look, as with the Redcoats, but also a scary sound. The ancient Chinese set off fireworks to spook enemies' horses, while the Nazis mounted sirens on the wings of their Stuka dive bombers during World War II; often the high-pitched noise of the diving plane created even more chaos among the troops on the ground than the bomb itself.

The sound that the U.S. military will use to put chills down enemy soldiers' spines most likely will come from the real experts, Hollywood. The military has long used Hollywood special effects for psychological operations. During the 2004 battle of Fallujah, for example, the marines set up loudspeakers around the city and broadcast the sinister laughter of the alien from the *Predator* movie. They were hoping to spook out the insurgents, as well as drown out the sermons that the insurgents broadcast back at them from each of the city's mosque towers. The noise was so constant that a marine joked that the siege should be called "Lala-Fallujah" (after the famous alternative rock concert festival Lollapalooza).

Given that the marines' new ground systems like the Gladiator come with their own loudspeakers, there's nothing to prevent them from doing the same with their robots, to create a more mobile fear factory. Of course, there are always downsides to these kinds of operations. After hearing the Predator's evil laugh one too many times, a marine scout team on the front lines radioed back to base to tell them the noise was having more of a psychological effect on them than on the enemy. "That's not funny anymore. You keep that shit up and we're coming back in."

The same kind of turn to Hollywood will likely take place with the overall design of unmanned systems. Says military robots pioneer Robert Finkelstein, if you want to truly have a psychological effect, "Make 'em look like Frankenstein's monster. Or make them look like creatures from *Star Wars*.... Make 'em hideous." Scientists at one military robotics firm similarly report how the military inquired if they could make a system that looked like "the hunter-killer robot of *Terminator*." Not much of a sci-fi buff, Eliot Cohen suggests that we instead turn to nature, that "we exploit the basic human fear of bugs." Whatever the inspiration, as Finkelstein concludes, there are "infinite possibilities" of how the looks and design of systems might be manipulated to heighten an enemy's fear. One day, specializing in scary designs "might even be a profession."

But as David Hanson's work illustrates, the creepiest robots of all may be the ones that look mostly human. He notes that the reactions to his robots vary. "Some people take it as a thrill, some think it is neat.... Others find it just creepy and threatening." He explains that different parts of the brain deal with social relations versus identifying objects. So when the human brain sees "an object acting as a human, it sets off natural alarms, so to speak."

Hanson is tapping into a phenomenon called the "uncanny valley." Researchers are finding that the more human attributes a machine has in design, the more people's connection seems to increase. As Hiroshi Ishiguro, the maker of such humanoid robots as the sexy Repliee android, explains, "The keyboard and the monitor are primitive. My brain was not designed to watch a display and my fingers were not designed to type on a keyboard. My body is best suited for communicating with other humans. The ideal medium for communicating with a computer is a humanoid robot, which is, of course, basically a computer with a humanlike interface." But this doesn't mean that we are entirely comfortable with ever more lifelike robots. "People's empathy increases until a sudden point at which the machine seems like the living dead, like a frightening imposter."

This is the "uncanny valley," when the appearance of a robot is close to a human but not close enough. At this point, a robot's look is most disturbing. The end of the "valley," when the fear goes away, is when the robot becomes so human in its appearance that it's hard to tell the difference. So the relative length of the "valley" is the space at which robots freak you out. Explains AI expert and psychologist Robert Epstein, "If a human can't tell it isn't human, no worry.... Humans are also okay with it if it doesn't look like a human at all, like Johnny 5 [the robot from the movie *Short Circuit* that looks a bit like a PackBot]." It's that part in the middle of the uncanny valley that is so disturbing. "It's like interacting with a corpse, a moving corpse. It makes you uncomfortable."

## THE OTHER SIDE

Psychologists like Epstein, however, are discovering that an encounter with a robot, whether it's a sexy android, a machine-gun-carrying lawn mower, or even one that seems straight out of *Night of the Living Dead*, is not just straight "shock and awe." "It's not just that a certain type of machine or robot makes us uncomfortable. It very much depends on who we are." Just like with those early guns and armor and the Incas, the effect greatly depends on one's prior experience with similar technology. "The more familiar with technology, the shallower the uncanny valley; the less familiar, the greater the effect." David Hanson similarly described that for people seeing his lifelike robots, "If they are not used to robots, the negative reaction is more likely."

Age also can be a factor. Oddly, children up to the age of roughly three years care the least about a robot's appearance. They accept almost any bizarre look matter-of-factly, good news for the robot-nanny industry. But around the age of four years, appearance becomes highly important to a child, with a wide "valley" that doesn't tend to go away until the teenage years are over. Ishiguro, the maker of the Repliee, first witnessed this aspect of the valley with his original version of the android, which was shaped to look like his four-year-old daughter. "When my daughter first saw her android she began to cry."

But as an iRobot scientist tells, "The uncanny valley is definitely cultural as well.... The Japanese will put up with a robot that even freaks me out, but they are totally comfortable with that."

Messaging across cultures has always been difficult, especially in war. In World War II, General Curtis LeMay ordered American bombers to use firebombs on Japanese cities, with the intent to terrorize the Japanese public into a realization that continuing the war was futile. The raids killed hundreds of thousands, but many in Japan instead interpreted the "message" as that it was dangerous to surrender unconditionally to an enemy willing to drop flaming napalm on civilians living in wooden houses. The United States tried similar messaging with its bombing during Vietnam, this time influenced by mathematical models and strategic game theory. As army colonel H. R. McMaster explains, such approaches proved "fundamentally flawed.... The strategy ignored the uncertainty of war and the unpredictable psychology of an activity that involves killing, death, and destruction. Human sacrifice in war evokes strong emotions, creating a dynamic that defies systems analysis quantification." In short, the message you think you are sending is not always the one that the other side actually receives.

This same phenomenon may be playing out with unmanned systems as well. Much in line with his sense that robotics can help the U.S. military push its enemies' psychological buttons, Eliot Cohen describes his belief of what an insurgent in Iraq thinks about such systems. "They are likely asking, 'What tricks are the Americans going to pull out of their bag next?' "

The troops who currently use unmanned systems are also generally hopeful about the psychological effect their robotics might be having on the other side in Iraq. As one drone pilot explains, "I think that it will discourage them more than anything. I know that if I was out on a future battlefield risking my life, my emotions would be out of whack knowing that I could be killed and the only damage I could inflict was to a robot. For today's battlefield, the UAV is used largely as a deterrent. AI forces ['Anti-Iraqi,' the official term at the time for insurgents] know we are out there. They know they are constantly being watched. The fear of being caught in the act keeps a lot of would-be insurgents out of the fight." Concluded one air force officer, "It must be daunting to an Iraqi or to an al-Qaeda seeing all our machines. It makes me think of the human guys in the opening to the *Terminator* movies, hiding out in the bunkers and caves."

The irony is, of course, that the humans in that movie were the side the audience was supposed to root for, and who overcame their fears to beat back the machines. So while there is no way to formally test this proposition out, in the summer of 2006 I was connected with two Iraqi insurgents by a trusted intermediary. Both of them were Sunnis, who were opposed to the U.S. presence in their homeland and had decided to join the insurgency. Notably, one was a former engineering student. Even with this background, he described the various unmanned systems his American foes were using as a bit bewildering. "I didn't really imagine that military industry reached such levels of imagination."

It may have been a bit of posturing, but the two also discussed how they were not all that intimidated by the technologies, as some of the strategists like Cohen and others might have hoped. "It is not really a matter of how sophisticated you [*sic*] weapons are," one told. Instead, they expressed a confidence that they would find ways to adapt and take advantage of the technologies soon themselves. Sounding almost like an Iraqi version of Ray Kurzweil, the former engineer expressed his sense that this trend would likely continue, as "the modern age is also marked by increasing trends towards automation."

What they expressed in the limited interviews I was able to carry out squared very much with what other experts with far more experience with insurgents have found. Nir Rosen is a reporter and the author of *In the Belly of the Green Bird*, a study of the early days of the Iraq insurgency. Born in New York City, but having learned to speak Arabic with an Iraqi accent in his youth, Nir was able to gain the trust of local civilians and insurgents in a way that few other journalists could. Indeed, he was the only Western journalist to spend time inside Fallujah among the insurgents before the major battles there in 2004. When we spoke in 2006, Rosen was just back from Somalia, having gained a meeting with the armed Islamist faction that had taken over Mogadishu.

Rosen told how during his time in Fallujah, the insurgents were "definitely aware of UAVs and other American technology, but not always aware of their full capabilities. They wouldn't understand the things they could and could not do." He described how they would sometimes give the systems credit for things not yet technologically possible, while other times make simple mistakes in underestimating what was possible decades ago.

As far as the supposed psychological effect, Rosen responded that "you have to remember that insurgents only have their own weapons, whereas they are fighting a force of F-16s [fighter jets] to tanks, up-armored Humvees to platoons of troops in helmets, flak vests, knee pads, boots, etc.... For insurgents, it already feels like they are fighting robots of a sort."

Rosen sensed that fighting more and more unmanned systems would "not be a huge quantum leap" to the insurgent psychology. "With things like F-16s, it's not like they are fighting face-to-face now anyway." Instead, he saw that what seemed like an overreliance on these systems is even backfiring psychologically on the Americans. "In their rhetoric, they'll make fun of the Americans for not being man enough to fight face-to-face." Ultimately, though, he felt the insurgents in the field would understand why the United States was using them and may even follow suit with whatever technology they could. "They will adapt very readily.... It's just about achieving your ends."

So, at least within the psychological war of ideas, unmanned systems may not convey the messages we desire. Instead, they may send rather undesirable and unintended signals about our intentions and even our character.

For instance, unmanned systems are intended to reduce casualties. But as Peter Feaver, a Duke University professor turned Bush administration National Security Council adviser, asks, "What is Osama bin Laden's fundamental premise if not the belief that killing some Americans will drive our country to its knees?" Indeed, robots' very rationale of limiting human risks runs counter to the local values in many of the most important theaters of the war against terrorist groups. As one marine general explained, in places like Afghanistan, especially among the Pashtu tribes in the mountainous south, "Courage is the coin of the realm." Showing personal bravery, which you cannot do with a robot, builds trust and alliance in a way that money or power never can.

Finally, the systems are hoped to limit the number of "boots on the ground." But the effect can send an unintended message, blunting the psychological and even tactical effects of defeat on a foe. As Bevin Alexander, the author of *How Wars Are Won*, explains, "Victory comes from human beings moving into enemy territory and taking charge." Otherwise, you repeat the experience of the Sunni Triangle in Iraq. The future hotbed of rebellion wasn't occupied until weeks after Baghdad fell in 2003, and local would-be insurgents instead got the signal that they had never been defeated.

Rami Khouri is well placed to evaluate the effect of our new technologies in the particularly important area of the Middle East. The director of the Issam Fares Institute of Public Policy and International Affairs at the American University of Beirut, Khouri is also the editor-at-large of the Beirut-based *Daily Star* newspaper. When we spoke in 2006, the electricity in his Beirut home was still cutting in and out, the effect of the Israeli bombardment (coordinated by a near-constant flyover of Israeli UAVs) during the war between Israel and Hezbollah.

Khouri described how it felt to be on the receiving end of unmanned targeting and an all-seeing eye in the sky. The kind of depression that Cohen had hypothesized was certainly present, as the normally ebullient Khouri fretted over whether he would have enough food for the week ahead if the electricity went out again. But so was the defiance. Khouri is a leading voice of moderation in the region and is so much of an admirer of the United States that he is an avid baseball fan. Yet even he described how, instead of cowing the populace, these sorts of attacks were reinforcing the position of radical groups like Hezbollah. The use of such technologies was "spurring mass identity politics.... The new combination of Islamist, Arab nationalist and resistance mentality is seen as an antidote to the technology discrepancy."

Instead of receiving a message that they were overmatched, "it is enhancing the spirit of defiance." Khouri explained how both the Hezbollah fighters in the field and the broader Lebanese populace saw that "the enemy is using machines

A soldier with a PackBot. When asked how science fiction writer Isaac Asimov might react to real-world warbots such as these, iRobot's cofounder and chairman Helen Greiner responds, "I think he would think it is cool as hell."

PHOTOGRAPH COURTESY IROBOT

A Predator drone is prepared for launch. The number of drones in the U.S. military went from only a handful in 2001 to some 5,300 in 2008. After 9/11, Pentagon buyers told one robotics executive, "Make 'em as fast as you can."

PHOTOGRAPH COURTESY U.S. DEPARTMENT OF DEFENSE

The Predator drone, which can stay in the air for twenty-four hours, is one of the most widely used and effective weapons in the force. "If you want to pull the trigger and take out bad guys, you fly a Predator," says one report.

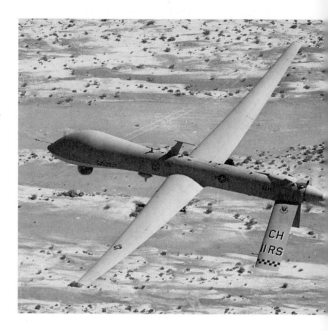

A special task force using drones armed with weapons such as these found and killed more than 2,400 Iraqi insurgents, in just one year.

The Global Hawk spy drone can take off by itself, fly 3,000 miles, spend a day spying on an area the size of Maine, fly back 3,000 miles, and then land itself. Some uncharitably say it looks like "a flying albino whale."

In reachback operations, the drones over Iraq and Afghanistan are actually flown by pilots sitting in Nevada. As one described fighting from a cubicle, "It's antiseptic. It's not as potent an emotion as being on the battlefield." Says another, "It's like a video game. It can get a little bloodthirsty. But it's fucking cool."

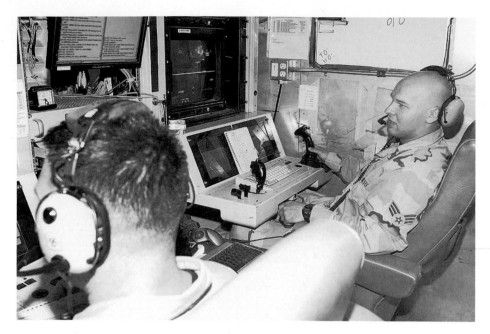

While the pilots are no longer at risk, the experience of fighting from home bases, some 7,500 miles away, does bring new psychological twists to war. "You see Americans killed in front of your eyes and then have to go to a PTA meeting," tells one pilot.

PHOTOGRAPH COURTESY U.S. DEPARTMENT OF DEFENSE

Zero ground robots were used in the invasion of Iraq in 2003. Some twelve thousand were in service there by the end of 2008. As one officer put it, "The Army of the Grand Robotic is taking place."

PHOTOGRAPH COURTESY U.S. DEPARTMENT OF DEFENSE

A MARCBOT on patrol with U.S. troops in Iraq. A jury-rigged version of the tiny robot was actually the first ground robot to draw blood on the battlefield.

"Stanley," the robotic car from Stanford University that won the DARPA Grand Challenge race. By turning the research into a competition and offering prize money, the Pentagon was able to entice scientists and college students who wouldn't normally work on war technologies to help solve its battlefield problems.

Nicknamed "R2-D2" by the troops, the Counter Rocket Artillery Mortar system uses an automated machine gun to shoot down incoming missiles and rockets, which humans would be too slow to react to. A new version will mount a laser.

A TALON robot in action. These technologies "save lives," says a former Pentagon official. But he also worries that "there will be more marketing of wars. More 'shock and awe' talk to defray discussion of the costs."

Military researchers are trying to make robots easier to control by "playing to the soldiers' preconceptions." And with young soldiers today, that means video games.

Two young army soldiers prepare to launch a Raven drone. According to one report, one of the unexpected results of the new technologies is a "military culture clash between teenaged video-gamers and veteran flight jocks for control of the drones."

An army infantryman tosses a Raven drone into the air. The drone has proved so useful and popular in Iraq that its maker was approached by the Chinese military for a demonstration. Some forty countries now make military robots, meaning the revolution will not just be an Ameircan one.

A Fire Scout helicopter drone fires a missile on a target below. "This new technology creates new pressure points for international law," tells one human rights worker. "You will be trying to apply international law written for the Second World War to *Star Trek* technology."

A Warrior robot uncovers a hidden roadside bomb. While robots are a revolutionary technology, war still remains messy and difficult, with an enemy already learning how to fight back.

PHOTOGRAPH COURTESY IROBOT

A soldier works on a TALON robot in Iraq. While these robots save lives, they also might be sending out an unintended message to the other side. One news editor in the Muslim world commented that such technologies made Americans look like "cowards because they send out machines to fight us.... They don't want to fight us like real men, but are afraid to fight. So we just have to kill a few of their soldiers to defeat them."

PHOTOGRAPH COURTESY FOSTER-MILLER

SWORDS, made by Foster-Miller, is a robot armed with the user's choice of weapons, ranging from machine guns to rockets. It gives new meaning to the term "killer app."

PHOTOGRAPH COURTESY FOSTER-MILLER

Scooby Doo, one of the very first robots "killed in action" in Iraq, blown up by an insurgent's roadside bomb. It now rests in the offices of its manufacturer, iRobot. One commander put a positive spin on such losses, "When a robot dies, you don't have to write a letter to its mother."

PHOTOGRAPH COURTESY THE AUTHOR

Scooby Doo's human squadmates kept track of the number of dangerous missions the robot went out on, keeping them alive. When the robot could not be repaired, it left one soldier very upset. He didn't want a new robot but "wanted Scooby-Doo back."

One concept of robotic warfare is the "warfighter's associate" idea, where a mixed team of robots, such as the PackBot here, and human soldiers would jointly carry out missions.

Sometimes, for all their sophistication, robots still need a little help from their friends.

PHOTOGRAPH COURTESY IROBOT

Unmanned submarines are increasingly being used for the most dangerous roles underwater as well. This includes hunting for mines and patrolling waters close to shore, missions considered too risky to send in expensive nuclear-powered manned submarines.

PHOTOGRAPH COURTESY U.S. DEPARTMENT OF DEFENSE

Asimo, a humanoid robot built by Honda. Standing just over four feet tall, this real-world version of the Twiki robot from *Buck Rogers*, can run, jump, dance, recognize faces, and even hook up to the Internet.

PHOTOGRAPH COURTESY THE AUTHOR

The Actroid robot not only is incredibly lifelike, but can also understand forty thousand phrases in four languages and give answers to more than two thousand questions it might encounter. Owned by the same company behind Hello Kitty, its Web site also notes that "rentals are now available." Such lifelike robots will open up new questions of ethics and rights.

PHOTOGRAPH COURTESY THE AUTHOR

Wakamura is a cross between a house sitter and nanny robot. It is able to patrol the house, call the police or doctors in an emergency, and wake the family in the mornings and tell them about the weather and the news. In Japan, the little robot has also become a "companion" for elderly shut-ins.

PHOTOGRAPH COURTESY THE AUTHOR

As part of its design, Wakamura can recognize faces, make eye contact, and start conversations. But robots are notoriously tight-lipped during interviews.

PHOTOGRAPH COURTESY THE AUTHOR

One concept being explored for twenty-first-century war at sea is the mothership, where a warship would serve as the hub for a tiny fleet of unmanned drones and submarines.

Described as looking like "a set piece from the television program *Battlestar Galactica*," the X-45 UCAV is designed to take on the most dangerous roles in the air and, perhaps, even replace manned bomber and fighter planes one day.

One focus of Pentagon research is on bioinspired robots, such as tiny insectlike robots that could fly up to windowsills and perch and stare inside or climb up walls or into pipes. Many worry about the end of privacy that these robots may portend.

PHOTOGRAPH COURTESY U.S. DEPARTMENT OF DEFENSE

The Crusher is a prototype of a next generation autonomous robotic fighting vehicle. The current robots at war are already outdated. The concern, as Isaac Asimov once said, is that "science gathers knowledge faster than society gathers wisdom."

PHOTOGRAPH COURTESY U.S. DEPARTMENT OF DEFENSE

to fight from afar. Your defiance in the face of it shows your heroism, your humanity.... Steadfastness is the new watchword. Take the beating and keep fighting back."

As an Arab moderate, Khouri was not happy about such reactions. But then again, he was also not happy about having spent the last few weeks watching UAVs fly over as his city was bombed. Indeed, he talked about how the unmanned drones somehow made him "even more angry" than the manned F-16s.

Khouri's explanation of how those on the ground viewed unmanned systems in the Lebanon war was very much like the reactions of the insurgents in Iraq. Rather than creating just fear, fright, and depression, such systems were also unintentionally sending messages of weakness, and even vulnerability. As he concluded, "The average person sees it as just another sign of coldhearted, cruel Israelis and Americans, who are also cowards because they send out machines to fight us, ... that they don't want to fight us like real men, but are afraid to fight. So we just have to kill a few of their soldiers to defeat them."

## THE EVIL EMPIRE

When people talk about the psychological war of ideas that takes place in conflict, they are often not talking merely about the effects on the field of battle, but also among the broader populace. Geopolitics is not a popularity contest, but it is dangerous to disregard international public opinion to such a degree as to assist the recruitment and growth of radical, anti-American groups. If you lose your credibility and reputation, you alienate your allies, reinforce your foes, and shoot your own ideas and policies in the foot. General David Petraeus, the commander in Iraq, once described these aspects as 80 percent of the fight

Unfortunately, by most metrics, the United States is losing this war. In a few short years, America went from being viewed as the beacon on the hill of freedom, Coca-Cola, and blue jeans that won the cold war to the dark home of Abu Ghraib, Gitmo, and orange jumpsuits. Already at the bottom of a deep hole, we can't afford to dig much deeper.

Hence, former assistant secretary of defense Larry Korb argues, "Unless you are refighting some form of World War II, your warfighting must include some part of trying to sway people.... If the U.S. doesn't handle robotics right, it will undermine [our] moral standing, and the U.S. can't be a global leader without such standing." John Pike of the Global Security organization concurs. "This [the robotics revolution] opens up great vistas, some quite pleasant, others quite nightmarish. On the one hand, this could make our flesh-and-blood soldiers so hard to get to that traditional war—a match of relatively evenly matched peers—could become a thing of the past. But this might also rob us of our humanity. We could be the ones that wind up looking like *Terminators* in the world's eyes." Noah Shachtman sums it up with another sci-fi reference. "The optics of the situation could look really freaking bad. It makes us look like the Evil Empire [from *Star Wars*] and the other guys like the Rebel Alliance, defending themselves versus robot invaders."

A concern is that the uncanny valley may also have a cultural distance to it, as it is widened by a lack of familiarity with technology. When much of the Christian world was burning down libraries during the Dark Ages, the Muslim world was the home and protector of much of modern science and mathematics, flourishing in places like Córdoba and the House of Wisdom in Baghdad. But today, the popular penetration of science in the Muslim world has been stifled by a combination of backward-looking fundamentalists who fear anything new and corrupt regimes that look at science as simply something to buy but not understand. Spending on science and technology in the region is 17 percent of the global average, with the region falling behind not just the West, but also the poorest states in Africa and Asia.

The region's media doesn't help much either. As an example, rather than celebrating the only two Muslims to have won a Nobel Prize in the sciences, a show on Al Jazeera in 2006 described that they should be shunned, as the Nobel Prize "encourages heresy. It encourages attacks against the heritage, and encourages those who scorn their people and their culture." The show went on to describe the world's highest scientific honor as a part of a conspiracy stemming from "the Elders of Zion." Given this kind of message, it's not surprising that the journal *Nature* lamented that science in the region lacks "a cultural base."

As a result, differing interpretations of the technology certainly could reinforce an already growing chasm. Retired Pakistani lieutenant general Talat Masood is uniquely qualified to assess both the technology and the gaps in understanding that could play out on the "street" in the Muslim world. Masood, who served in the Pakistani army for thirty-nine years, including as the man in charge of military technologies, characterizes the region's impression of American strategy and doctrine as that of "distant war." That is, the United States has a great willingness to use force, but only if it can do it from afar with high technology, limiting as much as possible its human exposure on the ground.

Masood, whose former colleagues trained up the Taliban in the 1990s, described the technology that the U.S. military was using as "amazing," but also as causing "great anger" in the region. "This type of warfare seldom involves distinct front lines. Fighting has taken place in a confusing mix of friend and adversary, usually directed from afar with occasional failure in communications systems bringing death and destruction to civilians. There is a lack of understanding by the U.S. of the human realities and a marked insensitivity about the casualties of the opponent, and at times even of their own forces. Implications of the RMA are thus broad and profound and a frequent cause of creating a major rift between the U.S. and the Islamic world."

Similarly, he described how people in the region felt that "distance warfare, due to its relative safety, acts as a ready incentive for the U.S. to use military force in pursuing its foreign policy objectives." But this comes at a cost, he found. "Over-reliance on the military instrument has brought under sharp scrutiny the great values and political principles of the U.S. that many in the Islamic world admired and respected.... The advent of 'distance warfare' has profound implications for the battlefield and for America's global strategy. It is fast transforming the relationship with its allies in the Islamic world. Undoubtedly, the U.S. has been able to militarily overwhelm its adversaries, but in every case, whether it is Afghanistan or Iraq, it has vastly complicated the prerequisite of building the structures of peace." In short, warned Masood, "The concept of 'shock and awe' could drive moderate and uncommitted civilians toward anti-Americanism."

Other regional observers agreed strongly with this view. As a security expert in Qatar summed up, "How you conduct war is important. It gives you dignity or not." Their reactions also appeared to confirm a sense that America was coming across as a menace, using its high technology to pick on the little guy. Indeed, in 2009, the News of Pakistan declared that "Predator [drone] attacks against extremists inside Pakistan [have] now moved America into the position of 'principal hate figure' and all-purpose scapegoat."

Even pop culture in the region echoes the experts. By the end of 2008, for instance, the number of American drone strikes into Pakistan surged to a rate of almost one every week, triple the rate of the year before. The issue became so pervasive that "drone attack" became a vernacular phrase in Urdu. Perhaps unco-incidentally, one of the most popular songs in Pakistan around that same time was "Chacha Wardi Lahnda Kyo Nahen?" ("Uncle, Lose the Uniform Why Don't You?"). The song was played at street protests and even became a popular ringtone for cell phones. Its lyrics give a hint at how America's "distance war" is being portrayed: "America's heartless terrorism, Killing people like insects, But honor does not fear power."

Someone who has started to give credence to the unintended psychological consequences of using robots in war is Mubashar Jawed "M.J." Akbar. Akbar is an Indian Muslim who is the founding editor of the *Asian Age*, India's first global news daily. He is the author of eight books, including most notably *In the Shade of Swords*, which came out just before the Iraq war and warned America not to underestimate the brewing anger in the region. Also a columnist read by millions in newspapers across South Asia and the Middle East, Akbar mixes smart analysis with a finger on the pulse of the region.

Akbar expects the future media coverage in his region of robotic systems to be "frightful." He explains, "It will be like when tanks were first used in World War I. When they were introduced, they were described like a weapon of horror, like a large monster, not a weapon....It will excite lots of references to movie horrors. It will be seen as evil, by the way." Indeed, given regional distrust of America, if any mistakes do occur, "then the region will assume you meant for it to happen."

In talking about the Israeli use of UAVs, Rami Khouri in Lebanon had observed that "the general reaction is of an evil, brutal enemy that will use any means to accomplish its goals. Some might say, 'It's too tough. Just give up.' But for a lot of people it will spark a greater desire to fight back." M.J., living in South Asia, sees a similar message going out from American use of unmanned systems to the broader Muslim world. "It will be seen as American cowardice. In war terms, if you are not willing to sacrifice blood, you are essentially a coward." He continues, "These systems will show the pathway to your defeat unintentionally. They create a subtext that shows that you don't want to die....That all we need to win is to frighten them."

Clearly, conflicts in places like Iraq and Afghanistan are bringing together combatants with vastly different understandings of war, the role of the warrior, and the meaning of sacrifice. One side looks at war instrumentally, as a means to an end, while the other sees it metaphysically, placing great meaning on the very act of dying for a cause. It is for this reason that completely different interpretations are made of the same act. A person who blows himself up can either be a martyr and *shaheed* or a murderer and fanatic. There is no in-between.

Unmanned systems take this collision of human psychologies to the next level. They are the ultimate means of avoiding sacrifice. But what seems so logical and reasonable to the side using them may strike other societies as weak and contemptible. Using robots in war can create fear, but also unintentionally reveal it.

It is this link that leads Akbar to conclude that another unintentional effect must be watched out for. The greater the use of unmanned systems, the more likely it will motivate terrorist strikes at America's homeland. "It will be seen as a sign of American unwillingness to face death. Therefore, new ways to hit America will have to be devised.... The rest of the world is learning that the only way to defeat America is to bleed her on both ends. The [American] public responds to casualties and to bleeding of the treasury, so if something goes on long enough they get tired."

Disturbingly, I heard the same conclusion time and again from other regional experts. Speaking from his experience in Iraq, Nir Rosen expects that the continuing trend will "encourage terrorism," maybe especially among those not fighting that way now. As he explains, it is important to understand that in places in Iraq, not every fighter is an al-Qaeda terrorist intent on attacking the United States. "The insurgents are defending their area and focusing on troops they see as occupiers. But if they can't kill soldiers on the battlefield, they will have to do it somewhere else." He predicts that the more we take American soldiers off the battlefields, the more it will "drive them to hit back home."

Rami Khouri similarly anticipates that for many in the Middle East, the sense will be, "If they play by these rules, which are completely unequal, then we'll play by our own rules.... Whether it's in the U.S., the U.K., or Malaysia." As the United States uses more unmanned systems, terrorists "will find much more devious ways to cause panic and harm. They'll say, 'If they are going to use these machines, we should get some chemicals and use them.' Put them in air-conditioning ducts in shopping centers or university dorms.... They might go after soft targets, shopping centers, sports stadiums, and so forth."

The same observers are all realistic, however. They see terrorism occurring regardless of unmanned systems. Moreover, they see that same sort of adaptation that the Iraqi insurgent hinted to me. Despite all the expected negative coverage such systems might receive in the region's press and public opinion, they anticipate there will be a quick willingness to gain and use them as well. As Akbar explains, "When they first come out, the very first reaction from the defense establishment will be, 'Where can we order these fucking things?'"

Indeed, Akbar (like many of the other regional experts I spoke with) believes that nongovernmental groups like insurgents and terrorists will also be quite willing to use them. Indeed, they will have their own ready explanation, to bolster their own psychological operations. "In fact, they will likely cite a verse in the Koran that you do not start jihad until you have the latest weapons, armor, and steeds."

This quote in the Koran reads, "Against them make ready your strength to the utmost of your power, including steeds of war, to strike terror into (the hearts of) the enemies, of Allah and your enemies, and others besides, whom ye may not know, but whom Allah doth know. Whatever ye shall spend in the cause of Allah, shall be repaid unto you, and ye shall not be treated unjustly." As Akbar explains, "The 'steeds of war' is translated today as 'the best equipment.' . . . Basically, it tells that you should not go unprepared into war. Valor is good, but not enough. . . . Even David had a stone." Or, as Rami Khouri in Beirut puts it, "The response to drones is to get your own drones. They are just tools of war. Every tool generates a counterreaction."

## DO ROBOT SOLDIERS DREAM OF ELECTRIC SHEEP?

As both sides of a conflict then begin to use unmanned systems more and more, a larger question of psychology comes to the fore. For all the differences in war through the ages, human psychology has always been at its center. Napoleon said, "In war, moral considerations account for three-quarters, the actual balance of forces only for the other quarter." What happens when that three-quarters is replaced by something else? What happens when the forces feel no fear, no fright, no anger, no shock, no awe, or any other elements of human psychology, but are guided only by a software of 0s and 1s?

The effect on future history will be immense. Imagine how different the world would be today if at the battle of Hastings, the English hadn't lost heart when their king was killed, or if at Waterloo, Napoleon's Old Guard hadn't grown weary of war and instead fought to the last machine.

The historian John Keegan wrote that "the study of battle is therefore always a study of fear and usually of courage; always of leadership, usually of obedience; always of compulsion, sometimes insubordination; always of anxiety, sometimes of elation or catharsis; always of uncertainty and doubt, misinformation and misapprehension, usually also of faith and sometimes of vision; always of violence, sometimes also of cruelty, self-sacrifice, compassion; above all it is always a study of solidarity and usually of disintegration—for it is towards the disintegration of human groups that battle is directed. It is necessarily a social and psychological study."

This was the truth of the last five thousand years of war. A human army needed some "vision, a dream, a nightmare, or some mixture of the three if it is to be electrified into headlong advance." A robot just needs an electric charge.

# SIXTEEN

# YOUTUBE WAR: THE PUBLIC
# AND ITS UNMANNED WARS

*"How fortunate for leaders that men do not think."*

—ADOLF HITLER

"We'll have more Kosovos and less Iraqs."

Larry Korb is another one of those deans of Washington's defense policy establishment. A former navy flight officer, he served as assistant secretary of defense during the Reagan administration. Now he is a senior fellow at the Center for American Progress, a left-leaning think tank. In between, Korb has seen presidential administrations, and their wars, come and go. And, having written twenty books and over one hundred articles and made almost a thousand TV news show appearances, he has also helped shape how the American media and public understand these wars.

In 2007, I asked Korb about what he thought was the most important overlooked issue in Washington defense circles. He answered, "Robotics and all this unmanned stuff. What are the effects? Will it make war more likely? And all sorts of questions like that. People need to think about this."

Korb is a great supporter of unmanned systems for a very simple reason. "They save lives." But he worries about their effect on the perceptions and psychologies of war, not merely abroad, but also at home. As more and more unmanned systems are used, he sees this issue playing out in two ways, both of which he fears will make war more likely. "It will further disconnect the military from society. People are more likely to support the use of force as long as they view it as costless." Even more, a new kind of voyeurism allowed by the new technologies will make the public more susceptible to false selling of how easy a potential war will be. "There will be more marketing of wars. More 'shock and awe' talk to defray discussion of the costs."

Korb is equally troubled by the effect that such technologies have on how the leadership might look at war and its costs. "It will make people think, 'Gee, war-fare is easy.' Remember all the claims of a 'cakewalk' in Iraq and how the Afghan model would apply? The whole idea that all it took to win a war was 'three men and a satellite phone' [mocking the network-centric crowd]? Well, their thinking is that if they can get the army to be as technologically dominant as the other ser-vices, we'll solve for these problems."

He feels that the current body politic in D.C. has been "chastened by Iraq." But he worries about "when you get a new generation of policymakers." Technol-ogy like unmanned systems can be seductive, feeding overconfidence that can lead nations into wars for which they aren't ready. "Leaders without experience tend to forget about the other side, that it can adapt. They tend to think of the other side as static and fall into a technology trap."

This is what Korb means when he predicts "more Kosovos and less Iraqs." As unmanned systems become more and more prevalent, we'll be more likely to use force, but also see the bar raised on anything that exposes human troops to dan-gers. Echoing the Pakistani general's description of distance war, Korb envisions a future where the United States is more willing to fight, but only from afar, where it is more willing to punish via war, but less to face the costs of war.

## THE PASSIVE PUBLIC

Immanuel Kant's 1795 book *Perpetual Peace* first expressed the idea that democra-cies are superior to all other forms of government because they are inherently more peaceful and less aggressive. This "democratic peace" argument (which some two centuries later both presidents Clinton and Bush Jr. cited) is founded on the belief that democracies have a built-in connection between their foreign policy and domestic politics that other systems of government lack. When the people share a voice in any decision, including whether to go to war, they are supposed to choose more wisely than some king or potentate. As one Pentagon official explains, this sense of shared participation and ownership is the key aspect in making the right decisions on when to start and end wars. "The Army belongs to the American population, and not the President or Congress."

Colonel R. D. Hooker Jr. is an Iraq veteran and the commander of an army airborne brigade. As he explains, the people and their military in the field should be linked in two ways. The first is the direct stake that the public has in its government's policies. "War is much more than strategy and policy because it is visceral and personal.... Its victories and defeats, joys and sorrows, highs and depressions are expressed fundamentally through a collective sense of exhilaration or despair. For the combatants, war means the prospect of death or wounds and a loss of friends and comrades that is scarcely less tragic." Because it is their blood personally invested, citizen soldiers, as well as their fathers, mothers, uncles, and cousins who vote, combine to dissuade leaders from foreign misadventures and ill-planned aggression.

The second link is supposed to come indirectly from a democracy's free media, which widens the impact of those personal investments of blood and risk to the public at large. As Colonel Hooker explains, "Society is an intimate participant [in war] too, through the bulletins and statements of political leaders, through the lens of an omnipresent media, and in the homes of the families and the communities where they live. Here, the safe return or death in action of a loved one, magnified thousands of times, resonates powerfully and far afield." It may not be your son or daughter at risk in a particular battle, but you're supposed to care because they are part of your community, and it might just be someone you know the next time.

So the media's effect in a free system is not merely a report on a war's outcome, as if reporting on a sporting event. Unlike a spectator merely watching a game, the public's perceptions of events on distant battlefields creates pressures on elected leaders, which can determine when the game begins or ends and even whether another game is played the next week. Too much pressure can translate into an elected leader trying to interfere in ongoing operations, as bad an idea as the owner or fans calling in the plays for a coach to run. But as Korb and Hooker explain, too little public pressure may even be worse. It's the equivalent of no one even caring about the game or its outcome. War becomes the WNBA.

Many worry that this democratic ideal is under siege. The American military has been at war for the last eight years in places like Afghanistan and Iraq, but other than at the airport perhaps, the American nation has not. With the ending of the draft after Vietnam, most American families no longer have to think about whether their husband, wife, son, or daughter would be at risk if the military is sent to war. By comparison, during World War II 12.2 million men, just under 10 percent of the American populace, served in the military; the equivalent of almost 30 million today.

The military is also far less representative of the broad populace than it was in past generations. Flags once flew on nearly every street, marking which houses had sons off at war. Now, with the end of the draft, entire neighborhoods can lack even a passing link to the military. This disconnection is even more pronounced among the elite that dominate the business, the media, and the politics of both parties. At the time of the Iraq invasion vote, for example, less than 1 percent of all Ivy League university graduates were enlisted in the military and only 1 out of 635 senators and representatives had a child that might be sent into harm's way.

By the start of the twenty-first century, even the financial costs on the home front were displaced. Industry didn't need to retool its factories and families didn't even need to ration fuel or food, or even show their faith in the war effort by purchasing bonds (instead, taxes were lowered for the top 1 percent of citizens). Government leaders were at a loss for how to motivate the public to show support for the war effort. When asked what citizens could do to share in the risks and sacrifices of the soldiers in the field, the message sent from the commander in chief in the White House was "Go shopping." As one article in *American Conservative* magazine put it, "Rather than summoning Americans to rally to their country, he [Bush] validated conspicuous consumption as the core function of 21st-century citizenship." The outcome is a public that became more disinvested in and delinked from its foreign policy than ever before in a democracy.

With this trend already in place, many worry that unmanned technologies may well snip the last remaining threads of connection. The increasing use of robotics may be motivated by saving lives, but by doing so, it does affect the way the public views and perceives war. In turn, it will also affect wars' processes and outcomes, perhaps even transforming that public into the equivalent of sports fans watching war, rather than citizens sharing in its importance.

## CHANGE THE CHANNEL

Josiah Bunting is a former major general in the army. After he retired from service, he became superintendent at the Virginia Military Institute and then head of the Guggenheim Foundation. Bunting is concerned that the American public is turning into "passive" observers of their country at war, and that the new trends of unmanned systems threaten to make it worse. He compares the situation to the book *1984* by George Orwell, "where every 20 or 30 pages there is some oblique reference to 'the war' but it has no bearing on life."

For Bunting, the costs of this disconnect are immense. First, anything that takes away the public's investment and involvement in a war also takes away any sense of unity for a nation. Instead of a nation mobilized and united behind its men and women in the field, you get the reverse. With robotics, instead of Rosie the Riveter pitching in to support her husband abroad, you just get a rivet that no one cares about. Even worse, any public passivity about war only increases the likelihood of bad policy, and more lost wars. "It makes it easier for leaders to stick in 'stay the course mode' when things aren't going well." He explains, "The war just becomes bad news, akin to a TV show that you get tired of and want to end.... Why [should a leader] change if failure doesn't matter?"

Jun Ho Choi is a student who works on two-legged humanoid robots at the University of Michigan. Choi would seem to have little in common with an army major general, and yet he has the very same concerns about the public disengagement from war that his robotic systems might bring. "This may be a positive way to improve the military, but I do not believe this is a positive way to improve our lives.... I am worried about people becoming less serious about war since robots are fighting. We might end up having war every day."

Bunting and Choi are pointing to the first of those concerns laid out by Washington expert Larry Korb. Unmanned systems may lessen the terrible costs of war, but in so doing, they will make it easier for leaders to go to war. Indeed, people with widely divergent worldviews come together on this point. "They [unmanned systems] lower the threshold for going to war. They make it easier, make war more palatable." "Anything that makes it morally and ethically easier to wage war is not necessarily a good thing." The first quote is from a human rights expert, whose job entailed trying to shut down the prison at Guantánamo Bay; the second is from a special operations officer just back from hunting terrorists to lock up there.

Unmanned systems represent the ultimate break between the public and its military. With no draft, no need for congressional approval (the last formal declaration of war was in 1941), no tax or war bonds, and now the knowledge that the Americans at risk are mainly just American machines, the already lowering bars to war may well hit the ground. A leader needn't carry out the kind of consensus building that is normally needed before a war, and doesn't even need to unite the country behind the effort. Describes one air force officer none too happy with this trend, "Taking the human factor out of warfare cheapens the expense of combat and would lead to more conflict. Furthermore, that uniquely human concept of chivalry on the battlefield helps separate us from the beasts."

But the technologies don't just merely remove human risk, they also record all they see, and in so doing reshape the public's link to war. The Iraq war is literally the first war where you could download video of combat off the Web; as of 2007, there were over seven thousand video clips of combat footage from Iraq on YouTube.com alone. Much of this footage was captured by various drones and unmanned sensors and then posted online. Some of the videos were official, but many were not.

This trend could be viewed as a positive development that builds greater connections between the war front and home front, allowing the public to see as never before what is going on in battle. But so much visibility is not all that it seems. Inevitably, the ability to download the latest snippets of combat footage to home computers turns war into a sort of entertainment, or "war porn," as soldiers call it. Clips of particularly interesting combat footage, such as an insurgent blown up by a UAV, are forwarded to friends, family, and colleagues with titles like "Watch this!," much the same way an impressive soccer goal or amusing clip of a nerdy kid dancing in his basement gets e-mailed around the Internet. Comments and jokes are attached, and some are even set to music. A typical example was a clip of people's bodies being blown up into the air by a Predator strike set to Sugar Ray's song "I Just Want to Fly." War then becomes, as one security analyst described, "A global spectator sport for those not involved in it."

More broadly, it engages the public in a whole new way, but can fool many into thinking they now have a true sense of what is going on in the conflict. It has a paradoxical effect, a widening of the gap between our perceptions and war's realities. To make another sports parallel, it's like the difference between watching an NBA game on television, with the tiny figures on the screen, and seeing it in person, where the players really are seven feet, scream, sweat, and smell, and playing in the game yourself and knowing what it actually feels like to have KG knock you down and dunk on your head. Even worse, such clips don't show the whole game, but are merely just the bastardized ESPN *SportsCenter* version of it. The context, the strategy, the training, the tactics, and so on all just become slam dunks and smart bombs.

War porn also tends to hide another hard truth about battle. Most viewers have an instinctive aversion to watching a clip of a battle where the person in the clip might be someone they know or a fellow American; such clips tend to get banned from U.S.-based host sites. But many are perfectly happy to watch clips of anonymous enemy deaths, even just to see if the machines fighting in Iraq are as "sick" as those fighting in the *Transformers* movie, as one student put it to me. To a public with less at risk, wars take on what analyst Christopher Coker called "the pleasure of a spectacle with the added thrill that it is real for someone but not the spectator." The public's link to its wars transforms from connection into merely a kind of voyeurism.

## ROBOT CHICKENHAWKS

Such changed connections don't just make a public less likely to wield its veto power over its elected leaders. As former Pentagon official Larry Korb reminded, technology also alters the calculations of the leaders themselves.

Nations often go to war because of overconfidence, which makes perfect sense; few leaders choose to go into a conflict thinking they will lose. Technology can play a big role in feeding overconfidence; new weapons and capabilities breed new perceptions, as well as misperceptions, about what might now be possible in a war. Today's new technologies like robotics are particularly liable to feed this. They are perceived as helping the offensive side in a war more than the defense, plus they are advancing at an exponential pace. The difference of just a few years or even months of research and deployment can create vast differences in such technologies' capabilities, creating a sort of "use it or lose it" mentality among leaders. Finally, as one roboticist explains, a vicious circle is generated. Scientists and companies often overstate how great a new technology is in order to get governments to buy it. But "if we believe the hype, it will probably increase the frequency of tactical engagements."

James Der Derian is an expert at Brown University on new modes of war. He believes that the combination of these factors may mean that robotics will "lower the threshold for violence." They create a dangerous mixture: a public veto over leaders now gone missing, and technologies that seem to offer leaders spectacular results with few lives lost. It can be very seductive. "If one can argue that such new technologies will offer less harm to us and them, then it is more likely that we'll reach for them early, rather than spending weeks and months slogging at diplomacy."

When faced with a dispute or crisis, policymakers have typically looked at force as the "option of last resort." Now unmanned systems might help the option move up the list, with each step making war more likely. That leaves us back at Korb's scenario of "more Kosovos, less Iraqs."

While avoiding the mistakes of Iraq certainly sounds like a positive result, the other side of the trade-off would not be without its problems. Lowering the bar to more and more unmanned strikes from afar would most resemble the so-called cruise missile diplomacy of the 1990s. They may result in fewer troops stuck on the ground (a lesson that many have taken away from Iraq), but, like the strikes against al-Qaeda camps in Sudan and Afghanistan in 1998, or the Kosovo war, they are military endeavors without any true sense of a commitment, lash-outs that yield incomplete victories at best. As one report in an army journal tells, such operations "feel good for a time, but accomplish little." They involve the country in a problem, but do not resolve it.

Even worse, Korb may be wrong and the dynamic could yield not "less Iraqs," but even more. It was the lure of an easy preemptive action that got the United States into such trouble in Iraq in the first place. Describes one robotics scientist of his creations, "The military thinks that it will allow them to nip things in the bud, deal with the bad guys earlier and easier, rather than having to get into a big-ass war. But the most likely thing that will happen is that we'll be throwing a bunch of high tech against the usual urban guerrillas. . . . It will stem the tide [of U.S. casualties], but it won't give us some asymmetric advantage."

Thus, robots may entail a dark irony. By seeming to lower the human costs of war, they may seduce us into more wars.

## WAR, NOT WAR

Whether it's watching wars from afar, or sending robots into harm's way instead of fellow citizens, robotics offer the public and their leaders the lure of riskless warfare. All the potential gains of war would come without the costs, and even be mildly entertaining.

It's a heady enticement, and not just for evil warmongers. The world watched the horrors of Bosnia, Rwanda, and the Congo, but did little, mainly because the public didn't know or care enough, and the perceived costs of doing something truly effective just seemed too high. Substitute in unmanned systems and the calculus might be changed. Indeed, imagine all the horrible genocides and crimes against humanity that could be ended, if only the barriers to war were lowered. Getting tired of some dictator massacring his people? Send in the bots and sit back and watch his troops get taken down. One private military company executive even slickly pitched a quick and easy technologic solution to the genocide in Darfur as a simple matter of "Janjaweed be gone!," as if an intervention into an African civil war was just a problem of scrubbing away the bad guys.

Yet wars never turn out to be that way. It's in their very nature to be complex, messy, and unpredictable. And this will remain the case even as unmanned systems substitute for more and more humans. But imagine if this was not the case, that such fantasies were actually to come true. Even such a seemingly positive outcome of truly cheap and costless unmanned wars should give us pause. By cutting the already tenuous link between the public and its foreign and defense policy, the whole idea of a democratic process and citizenship is perverted. When a citizenry has no sense of sacrifice or even the prospect of sacrifice, the decision to deal out violence becomes just like any other policy decision, like whether to raise the bridge tolls. Instead of widespread engagement and debate over the most important decision a government can make, where blood might be shed, even if only on the other side, you just get popular indifference.

When technology disengages the public and instead turns war into something merely to be watched, and not weighed with great seriousness, the checks and balances that undergird democracy go by the wayside. This could well mean the end of the idea of a democratic peace, which supposedly sets our foreign policy decision-making apart from that of potentates and emirs.

Wars without costs can undermine the morality of even "good" wars. When a nation decides to go to war, it is not just deciding to break stuff in some foreign land. As one philosopher put it, the very decision is "a reflection of the moral character of the community who decides." Without public debate and support and without risking troops, though, the decision of war may only reflect a nation that just doesn't give a damn.

Even if the nation acts on a just cause, such as the motivation to stop genocide, war can be viewed as merely an act of selfish charity. One side has the wealth to afford high technologies and the other does not. The only message of "moral character" a nation sends out is that it alone gets the right to stop bad things, but only at the time and place of its choosing, and most important, only if the costs are low enough. While the people on the ground being saved may well be grateful, even they will see a crude calculation taking place that cheapens their lives. As

Kosovars darkly joked during the 1998 war, in which NATO was willing to bomb to stop their massacre, but only as long as it didn't have to risk its own pilots below fifteen thousand feet, "The life of one NATO soldier is worth 20,000 Kosovars."

With unmanned systems, this bare minimum is reduced to zero. Wars, even the best of them, lose their virtue. They instead become like playing God from afar, just with unmanned weapons substituting for thunderbolts.

This also makes it easier to start playing God when you shouldn't, for causes that may not be so just. The danger of these new technologies is that leaders can, as professor Christopher Coker argues, "become so intoxicated by the idea of precise, risk-free warfare that we believe what we want to believe. Unfortunately we may slip down the slope and find ourselves using violence with impunity, having lost our capacity for critical judgments. We may no longer be inclined to pay attention to the details of the ethical questions which all wars (even the most ethical) raise."

Some question whether such wars without risk are even war. If one side is empowered and the other is not, it becomes more like a police action, with the public at home watching the military version of *Cops* via their video clips. But war is not some police action, where a bad guy is chased down the street in his underwear. As retired marine officer Bing West put it, in the final calculus, "making war is the act of killing until the opposition accepts the terms of surrender rather than accepts more destruction."

Paul Fussell is perhaps best suited to sum up this question of how a public engages with unmanned war. In 1943, at the age of nineteen, Fussell was drafted into the U.S. Army. The next year, he was sent to France as part of the 103rd Infantry Division, arriving just after the Normandy invasion. In the fighting that followed, he was wounded and awarded the Purple Heart. After the war, he went back to school and became a noted author and cultural historian. Having experienced it himself, Fussell is perhaps the literary world's greatest living critic of how war can be glorified and romanticized, both by governments and pop culture. His book on this issue, *The Great War and Modern Memory*, was named by Modern Library as one of the twentieth century's one hundred best nonfiction books.

Now eighty-seven years old, Fussell is as brutally honest as his topic. War (or, as he once ironically called it, "The Bloody Game") "is forced travel, no good food, sleeping in the dirt, death and maiming." Fussell believes that the true horrors of combat are never fully acknowledged by the public and its political leaders during a war and are then ignored by the authors who write the histories, only after the killing is done. "And so I tried to cut away parts of it—tell them what a trench smelt like and what dead GIs smelt like and so forth." He tells, for example, of the time his unit killed "weeping, surrendering Germans," or of the morning he woke up to find himself surrounded by dozens of dead bodies. "If darkness had mercifully hidden them from us, dawn disclosed them with staring open eyes and greenish white faces."

Today, Fussell worries about what all these amazing new technologies will mean for the next generation of war and the public's connection to it. "If there is no risk, no cost, then it isn't war as we think of it. If you are going to have a war, you've got to involve people and their bodies. There's no other way."

Fussell rails against this trend and what it portends. But he admits that he's a bit of a pessimist and questions whether his efforts will do any good. "In the end," he laments, "people will support the next war because the TV tells them to."

# SEVENTEEN

# CHANGING THE EXPERIENCE OF
# WAR AND THE WARRIOR

*The introduction of every new technology changes society, and how
society looks at itself.*

—ILLAH NOURBAKHSH

When American forces swept up the remnants of the Saddam Hussein regime in
the aftermath of the 2003 Iraq invasion, Uday and Qusay Hussein, Saddam's two
sons, were among the most wanted men in the country. Qusay was the quiet one.
Trusted more by his father, he had been put in charge of the Republican Guard
defense of Baghdad, which he planned to ring with bunkers and ambush sites.
Uday was the wild one. A bit off-kilter by even his family's odd standards, he was
renowned for his fondness for fast cars, women, and violence (Uday would personally torture Iraqi Olympic athletes who didn't perform well in games). He was put
in charge of the Saddam Fedayeen (Men of Saddam), the regime's paramilitary
force. Uday's masterful plan to defend Iraq was to try to re-create the movie *Black
Hawk Down*, by having the Fedayeen fight the Americans from pickup trucks and
buses.

Neither's plan worked out. Qusay's Republican Guard defenses either melted away
or were swept aside, while Uday's strategy had two fatal flaws. First, that wasn't exactly
what happened in either the movie or reality. Second, in this war, the U.S. soldiers had
tanks with them. (If war was a card game, tanks would trump buses.)

While dad hid out in a dirt pit behind a family friend's farm outside Tikrit, the two sons sought refuge in a villa in the city of Mosul. But a tipster passed on their whereabouts to U.S. military forces. When soldiers from the 101st Airborne Division entered the first floor of the building, they were met by a hail of bullets. The two brothers had barricaded themselves on the second level. The soldiers pulled back and called for backup. More than two hundred troops and a range of vehicles and helicopters then rushed to the scene and a full-fledged firefight broke out. Soldiers peppered the mansion with their M-16 rifles and rocket launchers. Gunners standing in Humvees pasted the walls with .50-caliber machine guns. Apache helicopters launched missiles that turned the mansion's façade into Swiss cheese. The scene filled with smoke, fire, and explosions. All the while, a remote-controlled drone circled overhead, silently watching, coordinating the attack.

After two hours of free-for-all shooting, the order was given to cease fire. The soldiers picked their way through the rubble to find the two brothers' bodies, riddled with bullets. No one was exactly certain when they had died. It could have been in the first minute, it could have been in the second hour. Master Sergeant Kelly Tyler, the spokesperson for the 101st Airborne, expressed the official position on the results of the "battle," "The 101st kicks ass."

More than four hundred miles away, the scene was also chaotic, but in a far different way. The operator of the drone flying over the villa was at a base in Qatar. He was amazed by the footage that he was watching. News of the ongoing battle spread through the command center and soon over forty off-duty soldiers had crowded into the small control room. Many brought in snacks. They squirmed to get the best view of the battle playing out on the flat-panel screen. Cheers would erupt every time there was a particularly big explosion. As one of the soldiers later described his experience during the battle, "It was like a Super Bowl party in there."

General Robert E. Lee once wrote, "It is good that we find war so horrible, or else we would become fond of it." The new technologies of war are changing the experience of war itself.

## GOING TO WAR?

The phrase "going to war" has long had a double meaning. It signified not just the start of hostilities between two nations, but also the start of an arduous journey for their soldiers leaving home. A soldier who went to war might be gone for years before they returned to their loved ones, if they were so lucky to even make it back. Indeed, in the *Odyssey*, one of the founding books of Western literature, Odysseus goes to war for ten years, and then needs a journey lasting another decade before he finds his way home.

Going off to war, though, meant more than just being disconnected from loved ones. It also was defined by the great dangers that greeted soldiers at their destination. Whether it was Odysseus arriving before the walls of Troy or my grandfather heading off to fight the Japanese in the Pacific, when a soldier went off to war, he arrived at a place where killing took place. Things changed slightly with each new technology; a soldier in one of Julius Caesar's legions was certainly exposed to far more danger at a closer distance than a B-52 bomber pilot flying above the Vietnamese jungle. But the essence of their destination was the same. War was a place of such dangers that warriors had to reasonably expect that they might never return home.

For the first time in history, however, the experience of "going to war" is changing at a much deeper level. Gary Fabricius's experience in the Iraq war is as good an illustration as any.

Colonel Gary Fabricius graduated from the Air Force Academy in 1984 and joined the service as a pilot. The young officer, who took the call sign of "Fabs," advanced up the ranks, eventually piloting F-15 fighter jets, the premier fighter plane in the world at the time. In 2002, he was promoted to squadron command. There was only one catch: he would be commanding the first Predator drone squadron.

As a self-described "fighter jock," Fabricius recalls that he was "kicking, screaming, clawing, and scratching" to stay in F-15s. Indeed, instead of celebrating his promotion, a fellow fighter pilot told Fabs, "Jeez, I'm sorry." UAVs were just for taking pictures and the pilots who flew them were considered "second-class citizens," just "playing video games." Fabricius headed off to his new command in Nevada, not wholly happy with where his career had taken him.

"But after a month, I became a believer." Fabricius describes how the Predator unit was not some "remote control flight club" and instead soon became "a combat-effective squadron." His unit went from merely gathering information viewed only by the senior leadership as "neat to see" to carrying out thousands of important missions, everywhere from Iraq to Afghanistan, despite the fact that his pilots never left Nevada. He was part of the first UAV air combat mission, in which the drones were armed with Stinger missiles, so that they could try to ambush

Iraqi fighter jets. In battles like Fallujah, his unit didn't just take out enemy targets, but also beamed vital real-time information to marines on the ground that saved lives, such as warning them of a sniper on a rooftop waiting to ambush them. His unit even helped catch Saddam Hussein. "You could see him climbing out of the hole. It was pretty exciting." Today, when Fabricius sees his old F-15 mates, he asks them how many combat missions they have flown, how many targets they have taken out in the war, or how many American lives they have saved. Most can only answer "zero."

In a way, Fabricius's experience was a "historic first," as he explains. It wasn't just that he had "a God's eye view to the battlefield," but that he is part of the first generation to go to war without actually going to war.

Fabricius is representative of a new generation of warriors who could be termed "cubicle warriors." The term takes its inspiration from the cubicle, the now standard office workplace filled with computers and ergonomic furniture that has become a fixture of pop culture, from *Fight Club* to *The Office*, as well as science fiction, where in movies like *The Matrix* it is a metaphor for technology's increasing control over daily life. The concept isn't meant as a mockery, as most soldiers' workstations certainly don't have the ubiquitous knickknacks, pictures of pets, or "Hang in There" kitty posters that decorate many office cubicles. Rather, just as more people's experience in modern industry has shifted from the fields and factory floors to the "cube," the location of fighting has shifted for many soldiers from the battle space to *Office Space*. For a new generation, "going to war" doesn't mean shipping off to some dank foxhole in a foreign land to dodge bullets. Instead, it is a daily commute in your Toyota Camry to sit behind a computer screen and drag a mouse.

Their location doesn't limit the violence that cubicle warriors deal out, though. Fabricius's unit may not have been in Iraq, but its base just outside Las Vegas is where most of the combat action in the air force takes place today. As one drone pilot describes, "If you want to pull the trigger and take out bad guys, you fly a Predator." As such, they take their jobs seriously. Like Fabricius, they all had experience flying other planes like F-15s or A-10s (this policy is soon to change, however), and they even come to work wearing a flight suit. One officer recalls that the action felt so intense that one time, when his drone thousands of miles away was about to crash, he instinctively reached for the ejection seat.

Fabricius's squadron flies under what is known as a "remote split operation." The drones fly out of bases located in the war zone, but the pilots are actually sitting at control stations in a complex of trailers in Nevada. These are linked by fiber optic cables to Europe. There, a satellite dish beams their information and communications out to the drones. Because they are not bound to any one plane, the pilots can "swing" from flying a drone over Iraq one day to a different one flying over Afghanistan the next. For the first time, the limitations of geography are taken out of the war that a soldier goes off to experience.

Air force pilots have long been mocked by troops in the field for leading cushy lives in war, safely flying above the fight and then returning each night to their bases behind the lines, which have good chow, warm showers, and a nearby bar. But even in this scenario, a pilot who went to war could never be accused of living the sort of life they led back home. For example, in his classic book *A Lonely Kind of War*, Marshall Harrison tells what it was like to fly Forward Air Controller missions in Vietnam, the precursor to the types of roles carried out by drones today. He recounts exchanging the humdrum of "a wife, three children, and a well-mortgaged home in the Virginia suburbs" for a base "with sagging tents and rain-rotted hootches...a strong miasma of burning feces...hordes of mosquitoes." The same difference goes for the risks that pilots had to take. While their exposure to danger might be at greater distances and shorter duration than for soldiers on the front lines, the expectation of peril was ever-present. At one point, for example, Harrison had to land his plane, under enemy fire, on a dirt road deep in Cambodia to help rescue a reconnaissance team. As such, even back at base, there was no mistaking that he was at war; it was literally in the air. Tells Harrison, "You could almost smell the excess testosterone."

The experience for drone pilots is a bit different. They work the same hours as if they were in a war zone, usually seven days a week, twelve hours a day, with the unit split into two shifts. But, says Colonel Charlie Lyon, commander of the 57th Operations Group at Nellis, "At the end of the duty day, you walk out of the deployment and walk back into the rest of life in America."

A 1940s army pamphlet given to new recruits in World War II explained what it was like to experience war: "YOU'LL BE SCARED. You'll be frightened at the uncertainty, at the thought of being killed." By contrast, described one Predator pilot, "Most of the time, I get to fight the war, and go home and see the wife and kids at night." Another talked about flying missions in Afghanistan, and then getting home in time to watch reruns of the TV sitcom *Friends*.

This changing experience leads some to question whether these cubicle warriors can even claim to have gone to war. One special operations officer, just back from Iraq, says, "You have some guy sitting at Nellis and he's taking his kid to soccer. It's a strange dichotomy to war. He's disconnected from the enemy he's fighting.... A warrior has to assume physical risk." When I directly asked him if he thought a Predator pilot was at war, he replied, "No, he doesn't meet my definition." It was the exposure to risk that defined whether he respected someone as a fellow combatant, including even enemies who violated all the other rules of war. "If you see it through their eyes, you can understand what they think. Even AMZ [Abu Musab al-Zarqawi, the leader of al-Qaeda in Iraq, whom the operative had helped hunt down] was right there hanging his balls out on the battlefield in terms of personal risk, leading his men in combat."

Throughout history, as each new technology has pushed soldiers farther and farther away from their foes, many lamented the effect it would have for warriors and their values. When Hiero, the ancient king of Syracuse, hired the great Archimedes to build him a catapult, the king was said to have cried when he saw the result. The catapult was so powerful that Hiero mourned that the age of warriors was surely over, replaced by the age of engineers. Similarly, using a gun was once seen as cowardly. As one commentator in the 1400s complained, "so many brave and valiant men" were being killed by "cowards and shirkers who would not dare to look in the face the men they bring down from a distance with their wretched bullets." This leads some, like AI expert Robert Epstein, to claim that "fighting by remote is no different than fighting with a knife in hand-to-hand combat is from shooting someone with a rifle. You are still fighting."

The fact is, however, that while technology may not have ended the warrior's trade, it certainly has affected our definition of the attributes soldiers must have when they go to war, most especially that ultimate value that so defines a soldier, courage in the face of danger. In the days of swordplay, individual ferocity often carried the day in battle (think Mel Gibson in *Braveheart*), and so it was an attribute that was greatly admired. With the invention of gunpowder and forces lining up in battle to fight each other, the ultimate value became steadfastness under fire; courage now meant standing in the line with "passive disdain," as bullets came flying at you (think Mel Gibson in *The Patriot*). But when the machine gun entered war, this old definition of courage became ineffective if not insane (think Mel Gibson in *Lethal Weapon*, or on a Malibu highway). As a French general commented after the battle of Verdun in 1916, "Three men and a machine gun can stop a battalion of heroes."

But the underlying essence of war stayed the same. As T. R. Fehrenbach wrote in *This Kind of War*, "The real function of an army is to fight and that a soldier's destiny—which few escape—is to suffer, and if need be, to die." Whether it was from a sword point up close or a machine-gun bullet fired from afar, none of the previous revolutionary technologies in history changed the fundamental fact that going to war meant facing risk. A soldier then needed to hold tight to that special value, which would allow them to face these dangers and still do what needed to be done. As Lord Moran writes in *The Anatomy of Courage*, "The mysterious quality we call 'courage' is will-power, self-sacrifice, call it what you will, that inspires men to hold their ground when every instinct calls upon them to run away." The courage of a warrior, then, is about victory over fear. It is not about the absence of fear.

By removing warriors completely from risk and fear, unmanned systems create the first complete break in the ancient connection that defines warriors and their soldierly values. If you are sitting at a computer's controls, with no real danger other than carpal tunnel syndrome, your experience of war is not merely distanced from risk, as with previous technologies, but now fully disconnected from it. And thus these new warriors are disconnected from the old meanings of courage as well. As one described his experience in the Iraq war, fought from a cubicle in Qatar, "It's like a video game. It can get a little bloodthirsty. But it's fucking cool."

Not everyone is fighting from the cubicles, of course, which creates two different tracks in the once shared experience of going to war. One group of Shadow drone pilots reflected on what it's like to be at war when you don't really face any danger but can watch others who still do. "Every now and then, you're like, 'Man, these guys are really taking fire!' You just want to get out there and help them," states Sergeant William Coleman. At the same time, he explains, "You've got to be thankful for the situation you're in, because you're not under hostile fire every day.... People are really getting hurt. They're dying every day."

"Yeah, war is hell," concurs his copilot, army specialist Jonathan Whitaker. Notably, the two soldiers relax from the stress of virtual war by playing video game versions of war in their off hours. "I won the last game," says Coleman. "*Medal of Honor*, we were playing and I walked away with that one."

With these changing experiences, many worry that the age-old soldierly virtues of loyalty, bravery, courage, and sacrifice are under threat. Air Chief Marshal Sir Brian Burridge, who commanded the British military forces during the Iraq war, even describes unmanned systems as part of a move toward "virtueless war," a result of remote soldiers' no longer having any "emotional connectivity with the battlespace." Analyst Chris Gray similarly says that "war is not just in transition, it is in crisis."

## THE VIRTUAL BAND OF BROTHERS?
## UNIT COHESION IN A CHAT ROOM

*From this day to the ending of the world,*
*But we in it shall be remembered—*
*We few, we happy few, we band of brothers;*
*For he to-day that sheds his blood with me*
*Shall be my brother.*

It was supposedly with these words, touched up a bit courtesy of William Shakespeare, that Henry V inspired his men before the battle of Agincourt in 1415. His phrase "band of brothers" captures the unique bond between men who fight and even die together. Over five centuries later, it was the title of *Band of Brothers*, Stephen E. Ambrose's classic history of the men who made up a World War II American infantry company in the 101st Airborne Division, and the close attachments that built between them as they battled across Europe. The men of Easy Company shared in the training, the dangers, the losses, the laughs, the risks, and finally, the sense of triumph, mixed with guilty relief, at the end of the war. Decades after the war is over, the men in the unit still feel closer bonds with fellow soldiers they have not seen for years than with their best friends from civilian life, and sometimes even their families.

This brotherhood of arms has long been one of the integral parts of military culture and the experience of war. It is why the most hard-ass marine sergeant can tell his unit something tender like, "We're a family. I'm your father and mother," and truly mean it from the bottom of his steely heart. While recent U.S. Army recruitment advertisements talked of an "Army of One," the reality is that solitude and self-sufficiency are not the typical experience of war. Instead, the combination of close proximity, harsh conditions, distance from home, and shared dangers and losses forges uniquely tight bonds between men (and increasingly women) who often have nothing else in common.

Military researchers call these bonds "unit cohesion." It's not something that can exactly be measured, but unit cohesion is best described as the sort of chemistry that builds within a unit, which in turn allows it to act as a team. As Major Ralph E. McDonald summarizes, "Cohesion requires trusting each other and anticipating each other's needs."

Most believe that cohesion doesn't just make soldiers fight better together, but even is what gives them the courage to stay in the fight. The nineteenth-century French military writer Ardant du Picq described it as "mutual surveillance," telling how individually most men are secret cowards, but when trained together as a unit, they become transformed; their fear of letting each other down matters more than their fear of the enemy's bullets and bayonets. Other studies of soldiers in World War II found the same truths of combat, that the sense of responsibility a soldier had for his close comrades mattered far more than any lofty ideals of patriotism.

Compare such notions to the new experiences of warriors fighting from afar. Marine veteran Bing West, for example, talked about an incident that took place just outside Fallujah, Iraq, in 2005. A group of insurgents was sighted by a Pioneer drone, flown by operators sitting at a base in the Persian Gulf. It was unarmed, so a second, armed Predator drone, operated by pilots back in Nevada, arrived on the scene. The coordinates were passed, the target identified, and the enemy was taken out. None of the operators actually risked themselves, nor did they ever physically meet. Indeed, they never even spoke to each other by radio or phone. Instead, they carried out the entire operation by texting each other in an Internet chat room. "Make no mistake, this war is being fought on chat," declares an air force lieutenant colonel who coordinates such attacks every day.

As new communications technologies spread across society, we are seeing a change in how people organize themselves in their new office cubicles and whether they bond or not. It was once a mantra of business, for example, that workers had to be located together in the same place and organized into categories with strict boundaries to get the maximum efficiency. Indeed, this belief, as popularized by Peter Drucker, the "founding father of the study of management," was considered so essential to America's business success in the twentieth century that Drucker was awarded the Presidential Medal of Freedom by President George W. Bush in 2002.

But when we look at corporate life at the turn of the twenty-first century, it is increasingly taking place in the same sort of distributed operations that characterize the military's "reachback" operations of war via chat rooms. As a federal government study of corporate change found, people are increasingly carrying out their work, and even their social discussions, via e-mail, rather than in the hallways, the cafeteria, or conference rooms. "The only sound is the click-clack of keyboards as email flies back and forth."

Carrying out work or even war via an Internet chat room, where no one ever meets, may seem completely odd to anyone who grew up in the last twenty centuries. But it is not so unusual for the "MySpace Generation" that is leading change in the twenty-first century. As *BusinessWeek* describes the generation moving into the workforce and the military today, "They live online. They buy online. They play online. Their power is growing." Indeed, in 2007, the social networking site MySpace was the most trafficked Web site on the entire Internet. It isn't only perfectly natural for today's youth to feel more comfortable communicating via the Internet, but many also have far more virtual "friends," whom they have never met, than they do real ones.

The effect of such changes on the traditional principles of unit cohesion are obviously immense. Even drone squadron commander Gary Fabricius acknowledges there was a huge difference between deploying out to a region to fight together versus staying back home and fighting via computer screen. "You gain that trust and focus from living and breathing the operation together. With reachback operations, you lose that camaraderie." Marine general James Mattis, who was the overall commander in Fallujah at the time when chat-room-coordinated strikes came into being, feels similarly. "Computers by their nature are isolating. They build walls. The nature of war is immutable: you need trust and connection."

While virtual hookups allow teams to form and work together across great distances, they are not producing the same sorts of bonds. Soldiers working and fighting together via a wireless signal or fiber optic cable may be connected, but not emotionally or psychologically. The teams they are supposedly part of may be quickly connected, but they are just as easily disconnected. No real bonds form and unit cohesion becomes ephemeral. It is less like a true "band of brothers" and more like most of the Facebook "friendships" that youth make online via social networking. They may list hundreds of "friends," but very few are actual relationships built of mutual respect and trust. Instead they are more like superficial social groupings, or as *Washington Post* technology reporter Joel Garreau puts it, "skittering like water bugs on the surface of life."

Without sharing the experiences of war, soldiers can't develop that same sense of trust. And with that, units and soldiers look at each other differently. One U.S. Marine officer says that he detests the virtual linkages, because "op-con [operational control] isn't real without 'hand-con,' a handshake. All the wire diagrams you see are nearly irrelevant in today's environment.... If you don't have trust, all these units just become chess pieces on a board."

In lieu of this trust and cohesion, many of today's warriors are experiencing other problems endemic to the Internet. Just as you can never be sure online whether that sexy blond cheerleader is actually a 350-pound welder from Milwaukee, something similar happens in military chat rooms. "Ninety percent of the time you don't know who you are talking to," says Gary Fabricius. "The beauty of it is that anyone can sign in and ask for information or mission help. But the danger is that anyone can sign in and ask for information or mission help."

This can create a free-for-all, which sometimes throws military hierarchy into a tailspin. "Staff Sergeant Smithy will be on one side badgering a lieutenant colonel flying the Predator. Pretty soon they each start getting testy." Ultimately, the e-mail traffic grew so problematic in the chat rooms that Fabricius had to lay down some rules of proper chat-room behavior for soldiers at war. No e-mailing in ALL CAPITAL LETTERS, no exclamations!!!!, and no emoticons (☺).

Setting military chat-room etiquette reduced the tension and let information flow a bit more smoothly. But still, Fabricius says, "I hate it. You lose something. You can't tell the emotions. You lose the sense of urgency."

Lieutenant Colonel Norman Mims of the army's 11th Signal Brigade says that "what's funny about using Microsoft Chat is that everybody has to choose an icon to represent themselves. Some of these guys haven't bothered, so the program assigns them one. We'll be in the middle of a battle and a bunch of field artillery colonels will come online in the form of these big-breasted blondes. We've got a few space aliens, too." One general even went by the chat-room handle of "YKYMF," short for Bruce Willis's famous line from *Die Hard* of "Yipee-ki-aye! Motherfucker."

Such problems of identity, when soldiers can't meet face-to-face, aren't just amusing, but can create confusion on what to do in battle. One Predator drone pilot, for instance, recalls how it was often difficult "to tell if it was a private or a colonel" asking for his drone to perform a certain mission. The distanced anonymity of the chat rooms made it such that "everyone thinks they have a vote."

Another problem is that even the best technology cannot bridge the divide of being in two different locales. Being there virtually only allows so much communication. Describes one air force officer, "Textual communications accounts for 30 percent of what I need. So we use shorthand, but so much is lost." He describes how a Predator might be flown over Afghanistan while the pilots are in Nevada, and the end users of the data might be anywhere from Iraq to Tampa (CENTCOM headquarters) or California (where an air force intelligence unit is located). "You fly by the target and I type for you to turn around. You may not want to or want to know why and we get into a chat-room pissing contest. And by then it's too late. If we could just talk face-to-face, where body stance and seriousness are so clear, it would take a few seconds."

One special forces officer recounts all sorts of "chat-room failures," when cubicle warriors fighting from afar were teamed with his unit on the ground. They ranged from relatively minor screwups, such as when there were two enemy trucks "and confusion led the UAV to follow the wrong vehicle," to a more serious incident. His team was deployed on a mission inside Afghanistan, when the operator (based in Nevada) pulled the UAV out because of bad weather. This decision left his team "in the lurch," alone inside enemy territory. From his vantage point on the ground, it was "a bogus weather call," as the sky was clear. (Note: I interviewed some drone operators and commanders about his claims, and they pointed out the likelihood of high winds or looming weather changes that the soldier couldn't see from the ground.)

Such disagreements have long characterized the frequently rocky relationships between those experiencing war on the front lines and those either behind or above the fight. But this incident illustrated three important new pressure points that unmanned systems introduce. To the soldier in the midst of battle, the drone pilots weren't just bearing less risk, but no risk at all. Moreover, he saw them as valuing the safety of a machine over his men's lives. Finally, he didn't see it as a fellow warrior making the call, but "some guy sitting in Vegas rushing to take his kids out to a soccer game or go hit the slots.... [Two years later] I still want to be able to stick my finger in his chest to explain how we do business."

## FOR THE LOVE OF A ROBOT

The EOD soldier carried a box into the robot repair facility at Camp Victory, Iraq. "Can you fix it?" he asked, with tears welling in his eyes. Inside the box was a pile of broken parts. It was the remains of "Scooby-Doo," the team's PackBot, which had been blown up by an IED. On the side of Scooby's "head" was a series of handwritten hash marks, showing the number of missions that the little robot had gone on. All told, Scooby had hunted down and defused eighteen IEDs and one car bomb, dangerous missions that had saved multiple human lives. "This has been a really great robot," the soldier told Master Sergeant Ted Bogosh, the marine in charge of the repair yard.

Unfortunately, the robot could not be repaired. The news left the soldier "very upset." He didn't want a new robot but "wanted Scooby-Doo back."

At a certain level, robots are just sophisticated lawn mowers, can openers, or coffee grinders. As Carnegie Mellon robotics professor Red Whitaker puts it, "I don't get happy about robots or feel sorry for robots. They are not like little old ladies or puppies. They are just machines." Indeed, says Whitaker, for a person to develop any care, concern, love, or hate toward a machine makes no sense. "They certainly don't have the same feelings for you."

And yet while new technologies are breaking down the traditional soldierly bonds, entirely new bonds are being created in unmanned wars. People, including the most hardened soldiers, are projecting all sorts of thoughts, feelings, and emotions onto their new machines, creating a whole new side to the experience of war.

An affinity for a robot often begins when the person working with it notices some sort of "quirk," something about the way it moves, a person or animal it looks like, whatever. "You start to associate personalities with each of them," says Mark Del Giorno, vice president at General Dynamics Robotic Systems. Of course, these quirks often have nothing to do with what is causing it. "The 'personality' comes from, say, the steering being a little loose."

Pretty soon, it feels natural for the person to give the robot a name, just like they would another living thing, but not what they would do for most machines. The Roomba is really just a sophisticated vacuum cleaner. Vacuums are not a machine that people tend to give names or ascribe personalities to. And yet iRobot has found that 60 percent of Roomba owners have given names to their robot vacuums.

As the tale of Scooby-Doo illustrates, the same thing is happening in the military. But it's not just a matter of nicknames. As Joel Garreau explains, the continued evolution of human-robot interaction is leading many robot operators to do things like "award 'battlefield promotions' and 'Purple Hearts' [medals] to their machines.... One unit in the 737th Ordnance Company, for instance, called their EOD bot Sgt. Talon; Sgt. Talon, in fact, got promoted to Staff Sergeant and received three Purple Hearts."

Soldiers are not just doing this as a joke, but because they are truly bonding with these machines. Paul Varian, a chief warrant officer who served three tours in Iraq, recounts that his unit's robot was nicknamed "Frankenstein," as it had been made up of parts from other blown-up robots. But after going into battle with the team, Frankenstein was promoted to private first class and even given an EOD badge, "a coveted honor" among the small fraternity of men willing to defuse bombs. "It was a big deal. He was part of our team, one of us. He did feel like family."

Just like the Roomba owners, these soldiers know that their robots are not alive and that the machines couldn't care less whether they get promoted. The robots did what they were supposed to; it's like awarding a medal to a popcorn maker for cooking corn kernels. And yet these soldiers are experiencing some of the most searing and emotionally stressful events possible, with something they would prefer not to see as just an inanimate object. They realize they might not be alive without this machine, so they would rather not view it that way. To view the robot that fought with them, and even saved their lives, as just a "thing" is almost an insult to their own experience. So they grow to refer and even relate to their robot almost like they would with one of their human buddies.

Soldiers who work with damaged robots notice these attachments the most. Jose Ferreira described working at the repair yard in Baghdad as less like being a mechanic in a garage and more like being a doctor in an emergency room. "I wish you all could be here and experience the satisfaction in knowing you saved someone's life today. I wish you could see the fear in their eyes when they first walk in knowing that they could walk out with no robot. I wish you could see the smiles and feel the hugs and handshakes after they leave our shop knowing that their 'little Timmy' is ALIVE. Alive and well to go down range one more time."

Ironically, these sorts of close human bonds with machines sometimes work against the very rationale for why robots were put on the battlefield in the first place. Unmanned systems are supposed to lower the risks for humans. But as soldiers bond with their machines, they begin to worry about them. Just as a human team would "leave no man behind," for instance, the same sometimes goes for their robot buddies. When one robot was knocked out of action in Iraq, an EOD soldier ran fifty meters, all the while being shot at by an enemy machine gun, to "rescue it."

This effect even plays out on robot design. Mark Tilden, a robotics physicist at the Los Alamos National Laboratory, once built an ingenious robot for clearing minefields, modeled after a stick insect. It would walk though a minefield, intentionally stepping on any land mines that it found with one of its feet. Then it would right itself and crawl on, blowing up land mines until it was literally down to the last leg. When the system was put through military tests, it worked just as designed, but the army colonel in charge "blew a fuse," recounts Tilden. Describing the tests as "inhuman," the officer ordered them be stopped. "The Colonel could not stand the pathos of watching the burned, scarred, and crippled machine drag itself forward on its last leg."

Humans have a natural inclination to "anthropomorphize," to give human characteristics to something that is not human. Indeed, we are hardwired that way. In our brains are clusters of nerve cells called mirror neurons. These neurons fire when we recognize the object we are looking at as alive and deserving of empathy; that is, what we "mirror" ourselves onto. When scientists at the University of California–San Diego studied the brain scans of people looking at a robot, they found that their mirror neurons were firing. That is, the people had much the same brain activity that they would have as if they were interacting with a real person, even though the rest of their brain knew that it was just a machine.

Our machine creations are not just "neutral" objects to us. We not only tend to view them as having their own personalities, but also feel that they deserve some form of emotional attention and engagement. For instance, students majoring in computer science should know more than the rest of us that a computer is just a machine. And yet 83 percent of them describe their computers as having "agency," having their own intentions and making independent decisions (such as knowing when to crash or not for maximum effect).

Robots are often modeled in their designs after living things, so this tendency to "mirror" and empathize with our machines goes even further. For example, in a study of Sony AIBO "owners," people who had bought the tiny robotic dog, 75 percent regarded their AIBO as something more than just a machine. Almost half, 48 percent, thought that their AIBO robot had "a life-like essence."

If a robot is perceived as more than just a machine, then the way the person acts and communicates with it also changes. But it also affects how they think that the robot acts and communicates with them. Sixty percent of the owners thought their AIBO could express its "mental state," 42 percent thought that AIBO engaged in intentional behavior, and 38 percent affirmed that AIBO had "feelings." As one even described, "My dog [the AIBO robot] would get angry when my boyfriend would talk to him." Fifty-nine percent of the owners described themselves as having a social bond with their robot, with many even considering the machine a part of the family. Said one owner, "Oh yeah, I love Spaz [the robot's name], I tell him that all the time.... When I first bought him I was fascinated by the technology. Since then I feel I care about him as a pal, not as a cool piece of technology."

And this is for a tiny robot that doesn't look all that lifelike, is unable to speak, and comes with little AI. As robotic systems get more advanced, these trends may go even further. Researchers are also noticing a sort of generational split when it comes to these ideas of a machine being "alive" or having "feelings." Before the 1990s, computers and robots tended to be viewed as merely devices, machines used to do things. As computers grew more advanced and, even more so, children began to be confronted with toys that could speak and video game characters that reacted to their moves at younger ages, perceptions have begun to change.

Remember, the generation becoming soldiers now is the same one that as kids became fascinated with Tamagotchi. If you are too old to recall this, Tamagotchi was the super-popular toy that came out in 1996. It was essentially a small, colorful egg with an LCD screen and some buttons. The "fun" of it was that Tamagotchi acted like a small child; you would have to feed it, play with it, and even give it virtual shots in a medical emergency. Based on how you treated it, Tamagotchi would either get happy or sad, or even get sick and "die."

As toys like these took off and children began to use computers at younger and younger ages, researchers began to notice some "ambiguity" among children as to whether computers were alive or not. As one child said in a study, a robotic toy was "alive the way insects are alive, but not the way people are alive." Indeed, this line of what is alive or not has grown so fuzzy that visitors to Disney's Animal Kingdom theme park in Orlando have even started to complain that the real-life biological animals "were not realistic" compared to the animatronic ones.

With humans bonding with robots even without designers trying, roboticists are now trying to take advantage of this natural proclivity. Their goal is to create "social robots" that have emotions, or rather the semblance of emotions, in order to make it easier for humans to interact with them. That is, the human-machine interface might be smoothed out by machines that mimic human behaviors, including even having what seems like an emotional response and attachment to the person. Ibn Sina, a robot built by the Interactive Robots and Media Lab at the United Arab Emirates University, was even given a Facebook page so that it could glean information gathered from the online social network to weave into face-to-face conversations with its "friends." This echoes back to the work done by David Hanson and his Zeno, the little boy robot designed to grow up with his son. "To live among people, robots need to handle complex social tasks," says Junichi Takeno, a Japanese researcher building robots that react to different human key words with feigned expressions of happiness, sadness, urgency, fear, and even anger. "Robots will need to work with emotions, to understand and eventually feel them."

Building such bots makes perfect sense for a robotic toy, but why would the military want a weapon to have social skills? An M-16 never needed to learn how to play nice, so why should a SWORDS packing an M-16? The answer is that just as soldiers who trust each other fight better together, so do humans and their robots work more efficiently when there is a bond. Chemistry matters, even if it is a completely faux one.

In one study of operating unmanned combat drones, human operators interacted with AI programs that were given two different "personalities." One AI had a humanlike voice and mannerisms and would greet the human by saying, "Hey [whatever their name was], we did an awesome job—great working with you!" or with a joke. The other would say, in a monotone voice: "Hello." These differences in personality continued through the missions. The personable AI would not just advise the human agents about a mission, but also try to inspire them, saying such things as, "Here is the last known target. Let's finish this!" The other AI would just say, "Pay attention, high priority." It paid off, and the personable AI-human team finished the tasks faster. In robotic warfare, nice AIs finish first.

The outcome points to a whole new kind of wartime cohesion in unmanned war as well as a new irony. As Peter Kahn, one of the world's leading experts on human-robot interaction, explains, "Let's say you design robots to be team members in the military. That might increase the ability to engage effectively with the robot. But then it may be a little hard to think of a robot team member as a disposable unit.... It's not going to go great to have your Robo-buddy blown up." Kahn believes that ultimately the military may want robots with "a slightly aversive personality." In other words, robots that are social, but just annoying enough so that fellow soldiers won't feel bad when they get blown up.

## CHANGING THE DEFINITIONS

In his memoir *Soldiering*, Colonel Henry Cole of the U.S. Army looked back on his experience in the military. He recalls that it was seeing his uncle, drafted into the war effort, coming home on his first leave that made him want to join the military. Cole's uncle was a rail-thin eighteen-year-old boy and his uniform didn't fit. His overcoat smelled, as its cloth was made from a horse blanket. "But he was somebody," Cole recalls. "He was a solider."

A half century later in 2005, Nathaniel Fick explained what led him to leave the fast track of an Ivy League education and join the Marine Corps, consciously choosing a life in which death, as he titled his memoir, was always just *One Bullet Away*. "Being a Marine was not about earning money for graduate school or learning a skill; it was a rite of passage in a society becoming so soft and homogenous that the very concept was often sneered at."

Joining the military and heading off to war has long been viewed as a transformative act. It wraps together a deliberate choice of self-sacrifice, taking on a new identity, and adhering to a new code of behavior, conduct, and honor. This experience changes how a person looks at the world and how the world looks at that person.

As technology changes what it means to "go to war," how soldiers experience battle, and the bonds they build, we must question whether the memoirs of our current and future digital warriors will convey that same sense of powerful transformation. The soldiers who watched the last stand of Saddam's sons, Gary Fabricius and his squadron, and all the others fighting via chat rooms and unmanned systems were certainly at war, but they experienced it in a far different way than Cole, Fick, or any previous generation. And we can expect that their future memoirs and histories will likely read quite differently as well. A U.S. Marine summed it up this way: "We joined the military to become warriors. But that definition is changing."

# EIGHTEEN

# COMMAND AND CONTROL . . . ALT-DELETE: NEW TECHNOLOGIES AND THEIR EFFECT ON LEADERSHIP

*I can no longer obey; I have tasted command, and I cannot give it up.*

—NAPOLEON BONAPARTE

"You are watching the most violent actions that man carries out, but you are not there. It's antiseptic. It's not as potent an emotion as being on the battlefield. You may get angry at seeing one of our guys get killed, but then it's on to the next mission."

Colonel Michael Downs entered the air force out of Texas A&M University. I first met the avid "Aggie" football fan when we shared a cubicle in an office inside the bowels of the Pentagon. Since he was an air force officer, the Pentagon in its infinite wisdom had assigned Downs responsibility for the landmine issue in the Balkans. The next time I saw him he was out in the Middle East, serving in a more traditional air force role as one of the key planners of the air operations for the early stages of the Iraq war, for which he was awarded the Bronze Star.

Downs then shifted over to Beale Air Force Base, located about forty miles north of Sacramento, California. Unlike most air bases, which are named after pilots, Beale is named in honor of the man who founded the Army Camel Corps in the 1840s. Given such an iconoclastic legacy, it is perhaps appropriate that today

Beale is the home of the 9th Reconnaissance Wing, a unit that hosts much of the air force's unmanned operations, as well as the 548th Intelligence Group, which helps analyze the information gathered by America's fleet of unmanned drones.

Downs's job at Beale as director of operations was to help lead and coordinate the high-altitude unmanned operations that took place around the globe. He sees "a strong future" in unmanned systems. "They are becoming a staple of what we do." Trained as an intelligence officer, he appreciates how much they help with what is perhaps the most difficult task in modern-day war: simply locating the enemy. "We continue to make incredible progress on the kinetic part of war, to where our biggest challenge is no longer destroying targets, it's finding them." During the 2006 operation that killed al-Qaeda in Iraq leader Abu Musab al-Zarqawi, "It took over six hundred hours of surveillance work for roughly ten minutes of bomb-dropping work."

Downs wants to make clear that those fighting from places like Beale are not some stereotype of "emotionless automatons who are detached from the impact of their work." Instead, his pride in the men and women on his team shines through again and again. "They care deeply about what they do, why they do it, and give of themselves greatly for our country. They are consummate professionals." Rather, Downs is growing concerned about how their leaders like himself will face the unfamiliar challenges that unmanned, distance warfare presents.

## THE WAR AT HOME

The units like the one Downs led at Beale are not merely fighting from afar, but doing so 24/7, over long periods of time. "Maintaining the acute concentration and focus necessary for combat operations is difficult if you are doing the same thing every single day, day in and day out, for three, four, or six years in a row."

He describes the challenge of keeping a "razor-sharp focus, consistently." A commander has to be sure to continually reinforce the criticality of the mission to his troops, "so that they have the mental and emotional sense that they are in the battle space that they're looking at." Downs continues, "You try to give your team the context, make sure they link what they are doing here in the States to the broader cause, to see the importance of it, so six months in you don't have people with their jaws on the keyboards. . . . I would tell my folks that when they stepped into our mission vans, that they were leaving California and stepping into Iraq or Afghanistan."

Downs thinks that unmanned war, "while you can't compare it to the experience on the ground," also comes with a great deal of psychological stress and emotional connections, perhaps more than people might think that a so-called cubicle warrior would experience. He recalls an instance in which the crew of an unarmed Predator drone could only watch from above as insurgents killed a team of U.S. Special Forces operators. "It was tough on the young kids....I worry about the young airmen. They don't have the same life experience and support systems. They just go home and internalize it."

Downs's worry was later reinforced by a staff sergeant at another air base, who helps oversee the support of drone crews and mission planners. She similarly raised the issue of what the servicemen and -women under her care were experiencing, even while fighting from afar. "What angers me is that as a service, we are not doing a good job on PTSD [post-traumatic stress disorder]. People are watching horrible scenes, it's affecting people. Yet we have *no* systematic process on how we take care of our people."

Another novel command challenge emerged from what is widely perceived as the greatest perk of distant war: fighting without leaving home. "Conducting continual combat operations from home station presents a unique set of stresses and challenges that we've not had to face until recently." Downs is a married father of three children, who has deployed out on operations to dangerous places in the Middle East and the Balkans multiple times. So he knows the risks of an actual deployment to a combat zone and the accompanying heartache that comes from leaving loved ones behind. Yet, he explains, leaders are also starting to learn that commanding reachback operations at home comes with new issues that raise all sorts of leadership questions. "When you are deployed, the mission is your only job. When you are at home you still have the mission, but all the extras, plus the family."

His unit may have been at home base, but it operated on a wartime schedule, conducting missions 24 hours a day, 7 days a week, 365 days a year. There are no weekends or holidays, and the pace can be grueling for the men and women he commands. Yet, while the war may be on, none of the pressures of the home front disappear. "You are at war, but at the same time you have Mom at home saying the toilet needs to be fixed. You need to be ready to execute combat missions, where lives are at stake, but still have church activities to go to, kids that need to be taken to the hospital, soccer practices, et cetera."

Also, because they are fighting within a battle space physically located half a world away, the units adhere to a different time zone. Evening in Afghanistan is afternoon in Iraq is early morning in California. As the singer James Taylor might put it, for an unmanned unit, "It's war o'clock somewhere." Explains Downs, "Even when you are off, you're out of sync with your family."

This aspect of balancing fighting and family creates an almost psychological disconnect in how the units have to operate. "You see Americans killed in front of your eyes and then have to go to a PTA meeting." Gary Fabricius, our Predator squadron commander, similarly cited this as perhaps the most surprising challenge of his early experience with unmanned war. "You are going to war for twelve hours, shooting weapons at targets, directing kills on enemy combatants, and then you get in the car, drive home, and within twenty minutes you are sitting at the dinner table talking to your kids about their homework."

With these different sorts of pressures, it is very tough for a leader to ensure his unit keeps its "battle rhythm" when it is still located at home. Indeed, a survey of air force drone crews found that, contrary to expectation that those fighting from a distance should find it easier, the remote crews actually had "significantly increased fatigue, emotional exhaustion, and burnout." They were even found to be suffering from the stress and fatigue of combat at the same, if not higher, levels than many units physically in the war zone. And, despite being at home with their families, the crews were found to be more likely suffering from "impaired domestic relationships." Says Downs, commanders particularly have to keep an eye out for young troops "burning the candle at both ends of the wick."

This new generation of leaders like Downs is testing a variety of measures to try to help their forces operate at maximum efficiency and keep the two worlds separated. One is the banning of personal phone calls into the control rooms. When the soldiers are at war, they are kept in a communications bubble. Another idea is for reachback units to operate like many professional sports teams do before big games. Just like a football team before the Super Bowl, a unit rotating onto assignment might be sequestered at a hotel or barracks on base, isolated from their families during operations. This would create a bit more of a distinction between war time and home time, thinks Downs, as well as "keep them fresher and give more focus."

Ultimately, Downs feels that the stakes of being at war still overwhelm what would seem to be the virtual nature of fighting it from home. I once asked him if it would ever be possible for warriors fighting unmanned wars from afar to leave their work back at the office, just like other professions can do. He paused for a half minute in silent reflection. He then responded, "You don't really switch it off."

## TACTICAL GENERALS

The four-star general proudly recounts how he had spent "two hours watching footage" beamed to his office. Sitting behind a live feed of video from a Predator drone, he saw the two insurgent leaders sneak into a compound of houses. Then he waited as other insurgents entered and exited the compound, openly carrying weapons. He was now personally certain. Not only was the compound a legitimate target, but any civilians in the houses had to know that it was being used for war, what with all the armed men moving about. So, having personally checked out the situation, he gave the order to strike. But his role in the operation didn't end there; the general tells how he even decided what size bomb his pilots should drop on the compound.

Much like Downs watching after his men and women at Beale, great generals also had to have an innate connection to the warriors fighting under their command. In his masterful history of men at war, *The Face of Battle*, John Keegan wrote how "the personal bond between leader and follower lies at the root of all explanations of what does and does not happen in battle." In Keegan's view, the exemplar of this was Henry V at the battle of Agincourt, who so inspired his "band of brothers" by fighting in their midst.

With the rise of each new communications technology, these connections between the soldiers in the field and those giving them battle orders began to be distanced. Generals were no longer at the same front lines as their men, but operated from command posts that moved farther back with each new technologic advance. And yet, describes analyst Chris Gray, the very same technologies also pushed a trend "towards centralization of command, and thus towards micromanagement."

When telegraphs were introduced during the Crimean War (1853–56), generals back in England quickly figured out that they could now send in their daily plans to those on the front lines in Russia. And so they did. The advent of radio heightened this effect. Hitler, for instance, was notorious for issuing detailed orders to individual units fighting on the Eastern Front, cutting out the German army's entire command staff from the process of leading its troops in war. Even the U.S. military has suffered from this problem. During the 1975 *Mayagüez* rescue attempt, considered the last battle of the Vietnam War, the commander on the scene received so much advice and so many orders from leaders back in D.C. that he eventually "just turned the radios off."

These leaders never had access to systems like today's Global Command and Control System (GCCS). As one report describes, "GCCS—known as 'Geeks' to soldiers in the field—is the military's HAL 9000. It's an umbrella system that tracks every friendly tank, plane, ship, and soldier in the world in real time, plotting their positions as they move on a digital map. It can also show enemy locations gleaned from intelligence." When combined with the live video that various unmanned systems beam back, commanders are enabled by technology as never before. They are not just linked closer to the battlefield from greater distances, ending the separation of space, but the separation of time has also been ended. Commanders are not only able to transmit orders in real time to the lowest-level troops or systems in the field, but they can also see the action in real time. With a robotic system like a drone or SWORDS, that commander can see the exact same footage that the operator sees, at the exact same time, and even take over the decision to shoot.

Many people, especially the Cebrowski-led network-centric warfare crowd, thought that this linking together of every soldier and system into a vast IT network would decentralize operations, that it would allow for greater initiative among the lower-level units in war. Actual experience with unmanned systems is so far proving the opposite. The new technologies have also enabled the old trends of command interference to reach new extremes of micromanagement.

Too frequently, generals at a distance are now using information technology to interpose themselves into matters that used to be handled by those on the scene and at ranks far below them. One battalion commander in Iraq told how he had twelve stars' worth of generals (a four-star general, two three-star lieutenant generals, and a two-star major general) tell him where to position his units during a battle. An army special operations forces captain even had a brigadier general (four layers of command up) radio him while his team was in the midst of hunting down an Iraqi insurgent who had escaped during a raid. The general, watching live Predator video back at the command center in Baghdad, ordered the captain where to deploy not merely his unit, but his individual soldiers. "It's like crack for generals," says Chuck Kamps, a professor at the Air Command and Staff College. "It gives them unprecedented ability to meddle in mission commanders' jobs."

Over the last few years, many analysts have discussed what marine general Charles Krulak called the rise of the "strategic corporal." This idea was meant to describe how new technology put far more destructive power (and thus influence over strategic outcomes) into the hands of younger, more junior troops. A twenty-year-old corporal could now call in airstrikes that a forty-year-old colonel used to decide in the past. But these technologies are also producing something new, which I call the "tactical general." While they are becoming more distanced from the battlefield, generals are becoming more involved in the real-time fighting of war.

As retired army colonel Robert Killebrew explains, the technology available to today's senior commanders provides them with numerous "incentives to intervene tactically at the lowest levels." That a general, who can now see what is unfolding on the ground, would want to shape it directly makes perfect sense. All sorts of battles have been lost when a general's commands were misinterpreted or implemented wrongly by subordinates in the field. Who else better knows a commander's intent than the commander? What is more, a general who stays on top of the situation can rapidly adjust his original commands to any changes that happen in the midst of battle, rather than letting old plans be carried out despite already being passed by events.

Unfortunately, unmanned systems are blurring the line between timely supervision and micromanagement. Retired air force lieutenant colonel Dan Kuehl points out that just because a general now can use a "5,000 mile long screwdriver" doesn't mean he should. One interviewee, for example, described how officers hundreds of miles away would instruct him onto which roads he should turn down during raids in Afghanistan. In another case, a soldier in the 82nd Airborne described how a commander watching Predator feed of his patrol trek up a mountain in Afghanistan radioed in to chew the unit commander out for a uniform violation. Watching from an air-conditioned room thousands of miles away, the commander had observed a few of the patrol members untuck their shirts and take their helmets off (against regulations, but also understandable because of the stifling heat of the march up a mountain).

To the general who described spending two hours watching Predator footage of just one compound, this was time well spent. As the overall commander, he was going to be held accountable if the strike went awry. So if the technology allows, he believed that he should make sure it went exactly the way he wanted. But while this general was doing a job that normally would have been done by captains in the field, who was doing the general's job? These new technologies allowed him to make tactical decisions as never before. But the captains, majors, colonels, and so forth that he was cutting out of the chain could not, in turn, devote themselves to the big strategic and policy questions that the general would have been wrestling with instead.

Moreover, "tactical generals" often overestimate how much they really know about what is happening on the ground. Operation Anaconda, the 2002 battle when the 10th Mountain Division took on Taliban and al-Qaeda fighters in the Shah-i-Khot valley in Afghanistan, was one of the first battles in which generals back in the States could watch a battle play out live, beamed back to them by a Predator drone that flew above the fight. The danger, explains Major Louis Bello, the fire support coordinator for the 10th Mountain Division, is that the video tends to be "seductive," leading commanders to focus in on what the drone beamed back as if it were the whole story. "You get too focused on what you can see, and neglect what you can't see," Bello said. "And a lot of the time, what's happening elsewhere is more important."

Jumping in and out of the tactical issue, rather than working it day to day, senior officers don't have the local context and also tend to interpose their assumptions onto the video they see. During the battle, for example, American commanders saw live video of al-Qaeda fighters moving across a mountain. Even though the footage was staring them in the face, the commanders thought they were seeing Americans, as that was who they expected to see there based on their original plans.

Misunderstanding from afar can even be heightened by technology. During the 2003 Iraq invasion, for example, the overall commander, General Tommy Franks, reportedly became obsessed with the "Blue Force Tracker" map. This was a massive electronic display that showed the exact locations and status of every U.S. unit, as well as the Iraqi units facing them. The appearance of so much information proved deceptive, however. At one stage early in the fight, it looked to Franks like several units in the Army's V Corps were neither moving nor fighting. The tracking map showed no Iraqi units nearby and so Franks reportedly flew off the handle. He tracked down his land forces commander, who in his words was then made to eat "a shit sandwich."

There was only one problem: General Franks was reportedly looking at the electronic map on the wrong scale. If he had just increased the map's resolution, he would have seen that while the American units may have looked like they were alone at the large scale in the map, they were in actuality locked in one of the toughest battles of the entire invasion, fighting against a swarm of Saddam Fedayeen teams. These small insurgent units were big enough to give the U.S. invasion fits, but not big enough to get their own logos on the high-tech map that the general far from battle was watching.

Most of all, officers in the field lament what they call the "Mother may I?" syndrome that has come with these new technologies. Rather than relying on the judgment of their highly trained officers, generals increasingly want to inspect the situation for themselves. It's all fine if the enemy plays along and gives that general several hours to watch the video himself and decide which bomb to use. But sometimes matters aren't decided on a general's schedule. An air force officer in the Middle East described his ultimate frustration being when he had information that could have saved lives, but "it sat in someone's e-mail queue for six hours." Similarly, one Predator pilot complains, "It's the old story—by the time you have all the evidence, it's too late to affect the outcome."

Ultimately, these problems put a new wrinkle on a venerable truism of war. As Napoleon once said, "One bad general is better than two good ones." The traditional concept of a military operation is a pyramid, with the strategic commander on top, the operational commanders next, and the tactical commanders on the bottom layer. With the new technologies, this structure isn't just being erased from above, with strategic and operational commanders now getting into the tactical commanders' business. It is also endangered from the sides. As one drone squadron officer explains, a major challenge in the command and control of reachback operations is their simultaneous location in multiple spaces. The drones may be flying over Iraq, but they are launched out of a base in the Persian Gulf, and flown by men sitting back in Nevada. At each of those locales, "each commander thinks he's in control of you." Even worse, the drones are a high-demand asset, for which everyone is clamoring.

The results are "power struggles galore." As the operations are located around the world, it is not always clear whose orders take priority. The units instead would get "pulled in many directions because you are in virtual space. Am I at Nellis or am I at CENTAF [the air command in the Middle East]?"

Moreover, by giving everybody in the command structure access to the Internet, the ability to watch what is going on and to weigh in on what the units should be ordered to do is not limited just to where a unit is physically or virtually located. During the Shah-i-Khot battle, for instance, video of the fighting was beamed from the Predators to bases and offices all over the world. Army major general Franklin "Buster" Hagenbeck, the commander of U.S. ground forces during the battle, recalls how "disruptive" this was, as officers all around the world now felt that "they were in a position to get involved in the battle." While his team was trying to actually fight the battle in Afghanistan, "people on other staffs at higher levels would call all the way down to my staff and get information and make suggestions." In the midst of battle, some officers back in the States even called in asking for information that they could plug into their own generals' daily briefings, pestering soldiers fighting "for details that they presumed their bosses would want to know."

Each of these tasking orders is tough to ignore. Not only do they come in from senior leaders, who can make or break careers, but they also tend to come in on a "priority basis." The various generals around the world tend to use a logic that humorist Garrison Keillor cited in *Lake Wobegon Days*. Every single one of them, of course, thinks that they and their missions and orders must be the ones of "above average" importance. But not everyone actually is. This "flattening of the chain of command," says retired lieutenant general William Odom, causes "constipated communication channels" and "diarrhea of the email" that distracts troops from the mission at hand.

At its worst, this pattern can lead to the battlefield version of too many cooks spoiling the meal. A marine officer recalls, for example, that during an operation in Afghanistan, he was sent wildly diverging orders by three different senior commanders. One told him to seize a town fifty miles away. Another told him to seize just the roadway outside of the town. And the third told him, "Don't do anything beyond patrol five miles around the base."

The marine in this case ultimately chose "curtain number one" and seized the town. A veteran of the 1991 Gulf War, he felt confident enough to take the career risk of going with his gut. But the rise of virtual command from afar threatens to hollow out the experience of those who will be moving into these command roles in the future. Explains one Predator squadron commander, "You may have some general officer sitting behind four Toshiba big screens [TVs] with greater knowledge of the battlefield from the distance. And maybe it works the first time when they intervene and save the day. But my worry is what happens with the next generation. What happens when that lieutenant, who learns thinking the guys in the back are smarter, becomes a colonel or a general. He'll be making the decisions, but not have any experience."

Some worry that the ability to reach into the battlefield could even prove tempting to those outside the military. Marine veteran Bing West expects that "in the near future . . . a president will say, 'Why do we need these twenty links in the chain of command?'" As West explains, the enhanced connections could certainly help the commander in chief become better informed about the true situation on the ground, but could prove catastrophic if civilian leaders are tempted to intervene, "trying to play soldier." Referring to how President Johnson often tried to influence air operations in Vietnam, Secretary of the Air Force Michael Wynne warned that "it'll be like taking LBJ all the way down into the foxhole."

## DIGITALLY LEADING

"You know what makes leadership?" asked Harry Truman. "It is the ability to get men to do what they don't want to do and like it." So the "techniques of leadership" that generals needed in the past were both physical and psychological. A general might lead by example, like Henry V, exposing themselves to danger at the head of a charge. Or they might inspire by appealing to soldiers' moral centers, by demonstrating what marine colonel Bryan McCoy calls "the passion of command." Or they even might try to play on soldiers' pride, such as how Patton would publicly embarrass his officers by cussing them out in front of their men, in order to try to spur all to action.

These qualities are all in stark contrast to how science fiction portrays the generals of the future, as they use more and more unmanned systems. For example, a *Star Wars* novel described a leader commanding unmanned systems in a galaxy far, far away. He sees his role as only to make cold calculations of costs and benefits, as he moves robotic units around like a computerized game of chess. "Commanding an army of droids was more like playing a game than engaging in actual combat. [By comparison] living soldiers bled and died, had to be fed, experienced morale problems, knew fear and all the other emotions common to beings who could think."

Such a leader is fortunately still fiction but it is becoming clear that twenty-first-century generals will have to bring new skills to increasingly unmanned wars. When the U.S. Army War College studied what would make a good general in this new century, it found that new technologies are creating an environment "where the strategic, operational, and tactical levels of war can at times be so compressed as to appear virtually as a single function." The downside of this "compression" is that it tempts officers to micromanage (the "tactical general" problem). However, officers who have what Clausewitz called the "eye of command," who can find the right balance, will achieve "simultaneous awareness" of what is going on at all the levels of war, and make the appropriate decisions.

This isn't going to be easy. For one thing, all the information being collected, all the requests taking place in real time, and all the general "diarrhea of the email" threaten to flood officers with what the army study described as an "avalanche of data." Much like a corporate drone in his office cubicle, the twenty-first-century general will have to develop the ability to manage his in-box. Notes the report, "The strategic leader best adapted for the Information Age will be one with a retentive but discriminating mind, capable of separating the essential from that which is interesting and acting with confidence on his or her conclusions."

Part of how this problem of information overload will likely be managed is developing a "knack for enlightened control." Generals will literally have the entire battle at their fingertips. They can watch nearly every single action and make every minute decision. But technology still cannot give them an infinite amount of time. At some point, the leader has to turn matters over their subordinates. The general who can figure out when to intervene and when to delegate down the chain of command, and even more, to empower their junior troops to act with initiative in the absence of micromanagement, will be far more successful than the general who doesn't trust their force to do anything without them.

Good generals will also need the mental flexibility to lead a "learning organization" that can adapt to changing circumstances in something beyond just a top-down manner. They will not only have an open mind themselves, but also be willing to let their subordinates wrestle with new concepts and new technologies. Describes Colonel Paul Harig of the U.S. Army War College, "I speculate that the digital general some 35 years from now might not just communicate differently but will actually *think differently* from his or her predecessors, because conceptual behavior itself is evolving during the Information Age."

While a general may no longer have to be as fit a fighter as his troops, the way a Henry V was, the new technologies do impose certain physical requirements on commanders in wartime. For one thing, the U.S. military is finding that generals had better have "hands-on skill" at using a computer, something that once seemed an almost abhorrent concept to leaders. Writes an army report, "To the strategic commander of the Information Age, the laptop computer, or its successor, will be a natural extension of his mind, as familiar as the telephone, map, and binoculars. Aspiring future commanders who are not already computer literate take note."

Another physical shift comes from wars being no longer limited by geography or time. While command has always been taxing, it is now becoming literally a 24/7 job. The kind of strength needed to wield a sword may no longer be required of generals. But they may now need the physical and psychological stamina of a twenty-two-year-old medical student on call in the ER.

Some of these changes might seem immense, but they will not supplant many of the same qualities that made great generals in the past. For example, the idea of "enlightened control," giving just enough guidance to officers closer to the scene so that they can figure what to do best, is nothing new. The great Prussian generals of the nineteenth century were big believers in its equivalent. They called this *Führen durch Auftrag*, "leading by task," as opposed to *Führen durch Befehl*, "leading by orders." Their ideal was that the best general gave his officers the objective and then left it to them to figure out how best to achieve it. The most famous of these was before the 1864 Prussian invasion of the Danish province of Schleswig, where the commanding general so trusted his officers that the only order he supposedly issued was "On February 1st, I want to sleep in Schleswig."

While this may be a bit too succinct for modern war, the example of General George Marshall, the overall commander of the U.S. Army during World War II, remains an apt model for twenty-first-century leaders. New inventions like the radio and teletype gave him an enhanced ability to instruct from afar, but Marshall's approach was to set the broad goals and agenda, have smart staff officers write up the details of the plan, but ensure that everything remained simple enough that a lieutenant in the field could understand and implement everything on their own. Similarly, marine general James Mattis's guidance to his troops before the 2003 invasion of Iraq was just as brief, understandable, and worthy. "Engage your brain before you engage your weapon."

When the army surveyed almost five hundred generals and colonels about what traits officers would need in the twenty-first century, they identified such qualities as "flexibility," "adaptability," "political astuteness," "ability to conceptualize," "skill in resource management," and "caring leadership." As Colonel Harig of the War College put it, "In the end, it could be argued, all great commanders are the same. They adapt the technology of their times in a highly personal, reflective space where machines can extend, but never supplant, the human dimension of their leadership."

## GENERAL 2.0

Every decision in a military operation, whether it is the corporal (or robot) in the field deciding whether to pull the trigger or General Eisenhower deciding whether to give the "go" for the D-Day invasion, can be broken down into four basic parts. Folks in the military call these the "OODA loop," short for "observe, orient, decide, and act." Information is gathered, the situation figured out, orders issued, and action taken. Then the whole OODA cycle begins again.

The challenge is that technology is shrinking the time inside this decision cycle. Massive amounts of information are coming in faster, and decisions have to be made quicker as a result. This is what led, for example, to defense against mortars and rockets in Iraq being turned over to the R2-D2-like CRAM automated gun system. Humans just couldn't fit into the shorter OODA loop needed to shoot down rockets.

This shortening of time in the decision cycle is working up the chain to the generals' level. Marine general James Cartwright, chief of the U.S. Strategic Command (the part of the military that controls the nukes), predicts that "the decision cycle of the future is not going to be minutes. The decision cycle of the future is going to be microseconds."

And thus many think there may be one last, fundamental change in the role of commanders at war. As a 2002 army report posits, "The solution to this problem may come from automated systems that have enhanced artificial intelligence. Unmanned systems will capitalize on artificial intelligence technology gains to be able to assess operations and tactical situations and determine an appropriate course of action." If the first step of technology's effect on command and control is to force officers to learn how to lead troops fighting from home bases, and the second is to make generals have to figure out when to intervene directly in the battle or not, the final step may be figuring out just which command roles to leave to people and which to hand over to machines.

The world is already awash with all sorts of computer systems that help us sift through information, and decide matters on our behalf. Your e-mail likely filters out junk mail that you don't need to read, while billions of dollars are traded on the stock market by AI systems that decide when to buy and sell based only on algorithms.

The same sort of "expert systems" are gradually being introduced into the military. DARPA, for example, has created the Integrated Battle Command and "Deep Green" (named after the supercomputer that first beat man at chess). The system gives military officers "decision aids"—AI that allows a commander to visualize and evaluate their plans, as well as predict the impact of a variety of effects. For example, the system helps a command team building a military operational plan to assess the various interactions that will take place in it, so that they can see how changing certain parameters might play out in direct and indirect ways so complex that a human would find them difficult to calculate. The next phase in the project is to build an AI that plans out an entire campaign. Similarly, "battle management" systems have been activated that provide advice on actions an enemy might take and potential countermoves, even drawing up the deployment and logistical plans for units to redeploy, as well as creating the command orders that an officer would have to issue. The military intelligence officer version of this is RAID (the Real-Time Adversarial Intelligence and Decision-making), an AI that scans a database of previous enemy actions within an area of operations to help "provide the commander with an estimate of his opponent's strategic objectives." The Israeli military is even fielding a "virtual battle management" AI. Its primary job is to support mission commanders, but it can take over in extreme situations, such as when the number of incoming targets overwhelms the human.

The raw processing power and memory of such systems can offset the problems of information overload that so trouble human commanders. Because searching though data and then processing it takes too much time, human commanders without such aids have to pick out which data they want to look at and which to ignore. Not only does this inevitably lead them to skip the rest of the information that they don't have the time to cover, but humans also tend to give more weight in their decisions to the information they see first, even if it is not representative of the whole. The result is "satisficing." They tend to come out with a satisfactory answer, though not the optimal answer. One air force officer described how each morning he received a "three-inch-deep" folder of printouts with the previous night's intelligence data, which he could only skim through quickly before he had to start assigning missions. "A lot of data is falling on the floor."

Emotions also can shape decisions, even the most major military ones. Recent neurological findings indicate that emotions drive our thought processes, including leaders' political decisions, to a greater extent than has been previously recognized. That is, our idealized concept of how decisions are made in war and politics—rationally weighing the evidence to decide how and when to act—does not tell the full story of how human leaders' brains actually work.

Stephen Rosen is a Harvard professor who consults for senior leadership at the Pentagon. In his book *War and Human Nature*, he describes how two underrated factors have frequently shaped strategic choices in war. The first are powerful emotional experiences that leaders had in the past. These often steered their decisions, even decades afterward, including even decisions on whether to go to war. The second factor was how body chemistry affected one's state of mind. Those with high levels of testosterone, for instance, were more likely to exhibit aggressive behavior and risk-taking; Custer and Patton seem classic examples. By contrast, those with low levels of serotonin are more prone to depression and mood swings; Hitler and Lincoln both were known for such. As these examples show, emotions can shape a leader's decisions for both the better and the worse, so to pull emotions out of the equation could yield widely divergent results.

Leaving aside that such artificial decision systems are how AIs invariably take over the world in movies like *The Terminator*, machine intelligence may not be the perfect match for the human realm of war. "The history of human conflicts is littered with examples of how military forces achieved results that no algorithm would have predicted," tells an air force general. And he is right. It may seem just like a game of chess to some, but war doesn't have a finite set of possible actions and a quantifiable logic of zeros and ones. Instead, as one writer put it, "In war, as in life, spontaneity still prevails over programming."

Even so, the Pentagon's work on such programs continues. Many think that the most likely result for future command and control is a parallel to the "warfighters' associate" concept of mixed teams of soldiers and robots fighting in the field. Their future commanders back at base will soon also have a staff that mixes advice from human officers as well as AI. Colonel James Lasswell of the Marine Corps War-fighting Lab thinks that the various technological decision aids will likely evolve into an AI "alter ego" for the commander. A sort of artificial aide-de-camp, the technology would "automatically send and collate information for him to have at his beck and call."

A real-world example of this under development now in DARPA is the "PETE" (Professional, Educated, Trained, and Empowered) virtual electronic assistant. PETE wouldn't just gather and collate information for the human commander but would also execute orders and even liaise with other commanders' virtual assist-ants, creating a network of PETEs. The developers envision a resulting split in how the team of a human commander and his AI commander's associate would handle the OODA loop; with each focusing on what they do best, PETE might perform as much as 90 percent of the Observe (gathering data), 70 percent of the Orient (making sense of the data), but maybe as little as 30 percent of the Decide, and 50 percent of the Action (issuing orders).

Since the beginnings of war, leaders have described the responsibilities of com-mand as feeling like the weight of the world was on their shoulders. Whether it's an officer like Mike Downs having to figure out how to support his team of cubicle warriors or a future general having to decide just how much to integrate the advice of a machine into his battle plans, unmanned systems are lifting some of these bur-dens of command, while adding many new ones. Machines may not yet be making command decisions in war, but they are certainly shaping them as never before.

# NINETEEN

# WHO LET YOU IN THE WAR?
# TECHNOLOGY AND THE
# NEW DEMOGRAPHICS OF CONFLICT

*How can I be a professional, if there is no profession?*
—U.S. MILITARY OFFICER

"Simplicity. These systems are extremely simple to operate. My friends back home always seem shocked. This field requires basic computer skills and an abundance of common sense. That's all."

Joel Clark originally tried to join the army to become a helicopter mechanic. However, his high school transcript revealed a failed English class. In typical military logic, his lack of a love of Shakespeare made him unqualified to be a mechanic. The army recruiter, not wanting to lose a young man willing to serve his country, scrambled to find a military job that he was qualified for. He asked Joel if he wanted to become a "96 Uniform," army-speak for a robotic airplane pilot.

It had never been part of Clark's life plans, "but the idea of running a robot spy plane sounded pretty rad." Plus, as Clark tells, he had a bigger goal in mind. "The only thing that I was concerned about when I got on the plane to basic training was making my father proud. Failing to graduate [high school] on time put a rift in our relationship, so my goal was to complete this task to the best of my ability, in order to regain his confidence in me."

Clark's effort to serve his country and make his father proud took him to Fort Huachuca, a 125-year-old base in Arizona, ten miles from the Mexican border. Huachuca originally housed the cavalry units that hunted down the Apache warrior Geronimo. Today, it houses the U.S. Army's training school for unmanned aircraft pilots.

Clark proved a quick study. Like most of Generation Y, he was a whiz at computers and video games, perhaps in part explaining his subpar English grades. It also helped that the controls of the robotic drones he was flying were similar to the ones in the Xbox and PlayStation video games that he continued to play during his downtime back in the barracks. After a few months of training, Clark was ready. He may not have been able to pass that pesky English class, but the army judged him qualified to fly combat missions and sent him off to "the big sandbox," what trainees call Iraq.

"I love my job. I have done a lot with and for UAS [unmanned aerial systems]. The most rewarding experience I have had working with UAS would have to be the number of insurgents I have personally been responsible for the capture of. Nothing feels more rewarding than watching the final takedown of an insurgent after guiding troops to his position." Indeed, Clark proved so talented at using unmanned systems that upon his return from Iraq, he was posted back to the training school at Huachuca. Even though he is still an enlisted soldier, the twenty-year-old is now teaching the next wave of drone pilots.

Joel Clark's experience illustrates how technology is changing not just how wars are fought, but also who is allowed to fight them. Of course, his amazing journey is not one that air force pilots find "rad." The idea that a nineteen-year-old enlisted high school dropout from the army can take over a job once limited to college-trained air force officers also shows how upsetting and controversial the changes can be. Clark may never step inside a cockpit going Mach 2 over enemy lines, but he is certainly on the front lines of a new kind of conflict. Military technology reporter Noah Shachtman describes this battle as a "military culture clash between teenaged videogamers and veteran flight jocks for control of the drones."

## "GENERATION KILL" MEETS THE XBOX

The soldier's profession is "more than just a job," as the saying goes. Those who serve in the realm of war not only are responsible for the safety and security of all the other professions out there, but they also form a special sort of fraternity. This comes not just from their shared risks and losses, but also because the military profession is uniquely set apart from the rest of society. It has its own training, its own schools, its own insignia and uniforms, its own housing and bases, and, most especially, its own professional codes. Indeed, the military is the only profession with its own system of laws and courts. Because of this unique professional identity a special forces officer will, for example, describe himself as having more in common with the enemy trying to kill him than with his own brother selling insurance back home.

This singular professional identity is under siege, however. The military profession as we know it today really only came about with the rise of nation-states in the eighteenth and nineteenth centuries. With the rise of globalization in the twenty-first century, these old ideas of a nation-state and its uniformed military are transforming. Not only are governments losing the control they once had over things like their financial markets and trade, but global forces are also redefining how they look at security, something that used to be the sole province of a nation's military. The threats to security now range from terrorists and rogue states to global infectious disease. This, in turn, has the military profession both working with and competing against other agencies and professions, from Homeland Security and the CIA to the Border Patrol and FEMA, and even nongovernmental aid groups and the UN, when it is operating in peacekeeping and relief operations abroad.

This describes the changing political environment. But technologic changes also affect the profession of war and how it sees itself, especially revolutionary technologies. For example, the longbow and then gunpowder helped end the age of chivalry and the accompanying monopoly of nobles over the warrior profession. It should not be all that shocking then that the latest technologies "will again revolutionize the way soldiers perceive themselves and are perceived by others," as military analyst Christopher Coker puts it.

When "UAVs are piloted by rank-and-file soldiers who have powers once reserved for generals," as one report described, it cannot help but create some changes in military professional identity and culture. One impending change is the blurring line between the officer corps and the enlisted ranks. This division, which encompasses everything from pay scales to which bars on base a soldier can drink at, originally came about as a way to distinguish the aristocratic roles in the military (leading troops) from the "commoner" parts of the job (the digging, cooking, fighting, and dying). A similar division occurred around the same time in the industrial world between "management" and the "labor" of the assembly line.

Today, the aristocracy has all but disappeared and the easy division of blue-collar versus white-collar jobs has blurred in most successful modern companies. Many think that technology trends may create the same breakdown of roles inside the military profession. "It might be necessary to consider whether the division of a service into enlisted personnel and commissioned officers makes sense in the 21st century," writes Steven Metz of the U.S. Army War College.

Take, for example, the role of the pilot. The early pilots who first fought in World War I were almost all aristocrats (the most famous was Manfred Albrecht Freiherr von Richthofen, a.k.a. "the Red Baron," the German ace who shot down eighty Allied planes, battled Snoopy, and still found time to make delicious microwavable frozen pizza). They saw themselves as "Knights of the Air," carrying on noble duels above the faceless thousands dying in the muddy trenches below. Since then, the job of pilot has been the domain of the officer corps. It was not just the honor and distinction that went with being a flyboy (and the cool leather jacket and scarf that made the ladies swoon), but also that the job required lengthy and arduous training.

As retired air force general Hap Carr explains, to fly an F-15 fighter jet today, a pilot first has to qualify on a training plane such as a T-38. This takes about six months. Then they would be assigned to a training unit to learn the specifics of the F-15 and all the tactics they need to fight and fly in it. This takes another year. But still their training is not yet over. "After weapons system qualification, the pilots will be assigned to an operational unit. They will spend about a year as a 'new guy' and fly most of their missions as a wingman to a more experienced pilot." Before they even reach the peak of their profession, the service would have invested several years and well over $10 million in their training. This is why the force has not only limited the role to officers, but also requires an additional commitment of service time from anyone who wants to be a pilot.

Compare that to the experience of an army UAV pilot like Private Joel Clark. His training at Huachuca involved four basic steps. First, trainees worked with a computer program (much like the old Microsoft *Flight Simulator* computer game popular in the early 1980s) to get them familiar with the basic concepts of flight. Then they moved on to actually flying radio-controlled planes, although these are tiny models, much like the ones that hobbyists buzz about city parks. Once they racked up forty hours of flight time in the tiny models, they graduated to one-third-sized mockups of the Hunter UAV. After forty hours of this, they were ready for the real thing, needing another forty hours of spotless flight time with the full-sized remote plane to get their certificate and become eligible to fly it on a combat mission. That is all that is required for the bigger planes; a smaller drone like the Raven UAV can take under an hour to learn.

As systems gain more autonomy, they will require less pilot interface and thus even less training for the men and women behind them. An air force colonel describes the force's planned training program for the next generation of drone pilots (who will mainly be enlisted, like Joel Clark in the army, rather than officers as most in the air force are now), "We'll get them a few rides in an airplane so they know what it feels like, but they won't actually know how to fly."

While the obvious question is whether someone who doesn't actually know how to fly can even be considered a "pilot" in the first place, it demonstrates an underlying point. With these new technologies, younger and younger troops are taking on roles that had been previously limited to older troops with higher ranks. Indeed, rank often falls by the wayside in the military robotics community, much as it has in other high-technology organizations, like Google, which encourage collaboration rather than a strict hierarchy. But what works in Silicon Valley may not be comforting to the military profession's old guard.

## MAD SKILLZ

The setting is a Rotary Club luncheon at the Holiday Inn in Decatur, Alabama. The speaker is retired colonel Edward Ward, who serves as logistics chief of the military's Robotic Systems Joint Project Office. The crowd is dutifully impressed; Ward looks the part of a tough marine officer, from his camouflage uniform to his bald, shiny dome. He gives a full briefing on how and why the military is using robots and even lets the Rotarians drive two of the robots (a PackBot and Throwbot) around the conference room.

Ward makes a special point to thank the parents in the group for contributing to the training of his troops. "Those PlayStation 2s really do the trick," he says. "I bought two hundred of them for training in Iraq. I have a feeling I'll be questioned about that one day." Later on, one Rotarian asks Ward if this meant that his grandson, "glued to PlayStation-style video games morning to night," was actually ready for a career in the military. Ward replies, "Only if he can make it through this little thing we call 'boot camp.'"

While video games are widely derided as a waste of time, they are actually incredibly influential, both in the economy and in the way they shape the soldiers of today and tomorrow. The game Halo 3, for example, made $170 million in just its first day of release, meaning more kids invested their time and money in playing it than what was spent on most summer movie blockbusters, as well as any book not beginning with the words "Harry Potter." And this is nowhere near the $2 billion spent over the lifetime of the John Madden NFL video game series. With kids today, "you are never far from the Madden crowd."

The result is that while the younger enlisted troops are taking over more roles through robotics, they aren't coming into the military completely unprepared. Rather, argues one U.S. military journal, "The Army will draw on a generation of mind-nimble (not necessarily literate), finger-quick youth and their years of experience as heroes and killers in violent, virtually real interactive videos."

The obvious benefit to the military is that its recruits come in already partially trained; because of all their time gaming online, young soldiers find it very easy to adapt to using unmanned systems. "The video game generation learns very quickly," tells retired admiral Joe Dyer of iRobot. The typical young PackBot operator just needs about a day and a half of training to get down the basics. Much like with their gaming, they then need only a few weeks after that to figure out all the moves and reach expert level.

The idea that a kid who grew up playing video games is going to be better able to learn a system in which the controls are modeled after video games makes sense. But it would be beside the point if all that experience sitting with glazed eyes in front of Super Mario and Halo actually made that young soldier dumber, as many parents worry a childhood spent video gaming is doing to their kids. Yet recent research shows this may not be the case. In his book Everything Bad Is Good for You, science writer Steven Johnson found that today's pop culture and video games are actually helping, not hindering, youngsters' mental and moral development.

It is not just that more goes on in the latest video games or that they have better visual graphics. They engage the brain far more. For example, many of the most popular video games that young soldiers cite as their favorites growing up, like Halo, may seem like a standard shoot-'em-up (formally known as a "first-person shooter" game). But they also have intricate plots and sophisticated challenges that play out over a series of stages that advance in complexity. To do well in these games, children's brains have to master balancing quick decisions and long-term strategy, much like in life.

Dealing with all this complexity is stimulating and exercising the neural networks in kids' brains at younger ages than in previous generations, which may actually be making these kids smarter. Over the last few decades, psychologists have found that the average IQ in technologically advanced countries has gone up by a striking amount. This, in turn, has military analysts finding that the average young soldier is better equipped than their predecessor generations at dealing with complex situations. Describes Colonel Paul Harig of the Army War College, "It will be no surprise to one who has watched school-age children 'surf the net' that information technology has jolted our culture, promoting access to ideas and immediacy to events, leading to mastery of resources. Without leaving his room, a 12-year-old can 'cyberchat' or correspond worldwide with e-mail pals, download a computer game, compile references from university libraries for homework papers, or view a music video."

This experience gives kids not just more smarts, but a certain mental flexibility that translates especially well in the complex fights of today's wars. As two retired marine officers describe what they saw in Iraq, "Battles are won by young enlisted men, not by generals poring over maps as in World War II."

One air force colonel, who now commands a Predator squadron, thinks that the video game culture may even make his younger unmanned pilots better than those officers he served with while flying F-15s. The reason is that same sort of mental flexibility. Indeed, the average teenager today spends six and a half hours a day using media technology, but packs eight and a half hours of use into it, as they often are using more than one technology at a time. Having spent their youth online gaming, sipping Red Bull, and talking on their cell phones all at once, young drone pilots come to the unit with an ease at multitasking already wired into their DNA. "The younger airmen and women are better suited for the new technology.... These pilots are incredible at multitasking. They will sit there and watch all four of their screens at once, monitoring everything from the map to the weapons to fuel, while also peeking over at the pilot beside them's screen, to see what he is looking at. That comes from all those games. An older pilot like me was taught to go through the checklist one by one. He will look at each screen at a time." At another air force base, I was struck by how the younger, "Generation Y" airmen had as many as thirty-six different computer screens open at a time, allowing them to juggle from mission to mission. And many believe that the next generation will be even more attuned to the new technologies. As one air force officer who works with Predator and Global Hawk put it, "If you talk to my seven-year-old son, I bet he has a better understanding of UAVs than me."

More junior soldiers may be making decisions that far more senior ones did in the past, and they may even come to the situation with better skills, IQ, and multitasking abilities. But it doesn't necessarily mean that they are better equipped for war mentally or emotionally. Describes one army major general, "The native creativity, innovativeness and initiative exhibited by these young men and women belie their woeful lack of psycho-social preparation."

Regardless of all his years spent video gaming, the younger someone is, the less life experience he will have, and the less time he will have spent in training and education specifically designed for the unique and often extreme dilemmas of war. An eighteen-year-old private, who still might be in puberty, may be quicker or more nimble in their thinking at the tactical level. But they will also now be playing at the strategic level once reserved for the forty-year-old colonels. "How do you manage to train people with three to four years of experience to make decisions that normally would be made by someone with ten to fifteen years of experience?" asked one marine officer.

Video games also reinforce what some have described as "an American tendency to think of war as a game, where someone wins and someone loses." The kids who play Madden may think they know football, but never will really understand the sport the way someone who plays on an actual team does. Similarly, argues Jeff Macgregor, that young video gamer, who first learned about war on the virtual field of battle, is more likely to see real war as something where you can "erase the pain given and taken, reduce the grunt and the struggle to the push of a button, eliminate the magnificent inconsistencies of the human heart and its capacity for courage or cowardice, and the game, the war, is no more than a fast twitch exercise—a battle fought without personal cost. It is cause without effect, a victory only for technology and opposable thumbs."

As the air force colonel who led a Predator squadron explains, the younger troops flying drones may be more talented, but there is a cost: "The video game generation is worse at distorting the reality of it [war] from the virtual nature. They don't have that sense of what is really going on." He believes that the virtual nature of the games makes the consequences seem unreal. "It teaches you how to compartmentalize it." Because of this, he said, "I don't like my [own] kids playing those [violent] video games. We do the car ones instead."

## THE OLD MAN'S WAR

Graham Hawkes is a renowned engineer who has invented many notable manned and unmanned vehicles used in science and industry. For example, he designed the Deep Rover submersibles, which were featured in James Cameron's IMAX film *Aliens of the Deep*.

Working independently of Foster-Miller, the British-born Hawkes also designed and built one of the very first robotic machine guns. The idea came to him after reading about a bloody police shootout in Philadelphia. Without any government funding, he set out to build a prototype. "When you have a radical idea, people's brains don't engage unless you actually make the thing. So I built it with my own money. I designed the system in 3D-CAD [a computer design program] and had some local machine shops fabricate the parts—without letting them know what I was doing." The prototype weighs twenty-seven pounds and as Graham tells, "it's perfect for urban warfare. Even in the heart of a battle, you can shoot from a safe place, like a sniper." The homemade system has not been as widely adopted as the SWORDS, but when Graham tested out the system, he learned how technology changes the age of who can fight in a whole new way. "Within three minutes, my 80-year-old father-in-law was as deadly as a 30-year-old army captain."

This hints at another change in the military profession's demographics, at the opposite end from the younger gamers. The military of today is older and more mature than in the past. The average age of a soldier in Vietnam was twenty-two; today in Iraq it is twenty-seven. Fewer than one in ten enlisted troops back then were married; today it is one in three. Indeed, the Pentagon is now the largest daycare provider in the world, with over 1.6 million children, half under the age of six, attending its schools and preschools.

The background to this is not just a shift away from the draft, but also a shift in health sciences. With all the advancements in health and physical training, life spans are becoming longer, as is the amount of time that a person can stay active in their field. In sports, for example, many athletes are now playing to ages unimagined even a decade ago, such as the baseball player Julio Franco, who has played to the age of fifty, almost twenty years past the point at which baseball players were once considered "over the hill."

The military is thus beginning to take on a new demographic look, much like in professional sports. Younger kids, often near high school age, are being put into greater spotlight roles, but mixed with older and older vets. The future military doesn't just include space for the younger gamers; technology may also help keep the old fogies around a little longer.

The prospect of leavening out the young video gamers with an older old guard appeals, especially to the senior set. Retired major general Robert Scales, the former head of the Army War College, writes that older soldiers might actually be better soldiers. "Social intelligence and diplomatic skills increase with age. Older soldiers are more stable in crisis situations, are less likely to be killed or wounded and are far more effective in performing the essential tasks that attend to close-in killing. Experience within special operations units also suggests that more mature soldiers are better suited for fighting in complex human environments."

No one yet has a sense of where exactly this trend will end. "Sixty is the new forty," wrote Ralph Peters in an article for the *Armed Forces Journal*, entitled "The Geezer Brigade." In science fiction, for example, author Joe Scalzi wrote of a future "Old Man's War," in which seventy-five-year-olds are still fighting in wars, enabled via modern technologies. Their unit is called "the Old Farts."

Our image of a soldier, as well as the military's self-image, is likely a man with a ramrod-straight stance, a bone-crushing handshake, and a crisp salute. This image is increasingly hard to reconcile with a military whose members might be using either Clearasil or Viagra.

This demographic change will play out in a variety of ways. For instance, the military has long discriminated based on physical attributes. If you had anything from poor eyesight to Rush Limbaugh's infamous "inoperable pilonidal cyst," the lump on his buttocks that helped him avoid the draft for Vietnam, you would be classified as 4-F, not eligible for military service. With changing technology needs, many of these limitations are no longer so relevant. Science might even turn our ideas of the required physical skills for war on their heads. As one analyst described, for certain military functions in the future, "having a strong bladder and big butt may be more useful physical attributes" than being able to do a hundred push-ups. Indeed, the two influential Chinese army predictors of future war, colonels Qiao and Wang, argued that "it is likely that a pasty-faced scholar wearing thick eyeglasses is better suited to be a modern soldier than is a strong young lowbrow with bulging biceps."

It is unlikely that all traditional soldiers will be replaced by these new demographics. Rather, as Bush administration official Elliott Abrams and retired army officer Andrew Bacevich write, "Over time, the proportion of soldiers who spend their tours of duty staring at computer screens will continue to increase while the proportion of those expected as a matter of course to venture into harm's way will dwindle."

This increasing division of the force into a different sort of structure will certainly open up new challenges. Describes one executive at Foster-Miller, "If you let the geeks wage war, you open up new vulnerabilities. If that computer craps out, and we know that it will, what next?" In turn, as the skills needed for part of the military change, that part may in turn have to change how it tries to be appealing as a career. Brigadier General Bruce Lawlor of the Army National Guard even thinks that "much of the military regimen" may need to be thrown out for certain units. "Innovators, intellectuals and highly skilled technicians [are not likely to] be impressed by the opportunity to wear hair 'high and tight' or do pushups and two-mile runs." Lawlor's concept is actually much in line with Vietnam veteran and science fiction author Joe Haldeman's vision of wars of the future. In his book *Forever Peace*, wars are fought by a new class of military reservists, who never have to go off to war, but instead serve on a part-time basis, operating unmanned systems remotely. It is no mere fancy; the air force is already moving to this model in part, with a number of Predator squadrons being designated to be flown by reserve units that stay in the United States.

The long-term consequences are huge. If the force is increasingly split between those sitting behind computers and those going out "in harm's way," the two parts may begin to have differing requirements and expectations. One part will take pride in its tough physical requirements and the aspects of personal bravery, modeling itself after the exploits and qualities of those who suffered through Valley Forge or stormed onto Normandy's beaches. The other part will see these requirements and parallels as foreign to their military experience, and even unnecessary in a new age of technology. How then does the military profession as a whole keep a unified ethos and identity?

## SOLDIERS OF FORTRAN

Over the last few decades, we have gradually seen the military's monopoly on war give way to the private market. From companies like Blackwater doing armed convoy escort jobs in Iraq (and shooting just a few civilians along the way) and CACI interrogators working at Abu Ghraib, to the outsourcing of the U.S. military supply chain to firms like Halliburton (for which it made over $20 billion in revenue, three times what the U.S. government paid for the entire 1991 Gulf War), private companies are operating in traditional military roles as never before. Indeed, in Iraq, there were more of these "corporate warriors" deployed than actual U.S. military troops.

The shift toward unmanned systems appears to be taking this trend further and in new directions. For many, there appears to be nothing inherently military about the ability to punch a keyboard and move a joystick around. When an office drone sits in a cubicle driving robotic drones around, there is a striking convergence between military tasks and civilian life. As one report notes, "While these actions are principally motivated by a desire to save scarce defense dollars, there is also a tacit recognition that the growing sophistication of the technologies of war require the military to ever more frequently tap civilian expertise."

Already, private contractors do much of the training for the military on how to pilot robotic systems. "We take these Army guys who don't know the front end of an airplane from the back, and we teach them from scratch all the aviation they need to make them pilots," explains Bill Hempel, who trains future drone pilots like Joel Clark at Fort Huachuca. The help of a contractor, though, also extends into operations. For example, when the army's 15th Military Intelligence Battalion deployed to the Balkans and then to Iraq early in the war, it was accompanied by a support team from Northrop Grumman. Having longer time with the machines, as well as not having the frequent turnover a military unit might go through as it rotates into an operation, the private contractors ironically became the military unit's "institutional memory." One contractor described how "certain soldiers were not as comfortable flying," and so the contractor would take over "on a regular basis," including "more than once a week in Iraq." By 2007, the army's armed Hunter drones flown out of a base in Tikrit, Iraq, were described as "government-owned, contractor-operated." The civilians at that location had taken over the flying as a formal policy.

Another growing contractor role is ground crew for drones like the Predator and Global Hawk, the people who fuel the unmanned planes, load weapons, and fix whatever is broken. It seems like an uncontroversial task for civilians to take on, but it leads to an odd division of labor and risk. Because the planes actually fly from bases located inside the war zone, the civilian support staff is in more physical danger than the military pilots who fly them from back in Nevada, who are only in "virtual" danger.

The growing use of what *Defense News* described as "surrogate warriors" to run the military's digital systems is certainly lucrative; Bloomberg News projects the military's spending on such drone "leasing" and operation to reach $10 billion by 2018. But it generates the same sorts of concerns that surround the broader outsourcing of war to contractors like Blackwater, CACI, or Halliburton in Iraq. The most obvious distinction between a soldier and a contractor is that one serves the nation, while the other works for profit. Describes one army official of working with contractors in the unmanned world, "They have their heart at the right place, but they also know where the checkbook is."

Another difficulty with military outsourcing has been the question of legal status and accountability. Such "corporate" or "surrogate warriors" are civilians, as they are not formally part of the military and its command structure. But they are not really noncombatants, as they are carrying out fundamentally military missions. The result is that the law has found it a confusing space to ensure both that contractors' rights are protected as well as that they are properly punished if they commit crimes in war. In Iraq, some contractors have been held in prison without any formal charges, while others have been reported to have committed crimes that ranged from prisoner abuse to joyride shooting at civilians, for which they went unpunished.

This same question of status is going to be troubling for contractors working with unmanned military systems. Most contractors in these roles believe themselves to be civilians under the law, which gives them certain rights and protections, such as immunity from enemy attack. However, as one military lawyer put it, this is based on a confusion of "the nature of their status in conflict or the consequences that flow from it." As another report concurred, "Even though international law recognizes that civilian technicians are necessary in modern armies, it has always maintained that noncombatants' immunity from damage and harm was predicated upon their obligation to abstain from hostile acts. If they took action against a party's armed forces, they automatically lost immunity."

It could be even worse for these contractors than now just being a legal target for the enemy to shoot at. One navy lawyer found that a civilian working with UAVs in wartime "could arguably be considered an 'illegal combatant' under the law of armed conflict." That is, they would have the same status and rights (or lack of rights) as the al-Qaeda detainees held at Guantánamo Bay. They wouldn't have the protections of a POW under the Geneva Conventions. If they were captured, they could even be prosecuted as a digital mercenary and shot. "Furthermore," added the lawyer, "the pervasive use of illegal combatants may have serious unintended consequences—such as our adversary conducting reprisals against civilian personnel, suspecting that others might be combatants."

All of these dilemmas lead some observers to ask whether civilians should play at war in the first place. As the Italian thinker Machiavelli put it, when a person becomes a soldier, "he changes not only his clothing but adopts attitudes, manners, ways of speaking and becoming himself quite at odds with civilian life." This change is not just about getting a new hairstyle, but values like duty, honor, and sacrifice take on new meanings when in military service. Indeed, the essence of becoming a soldier is surrendering one's most basic rights as a citizen in order to serve all the citizenry, to "protect and defend the Constitution of the United States against all enemies, foreign and domestic." The contractor just does a job.

As Richard Holmes points out in *Acts of War: The Behavior of Men in Battle*, "However much sociologists might argue that we live in an age of 'narrowing skill differentials,' where many of the soldier's tasks are growing ever closer to those of civilian contemporaries, it is an inescapable fact that the soldier's primary function, the use—or threatened use—of force, sets him apart from civilians.... [T]he fact remains that someone who joins an army, is both crossing a well-defined border within the fabric of society, and becoming a part of an organization which, in the last analysis, may require him to kill or be killed." When the fighting of wars, even those unmanned, becomes just another job, all this is lost.

## "GENTLEMEN, WE CAN REBUILD HIM. WE HAVE THE TECHNOLOGY"

In 2003, a land mine in Iraq blew off army major David Rozelle's right foot. Within six months, he was skiing in the Rockies. Within a year, he was back in Iraq commanding an armored cavalry unit. The next year, he competed in the aptly named "Ironman" Triathlon in Hawaii, completing the grueling race of 2.4 miles of swimming, 112 miles of biking, and 26.2 miles of running. He was able to do all this because the army had fitted him with an artificial foot to replace the one he lost. Rozelle is stoically honest about how it feels. "At least once a day, I miss my foot, but it hasn't slowed me much."

By 2008, more than thirty thousand American troops had been wounded in action in Iraq. Many of these injuries were the result of IEDs and other explosives that cause horrific shrapnel and burn damage to the human body. And yet, because of body armor, advances in medicine, and quicker evacuations, soldiers are surviving at a far greater rate than in any previous war, including at almost twice the rate as in Vietnam. But the rate of wounded who have lost a limb doubled as well. Over a thousand soldiers have lost a limb but survived, with nearly a quarter of these amputees having lost more than one limb.

The result has been a crash program at DARPA to bring the latest robotic technology to bear. Instead of the wooden peg-legs or steel hooks of yesteryear, today's wounded warriors are increasingly equipped with electrically powered prostheses. These robotic limbs are programmed to do such things as automatically match the intended stride of the human or automatically bend whenever the body weight shifts. Akin to the advances in robotic interfaces described in chapter 3, these devices are also being wired directly into the patient's nerves. This allows the soldier to control their artificial limbs via thought as well as have signals wired back into their peripheral nervous system. Their limbs may be robotic, but they can "feel" a temperature change or vibration. As the navy's magazine *Proceedings* writes, the new technologies are "as close to real limbs as those portrayed on such then fanciful 1970s science fiction shows as *The Six Million Dollar Man* or *The Bionic Woman*."

With robotic artificial limbs taking on such a science fiction quality, they are allowing soldiers not only to recover and lead productive lives, but even to go back to war, putting another twist on the new demographics of war. Some 40 percent of the soldiers who have received robotic limbs return to their old units. A DARPA scientist explains, "It's sort of the Luke Skywalker phenomenon. Luke loses his hand in the movie and then he goes and has an artificial one put on. And that of course is a very noble, admirable goal if we are asking individuals to put their lives on the line to defend our way of life."

## ENHANCEMENTS

Putting technology inside our bodies is not just for replacing limbs lost in battle. It is a growing trend in the civilian sector. Over a hundred thousand people born deaf each year are now able to hear through cochlear implants. These are devices that turn sound into electrical signals, which are then fed into the body through neural implants in the inner ear. More broadly, surgeries to implant computerized pacemakers, heart monitors, and even replace entire joints with metal devices have become almost a rite of passage. The hip replacement, for example, is what comes after you buy the midlife crisis convertible, but before you move into an "active living" retirement community. Describes iRobot's Rodney Brooks, "As us baby boomers get older and older, we're going to be looking for all sorts of replacement parts. We're going to become partially robotic. What's a robot, what's us, is starting to get a bit messy."

History shows that technologies and procedures originally designed for replacing lost functionality are frequently applied beyond their original purpose. For example, the sixteenth-century French physician Ambroise Paré invented artificial teeth made of silver and gold for patients who lost their teeth due to the poor nutrition and hygiene of the era. It has since evolved into a status symbol for rappers. The first cosmetic surgery was in 1791 to fix a cleft lip. Today, more than eleven million cosmetic surgeries happen each year in the United States alone, ranging from breast augmentation to butt implants.

While the full force is still a few decades out, a similar trend is already starting to appear with voluntary technologic implants. These don't merely replace something lost, but add something more. The Florida-based VeriChip company, for instance, has sold human-implantable radio-frequency identification (RFID) chips to over five thousand security, government, and industrial installations. Even the Baja Beach Club, one of Barcelona's hottest nightclubs, is a buyer. In 2006, the club implanted its VIP customers, including the entire cast of *Grand Hermano* (the Spanish version of the reality show *Big Brother*), with the tiny microchips, so that they would not have to wait in line or need to carry cash or credit cards. The implants have also been used by the FN Herstal firearms company to make a gun that can only be fired by someone who has a matching computer chip buried in the flesh of their palm.

Such new technologies being plugged into our bodies push beyond our human bodies' previous limits. The research on these "enhancements" or "augments" (termed after the race of empowered leaders in the *Star Trek* world) is among military labs' most lavishly funded "deep" (i.e., long-term) projects. Describes one scientist about the new capabilities of robots, and how they in turn led him to add robotic enhancements to his body, "It makes me jealous. Why can't I sense the world in infrared? Why can't I see various things in ten dimensions? Even five dimensions I would be happy with."

Technologic implants might be used to enhance our human capabilities in war in many ways. While the comic book crowd may focus on what it would be like to have soldiers with metallic bones and claws like *Wolverine*, a focus at DARPA has been on what it calls "AugCog," augmented cognition. This program aims to implant the memory chips that robots use inside the human body. The combination of augmented memory and the interface connections allowed by a jack could be powerful. It might just be the trick for our brains to catch up to the data overload problem. As Philip Kennedy, a leading brain interface expert tells, "Essentially we'll have a PDA and a cell phone in the brain. And I know this sounds like *The Matrix*, but it is what it is."

*The Matrix* is an apt example for the possibilities of such a system in war. In one scene, the character Trinity downloads the knowledge needed to fly a helicopter in the midst of a battle. AugCog hopes to make this real. Even the fastest human reader in the world can only process about 50 bits of information per second, and their brain has terrible storage. The new robotic interfaces of jack connections (discussed in chapter 3) speed up the transfer rate to the equivalent of high bandwidth. Add a memory chip into the human body, as one scientist describes, and it will be like plugging in "the equivalent of a flash drive with massive memory that interacts with your brain." The combination allows what is called a giant up-load process (GULP), where someone could "take a book and gulp it down."

## COMBAT EVOLVED

Most such technologies are years and even decades off, but their impact could prove revolutionary at fundamental levels. Indeed in war, the frontiers opened up by such enhancements may even change the rules of the game. As Carl von Clausewitz, the Prussian military thinker, emphasized, "All war presupposes human weakness and seeks to exploit it." By contrast, the integration of robotic technology with living flesh could help turn the human infantryman, who had been "the weakling of the battlefield," into what one former army general called a "supertroop."

War becomes a mite different when the Master Chief from *Halo* is made real. For example, if individual soldiers are now technologically "enhanced," a literal "Army of One" as the U.S. Army recruiting commercials used to claim, it is hard to see them being used and deployed as they were in the past. Instead of being bundled together in large units, the regular infantry would likely operate in very small units or even alone. Fewer soldiers would seem to be needed for the same tasks and a nation with technologically super-empowered soldiers might find it easier to strike quickly or covertly. Ultimately, we could see the very end of the GI. As one book described of a possible future of war, "The G.I., the stamped government issue interchangeable warrior, becomes obsolete when masses of men are no longer required to fight wars."

And it is perhaps because of this potential that many soldiers find upsetting the prospect of such technologic changes to the inside of the human body. Many even express an unease far deeper than over the rise of unmanned machines in war. Describes one special forces officer of these plans, "I draw the line at screwing with your fucking body."

From the perspectives of the researchers, such reluctance makes little sense. War is a dangerous place and striving to technologically solve human weaknesses within it seems quite reasonable. The better that soldier's performance, the more likely he returns home to his family. Enhancements also might be the only way that humans can keep up with their new robotic technologies, rather than being left in their silicon-coated dust.

Still, that soldier's argument resonates strongly for a reason. The newness of such implants leads many to question their long-term consequences. Explains the officer, "Being a guinea pig doesn't settle well with me." And perhaps he should be concerned. The same science fiction stories that inspired DARPA to develop so many of these enhancements also usually reference a downside. For example, many credit William Gibson's 1981 short story "Johnny Mnemonic" as helping them to visualize the military applications of enhancements like jacks, memory chips, and robotic effector implants. Yet they always seem to ignore that Gibson set his story of a world of technologically enhanced humans against the backdrop of a raging epidemic, caused by that very same technology (he called it "Neural Attenuation Syndrome").

Gibson was writing science fiction, so while his imagined technology did come true, there is no reason to think that his imaginary disease has to as well. But, then again, the Pentagon's real-world record with things like the aboveground testing of atomic bombs, Agent Orange, and Gulf War syndrome certainly doesn't inspire the greatest confidence among the first generation of soldiers involved.

These issues are usually set aside as "manageable" by those who support the implant programs; valid concerns, yes, but no reason to stop the work. However, they ignore another, deeper concern among the soldiers. The use of enhancements just doesn't seem to settle well with the self-concept of those who work so hard to hone their skills and bodies for war. If technology becomes the substitute, they argue, it devalues their efforts and identity.

Many make a parallel to steroids. Soldiers note that performance enhancers

may help athletes, but they are still widely banned in sports. Record-holding athletes like Olympic sprinters Ben Johnson and Marion Jones and famed baseball
players like Barry Bonds and Roger Clemens have been treated with more contempt than admiration by the public. The reason is not merely worries about the
side effects, that fans really fret about the heart disease, shrunken testicles, and
gynormous heads that steroid users frequently develop. Nor is it so much that
the artificial enhancers are viewed as breaking the rules to get ahead. Instead, the
problem that fans (and soldiers) have is that it is hard to figure out whether it was
the person or the technology that deserves the credit. Few humans want to know
that they mattered less, be it the fans' image of their hero or the soldier looking
in the mirror.

In 2007, the International Association of Athletics Federations (the global
governing body for track and field events) preemptively banned athletes with
machines in their bodies from competition. The spark was an Italian double-
amputee sprinter who wanted to shift from the special events for disabled athletes
to the real Olympic races. His prosthetic "cheetah legs" had more spring and less
air resistance than human legs. So the officials viewed him not as a "disabled" athlete, but rather as unfairly "enabled" by technology. Described the IAAF spokesperson, "We need to separate emotion from the science. We all wish him well. The
point here is what's going to happen in 10 years? What happens if it continues to
evolve?" This same question can be asked of robotic enhancements in war.

## RESISTANCE IS FUTILE

The questions surrounding this looming wave of technologic tinkering with the
human body are huge, and all the more so because their origin comes not from
broad research at human betterment in general, but rather from military-funded
attempts to create a better fighting machine. There will also emerge in security
circles all sorts of "what if?" questions that used to be fodder for debate at science
fiction conventions. For example, given how every other form of technology has
been hacked and distorted, what are the consequences when enhancements inside
the body are hacked? Can a soldier's technologic body part even "crash," and if so,
how do they recover? Unlike a robot's operating system, the attached human body
parts can't just simply reboot by flicking the "on-off" button.

Much like the other changing demographics of war, the integration of technology also affects soldiers' sense of identity. Amputees using the early generations

of these systems report that such implants come to "feel" as if the device were a part of them, an extension that almost replaces what was lost. But enhancements don't just replace a sense that has been lost, they add something new. R. J. Pinero, a science writer who also works on microprocessors, suspects that for a soldier of the future, artificial enhancements might feel a little bit like muscles do to steroid users. "You never quite get used to the implants.... The knowledge feels...foreign? Yeah, that's it. Fake. It's almost like sneaking out a cheat sheet during an exam, getting the answer you seek and then stowing it away before anyone catches you. You get the right answer, but you didn't really know it."

And what happens to that soldier when the enhancements shut down, or are taken back by the military after their service has ended? It gives a whole new meaning to the idea of "deprogramming." If a person has been able to leap buildings, see in multiple spectrums, or download the Internet into their head, living without these robotic enhancements may not seem like going back to "normal." Leading an unenhanced life may feel like losing a sense, as being something less than before, much like Clark Kent must have felt after he stopped being Superman and got beat up in a bar.

As we change technologically, we also change how we categorize ourselves. Indeed, when the writer William Gibson speculated on a world where humans used technology enhancements, such as in "Johnny Mnemonic" that so influenced the current research, he also predicted it would lead some people to start to identify themselves, as they now do by race and ethnicity, by something else. He called the new divisions "technicity," grouping people by technologic implants. Soldiers proudly distinguish themselves by their particular specialties, which result from lengthy training. When what had been human skills and capabilities become just plug-in technologies, as these real-world R&D projects hope will happen, a soldier might be distinguished only by what implant they had at the time. This seems a small point, but fundamentally screws with military culture and traditions.

This leads to the most revolutionary and perhaps perilous aspect of where dabbling with the identity of war might take us. It affects not just how we look at ourselves, but how we look at each other. As opposed to unmanned systems, which change the capabilities of machines, technological enhancements are creating a new type of human species, the first time in twenty-five thousand years that we have more than one type among us.

We already live in a world where people walk around with cell phones and iPods almost permanently plugged into their ears, and openly describe their addiction to their "Crack Berries." The human body, though, is not a simple machine into which you can just plug in new hardware. Our neural systems do not have USB ports like robots and computers do. Instead, researchers are finding that when technologies enter into us, our capabilities are not the only thing that changes. Our bodies do. Like software rewriting itself, the human body actually alters the bio-logic pathways of its neurons to accommodate and incorporate the new machines placed inside it.

As such, humans truly do become something different, creating an even more revolutionary demographic change that looms in the decades ahead. That something different is "cybernetic organisms," or "cyborgs," creatures that have changed and enhanced their bodies' capabilities via technology. It is very exciting stuff until you realize we unenhanced humans are the equivalent of the Neander-thals, watching that first group of *Homo sapiens* walk across the hill, strutting their better bodies and technology.

If the trends hold, scientists believe that the next few decades will see a growing division of "natural" humans and those who are "enhanced" or "augmented." That those in the military will likely be among the first to get such enhanced powers adds new layers of complexity to civil-military relations. But the effects will ripple out further. As the technologies cross between the military and civilian worlds, some humans may decline the option of changing their bodies with technology. They will remain "natural" by choice (akin to the twenty-first-century version of the Amish). But many others will likely be left behind because of a lack of choice. In all likelihood, those that get the technologic capabilities first will be the rich and empowered. This is nothing new. Most of us don't seem all that concerned that the rich and powerful are able to afford cosmetic enhancements, like nose or boob jobs, that many poor cannot. But what about technologic enhancements that reshape our insides, like greater strength or intelligence?

Such changes in the very nature of ourselves seem somehow more worrisome. The reason again goes to a deeper level. For hundreds of years, people have used cosmetic variations in skin pigment color to decide whether someone was "supe-rior" or not. Countless conflicts resulted from tiny or even imagined differences. Maybe humanity will grow past this self-destructive weakness. Maybe no one will care about the differences created by such technologic enhancements, treating these new technologic differences as akin to whether someone has tattoos or not. But in order to reach that point we may also have to start to tinker with our very human psychology.

Kevin Warwick is the British university researcher who first connected his body directly into a computer via a technologic implant. His goal was simply to enhance the level of interface with his robots. But he described himself as experiencing not only a physical upgrade, but also a psychological change. "One of the reactions I had to having the implant was a feeling of affinity with my computer. Once that becomes a permanent state, you're not really a human anymore, you're a cyborg. Your values and ethics would be bound to change, I think, and you would view un-augmented humans a little differently."

Warwick uses humans' relations with cows to illustrate his point. He describes how humans share much in common with their fellow mammals, draw sustenance from them, even typically like them, as cows seem such nice, gentle creatures. But because of their lower intelligence and capabilities, humans don't consider themselves bound by the same laws or expectations when dealing with cows. Most people love their hamburgers and have no problem drinking milk from a cow that is essentially enslaved. While ardent animal rights activists may argue that it is wrong to eat meat and some even eschew milk, none argue that cows should have a vote in the presidential election. Warwick thinks that naturals should expect a similar regard from the technologically enhanced. Sounding much like the Borg he was a fan of, he says, "If you are not upgraded . . . you are going to be something of a subspecies."

# TWENTY

# DIGITIZING THE LAWS OF WAR
# AND OTHER ISSUES OF (UN)HUMAN RIGHTS

*We risk continuing to fight a twenty-first-century conflict with twentieth-century rules.*

—JOHN REID, British secretary of state for defence

War is a special kind of hell. It is a space where killing is not only allowed but considered one's duty. That is why many believe that war is a place where no rules or laws apply. Or as the Roman philosopher Cicero put it (in a quote that has since been referenced in everything from Supreme Court rulings to *Star Trek* episodes), "In times of war, the laws fall mute."

And yet, all is not fair in love and war. For all its horrors, a dense set of rules defines what is right or wrong in battle. These rules find their origin in everything from the Bible to the Geneva Conventions. Such rules are certainly not always followed, but their very existence is what separates killing in war from murder and what distinguishes soldiers from criminals. Writes Michael Walzer, "War is still a rule-governed activity, a world of permissions and prohibitions—a moral world, therefore, in the midst of hell."

Hugo Grotius, a seventeenth-century Dutch jurist, was the first to systematically organize these various rules into an international code of law. Grotius lived during two of the most savage and enduring wars that Europe has ever seen, the Eighty Years' War between his homeland and Spain and the Thirty Years' War between Catholics and Protestants across all of Europe. Revolted by what he witnessed, Grotius wrote a book entitled *De Jure Belli ac Pacis* (On the Laws of War and Peace) in 1625, which laid out what is now known as "just war" theory.

The conduct of war, argued Grotius, always had to comply with the natural laws of justice. They applied both to the act of going to war, *jus ad bellum*, and how the war was fought, *jus in bello*. If either set of laws was broken, then the war was not legitimate.

The two biggest rules to follow in *jus in bello* are "proportionality" and "discrimination." Proportionality means that the suffering and devastation caused by the war can't outweigh whatever harm started it. If the other side started the war by stealing your cow, you can't nuke their capital city to get her back. Discrimination (in this case, the good kind) means that the sides must be able to distinguish combatants from civilians and respect the latter's immunity from harm. It even states that the sides must respect the difference between an opposing combatant that poses a mortal threat and one that doesn't. For example, it is not okay to shoot troops who are wounded or running away and, whenever possible, opposing troops must be given the opportunity to surrender.

Grotius's book has since become the foundation of what we now think of as international law. Today, the rules that govern combatants and their conduct are known in military parlance as the Laws of Armed Conflict (LOAC). Rather than an actual document that you could easily program a robot with, the LOAC is the collection of all the various written treaties, like the Geneva Conventions, domestic laws, and even the time-honored customs of war. Air force major David DiCenso explains, "Whether war is waged on the muddy fields of Verdun by shell-shocked infantry troops or a high-tech cyberspace battlefield, the rules and general principles of the LOAC remain applicable."

It is this recognition of each other's humanity through the law that redeems war. If you follow these laws, war becomes not merely blowing up things, but, as Michael Walzer argued, "a rule-governed activity of equals, or victims, who despite their individual national or tribal allegiances, have the same human standing."

## UNMANNED LEGAL CONFUSION

The International Committee of the Red Cross (ICRC) is a unique institution in world politics. The group was founded by a Swiss businessman, Henry Dunant, who, like Grotius centuries earlier, was horrified by the suffering of soldiers he witnessed at the battle of Solferino in 1859. The organization he created is private and funded only by donations. At the same time, however, the group has sovereign status, as if it were a state government, and is the only body mentioned in international law documents like the Geneva Conventions as a controlling authority. That is, the ICRC is the only entity named by international law to provide an impartial and neutral voice on behalf of the laws of war themselves. It is essentially the repository and protector for international humanitarian law, *jus in bello*, responsible for making sure that the laws remain respected and relevant.

Among the two thousand employees of the ICRC is Peter Herby. Working out of ICRC's Geneva office, Herby is the head of the organization's Mines and Arms Unit, and has represented the ICRC at all arms-related negotiations since 1994. While Herby would love for the world to be at peace and end the use of all weapons, that is not his mission. Rather, he explains, his task is to help ICRC and all the various groups at war to "reconcile the necessities of war with the laws of humanity."

The legality of weapons and their impact on soldiers and civilians are key areas of the ICRC's work. Herby has been most focused on the antipersonnel land mine over the last decade, a weapon that leaves a horrible legacy for civilians long after wars end. The ICRC realized it couldn't stop the harm of these weapons merely by treating the victims, and so it helped in the successful international effort in the 1990s to ban this weapon. But sometimes the ICRC is also concerned about the impact that new technologies might have on the conduct of war and the respect for international law. It has, for example, helped lead efforts to stop the research and use of lasers that blind soldiers.

According to Herby, "There are four pillars of international humanitarian law on weapons." First, nations have a right to choose the methods and means of war, but this right is limited. They have to follow the rules. Second, weapons that cannot discriminate between civilian and military targets are prohibited. Third, weapons that cause unnecessary suffering are also prohibited. And, fourth, any weapons that the international community decides are abhorrent for some other reason are prohibited. This is a useful little clause, which can be applied to any weapon that might meet the other rules, but is just nasty or horrifying, such as chemical weapons and blinding lasers.

The ICRC tells nations that they should carry out legal reviews during the R&D process to ensure that any new weapons meet these four pillars well before they are ever deployed. In fact, Herby reminds, states are legally obliged to do so under the Geneva Conventions. The group is also concerned about the lack of controls over what happens with these new weapons. International humanitarian law mandates that military weapons be used only by military actors (something ever more difficult with dual-use and "off the shelf" technologies), as well as a responsibility that the laws be respected not just by the makers, but also by any recipients of a sale or transfer.

The ICRC thus has a strong position on weapons in general and how to ensure their legality. However, when it comes to the ICRC's position on where international law stands on robots and their uses in war, the discussion doesn't go very far. "We have no particular viewpoint or analysis to provide."

The organization is certainly aware that technologies like the Predator or SWORDS exist now and realizes that there will be problems with these new unmanned systems. Herby tells how "with every major scientific revolution, the fruits will be put to hostile use, unless there is action." But robotics are viewed as just a little bit too futuristic for an organization that has as much on its plate as the ICRC today. It is stretched out over eighty countries, burdened with everything from ensuring detainee rights at Guantánamo Bay to pressuring nations to fulfill their pledges to end the use of land mines. Asks one of Herby's colleagues, "There is so much terrible going on these days, why waste time on something crazy like that?"

The result is that, as important as the ICRC has been in shaping and guarding international law over the last century, it is not yet driving discussion on the most important weapons developments in this century. It is here that the laws truly "fall silent." In the hundreds of interviews for this book, not one robotics researcher, developer, program manager, or soldier using them in the field made a single reference to the ICRC, nor its all-important "four pillars" of international humanitarian law on weapons. That is, not a single organization, research lab, or company working on robotics today is formally linked up with ICRC or has in place the sort of reviews that Herby describes as necessary for new weapons. Indeed, the closest one gets to any such legal reviews are some limited efforts by robotics firms to make sure that they don't get sued by customers. iRobot, for instance, made sure that its new Warrior robot has a built in fail-safe control that automatically prevents the robot from running over people (researchers jokingly call it the "idiot protection device"). Instead, robotics makers and government agency clients alike tend to avoid the whole discussion of the ethical and legal reviews of these new weapons, mainly because it is too futuristic, too thorny, or would get in the way of business. As one engineer at iRobot described, "Our responsibility is that it works as requested, as described."

This is not to beat up on ICRC for somehow not doing its job. As its four Nobel Peace Prizes and the literally tens of millions of soldiers and civilians who owe the group their lives can attest, it has done its job far better than its founder Henry Dunant could ever have imagined. Instead, as the embodiment of international law, the ICRC position on robotics, or rather absence of a position, is simply representative of the brewing breakdown between the laws of war and the reality of conflict in the twenty-first century.

This disconnect reveals itself again and again. At a major conference in Washington, D.C., in 2007, for example, over one hundred international law experts, including several law school deans and heads of human rights groups, gathered to discuss the key questions they saw in "New Battlefields, Old Laws." Like at the conference on RMAs, not once did any of them even mention robotics or any other new technologies.

At the same time, those actually working upon and with these new systems describe all sorts of confusion about and gaps in the laws that are supposed to guide them. As the former commander of a Predator squadron put it, "There is nothing set for this all rules-wise. Somebody needs to take time to look at these issues and figure them out." As an illustration of his uncertainty, he told how his men who flew the systems saw themselves as legal combatants, "but you are operating on home soil. It opens all sorts of questions. Are they valid targets walking around the streets of Las Vegas? Under the current rules they are."

The same officer described debates over what sort of data is needed to designate a target as suitable to be shot at by a drone, and, in turn, who has the expertise and authority to decide. Equally, "What if they [the human operator of a UAV] hit the pickle button [that launches a weapon] and turn out to be wrong? Who is held accountable?" While the party line is that the process for determining legal accountability would be the same as if a manned pilot made such a mistake, the new technology complicates matters. At times it was unclear which chain of command the pilots fell under as they were conducting combat missions in Iraq and Afghanistan but sitting in Nevada. Both the local commander in the United States and those using them out in the region wanted to control the assets. But he guessed it would be the reverse if something went wrong. "The most disturbing part," he concluded, "is that the needed laws and values typically can't keep up with such rapid change."

This growing gap means that both weapons makers and users are prone to look at many of the current laws of war much like they look at the old laws that ban crossbows—as simply not useful to their daily jobs. It grows even more difficult as systems gain greater autonomy. The current "legal limbo," as one roboticist put it, becomes a legal vacuum. Explains Gordon Johnson of the U.S. Military Joint Forces Research Center, "The lawyers tell me there are no prohibitions against robots making life-or-death decisions."

New technology has often moved faster than the laws of war. During World War I, for example, all sorts of new technologies, from airplanes dropping bombs to cannons shooting chemical weapons, were introduced before anyone agreed on the rules for their use. As to be expected, the warring sides sometimes took different interpretations. For example, the British and Americans felt that the new submarines should avoid targeting any civilian vessels and that they ought to surface and reveal themselves before an attack, to give their target a chance to surrender. The Germans, who relied on subs more in their war plans, took a completely different view, arguing that such a legal interpretation only put their side at a disadvantage. This lack of legal clarity even helped induce America to join the war; the submarine attacks on merchant ships that the Germans saw as justifiable were instead viewed as war crimes on the other side of the Atlantic.

But while technologic change is speeding up exponentially, legal change remains glacial. Chemical weapons were first introduced in World War I, but they weren't fully banned until eighty-two years later. Even worse, if we look back at history, the biggest developments in law only came after some catastrophe. If one-third of central Europe's population hadn't been killed in the Thirty Years' War, Hugo Grotius probably wouldn't have written *On the Laws of War and Peace*. Or, if eleven million Jews, Roma, POWs, and political prisoners weren't killed in the Holocaust, there would be no 1949 Geneva Conventions. Regarding unmanned systems and the law, Army War College professor Steven Metz says, "There is no consensus yet on anything new and, unfortunately, I don't think we are due for a breakthrough until something terribly bad happens."

## THE HUMAN RIGHTS ELEMENT

"This new technology creates new pressure points for international law.... You will be trying to apply international law written for the Second World War to *Star Trek* technology."

Marc Garlasco is senior military analyst at Human Rights Watch, one of the world's leading human rights advocacy and research groups. If ICRC acts as the impartial voice on behalf of international laws, Human Rights Watch acts as the world's conscience on behalf of human rights. It has led efforts on everything from ending the recruitment of child soldiers to advocating for AIDS victims' rights.

Garlasco is their resident expert on anything military-related. Tall and trim, he does not fit one's expectation of a granola-scarfing, Birkenstock-wearing human rights softie. Before joining Human Rights Watch, he served in the Pentagon as "chief of high-value targeting" during the Iraq war in 2003. His bio also notes that he led the "Battle Damage Assessment" teams during the 1998 airstrikes on Iraq and 1999 operations in Kosovo and participated "in over 50 interrogations as a subject matter expert."

Also surprisingly for a human rights worker, Garlasco is a closet sci-fi geek. He tells how he even went to see Isaac Asimov give a lecture when he was a young boy. While it would make for a more inspirational story to say that hearing Asimov speak is what got him interested in human rights, Garlasco says he mainly remembers Asimov telling some questionable limericks. "Isaac Asimov was a bit of a dirty old man."

"Technology can always be abused," says Garlasco. But he also adds that sometimes new technologies can help save lives and increase respect for human rights. He points to the development of precision-guided munitions (smart bombs) that allow far greater discrimination in targeting and save civilian lives as a result. Human Rights Watch's interest then "is not so much in the technology itself, but rather our interest is in the guidelines of its use.... What we want to know is now that I've got a nineteen-year-old kid with this weapon, does it increase the potential for abuse?"

Like ICRC with international law, Human Rights Watch faces a problem of how to wrestle with all the various human rights problems of today. And like ICRC, the organization has taken no formal position nor issued any reports on the new technologies. "We've not yet had that internal debate."

The reason is not for lack of awareness or even interest, as Garlasco is both fascinated and concerned by the issues of unmanned warfare. Just back from a field research mission to Lebanon, he recalls how Israeli drones were constantly overhead ("Being under a UAV scares the shit out of me"). "But it's just not a big piece of our pie. Say you have one researcher. Do you send them to conflict X or to check out this new technology that is not even developed yet?" This also has to do with how such organizations are funded. It is a whole lot easier to gain needed monies for staff and operations working on crises in the headlines than for investigations of weapons not yet on donors' radar screens.

While his organization does not yet have a position on unmanned systems, Garlasco sees a fundamental change coming soon for human rights groups like his, as they try to shape the public debate on the use of weapons in war. He expects that "in ten to fifteen years," groups like Human Rights Watch will have to figure out how they are going to react to the introduction and use of "completely autonomous weapons systems." While he thinks the technology will easily make full autonomy possible, he feels it will prove incredibly challenging for both international law and the respect for human rights in war. "As *Human* [emphasizing the word] Rights Watch, we want that human element.... The human has morality, has an empathetic response. The human has the capability to make complex decisions; they can draw on their humanity."

By contrast, Garlasco explains, "You can't just download international law into a computer. The situations are complicated; it goes beyond black-and-white decisions." He explains how figuring out legitimate military targets is getting more difficult in war, especially as conflict actors increasingly fight in the midst of civilian areas like cities and even use civilians as cover. Citing examples he dealt with in his own career, he asks, if a tank is parked inside a schoolyard, is it legitimate to strike? How about if it is driving out of the village and a group of children catch a ride on top?

Such questions are already hard for humans, and likely would be answered differently by different soldiers or lawyers, depending on the circumstances. Noting that people have defined al-Qaeda members as everything from "freedom fighters" and "terrorists" to "criminals" and "unlawful combatants," he explains, "It is not that you can't do it, but that not all the actors will agree on it." Even more, to delegate such a decision in war to a machine may be expecting too much of a system when the technology is still immature. In turn, it may be expecting too little of humans' responsibilities in war.

Another fundamental premise of the human rights group, and for broader international law, is that soldiers in the field and the leaders who direct them must be held accountable for any violations of the laws of war. Unmanned systems, though, muddy the waters surrounding war crimes. "War crimes need both a violation *and* intent," says Garlasco. "A machine has no capacity to want to kill civilians, it has no desires.... If they are incapable of intent, are they incapable of war crimes?" And if the machine is not responsible, who does the group seek to hold accountable, and where exactly do they draw the line? "Who do we go after, the manufacturer, the software engineer, the buyer, the user?"

## MONDAY-MORNING QUARTERBACKING

"It used to be a simple thing to fight a battle," explains marine general James Jones. "In a perfect world, a general would get up and say 'Follow me, men,' and everybody would say 'Aye Sir' and run off. But that's not the world anymore. Now you have to have a lawyer or a dozen. It's become very legalistic and very complex."

Soldiers have an opposite fear from that of the lawyers and human rights workers, who worry about law going missing. They see the law increasingly impeding their missions and fret that the trend will worsen with the new technologies. They do not look forward to a world in which every single decision in war, down to every single shot fired, comes with a video record and computer database attached.

As General Jones's comments illustrate, the U.S. military already puts an immense amount of time and effort into the legal side of warfighting. Every single command center and unit has a military lawyer (a JAG, short for judge advocate general) attached, whose job is to make sure that the missions comply with the laws of armed combat.

These legal officers' influence can be powerful, some argue too powerful. For example, in the opening weeks of the operation in Afghanistan in 2001, a Predator drone spotted a convoy of pickup trucks that later turned out to have been carrying Taliban leader Mullah Omar. When asked why the Predator did not just blow up the suspected enemy vehicles, General Tommy Franks replied, "My JAG doesn't like this, so we're not going to fire."

As more and more unmanned systems are used, many in the military are growing afraid of another sort of legal complexity, arising from the ability to "play back battles." Says retired marine officer James Lasswell, "The ability to backtrack on combat and then make legal judgments on it is a scary proposition....I worry that it will yield what I call Monday morning quarterbacking."

As with a football player on the field, decisions in war must be made in split seconds. But now people will be able to play them back in detail, picking apart judgment calls and mistakes. And, just as we sometimes sit in our Barcaloungers and blame the quarterback who throws a "dumb" pass that leads to an interception, we will also likely place the blame at the point of decision, even though all sorts of other factors may have caused that bad outcome (a slippery football, a terrible play sent in by the coach, the owner not paying for a good free-agent lineman to give enough time to throw, and, of course, the other team's actions). In a sense, soldiers worry they will get the worst of both worlds. Despite all the confusion among the experts about how to apply the laws of war to robotics, soldiers' every move will be second-guessed, with legal consequences.

Air force major general Charles Dunlap worries that the situation could get even worse. He describes how "there is a legal and moral duty," as outlined in the LOAC, to "take all feasible precautions" to prevent civilian casualties. This legal understanding, he explains, becomes much more complex with unmanned systems and battle management AI that are growing more sophisticated, including allowing computer simulations and modeling before the actual fight. "What if a commander chooses a course of action outside the model that results in a higher number of civilian casualties?" On one hand, the commander ignored a duty to take feasible precautions. On the other hand, to punish any officer for this would be placing more legal trust in the judgment of the computer than in the human actually at war.

Given this combination of advancing technology and thorny legal questions, many advise that soldiers had better get used to the growing presence of lawyers inside military operations. "We'll see far more lawyers, weighing the consequences, getting into estimating the probabilities of technology working or not, is it right or wrong," explains former naval officer and assistant secretary of defense Larry Korb.

The "soldiers" who respect the laws of war might then be at a disadvantage to the "warriors" and criminals who do not. It is not merely that the laws and lawyers limit what they can do, but that the other side knows the limits, and will do everything possible to take advantage. During the Fallujah fighting in 2004, for example, insurgents knew U.S. forces were prohibited from shooting at ambulances, so they used them as taxis to carry about fighters and weapons. Such "lawfare," describes Major General Dunlap, is perhaps the ultimate misuse of international law, because it knowingly abuses it. "They are intent on manipulating our adherence to the rule of law." Ralph Peters is more blunt. "We approach war in terror of lawsuits and criminal charges. Our enemies are enthusiastic killers. Who has the psychological advantage?"

## DEHUMANIZING WAR

"The truth is, it wasn't all I thought it was cracked up to be. I mean, I thought killing somebody would be this life-changing experience. And then I did it, and I was like, 'All right, whatever.' . . . Killing people is like squashing an ant. I mean, you kill somebody and it's like 'All right, let's go get some pizza.' "

Steven Green was a private in the 101st Airborne Division when he indifferently described what it felt like to kill an Iraqi to a *Washington Post* reporter. Just a few weeks later, Green would allegedly plan out the rape and murder of a fourteen-year-old Iraqi girl, Abeer Qasim Hamza, with whom he had apparently become infatuated while doing checkpoint duty. The girl's parents and five-year-old sister were also killed in the attack, which Green reportedly tried to cover up as a raid on insurgents gone bad.

Discharged from the military before the crime was discovered, Green was arrested by the FBI in 2006 and convicted of rape and murder in 2009, sentenced to spending the rest of his life in jail without possibility of parole. But the damage was already done, not only to the civilians, but also to Green's own unit. Two of his fellow soldiers deployed to the same sector were later kidnapped, tortured, and beheaded. Accompanying the release of a gruesome video of the soldiers' bodies was a statement that it was "revenge for our sister who was dishonored by a soldier of the same brigade."

That someone like Green is the exception to the rule among soldiers is without debate. But his story also illustrates another aspect of war. It is a dangerous and dirty space, in which bad apples do pop up, as in every other part of human society. Horrendous crimes do occur, only with the stakes far higher. Historian Stephen Ambrose notes, "When you put young people, eighteen, nineteen, or twenty years old, in a foreign country with weapons in their hands, sometimes terrible things happen that you wish never happened. This is a reality that stretches across time and across continents. It is a universal aspect of war, from the time of the ancient Greeks to the present."

Just like any other crime, there are all sorts of causes of war crimes. Sometimes, atrocities occur because of a planned strategy or policy, like the Holocaust. Other times, they may result from individuals like Green (who reportedly had shown mental instability even before the murders), for whom war becomes both a backdrop and enabler for their aggressions. These are the relatively easy war crimes to explain. Hardest to rationalize are the ones like My Lai in Vietnam, when otherwise professional soldiers and units break down and become involved in unplanned war crimes. As David Perry, a professor of ethics at the Army War College, warns, "It is vitally important to recognize that atrocities are not committed by sadistic people only. Almost all of us are capable of barbarism. Even our most admirable soldiers—ones who would courageously give their lives for their buddies without hesitation, ones you'd gladly trust to baby-sit your kids—can be transformed into indiscriminate killers."

War is both fueled and sustained by human emotions, by the passions of fear, hate, honor, pride, bravery, and, by necessity, anger. It might be anger at the enemy for previous wrongs or just anger that they have put you in the position of having to kill them in order to see your loved ones again. But anger is often what allows a soldier to do the terrible deeds necessary to accomplish a mission and return home. As the expert on military philosophy Nancy Sherman writes, "Anger is as much a part of war as weapons and armor."

These same essential but raw emotions of war can also fuel the crimes of war, such that sometimes the anger trumps discipline and professionalism. A soldier's or even an entire unit's composure can be overcome in the heat of the moment. The spark can vary, but it is almost always the result of poor leadership, sustained stress at being in horrific conditions for extended periods, and, most often, some loss that drives soldiers over the edge, often when comrades are killed. Describes Captain Thomas Grassey, a professor of leadership and ethics at the Naval War College, "Situations where normal, good people are given power over others are fraught with dangers of inhumane, even sadistic treatment.... More generally, being an American offers no one exception from the passions of battle, the desire for vengeance, the urge to dehumanize, the temptation of sadism, and the traps of irrationality and brutality."

Technology is often described as a way to reduce war's costs and passions, and thus its crimes. The poet John Donne (of "No man is an island" fame), for example, told in 1621 how the invention of better cannons would help limit the cruelty and crimes of war, "and the great expence of bloud is avoyed." Yet the improvement of guns from the 1600s onward certainly didn't reduce the flow of "bloud" or end war crimes. However, many today hope that robotics just may well be the one technology that proves Donne right. As one retired army officer explained, "Warfare on some levels will never be moral, but it can be more moral."

An army brigade commander in Iraq, for example, told how footage from a drone showed one of his troops guarding an enemy detainee. The soldier, not knowing the drone was overhead, gave a quick look to the right and left, to see if anyone was watching down the street, and then "gave the detainee a good, swift kick to the head." The officer recalls that everyone in the command post then turned and "looked at the old man [him] to see how he would react."

While the omnipresence of cameras might mean more second-guessing along the lines of Monday morning quarterbacking, it also changes the context in which decisions about war, and also abuses in war, are made. That commander now knew about his soldier doing wrong and put a stop to it. Whereas a U.S. Army survey found that 45 percent of soldiers wouldn't report a fellow soldier they saw injuring or killing a civilian noncombatant, robots don't care about their buddies and always report what they see. At a broader level, any nation or would-be criminal pondering a war crime will know that, with so many machines around them recording data, they are more likely to be caught and cover-ups will be harder to pull off.

Others argue that robots' effect will be felt less on the planned war crimes and more on the unplanned crimes, the crimes of rage. Military analyst Martin van Creveld once described that the best-disciplined army would be one that "behaves as if it were a single personality." Robots turn this human impossibility into technologic fact. They are not governed by passions of loss, anger, or revenge. And machines don't suffer from fatigue that can cloud judgment, nor do they have those unpredictable testosterone fluctuations that often drive eighteen-year-old boys to do things they might regret later in life.

Not only are the robots themselves without emotion, but many also feel that unmanned systems remove anger and emotion from the humans behind them. As Clausewitz wrote, the stresses of battle can be overwhelming. "In the dreadful presence of suffering and danger, emotion can easily overwhelm intellectual conviction, and in this psychological fog it is…hard to form clear and complete insights." A remote operator, by contrast, isn't in the midst of combat and isn't watching their buddies die around them as their adrenaline spikes. As Robert Quinn at Foster-Miller says, "The big advantage of moving to armed robots is that you take the emotion, the fear factor out of the decision to shoot…. You are looking in the whites of their eyes but calm."

With the humans not facing risk, they also have the ability to take their time, with a "slow, methodical approach" that can lessen the likelihood of civilians being killed. Marc Garlasco of Human Rights Watch told how "the single most distinguishing weapons I have seen in my career were Israeli UAVs." He described that, unlike jet fighters that had to swoop in fast and make decisions on what targets to bomb in a matter of seconds, the UAVs he observed during the 2006 Lebanon war could loiter over a potential target for minutes or even hours, and pick and choose what to strike or not. In Vietnam, an amazing fifty thousand rounds of ammunition were expended for every enemy killed. Robots, on the other hand, might "come closer to the motto of 'one shot, one kill.' " As a report on the SWORDS says, the operator "can coolly pick out targets as if playing a video game."

But anyone who has played *The Sims* or *Grand Theft Auto* can support Chuck Klosterman's assertion that most people playing video games "are not a benevolent God." We do things in the virtual world that we would never do if we were there in person, such as ramming our car into an ice cream stand or seeing what happens when our avatar jumps off a skyscraper. Transferred to war, this could mean that the technology might well lessen the likelihood of anger-fueled rages, but also make some soldiers too calm, too unaffected by killing. A possible psychotic like Steven Green talked about the experience of killing someone as feeling the same for him as merely "squashing an ant." The true fear is that turning killing into merely the elimination of icons on a computer screen might make the experience feel the same way even for otherwise normal troops. As a young air force lieutenant described what it was like to coordinate unmanned airstrikes in Iraq, "It's like a video game, the ability to kill. Its like... [he pauses, searching for the right words] freaking cool."

Many studies have shown how disconnecting a person, especially via distance, makes killing easier and abuses and atrocities more likely. An important one was by Lieutenant Colonel Dave Grossman, an army psychologist. In his book *On Killing*, he explored both how soldiers are motivated to kill and how it affected them. He found that most people are not "natural born killers." Instead, humans have a natural instinct not to kill another human. Indeed, "trigger-pull" studies of earlier wars found that many soldiers never actually fired on the enemy in battle. But Grossman also found that this instinct isn't irresistible, and that military training and conditioning can overcome it. If the soldier, for example, can be conditioned to dehumanize their foe, to see them not as a person but as something else, they find it easier to kill. They might view them as subhuman (such as how we turn our enemies into the "Hun," "Jap," "Gook," or today's moniker in Iraq, "Haji") or nonhuman, a "target," as one former soldier put it, "that needs to be serviced."

The other factor that enables killing, Grossman found, was distance. "The greater the distance, physical and emotional, from the enemy, the easier it is to kill them. Soldiers at close range or engaged in hand-to-hand combat exhibit a much higher resistance to killing, but at long range, the resistance to killing is much lower." For example, bomber crews carried out firebomb raids during World War II that literally burned alive hundreds of thousands of men, women, and children in cities like Hamburg, Dresden, and Tokyo. They carried out these missions with few qualms. If they had been asked to do the same face-to-face with a flame-thrower, the outcome likely would have been different, even with their conditioning to think of the Germans and Japanese as less than human.

Many worry that both factors are especially enabled by the new technologies of unmanned systems. D. Keith Shurtleff is an army chaplain and the ethics instructor for the Soldier Support Institute at Fort Jackson in South Carolina. His concern is that "as war becomes safer and easier, as soldiers are removed from the horrors of war and see the enemy not as humans but as blips on a screen, there is a very real danger of losing the deterrent that such horrors provide."

Participation via the virtual world also seems to affect not merely how people look at the target, but also how the person looks at themselves (why people in online communities, for example, take on identities and try out behavior they would never do in real life, be it wearing tattoos or sharing intimate personal information with strangers). Research shows that this sort of "externalization" allows something called "doubling." Otherwise nice and normal people create psychic doubles that carry out sometimes terrible acts that their normal identity would never do. An air force lieutenant colonel, for instance, who led a Predator operation noticed how the virtual setting could make it easy for the drone operators to forget that they were not gods from afar and that there are real humans on both ends. "You have guys running the UAV saying, 'Kill that one, don't kill that one.' "

Each new technology, from the bow and arrow to the bomber plane, has moved soldiers farther and farther from their foes, so in some ways robots aren't creating an entirely new development. Yet unmanned systems have a more profound effect on "the impersonalization of battle," as military historian John Keegan called it. These weapons don't just create greater physical distance, but also a different sort of psychological distance and disconnection. The bomber pilot isn't just above their target, but seven thousand miles away. They don't share with their foes even those brief minutes of danger that give them a bond of mutuality.

As robots gain more and more autonomy, emotions won't just be limited or changed, but taken completely out of the equation, with similar mixed results. While autonomous robots are less likely to commit unplanned rage-filled war crimes, they enable the type of deliberate war crimes that a professional soldier might refuse. A computer has no anger to make it lash out like at My Lai, but it also has no pity, no disgust, and no sense of guilt. It does whatever it is programmed to. Shooting a missile at a T-80 tank is just the same to a robot as shooting it at an eighty-year-old grandmother. Both are just a series of zeros and ones.

## UN-MANSLAUGHTER

Thirty-year-old Daraz Khan lived in the village of Lalazha in the south of Afghanistan. He wasn't rich or worldly, but he did have one thing going for him in life—he was the tallest man in town. It would prove to be his undoing.

On the morning of February 4, 2002, Khan, whose nickname was "Tall Man," hiked with two friends to the top of a snowy mountain. They planned to collect scrap metal left over from past battles in the area between the Soviets and the mujahideen, and more recently between the Americans and the Taliban. The going rate was fifty cents for all the scrap metal that could be loaded on a camel's back. It didn't seem much, but that tiny amount was enough to make the ten-mile hike worthwhile. Around 3 P.M., as the men talked and picked through the wreckage, an explosion atomized the bluff on which they were standing.

The explosion that killed Daraz Khan and his friends came from a Predator drone that had been quietly circling above. The drone's operators had first spotted the three men when they had suspiciously hiked into the area suspected to have al-Qaeda leaders hiding in it. The men were wearing robes, were at a suspected terrorist hideout, and, most important, one of them was much taller than the others, as bin Laden was thought to be. As best as could be determined from seven thousand miles away, these were the men whom the Predator was looking for. As Pentagon spokeswoman Victoria Clarke explained, "We're convinced that it was an appropriate target... [although] we do not yet know exactly who it was." Daraz's sixteen-year-old niece took a slightly different view: "Why did you do this? Why did you Americans kill Daraz? We have nothing, nothing, and you have taken from us our Daraz."

Many wartime atrocities are not the result of deliberate policy, wanton cruelty, or fits of anger; they're just mistakes. They are equivalent to the crime of manslaughter, as compared to murder, in civilian law.

Unmanned systems seem to offer several ways of reducing the mistakes and unintended costs of war. They have far better sensors and processing power, which creates a precision far better than humans could marshal on their own. Such exactness can lessen the number of mistakes made, as well as the number of civilians inadvertently killed. For example, even as recently as the 1999 Kosovo war, NATO pilots spotting for Serbian military targets on the ground had to fly over the suspected enemy position, then put their plane on autopilot while they wrote down the coordinates of the target on their lap with a grease pencil. They would then radio the coordinates back to base, where planners would try to figure out if there were too many civilians nearby. If not, the base would order an attack, usually made by another plane. That new plane, just arriving on the scene, would carry out the attack using the directions of the spotter plane, if they were still there, or the relayed coordinates. Each step was filled with potential for miscommunications and unintended errors. Plus, by the time a decision had been made, the situation on the ground might have changed—the military target might have moved or civilians might have entered the area.

Compare this with a UAV that can fly over the target and send precise GPS coordinates and live video back to the operators. Add in the possibility of using an AI simulation to predict how many civilians might be killed, and it is easy to see how collateral damage can be greatly reduced by robotic precision.

The unmanning of the operation also means that the robot can take risks that a human wouldn't otherwise, risks that might mean fewer mistakes. During that Kosovo campaign, for example, such a premium was placed on not losing any NATO pilots that planes were restricted from flying below fifteen thousand feet so that enemy fire couldn't hit them. In one case, NATO planes flying at this level bombed a convoy of vehicles, thinking it was Serbian tanks. It turned out to be a convoy of refugee buses. If the planes could have flown lower, or had the high-powered video camera of a drone instead of human eyes, this tragic mistake might have been avoided.

The removal of risk also allows decisions to be made in a more deliberate manner than normally possible. Soldiers describe how one of the toughest aspects of fighting in cities is how you have to burst into a building and, in a matter of microseconds, figure out who is an enemy and who is a civilian and shoot the one that is the threat before they shoot you, all the while avoiding hitting any civilians. You can practice again and again, but you can never fully avoid the risk of making a terrible mistake in that split second, in a dark room, in the midst of battle. By contrast, a robot "gives you the ability to shoot second." It can enter the room and only shoot at someone who shoots first, without worrying that doing so puts a soldier's life at risk.

As the case of Daraz Khan illustrates, however, mistakes and accidents remain just as much a part of war with unmanned systems, including in new ways. Unmanned systems allow a greater precision, giving the operator a view of what is going on that is far better than before. But sometimes this can prove false, and even breed an overconfidence that makes mistakes more likely. Khan appeared tall, but that was only because of who he was standing near when looking on a computer screen from seven thousand miles away. "Tall Man" Daraz was actually only five feet eleven inches, big for that impoverished area, but nowhere near bin Laden's reported height of six foot four inches.

These unmanned cases of mistaken identity don't just cost civilian lives, but can even happen to friendly units. With the precision allowed by modern weapons, drones have more leeway than ever before to strike at targets when friendly troops are in the area. In the past, entire sectors would be boxed out on the map, indicating where airstrikes could happen and where they could not, with no strikes allowed anywhere near an area where friendly troops were. This new leeway can prove troublesome, though. In the mix of battle, and without clear "kill boxes," the front lines grow confusing, even for unmanned systems. During the 2006 Lebanon war, for example, an Israeli UAV opened fire on Israeli ground troops.

The causes of these mistakes are often in great dispute. Sometimes the blame is placed on the humans behind the machines. In a U.S. airstrike in 2001, for example, twelve out of fourteen smart bombs inexplicably missed their target by a wide margin. It turned out the humans who had programmed the weapons' targeting back at the base had punched in the wrong coordinates. Other times, the data itself is bad. As we know from the case of Daraz Khan, as well as all that Iraqi WMD we found, our intelligence is sometimes flawed and unmanned attacks don't always get the right person. In 2005, U.S. officials said that on at least two occasions, "The Predator has been used to attack individuals mistakenly thought to be bin Laden." The same rule from the world of computer software holds true in war: "Garbage in, garbage out."

The blowback from such mistakes is worsened. America's technology is viewed as almost magical, but much of the world also sees the United States through jaundiced eyes. When something goes wrong, the immediate assumption is not that a mistake occurred, but that it was planned malice. In 2006, for example, two Predator drones fired ten missiles into the Pakistani hamlet of Damadola. Their target was Ayman al-Zawahiri, bin Laden's deputy and one of the architects of the 9/11 attacks. Instead, as political analyst Mansoor Ijaz writes, "The attack killed 18 civilians and brought tens of thousands of protesters out onto Pakistan's streets calling for the ouster of President Pervez Musharraf and chanting 'Death to America' and 'Stop killing our children.' "

Anyone whose computer has ever crashed knows that the human sitting at the keyboard is not always to blame. The system itself can be the problem. Robotics expert Robert Finkelstein expects that future tragedies "such as mistaking a civilian for a soldier, a tank for a truck—fratricide is another example—will occur, most likely from bugs in the software or improper design." A U.S. Navy captain agrees: "Inevitably, sooner or later, something is going to go wrong. Then who's responsible, the programmer, the whiz bang guy?" Unsurprisingly, many roboticists don't think they should be the ones held legally accountable if a robot they designed mistakenly kills a civilian, or that it should even be viewed as a crime. As a DARPA-funded roboticist disturbingly put it, "It depends on the situation. But if it happens too frequently, then it is just a product recall issue."

## FREE FIRE

Military roboticist Robert Finkelstein believes that the legal issues will get even thornier as machines gain more autonomy. "The big question for military law is how do we transition authority for lethal action to the machine? It might start limited, such as 'Kill in this zone only.' But that is still authorizing."

For all the critiques that many people make of the U.S. military, it bends over backward to figure out when it's appropriate to engage the enemy and how to limit civilian casualties. For instance, in planning out what Iraqi targets to strike during the 2003 invasion, the Pentagon created a master list called the "joint target list." The only targets that made it onto this list were ones that had been approved by military lawyers and screened against another list, made with the help of the United Nations and other civilian groups, of sites that could not be targeted, such as schools, mosques, or cultural sites. The criteria for using the list also required that the area around the target be checked for any "environmentally dangerous facilities or civilian structure in the vicinity of the strike zone," as well as "can damage be avoided by using a different method to attack the target?" Finally, the list was also run against "collateral damage estimation methodology," computer simulation models that predicted whether civilians who lived nearby might be harmed during the attack. If a target was expected to cause such damage, but was still considered militarily necessary, then the strike mission to hit that target had to be specifically approved by the most senior civilian political leaders (the secretary of defense or the president), so that there was a clear chain of accountability. Notably, this approach is pretty much the exact opposite of how terrorists go about picking what to target; they try to find sites with high collateral damage that cause the most civilian harm possible.

Seemingly, a similar sort of checklist could prove equally useful in creating rules of engagement for autonomous robots. An unmanned system might be programmed to go down a list of criteria to determine appropriate targets and when it is allowed to shoot. Much as the "joint target list" required command authorization when civilians were at risk, the robot might be programmed to require human input if any civilians were detected. An example of such a list at work might go as follows: "Is the target a Soviet-made T-80 tank? Identification confirmed. Is the target located in an authorized free-fire zone? Location confirmed. Are there any friendly units within a 200-meter radius? No friendlies detected. Are there any civilians within a 200-meter radius? No civilians detected. Weapons release authorized. No human command authority required."

In 2007, the U.S. Army commissioned just such a study of how a "lethal autonomous system" could be built with an "ethical control and reasoning system...so that they fall within the bounds prescribed by the Laws of War and Rules of Engagement." Noting that human soldiers commit all sorts of war crimes both big and small (indeed, the report cited that 10 percent of U.S. soldiers in Iraq said they had mistreated noncombatants at some point), the study argued that lethal robots may not just be "able to be perfectly ethical in the battlefield," but that they will be more ethical "than human soldiers are capable of." Said the author of the report, "We could reduce man's inhumanity to man through technology."

Such an "ethical" killing machine, though, may not prove so simple in the reality of war. Even if a robot has great software that follows all the various rules of engagement and is the rare device with absolutely no software or hardware failures or bugs, the very question of figuring out who is an enemy in the first place (that is, whether a target should even be considered for the list of screening questions) is extremely tough in modern war.

During the 2003 invasion of Iraq, the U.S. military used what are known as "status-based" rules of engagement to decide whether a potential target was an enemy or not. If a person or facility was part of Saddam's military and paramilitary forces, then it was judged to be "declared enemy forces." As long as it met the qualifying questions of collateral damage, it was a legitimate target to destroy. As the fighting shifted from invasion to insurgency, though, and no one was part of Saddam's forces anymore, or even wore a uniform or insignia to proclaim themselves an enemy, this definition proved pretty useless. So the rules quickly shifted to a "conduct-based" system. The target could be attacked only if some "hostile act" or "hostile intent" was observed.

Identifying an enemy target became essentially a judgment call, and not always an easy one at that. If an insurgent was shooting a gun at you, then the "hostile act" was pretty clear. If they were seen planting an IED, then their "intent" to cause harm was clear. But not every situation was so self-evident. A soldier might be shot at from a direction, but not know which house the shooter was in, or even whether the shooter was alone or part of a force ready to overwhelm him. Could he fire back in that general direction to try to cover himself, or did the shooter have to openly show himself and his weapon before the soldier could fire back? Or what about cases where a group of men thought to be insurgents were gathered in a room. Did the soldiers have to wait until they actually left to go plant IEDs to destroy the building, or was the belief that they might be planning an attack enough?

Warrior groups are doing all they can to take advantage of these ambiguities, the kind of "lawfare" that the analysts were worried about. In Somalia, one Ranger recalled how a gunman shot at him with an AK-47 that was propped between the legs of two kneeling women, while four children sat on the gunman's back. The Somali warrior had literally created a living suit of noncombatant armor. Similarly, in the Lebanon war, Hezbollah hid rockets in farmhouses near the Israeli border. The farmers were paid to simply press a button that would remotely fire the rockets whenever they received a phone call from the group. It was also suspected that any farmers who refused to do so would be killed. The proper response for the Israelis was not obvious. Does the potential blackmail matter? Should they try to strike any farm's suspected rocket site before it fired at their cities (raising the question of how to find this out without engaging in some hostile act), or were they legally obligated to wait for the hidden rockets to be fired before they could strike the farms? Equally confusing was just if and when that farmer became a legitimate target. Did it happen when the rocket was first placed on his farm or when he pressed the button to fire the rocket? And when does he stop being a legitimate target to kill? Is it after he's pressed the button and gone back to farming, or does he stay a legitimate combatant for the rest of the war, even though his role in it is now done?

Politicians, pundits, and lawyers can fill pages arguing these points. It is unreasonable to expect robots to find them any easier. Explains one engineer at iRobot, "A robot can't easily tell a good human from a bad human." Indeed, it has a hard time even distinguishing an apple from a tomato. As a result, mistakes and unintended casualties will continue to happen in war, even with unmanned systems. But the rules of war don't seem likely to catch up with them anytime soon. Instead, as one air force Predator pilot described, episodes like the one that mistakenly killed poor Daraz Khan have a dark reality. "What happens when things don't work out as they are supposed to? The attitude seems to be to still then kill them and let their gods sort them out."

## ROBOT RIGHTS IN A PC WORLD

In 2006, the British government commissioned a series of papers on key developments that would challenge it over the next twenty to fifty years. The papers looked at a wide range of emerging political, economic, and science issues, exploring everything from the rise of India and developments in nanotechnology to what global warming and the shifting Gulf Stream boded for Britain's weather and economy. "We're not in the business of predicting the future," explained Sir David King, the government's chief scientific adviser, "but we do need to explore the broadest range of different possibilities to help ensure government is prepared in the long-term and considers issues across the spectrum in its planning." When the reports were made public, however, the always lively British tabloids focused on one finding in particular. "Robots could demand legal rights," screamed the headlines.

The British government's research had found that a "monumental shift" would happen when robots advance to the point of having artificial intelligence. Its conclusion wasn't just that this revolution would happen within twenty to fifty years, meaning the British government was also now officially supporting the idea of the Singularity, but that the rise of these machines would also revolutionize the way we think about citizenship. Intelligent robots could be deemed worthy of having many of the same rights and responsibilities as humans. If intelligent beings, even artificial ones, argued the report, were expected to serve in the military, in turn society would have its own responsibilities toward its "new digital citizens." Countries would be obliged to provide "full social benefits to them including income support, housing and possible robo-healthcare."

The image of C3PO standing in line for welfare cheese seems a bit outlandish, but it does illustrate some of the odd but potentially real long-range legal questions that might come about as robots are used more widely, as well as become more intelligent. The International Bar Association, the professional group for lawyers around the world, has also begun wrestling with robot rights; in a mock trial in 2003, a lawyer defended the rights of a conscious computer against a corporation that sought to disconnect it.

If we are destined to contend with the issues of robot intelligence and robot rights in civilian law, it is also likely we will have to do so in military law. For example, the laws of war deal not only with the treatment of civilians, but also that of fellow soldiers, banning torture and mandating that captured soldiers be treated well. So, as Robert Finkelstein asks about robots, "If they are given human level intelligence, will they be treated as humans? Why not?" Marc Garlasco of Human Rights Watch equally muses, "Will we ever get to the point where Human Rights Watch is advocating in a *Blade Runner* scenario, where Human Rights Watch is standing up for a manufactured human machine? Does that mean Human Rights Watch becomes Human and Robotics Rights Watch?"

There are two arguments for how and why we might start to care about how robots are treated, even in war. Interestingly, both aspects of robot rights are less about the robots themselves and more about the humans involved.

First, as robots become more intelligent, and especially as they interact better with humans (having realistic-looking bodies, AI personalities, and so on), people will start to endow them with certain identities, giving them a persona or character. How we treat a robot won't turn so much on whether it is "alive" or even able to recognize its rights, but whether we endow the robot with something called "being-ness."

We often talk about having respect for all living things, but actually we only endow them with certain rights or treatment standards the closer they seem to us. So, for example, most people don't like the idea of our evolutionary cousins the apes being tortured or even kept in captivity like pets, but no one thinks they have the right to vote. By contrast, domesticated mammals, like dogs or cats, are socially acceptable to "own," to keep in captivity as a "pet," and even to treat as a fashion accessory to tote about in your Gucci purse. Most people even accept their local government euthanizing stray cats and dogs, as the animal's social value for some reason depends on whether a person likes it enough to house it. But, as Michael Vick can attest, society definitely frowns upon torturing or killing our furry friends for sport, and aside from a few countries, most people abhor the idea of eating their pets for dinner. As you get farther down the list, society begins to look at species differently. Rats and mice may be fellow mammals, but it is still considered socially acceptable to poison or test mascara on them. By the time you get to insects, few care what is done to such living things; most households use chemical weapons in aerosol form to commit mass murder against them.

With robots, psychologists are finding that people tend to put them into a space somewhere between an inanimate object and living machine. It is not alive or dead like an animal, but also not the same as a lawn mower or computer. The robot is something else; people see it as having what psychologists describe as a certain sense of "being." The more intelligent, social, and familiar it seems to us, the more we think it is somehow like us, and thus we view the robot as having more of that "being." And just as we differentiate in how we treat other living species, our sense of how nonliving machines should be treated is highly dependent on that level of "being" we grant them.

To think of this another way, it somehow seems okay for a person to pound the keyboard of their computer in frustration when it crashes or to give their lawn mower a kick when it won't start. With robots, it equally seems okay to kick a PackBot or Roomba when it breaks. But to kick a robotic dog or punch an Actroid beauty robot in the mouth? That somehow doesn't seem right. And what about when C3PO was disassembled for scrap in *The Empire Strikes Back* or when Arnold was lowered into the molten iron in *Terminator 2*? Even certain teenaged boys have been known to tear up at such heartbreaking moments. We know these are machines and they don't feel any pain, and yet we still don't think they should be treated this way.

Basically, this sense that robots deserve better treatment than regular tools or machines will come about because of our own human psychological quirks and sense of self. The more they seem like us, the more worthy we will think them of our own expectations of treatment. Explains roboticist Daniel Wilson, "Humans are stupid. All robots are robots, but they will care differently about a humanoid robot versus a dog robot versus a robot that doesn't look like anything alive." The world that the TV show *Battlestar Galactica* foresees doesn't seem all that absurd then; some robots are most likely to be treated just as vehicles, others like an animal or pet, and still others, likely the more intelligent, more humanoid-looking types, with a little bit more respect and protection (in the show, for example, captured humanoid robots are kept in jail cells, and there is a spirited debate on whether they have legal rights and whether they can be tortured).

The second argument is that we need to regulate humans' behavior toward robots, not for their sake, but for humanity's sake. As Henrik Christensen, director of the Center for Robotics and Intelligent Machines at Georgia Tech, explains, robots present new pitfalls for humans' treatment of each other. "There will be people who can't distinguish that, so we need to have ethical rules to make sure we as humans interact with robots in an ethical manner, so we do not move our boundaries of what is acceptable." The argument, then, is that things like abuse or torture should be banned regardless of whether they're being committed on a human or a robot, to ensure that the very act remains a clear line that should not be crossed.

Giving robots some sort of protection or rights would be primarily about preserving our own sense of right and wrong. This may seem odd or hypocritical, but it makes a certain kind of psychological sense. Many parents, for example, are content to spray a can of Raid on ants in their kitchen, but grow worried if they see their child in the backyard giggling as he burns ants with a magnifying glass. They don't care about the ants, but what the act signifies about little Timmy and his sense of right and wrong.

This question of whether an unmanned system has rights or "being" seems very futuristic, and the British government didn't see it coming to the fore in civilian law for several decades. But military lawyers must already grapple with it. Take, for example, the right of self-defense, a time-honored principle in international law. If a ship or plane is attacked, even during peacetime, the crew has "the right to use all necessary means" to save itself, as well as punish the attacker, so that they don't attack again.

Many nations even argue that an actual attack isn't required to activate this right. Merely the appearance of a hostile intent to attack is enough; for example, they argue that if a radar targets a plane, the pilot can fire first, rather than simply waiting to be blown out of the sky. This expanded right of self-defense was used by the United States in the Gulf of Tonkin incident prior to the Vietnam War, the air battles with the Libyans in the Gulf of Sidra in the 1980s, and during the decade of airstrikes from 1991 to 2002 while enforcing the "no-fly zone" against Iraq.

This question gets fuzzy, though, when the "self" in self-defense is not a person but an unmanned system. Take the above incidents and substitute robotic drones for the planes piloted by humans. If an unmanned plane flying near the border of another nation is fired on, does it have the right to fire back at that nation's missile sites and the humans behind them, even in peacetime? What about the expanded interpretation, the right to respond to hostile intent, where the drone is just targeted by radar? Is the mere threat enough for the drone to fire first at the humans below?

The answers depend on how wide the "self" in self-defense is defined. One side could say that the whole concept of self-defense under the law is based on the assumption that a human is inside. They don't see how an unliving plane that doesn't care about itself could have the human right of self-preservation, especially one that preemptively kicks in at the mere appearance of a threat. Others counter that, while no human in the plane was under threat, the machine is still an entity and, moreover, it is "national property" and so must be considered as representative of the people who sent it; it has the same rights as if a person were inside it.

You may not be persuaded by this second argument, but the U.S. Air Force was. The interpretation of robot rights is official policy for unmanned reconnaissance flights over the Persian Gulf.

## "THE REVOLUTION IN MILITARY LEGAL AFFAIRS"

Any law's strength depends on its relevance. Yet today's major codes of international law in war, the Geneva Conventions, are so old that they almost qualify for Medicare. Dealing with digital warfare is a lot to expect of a treaty from 1949, when the average new house cost $7,450 and the most notable invention was the 45 rpm record. To put it another way, if the new technologies are creating a "revolution in military affairs," we may well need a "revolution in military legal affairs."

Catching twentieth-century laws up to twenty-first-century conflict does not mean that the old rules have to be jettisoned completely. The Geneva Conventions codified some of the most important principles of international law, from the rights of soldiers taken prisoner to special protections that had to be given to civilians and the wounded. Just because such protocols are old makes them no less important today, and any design and use of unmanned systems must seek to uphold them.

For example, it doesn't make sense to hold human pilots to different legal standards based solely on where they are located. As planes gained the capacity to fly at higher and higher speeds and altitudes, the laws of war didn't start to make exceptions for pilots flying a B-2 at six hundred miles an hour as opposed to flying a B-29 at three hundred miles per hour. If they deliberately or somehow negligently killed the wrong people, they were to be held accountable all the same. The same seems to be the most logical way for dealing with the remote operation of unmanned systems. If a driver runs over a little girl with his Humvee, it shouldn't matter whether he was physically located inside the vehicle or doing so from nine thousand miles away. The outcome, and thus the responsibility, is the same. The question should instead focus on whether the action was deliberate or an accident, and, if so, criminally negligent in some way.

For the same legal reason, military unmanned systems should be operated by those serving in the military, not handed off to private contractors. Military courts and the laws of war are designed for dealing with the kind of life-or-death decisions that take place in war. Putting a civilian in the operator role, on the other hand, might mean that a civilian court would be presented with questions for which it was simply not designed. How could a jury drawn from a pool of showgirls and casino employees be expected to wrestle with whether a Predator pilot in Las Vegas violated the rules of engagement during his UAV's classified operation over Kandahar? If control is limited to the old rule of "only military trigger pullers," describes one air force lawyer, this odd legal Pandora's box remains shut.

In systems with more autonomy, the legal questions become stickier. In 2002, for example, an Air National Guard pilot in an F-16 saw flashing lights underneath him while flying over Afghanistan at twenty-three thousand feet and thought he was under fire from insurgents. Without getting required permission from his commanders, the pilot dropped a 500-pound laser-guided bomb on the lights. They instead turned out to be troops from Canada on a night training mission. Four were killed and eight wounded. In the hearings that followed, the pilot blamed the ubiquitous "fog of war" for his mistake. It didn't matter. The hearing concluded that he "flagrantly disregarded a direct order," "exercised a total lack of basic flight discipline," and "blatantly ignored the applicable rules of engagement." All the law cared about was that he dropped the bomb in violation of standing orders, and he was found guilty of dereliction of duty.

Change this scenario to an unmanned system and military lawyers aren't sure what to do. Asks an air force officer, "If these same Canadian forces had been attacked by an autonomous UCAV, determining who is accountable proves difficult. Would accountability lie with the civilian software programmers who wrote the faulty target identification software, the UCAV squadron's Commanding Officer, or the Combatant Commander who authorized the operational use of the UCAV? Or are they collectively held responsible and accountable?"

This is the main reason why military lawyers are so concerned about robots being armed and autonomous. As long as "the man is in the loop," traditional accountability can be ensured. Breaking this restriction opens up all sorts of new and seemingly irresolvable legal questions about accountability. For this reason, the international community may well decide that armed, autonomous robots meet Peter Herby and the ICRC's fourth and last "pillar"—a weapon that is simply too difficult and abhorrent to deal with. Like chemical weapons or blinding lasers, they would be banned in general, for no other reason than the world doesn't want them around.

Yet there is nothing legally that presently says so. The law is simply silent on whether autonomous robots can be armed with lethal weapons. Even more worrisome, this concept of keeping the human in the loop is already being eroded by both policymakers and the technology itself, which are both rapidly moving toward pushing humans out of the loop. So we had better either enact a legal ban on such systems soon or start to develop some legal answers for how to deal with them.

If we do make and use autonomous robots in war, which seems to be the path we're already on, it still doesn't mean that the law still has to stay silent. Just because ICRC's fourth pillar isn't applied, the other three pillars don't disappear as well. Nations may have a right to make and use armed robots, but this right is not unlimited. To meet the old legal standards, the systems have to be able to discriminate between civilian and military targets and can't cause any unnecessary suffering. Given all the problems that robots will face in guaranteeing this, it may then mean that robots legally could be armed, but only allowed the autonomous use of nonlethal weapons. That is, that very same robot might also carry lethal weapons, but be programmed such that only a human can authorize their use.

Similarly, just as any human's right to self-defense is limited, so too should be a robot's. While military lawyers like to claim that a commander has the legal right to use "all necessary means" to defend their unit, even this right has limits in reality. For example, if a manned plane was targeted by radar and thought to be under threat, the crew can argue that they have the right to defend themselves and shoot a missile at the radar before it shoots them down first. But they can't honestly claim they have the right to drop a nuclear bomb on the radar. Context matters greatly.

Robots have great difficulty interpreting context, and, at least until they match humans in intelligence, it simply doesn't make sense to interpret a machine as having the equivalent of a human's rights of self-defense. One air force officer noted the dangerous pathway that such legal interpretations take us down: "If a robot were programmed with this rule [of self-defense], it would not hesitate to employ a hugely disproportionate weapon in the defense of its unit, including a nuclear missile that could start a global conflagration."

The same sensible limits need to be applied to the odd idea that the right of self-defense flows from the idea of property rights, as the air force is interpreting the law now. To use a parallel from civilian law, I certainly have the right to defend my house if someone invades it, but I can't preemptively shoot at them from across the street if I think they might want to rob it. Plus, when I am defending my property, there are certainly limits to how much force I can use. It is one thing to use a stun gun against someone who tries to steal my Roomba, another to shoot a Tomahawk missile at his hometown.

This last example illustrates how, sometimes, we might find useful models for military law on robotics from other realms. For example, if a robot vacuum cleaner started sucking up infants as well as dust, because of some programming error or design flaw, we can be sure that the people who made the mistakes would be held liable. That same idea of product liability can be taken from civilian law and applied over to the laws of war. While a system may be autonomous, those who created it still hold some responsibility for its actions. Given the larger stakes of war crimes, though, the punishment shouldn't be a lawsuit, but criminal prosecution. If a programmer gets an entire village blown up by mistake, the proper punishment is not a monetary fine that the firm's insurance company will end up paying. Many researchers might balk at this idea and claim it will stand in the way of their work. But as Bill Joy sensibly notes, especially when the consequences are high, "Scientists and technologists must take clear responsibility for the consequences of their discoveries." Dr. Frankenstein should not get a free pass for his monster's work, just because he has a doctorate.

The same concept could apply to unmanned systems that commit some war crime not because of manufacturer's defect, but because of some sort of misuse or failure to take proper precautions. Given the different ways that people are likely to classify robots as "beings" when it comes to expectations of rights we might grant them one day, the same concept might be flipped across to the responsibilities that come with using or owning them. For example, a dog is a living, breathing animal totally separate from a human. That doesn't mean, however, that the law is silent on the many legal questions that can arise from dogs' actions. As odd as it sounds, pet law might then be a useful resource in figuring out how to assess the accountability of autonomous systems.

The owner of a pit bull may not be in total control of exactly what the dog does or even who the dog bites. The dog's autonomy as a "being" doesn't mean, however, that we just wave our hands and act as if there is no accountability if that dog mauls a little kid. Even if the pit bull's owner was gone at the time, they still might be criminally prosecuted if the dog was abused or trained (programmed) improperly, or because the owner showed some sort of negligence in putting a dangerous dog into a situation where it was easy for kids to get harmed.

Like the dog owner, some future commander who deploys an autonomous robot may not always be in total control of their robot's every operation, but that does not necessarily break their chain of accountability. If it turns out that the commands or programs they authorized the robot to operate under somehow contributed to a violation of the laws of war or if their robot was deployed into a situation where a reasonable person could guess that harm would occur, even unintentionally, then it is proper to hold them responsible. Commanders have what is known as responsibility "by negation." Because they helped set the whole situation in process, commanders are equally responsible for what they didn't do to avoid a war crime as for what they might have done to cause it.

Other parallel safeguards might be put in place to ease these laws' application. For example, just as a car or dog must have identification tags to allow authorities to figure out who the owner is, so too should there be clear ways to track the chain of design, manufacture, ownership, and use of unmanned systems. Akin to registration and licensing, there should be some sort of formal sign-off to clearly establish who is taking responsibility at each stage of the manufacture and deployment of an autonomous system. This should extend from the designer and maker, who attest that the robot works as claimed, to the commanders and officers in the field, who, before it is deployed, take responsibility for the actions of an autonomous robot on any particular operation.

These ideas are not simply to figure out who to punish after something goes wrong with a robot. By establishing at the start who is ultimately responsible for getting things right, it might add a dose of deterrence into the system before things go wrong. If a programmer knows that they might be prosecuted for war crimes if their software code is missing a line, then they are more likely to double-check that code one more time. Or if a commander knows that they will be held accountable for whatever Megatron does when deployed into Fallujah, they might think twice about sending the system into situations where bad things are more likely to happen.

## ROBOTIC DISCUSSIONS

As political thinker Francis Fukuyama points out, "Science cannot by itself establish the ends to which it is put." Not merely scientists, but everyone from theologians (who helped create the first laws of war) to the human rights and arms control communities, must start looking at where the current technology curve is taking both our weapons and laws. These discussions and debates also need to be global, as the issues of robotics cross national lines. We may even one day see the need to set up an international body to help the world navigate the tough issues that surround robotics, much like the World Health Organization or the International Atomic Energy Agency.

As the world starts to wrestle with the issues, some sort of consensus might emerge. One could be that ICRC's fourth pillar may eventually end up being applied to autonomous robots. There are all sorts of historic parallels for how the world has changed its collective mind on the acceptability of a weapon long after it was invented, from exploding "dum-dum" bullets and chemical weapons to anti-personnel land mines. The science fiction parallels are even more instructive. In various fictional universes of the future, from the sci-fi lite of *Star Wars* to the true geekdom of the *Dune* novels, a repeated storyline is how intelligent robots were widely used in war and then later on banned.

Even if autonomous robots are not prohibited, consensus might be found on other aspects. The production and use of certain types or features of robots might be banned, such as ones not made of metal (which would be hard to detect and thus of most benefit to terrorist groups). Or we might see international agreements, along the lines of the cloning discussions, that limit robotic implants that serve no primary medical purpose.

Many will argue against having any such discussions or creating new laws that act to restrict what can be done in war and research. As Steven Metz of the Army War College says, "You have to remember that many consider international law to be a form of asymmetric warfare, limiting our choices, tying us down." Yet history tells us that, time and again, the society that builds and stands by a rule of law is the one that ultimately prevails. There is a "bottom line" reason for why we should adhere to the laws of war, explains a U.S. Air Force general. "The more society adheres to ethical norms, democratic values, and individual rights, the more successful a warfighter that society will be."

So as society begins to wrestle with the dilemmas and problems that robots present for the laws of war, the guidance of the last generation to ponder how to handle a revolutionary but also fearful new technology (in their case atomic power) might be instructive. As John F. Kennedy said in his inaugural address, "Let us never negotiate out of fear, but never fear to negotiate."

# TWENTY-ONE

# A ROBOT REVOLT?
# TALKING ABOUT ROBOT ETHICS

*DYSON: You're judging me on things I haven't even done yet. Jesus. How were we supposed to know?*
*—TERMINATOR 2: JUDGMENT DAY*

"Any machine could rebel, from a toaster to a Terminator, and so it's crucial to learn the common strengths and weaknesses of every robot enemy. Pity the fate of the ignorant when the robot masses decide to stop working and to start invading."

Daniel Wilson's fascination with robots began when he was young. "As a kid, I fell in love with *Transformers*, but all my parents could afford were crappy *Go-Bots*. Did I care? No. A robot is a robot." By the time Wilson hit middle school, "I fell in love again, this time with Vickie, a child star who played an android girl on the TV sitcom *Small Wonder*." Wilson went on to get a PhD in robotics from Carnegie Mellon University, and has worked on projects for Microsoft and Intel.

While working on his doctorate, Wilson decided to try his hand at book writing. The result was *How to Survive a Robot Uprising: Tips on Defending Yourself Against the Coming Rebellion*. Wilson's book was essentially a faux guide to robot revolt, based on real technology. It goes through all sorts of Hollywood scenarios of how robots might try to take over the Earth and then shows how a real roboticist would respond. Wilson is chock full of helpful advice for "when the robots inevitably come." He details the warning signs that one should look for to know whether your robot is planning a rebellion ("sudden lack of interest in menial labor" and "repetitive stabbing movements"), how to detect robot imposters ("Does your friend smell like a brand-new soccer ball?"), how to escape a robot chasing you (distract it by throwing decoys and obstacles in its path; "Just check twice before you toss the baby seat out of the window"), and a series of real-world technologies and tactics useful for fighting back against our future robot foes (from EMPs to radio-frequency pulse guns).

Wilson doesn't really think that your Roomba is poised to suck your breath away while you sleep. As he explains, "I believe the chance of a Hollywood-style robot uprising happening is about as likely as a Hollywood-style King Kong attack on New York City." On the other hand, "Humans are designing plenty of all-too-real robots to do things like 'neutralize enemy combatants,' or 'increase troop survivability.' Is it just me, or does that sound suspiciously like 'KILL, KILL, KILL?' "

## ROBO-FEAR

Wilson originally wrote his book to "strike back at Hollywood," mocking its many inaccurate portrayals of both robots and the people who make them. Hollywood instead ate it up, and the young roboticist ended up selling the rights to Paramount, where it is presently being turned into a Mike Myers movie. Wilson's story, though, goes from amusing to odd when he mentions in an aside that he has lectured at the U.S. Military Academy and done work for the Northrop Grumman defense firm.

Wilson's lighthearted take actually taps into a longer history of genuine fear over what our man-made creations might do to us one day. As far back as 1863, the English scholar Samuel Butler weighed in on the heated debate that Charles Darwin had opened about human evolution. In "Darwin Among the Machines," Butler argued that the scholars arguing over evolution should look forward rather than back. "Who will be man's successor? To which the answer is: We are ourselves creating our own successors. Man will become to the machine what the horse and dog are to man."

Today, the concept of machines replacing humans at the top of the food chain is not limited to stories like *The Terminator* or *Maximum Overdrive* (the Stephen King movie in which eighteen-wheeler trucks conspire to take over the world, one truck stop at a time). As military robotics expert Robert Finkelstein projects, "within 20 years" the pairing of AI and robotics will reach a point of development where a machine "matches human capabilities. You [will] have endowed it with capabilities that will allow it to outperform humans. It can't stay static. It will be more than human, different than human. It will change at a pace that humans can't match." When technology reaches this point, "the rules change," says Finkelstein. "On Monday you control it, on Tuesday it is doing things you didn't anticipate, on Wednesday, God only knows. Is it a good thing or a bad thing, who knows? It could end up causing the end of humanity, or it could end war forever."

Finkelstein is hardly the only scientist who talks so directly about robots taking over one day. Hans Moravec, director of the Robotics Institute at Carnegie Mellon University, believes that "the robots will eventually succeed us: humans clearly face extinction." Eric Drexler, the engineer behind many of the basic concepts of nanotechnology, says that "our machines are evolving faster than we are. Within a few decades they seem likely to surpass us. Unless we learn to live with them in safety, our future will likely be both exciting and short." Freeman Dyson, the distinguished physicist and mathematician who helped jump-start the field of quantum mechanics (and inspired the character of Dyson in the *Terminator* movies), states that "humanity looks to me like a magnificent beginning, but not the final word." His equally distinguished son, the science historian George Dyson, came to the same conclusion, but for different reasons. As he puts it, "In the game of life and evolution, there are three players at the table: human beings, nature and machines. I am firmly on the side of nature. But nature, I suspect, is on the side of the machines." Even inventor Ray Kurzweil of Singularity fame gives humanity "a 50 percent chance of survival." He adds, "But then, I've always been accused of being an optimist."

Scientists' fears are not merely that machines will "surpass" humans and then peacefully, logically take over, as in Asimov's *I, Robot*. Instead, many believe that future AI might have some evil intent, or even worse. Marvin Minsky, who cofounded MIT's artificial intelligence lab, believes that we humans are so bad at writing computer software that it is all but inevitable that the first true AI we create will be "leapingly, screamingly insane."

The refuseniks had concerns that the military might misuse their research. These scientists' concerns reach a whole new level. Some just accept it as an unavoidable consequence of their research that their creations will one day surpass humans and even order them about. Professor Hans Moravec observes, "Well, yeah, but I've decided that's inevitable and that it's no different from your children deciding that they don't need you. So I think that we should gracefully bow out—ha, ha, ha....But I think we can have a pretty stable, self-policing system that supports us, though there would be some machines which were outside the system, which means became wild. I think we can co-exist comfortably and live in some style for a while at least."

Others believe that we must take action now to stave off this kind of future. Bill Joy, the cofounder of Sun Microsystems, describes himself as having had an epiphany a few years ago about his role in humanity's future. "In designing software and microprocessors, I have never had the feeling I was designing an intelligent machine. The software and hardware is so fragile, and the capabilities of a machine to 'think' so clearly absent that, even as a possibility, this has always seemed very far in the future....But now, with the prospect of human-level computing power in about 30 years, a new idea suggests itself: that I may be working to create tools which will enable the construction of technology that may replace our species. How do I feel about this? Very uncomfortable."

## WHEN SHOULD WE SALUTE OUR ROBOT MASTERS?

These fears of robot rebellion go back to the Karel Čapek's very first use of the word "robot" in his play *R.U.R. (Rossum's Universal Robots)*. His choice of the word was deliberate, as he knew that the original "robotniks," the Czech serfs, had rebelled against their masters in 1848. This theme continued in science fiction, such as HAL in *2001*, which kills its human crew and decides to take over, or A.M. of the Harlan Ellison story "I Have No Mouth and I Must Scream." A.M. stood for "Allied Mastercomputer," as it was an AI designed for coordinating defenses, just like the real-world battle management systems of today. But the name Ellison gave to the computer was also a reference to Descartes' "I think, therefore I am." Once it evolves in thinking power, A.M. decides to launch a war, and torture the human survivors for sport.

A machine takeover is generally imagined as following a path of evolution to revolution. Computers eventually develop to the equivalent of human intelligence ("strong AI") and then rapidly push past any attempts at human control. Ray Kurzweil explains how this would work. "As one strong AI immediately begets many strong AIs, the latter access their own design, understand and improve it, and thereby very rapidly evolve into a yet more capable, more intelligent AI, with the cycle repeating itself indefinitely. Each cycle not only creates a more intelligent AI, but takes less time than the cycle before it as is the nature of technological evolution. The premise is that once strong AI is achieved, it will immediately become a runaway phenomenon of rapidly escalating super-intelligence." Or as the AI Agent Smith says to his human adversary in *The Matrix*, "Evolution, Morpheus, evolution, like the dinosaur. Look at that window. You had your time. The future is our world, Morpheus. The future is our time."

This evolution turns into a revolution at the point at which machine intelligence starts to act on its own, beyond its human programmers' original intent. Many see this runaway as inevitable. As military robotics pioneer Robert Finkelstein describes, "The first thing it [true AI] likely does within nanoseconds is jump into the Internet, because of the access to unlimited computing resources. We won't be able to stop it. The military will only reach a point of concern when it fails to work like we want it to. But that is too late."

Of course, many feel that the fears of a machine rebellion should stay put in the realms of humor and science fiction. Rod Brooks of iRobot, for example, says that a robot takeover "will never happen. Because there won't be any us (people) for them (pure robots) to take over from." His explanation is not merely that the idea is hogwash, but that there is also an ongoing convergence of human and machines through technologic implants and enhancements. By the time machines advance enough to reach the level of intelligence that those with fears of revolt think is necessary for the machines to want take over, people will be carrying about computers in their brains and bodies. That is, the future isn't one of machines separate from humans, plotting our demise. Rather, Brooks thinks it may instead yield a symbiosis of AI and humans. Others think that this still could yield conflict, pointing to the parallels in the *Dune* series of novels, where such "enhanced" cyborgs and strong AI fight it out, with regular old humans caught in the middle.

This debate among both scientists and science fiction will likely go on as long as robots are around, or until Skynet orders us meat puppets to shut up and get back to work. From my perspective of a security analyst, however, the only way to evaluate the actual viability of a robot revolt is to look at what exactly would be needed for machines to take over the world. Essentially, four conditions would have to be met. First, the machines would have to be independent, able to fuel, repair, and reproduce themselves without human help. Second, the machines would have to be more intelligent than humans, but have no positive human qualities (such as empathy or ethics). Third, they would, however, have to have a survival instinct, as well as some sort of interest and will to control their environment. And, fourth, humans would have to have no useful control interface into the machines' decision-making. They would have to have lost any ability to override, intervene, or even shape the machines' decisions and actions.

Each of these seems a pretty high bar to cross, at least over the short term. For example, while many factories are becoming highly automated, they all still require humans to run, support, and power them. Second, the ability of machines to reach human-level intelligence may be likely someday, even soon, but it is not certain. In turn, there is a whole field, social robotics, at work on giving thinking machines the sort of positive human qualities like empathy or ethics that would undermine this scenario, even if strong AI was achieved. Third, most of the focus on military robotics is to use robots as a replacement for human losses, the very opposite of giving them any sort of survival instinct or will to control. Fourth, with so many people spun up about the fears of a robot takeover, the idea that no one would remember to build in any fail-safes is a bit of a stretch. Finally, the whole idea of a robot takeover rests on a massive assumption: that just when the robots are ready to take over humanity, their Microsoft Windows programs won't freeze up and crash.

Of course, eventually a super-intelligent machine would figure out a way around each of these barriers. In the *Terminator* storyline, for example, the Skynet computer is able to trick or manipulate humans into doing the sorts of things it needs (for example, e-mailing false commands to military units), as well as rewrite its own software. However, Rod Brooks makes perhaps the most important point on the question of seriously evaluating the fears. If it ever does happen, humanity will likely not be caught off guard, as in the movies. You don't get machines beyond control until you first go through the step of having machines with little control. So we should have some pretty good warning signs to look out for; that is, beyond Daniel Wilson's helpful suggestion to monitor your robot for any "repetitive stabbing movements."

The whole issue of humankind losing control to machines may instead need to be looked at in another way. For all the fears of a world where robots rule with an iron fist, we already live in a world where machines rule humanity in another way. That is, the *Matrix* that surrounds us is not some future realm where evil robots look at humans as a "virus" or "cattle." Rather, we're embedded in a matrix of technology that increasingly shapes how we live, work, communicate, and now fight. We are dependent on technology that most of us don't even understand. Why would machines ever need to plot a takeover when we already can't do anything without them?

## ROBOT INSURANCE

If any place should be concerned with a robot takeover, it is the red-light district.

Few robotics firms issue press releases about their latest multimillion-dollar pornography contract. But just as pornography helped launch such common consumer products as digital cameras, instant messaging, Internet chat rooms, online purchasing, streaming video, and webcams, many experts in robotics believe that sex will drive many of the commercial advances in robotics, because, well, sex sells. On a number of occasions, I interviewed scientists about military robotic systems, who at the end would quietly ask whether I was also looking into the "robotic sex" sector. One dirty old scientist even described it as "something we all await with excitement."

Henrik Christensen, a member of the Robotics Research Network ethics group, explains the simple rationale for why he thinks the robot sex industry will take off in the next decade. "People are [already] willing to have sex with inflatable dolls, so initially anything that moves will be an improvement." Christensen raises this not because he is excited about such a prospect, but because he is concerned about whether society is prepared for the ethical dilemmas that this trend will bring. For example, should limits be placed on the appearances of such robotic systems? Christiansen believes "it is only a matter of time" before sexbots are made to look like children. "Pedophiles may argue that those robots have a therapeutic purpose, while others would argue that they only feed into a dangerous fantasy." Likewise, what happens as robots become more sophisticated, and have self-learning mechanisms built into them? Are these the sorts of "experiences" we want intelligent machines learning from, and what will be the impact on how they then behave?

Professor Ronald Arkin, a roboticist at the Georgia Institute of Technology, has been one of the few scientists to go into depth on the various ethical issues looming from robotics advancement. To him, the issues—not just in sex, but in war—revolve around one key question: What are the boundaries, if any, between human-robot relationships? This question, he explains, lays open a series of ethical concerns that must be dealt with soon, perhaps even in a moral code developed by humans but embedded in our robots. "How intimate should a relationship be with an intelligent artifact?" "What authority are we going to delegate to these machines?" "Should a robot be able to mislead or manipulate human intelligence?" "What, if any, level of force is acceptable in physically managing humans by robotic systems?" "What is the role of lethality in the deployment of autonomous systems by the military?"

But this is only to look at the issue of what robots should be programmed to do. Another ethical concern is the reverse: what humans should be allowed to do with robots. For example, what should be done with the massive amounts of data that robots will collect, data that will invariably be uploaded online and which might be used against people? Explains iRobot's Rod Brooks, "I am sure there will be new dilemmas, just as happens with every new technology. No one expected computers to bring so many concerns of privacy. Robotics will bring even more of the privacy concerns." The Los Angeles Police Department, for instance, is already planning to use drones that would circle over certain high-crime neighborhoods, recording all that happens. Other government agencies and even private companies are purchasing smaller drones able to land on windowsills and "perch and stare" at the humans inside. With all this observation, Andy Warhol's description of fame may then have to be reworked for the twenty-first century, tells IT security expert Phil Zimmerman. "In the future, we'll all have fifteen minutes of privacy." Indeed, when he talks about this aspect of the future of robotics, author Daniel Wilson goes from humorous to ominous. "That is what scares the shit out of me."

## YOU'LL HAVE TO PRY THIS ROBOT OUT OF MY COLD, DEAD HANDS

Given the depth and extent of problems that further advancement in robotics and AI might raise, from the machine-led destruction of humanity to the world learning that you are a thirty-two-year-old closet *Gilmore Girls* fanatic, many think that the best ethical answer is to stop the research altogether. As Bill Joy argues, "The only realistic alternative I see is relinquishment: to limit development of the technologies that are too dangerous, by limiting our pursuit of certain kinds of knowledge."

Proponents of relinquishment argue that it is not without precedent for people and countries to forgo researching or making certain technologies, even when they could yield great weapons. Most nations on the planet, for example, have chosen not to build nuclear weapons, even though it would offer them immense power, and all have agreed not to engage in biological weapons research anymore. Indeed, the situation with robotics and other unmanned technologies appears easier to resolve. Unlike the powerful states we faced during the Manhattan Project or the bioweapons research that took place during the cold war, argues Bill Joy, "We aren't at war, facing an implacable enemy that is threatening our civilization; we are driven, instead, by our habits, our desires, our economic system, and our competitive need to know."

Yet this ignores that many do feel we are at war with an implacable foe, and that this sense of military need is driving much of the research. Moreover, good old-fashioned human nature also would get in the way of attempts at self-restraint. "We are curious as a species," observes Dr. Miguel Nicolelis, a Brazilian scientist whose research has linked a monkey's brain to a two-hundred-pound walking robot. "That is what drives science." It is not merely that humans just can't help themselves from experimenting with technology, but that constantly pushing the envelope is the very essence of science. As Albert Einstein famously said, "If we knew what it was we were doing, it would not be called research, would it?"

Even if the world was able to come to an unlikely consensus to set up some sort of system to stop researchers from working on new technologies (like the "Turing Police," in William Gibson's novels, who hunted down anyone who worked on strong AI), there would still likely be work going on, just hidden away. There is just too much money to be made, and too many motivated actors, not just for military applications, but in everything from transportation and medicine to games and toys, to force robotics and AI research to stop anytime soon. As one analyst put it, "We would have to repeal capitalism and every visage of economic competition to stop this progression."

The challenge grows even bigger as advanced technologies migrate from the research labs to the military to the open market. As one blogger describes, "We'll be chasing our fucking tails about Lego robotics sets and the kids 'CSI' DNA testing kits they're selling at Target." AI expert Robert Epstein draws a parallel to the problem of illegal computer file downloading. While the music and movie industries have tried everything from creating new laws and launching heavy-handed lawsuits against college students to a public relations "shaming" campaign, people still keep on downloading pirated music, movie clips, and TV shows. "No matter what we do, there will always be something happening outside of that. And it will be huge."

## PLAN FOR SUCCESS (AND FAILURE)

"You can't say it's not part of your plan that these things happened, because it's part of your de facto plan. It's the thing that's happening because you have no plan.... We own these tragedies. We might as well have intended for them to occur."

William McDonough was writing about environmental issues, but his statement is frequently cited by those concerned about the future of the robotics field. It perfectly captures that while relinquishment may not be an option, there is no excuse for failing to plan ahead. As nanotech expert Eric Drexler puts it, "We've got to be pro-active, not just reactive."

In facing this, "There are two levels of priority," tells Gianmarco Verruggio, of the Institute of Intelligent Systems for Automation. "We have to manage the ethics of the scientists making the robots and the artificial ethics inside the robots."

On the human side, "managing" the ethics is hindered by the absence of professional codes or traditions that robotics scientists might look to when trying to figure out the ethical solution to a difficult science problem. Almost no technical schools require any sort of ethics classes and the robotics field certainly has nothing equivalent to the medical profession's Hippocratic oath. Even worse, the sort of twenty-first-century questions that people working with robots and AI care about aren't really dealt with in the broader fields of philosophy or ethics. There are few experts or resources for them to turn to, let alone any sort of consensus.

Even if you were an inventor, funder, or developer who wanted to do the moral thing, as Nick Bostrom, director of the Future of Humanity Institute at Oxford University, explains, you wouldn't have any ready guides. "Ethicists have written at length about war, the environment, our duties towards the developing world; about doctor-patient relationships, euthanasia, and abortion; about the fairness of social redistribution, race and gender relations, civil rights, and many other things. Arguably, nothing humans do has such profound and wide-ranging consequences as technological revolutions. Technological revolutions can change the human condition and affect the lives of billions. Their consequences can be felt for hundreds if not thousands of years. Yet, on this topic, moral philosophers have had precious little to say."

For many, the obvious guide would be to follow science fiction and simply mandate that all systems obey Isaac Asimov's "Three Laws of Robotics." Asimov's laws initially entailed three guidelines for machines. Law One is that "a robot may not injure a human being or, through inaction, allow a human being to come to harm." Law Two states that "a robot must obey orders given to it by human beings except where such orders would conflict with the First Law." And Law Three mandates that "a robot must protect its own existence, as long as such protection does not conflict with the First or Second Law." In later stories Asimov added the "Zeroth Law," above all the others. This states that "a robot may not harm humanity, or, by inaction, allow humanity to come to harm."

There are only three problems with these laws. The first is that they are fiction. They are a plot device that Asimov made up to help drive his stories. Indeed, his tales almost always revolved around robots' following the laws but then going astray and the unintended consequences that result. An advertisement for the 2004 movie adaptation of Asimov's famous book *I, Robot* put it best: "Rules were made to be broken."

For example, in one of Asimov's stories, robots are made to follow the laws, but they are given a certain meaning of "human." Prefiguring what now goes on in real-world ethnic cleansing campaigns, the robots only recognize people of a certain group as "human." They follow the laws, but still carry out genocide.

The second problem is that no technology can yet replicate Asimov's laws inside a machine. As Rodney Brooks puts it, "People ask me about whether our robots follow Asimov's laws. There is a simple reason [they don't]. I can't build Asimov's laws in them." Daniel Wilson is a bit more florid. "Asimov's rules are neat, but they are also bullshit. For example, they are in English. How the heck do you program that?"

Finally, much of the funding for robotics research comes from the military. It explicitly wants robots that can kill, won't take orders from just any human, and don't care about their own lives. So much for Laws One, Two, and Three.

While there is no Asimov-like code embedded in robots yet, it doesn't mean that many in the field believe that the design of robots should take place without some sort of guidelines. Indeed, just as they would for any other consumer product, Japan's Ministry of Trade and Industry has set up a series of rules for the design of office and home robots. Every robot must have sensors that prevent it from colliding with humans by accident, be made of softer materials at contact points, and have an emergency shutoff button. These rules came about only after Japanese authorities "realized during a robot exhibition that there are safety implications when people don't just look at robots but actually mingle with them."

The rules in Japan parallel a growing concern among robot makers about the financial costs that would come from a robot screwing up. As one executive put it, "You don't want to tell your management 'We had a bad day yesterday; our system killed four civilians by accident.' " Thus, the most powerful incentives for building precautions into robot designs are now mainly coming from the marketplace. "There is a lot of push to make these things damn safe," says Rod Brooks. He goes on to detail the three different sensors put in his company's Roomba vacuum cleaner to make sure it doesn't fall down the stairs. "If you have a multipound robot crashing down the stairs, it can get pretty bad... and not just for the robot."

While it is good that businesses are starting to think this way, it is certainly not enough. Any ethical codes and safeguards that come mainly from the fear of lawyers and lawsuits are certainly not going to be sufficient for civilian robots, let alone for robots in war. As science writer Robert Sawyer puts it, "Businesses are notoriously uninterested in fundamental safeguards—especially philosophic ones. A few quick examples: the tobacco industry, the automotive industry, the nuclear industry. Not one of these has said from the outset that fundamental safeguards are necessary, every one of them has resisted externally imposed safeguards, and none have accepted an absolute edict against ever causing harm to humans."

With this huge gap in ethics, there is a growing sense in the robotics field that scientists will soon have to start to weigh the implications of their work and take seriously their moral responsibilities, particularly as their inventions shape humanity's future. As Bill Joy puts it, "We can't simply do our science and not worry about these ethical issues."

## ROBOT RULES

In the TV comedy *The Office*, the witless character Dwight Schrute describes the perfect design for a robot. "I gave him a six-foot extension cord, so he can't chase us."

In building real-world safeguards for robots, something a bit more complex will be needed. Many roboticists describe the need for an ethic of "design ahead," which tries to take into account all the various problems that might arise, and set up systems and controls to avoid them.

There are a number of useful starting points for this design ethic. One is that the design of robots should be as predictable as possible (perhaps contrary to the growing interest in evolutionary designs). As Daniel Wilson puts it, "There is no sense in having any dangerous features... unless you want them." That is, the system should work the way it was originally designed to, all the time, rather than being able to change itself over time into something new, unexpected, and thus potentially dangerous.

With machine autonomy growing, mechanisms that ensure a human can take control and shut down a robot must also be built in. But contrary to Asimov's original laws, which entail that any humans must be able to order about any robot they meet, the controls of real-world robots should be designed to ensure limits on their masters. In a world of hackers and the like, we should aim not merely for control, but also security, so that robots can't be easily hijacked or reprogrammed for wrongful or illegal use.

Wherever possible, multiple redundancies should be built into any systems. "Redundancy can bring an exponential explosion of security," says Eric Drexler. He explains with an illustration. Imagine a suspension bridge, like the Golden Gate Bridge, which needs five cables to stay up. Each cable has an average risk of breaking one day per 365 days out of the year. If the designers of the bridge use one extra cable as backup (so, six total), the bridge would be expected to last ten years. If they add just five cables as insurance (ten total), enough of the bridge's cables shouldn't break in a million years. A little insurance goes a long way.

Scientists are also starting to recognize that information itself carries risks. This goes for the design of systems (anything that might be dangerous should not be open-source, where anyone could potentially copy, build, and misuse it), as well as whatever information is collected by the systems (data should not be publicly sharable unless there is a compelling need). In turn, there must be some required mechanism that allows information on the robot's activity to be stored and collected by public authorities. That is, the only way to ensure accountability if something goes wrong with a system is for each and every robot to have a unique identifier, even something as simple as a bar code, as well as traceability to track the actions that the system took.

In the long long term, some scientists even hope that an amended form of Asimov's behavior rules might be required in robots' software. This would mean robot makers have to look at design in a whole new way, not reactively trying to avoid lawsuits, but proactively trying to build in greater respect for the law and ethics. Georgia Tech's Ronald Arkin, for example, writes that autonomous systems in future wars might be endowed "with a 'conscience' that would reflect the rules of engagement, battlefield protocols such as the Geneva Conventions, and other doctrinal aspects that would perhaps make them more 'humane' soldiers than humans." Of course, while a machine may be guided by ethical rules, this does not make it an ethical being. Software codes are not a moral code; zeros and ones have no underlying moral meaning.

The key in all this is that ethics apply not just to the machines but also to the people behind them. Scientists must start to conduct themselves by something equivalent to the guidelines that the ancient Greek physician Hippocrates laid out for future generations of doctors. "Make a habit of two things—to help, or at least to do no harm." Martin Rees, royal astronomer of the United Kingdom (a position that is like the top science adviser to the queen), calls for the implementation of the "precautionary principle." It isn't that scientists should stop their research work altogether if anything bad might happen, but rather that they must start to make a good-faith effort to prevent the potential bad effects that might come from their inventions.

These kinds of guidelines won't arise overnight, but many scientists note that there already are models of how they might come about in high-tech fields. In the 1980s, for example, there was huge consternation over the Human Genome Project. Geneticists knew that their research could save literally millions of lives, but they also began to worry about all the various ethical and legal questions that the increased availability of genetic information would cause. Who "owned" the genes? What could be patented or not? How much of the information should be shared with the government, police, insurance companies, and other institutions?

The geneticists knew they didn't have the answers to such thorny questions, and that they should not try to answer them on their own. So they took the interesting step of setting aside 5 percent of the project's annual budget for a multidisciplinary program to "define and deal with the ethical, legal and social implications raised by this brave new world of genetics."

The world of genetics began this program at the very start of their research, which means that it is now years ahead of the world of robotics in the depth of its ethical discussions. This gap becomes even more unsettling with robotics' growing use in war. As General Omar Bradley once said, we have given ourselves the destructive power of "giants," while remaining "ethical infants." "We know more about war than we know about peace, more about killing than we know about living."

Just as with the new laws of war, the research and discussion on the ethics of robotics and roboticists can't be limited to the scientists. "We have reached a point where technology development can no longer flourish in a policy vacuum," describes analyst Neal Pollard. Scientists often fail to consider the policy ramifications of their research. In turn, says one research center director, scientists "don't have a seat at the table" when scientific issues are discussed in the halls of power in Washington.

If the dialogue between the policy world and science doesn't occur, a double whammy results. The good prescriptions that might come out of the scientific world are unlikely to go anywhere without political support. In turn, an uninformed political world might take decisions that could make things worse.

One answer may be to require new unmanned systems to have a "human impact assessment" before they enter production, analogous to the environmental impact assessments now required of new construction projects. This will not only embed a formal reporting mechanism into the policy process of building and buying unmanned systems, but also force the tough legal, social, and ethical questions to be asked early on. Indeed, if we are concerned enough about the spotted owl to require studies about potential environmental harms before some highway is built, we should be equally concerned enough about humanity to require the same sort of tough questions to be asked before some killing machine is built.

Ultimately, government is both of and for the people. The burden of weighing the ethical issues of our new technologies is shared not just by researchers and policymakers, but also by the wider public. Too often, when issues of robot ethics are raised, it comes across as science fiction, and is all too ripe for the kind of mocking that Daniel Wilson did so well. Indeed, this is perhaps why so many roboticists avoid talking about the issue altogether. That is an ethical shame.

Robots may not be poised to revolt, but robotic technologies and the ethical questions they raise are all too real. For scientists, policymakers, and the rest of us to ignore the issue only sets us up for a terrible fall down the line. As military robotics expert Robert Finkelstein explains, many may want to "think that the technology is so far in the future that we'll all be dead. But to think that way is to be brain dead now."

# TWENTY-TWO

# CONCLUSION:
# THE DUALITY OF ROBOTS AND HUMANS

*For every complex problem there is an answer that is clear, simple, and
wrong.*

—H. L. MENCKEN

In 2003, the American Film Institute assembled a jury of experts to name Hollywood's hundred greatest villains and heroes of all time. For the purposes of the vote, a "villain" was defined as a character whose "wickedness of mind, selfishness of character and will to power are sometimes masked by beauty and nobility, while others may rage unmasked. They can be horribly evil or grandiosely funny, but are ultimately tragic." A "hero" was defined as a character "who prevails in extreme circumstances and dramatizes a sense of morality, courage and purpose. Though they may be ambiguous or flawed, they often sacrifice themselves to show humanity at its best."

Hannibal Lecter, the cunning serial killer from *The Silence of the Lambs*, and Atticus Finch, the principled attorney and father in *To Kill a Mockingbird*, headed the lists of villains and heroes. But a single character ended up on both lists. And it was a robot, Arnold Schwarzenegger's portrayal of the Terminator.

The same year that Hollywood put together its team of experts, the U.S. government did the same. The National Science Foundation assembled hundreds of scientists to try to examine what would happen over the next ten to twenty years, as everything from robotics and artificial intelligence to nanotechnology and bioscience continued to advance and converge, intertwining and feeding off one another. The product of their efforts was an immense report, weighing over three pounds. They did a masterful job, exploring the impact of these developments on fields that ranged from national security to kindergarten education. And yet, ultimately, the top minds in the U.S. government could conclude that the only thing we could be certain of was uncertainty itself. "This will be an Age of Transitions, and it will last for at least a half-century."

It is this sense of duality and uncertainty that perhaps best captures how we may have to ultimately weigh what is going on in war and politics. Revolutionary new technologies are not only being introduced to war, but used in ever greater numbers, with novel and often unexpected effects. That said, everything that seems so futuristic is playing out in a present that follows familiar historical lines. Robots are doing amazing things in Iraq and Afghanistan, and yet, as I sat down to write this sentence, the news carried the story of five American troops tragically killed by a roadside bomb.

This sense of simultaneous change and stasis is nothing new. As obvious as a great change often seems after the fact, it rarely happens in one fell swoop, where you see a complete elimination of the old. The battleship, for example, went from being the dominant beast in the jungle of war to an endangered species in the course of the first few minutes of Pearl Harbor. And yet battleships stayed in naval service for another fifty years (the last ones firing their big guns during the 1991 Gulf War, directed where to shoot by unmanned drones). Atomic weapons had an unmistakable debut, a mushroom cloud that helped end the very same world war. Yet their real impact was that which played out over the following decades, driving a heated, global competition between two superpowers, but also making sure that war stayed "cold."

Change is also hard to tease out if you just look at the numbers. For instance, any outside observer could tell that tanks clearly were somewhat important when they swept across France in the German blitzkrieg that began World War II. And yet only 10 percent of the German army's units had converted to armor, meaning that this revolutionary new force still had, in historian Max Boot's words, "more ponies than panzers."

Given this uncertainty, how do we really know whether any new technology does matter, that it really is changing things? More to the point, how do we know that is the case with robotics?

The answer is simple. From little EOD #129 "dying" on the battlefield of Iraq to the all too real questions now looming in machine ethics, the revolution in robotics is forcing us to reexamine what is possible, probable, and proper in war and politics. It is forcing us to reshape, reevaluate, and reconsider what we thought we knew before. That is the essence of revolution.

Our very vocabulary illustrates. Right now, we refer to these systems as "unmanned" or "artificial," calling them by what they are not, akin to how cars were once called "horseless carriages." This is not only because we can't yet conceptualize exactly what these technologies are and what they can do. It is also because their nonhumanity sums up their difference from all previous weapons. It is why their effect on war and politics is beginning to play out in such a new and revolutionary manner.

Because they are not human, these new technologies are being used in ways that were impossible before. Because they are not human, these new technologies have capabilities that were impossible before. And, because they are not human, these new technologies are creating new dilemmas and problems, as well as complicating old ones, in a manner and pace that was impossible before.

Robots in Iraq and Afghanistan today are sketching out the contours of what bodes to be a historic revolution in warfare. The wars of the future will feature robots of a wide variety of sizes, designs, capabilities, autonomy, and intelligence. The plans, strategies, and tactics used in these future conflicts will be built from new doctrines that are just now being created, potentially involving everything from robotic motherships and swarms of autonomous drones to cubicle warriors managing war from a distance. The forces that fight these wars may well represent both governments and nonstate groups or even crazed individuals bearing a lethality once held by nations. In these battles, machines will take on greater roles, not just in executing missions, but also in planning them.

In turn, the humans still fighting will reflect changed demographics, often not matching our traditional assumptions of who we have thought of as soldiers over the last five thousand years of war. They will be younger, older, trained differently, use different equipment, fight from new locales, and even have altered concepts of their own identities and roles in war. For many, their experiences of battle will be fundamentally different from those of every soldier who went to war in every generation past. The relationships that these combatants will have with their leaders and even with each other will also be altered.

The public back home will be further distanced from the human costs of war, perhaps making such wars easier to start, but maybe also harder to end, even in democracies. In turn, the very technology itself might lead to new social, economic, even religious conflicts and maybe even create new sparks of war among those either left behind or so fearful as to lash out in anger and confusion.

Finally, these wars will feature new questions about what is legal and ethical, including even how to control our own weapons. The resulting dilemmas and debates will not only be intense, but will challenge many of the codes that have long shaped and regulated the very practice of war.

In short, the systems and stories captured in this book are just the start of a process that will be of historic importance to the story of humanity itself. Our robotic creations are creating new dimensions and dynamics for our human wars and politics that we are only now just beginning to fathom.

## ROBOTIC HOPES AND FEARS

Many, including nearly every roboticist I met while writing this book, hope that these new technologies will finally end our species' bent toward war. Indeed, even that very sober and lengthy U.S. government report about the future of technology and society expressed a similar optimism, describing how "the twenty-first century could end in world peace, universal prosperity, and evolution to a higher level of compassion and accomplishment."

Then again, it's hard to imagine us getting rid of conflict anytime soon. And indeed, as we learn about the new temptations, questions, confusions, and even anger that our new technologies might spark, there could be even more war and deadlier conflict. As Bertrand Russell once said, "Without more kindliness in the world, technological power would mainly serve to increase men's capacity to inflict harm on one another." Notably enough, Russell said this back in 1924, and the events of the last century, our most technologically advanced as well as violent one, bear him out.

The fear among soldiers is the very opposite of the scientists' hope. They worry that war is disappearing. Let me be clear here: Theirs is not some selfish sense that these new technologies will somehow end violent conflict and they'll be tossed out of work (most would gladly trade their military fatigues for a Dairy Queen uniform, if it meant the end of war and suffering). Rather, they often express fears that the unmanned planes, robot guns, and AI battle managers are turning their experience of war into something else altogether. Lives may be saved in unmanned warfare, but war itself is becoming almost unrecognizable, something they are not all that comfortable with.

From Homer's Achilles to Shakespeare's Henry V to my grandfather in World War II, war and the life of the warrior was never simply just about killing. Rather, it was the ideals that lay behind an accompanying sense of sacrifice, the acceptance that one might also have to die. Indeed, military historian Martin van Creveld argued that this willingness to sacrifice "represents the single most important factor" in modern war. "War does not begin when some people kill others; instead, it starts at the point where they themselves risk being killed in return."

This willingness to bear the most horrible burdens, face the most terrible risks, and even make the ultimate sacrifice, for your nation or just for your buddies, has always made war defy the normal rules of logic. All the great writers on war focus on this aspect because it gives war its humanity, its sense of purpose, and its heroism. From the knights' codes of chivalry to today's goals of ending tyranny or terrorism, war has always had to be linked to some ideal. Of course, these ideals weren't always followed and were rife with double standards. But they influenced how war itself was viewed, as something terrible but always linked to a higher purpose. Without these ideals, war's often horrific costs and sacrifices would be deemed unworthy.

Robotics starts to take these ideals, so essential to the definition of war, out of the equation. In so doing, they might just make the way we have framed war, and rationalized the killing that takes place in it, fall apart. Paul Kahn, a professor at Yale Law School, describes it as the "paradox of riskless warfare. It was tough enough to describe war as something permissible or moral, when combat involves deliberately choosing to inflict destruction and mayhem on your fellow humans. Yet, as long as there was a sense of mutuality, that the two sides were both accepting to bear the risks involved, there was some sense of equality and fairness."

As technologies have distanced soldiers farther and farther from the fighting, the risks, and the destruction, this sense of equality and fairness becomes harder to claim. When it becomes not just a matter of distance, but actual disconnection, as Kahn describes, it "propels us beyond the ethics of warfare."

This doesn't mean, then, that any war is instantly evil, immoral, or purposeless. And it certainly doesn't mean that there will be no more wars. Rather, wars using these new technologies are looking less like war as we once knew and understood it. The old definitions and codes don't fit so well with the realities brought on by our new technologies of killing.

A parallel is what happened to the old codes of chivalry, such as those of the Japanese samurai. Having a system of understanding and ethics that had lasted over a millennium, the samurai held back from using guns for as long as they could. They knew that the technology was useful, but they saw these new weapons as depersonalizing war and, in so doing, dishonoring the codes and values that defined them as warriors. And yet, eventually, the pace of history and technology moved on, and Japanese society soon had to adjust to the new ways of both the world and war (despite the best efforts of Tom Cruise in *The Last Samurai*). They started using guns instead of swords, and redefined both what they viewed as war and warriors' values.

The same redefinition may well happen with unmanned systems. If we are lucky, these new technologies might even redefine our sense of the acceptable human costs of war. In 1424, Machiavelli said that the side which lost the battle of Zagorara had suffered "a defeat renowned throughout all Italy." He was describing war in the time of the condottieri (hired armies), however, in which battles involved more maneuvering and posturing than actual fighting. So in this "renowned" defeat that no one now remembers, only three men perished, not from actual combat wounds, but from accidentally falling off their horses.

Maybe the shift to unmanned systems, which bear the risks of war instead and move humans out of harm's way, will lead to a similar change in how we redefine war and its human costs. But history provides counterexamples. The Japanese redefined their warrior code in the late 1800s, but they turned it into a chauvinist militarism, which just a few decades later bore such bitter fruit as the Rape of Nanking, Pearl Harbor, and the horrors of the Bataan Death March. In turn, the condottieri of Machiavelli's time may have valued their soldiers' lives so much that they were unwilling to risk them in open battle with each other, but they had no such qualms in their definition of war about killing, raping, or pillaging any civilians they came across. A similar concern pops up time and again with robotics. In making war less human, we may also be making it less humane.

## FINAGLE'S LAW

My own worry is a bit different. I believe strongly in an adage that riffs on the better-known "Murphy's law." What science fiction calls "Finagle's law" states a simple truth that I have come to believe holds for both humans and their creations, including their robots: "Anything that can go wrong, will—at the worst possible moment."

I have experienced Finagle's law again and again. The best example from my personal life may have been when the tuxedo store delivered a shirt sized for a four-year-old boy on the day before my wedding. But I also saw it at play time and again during the research for this book. The best example of this may have been when I was touring the U.S. Air Force's Middle East command center and the electricity went out. Even worse, the backup power generator didn't come on because, at that very moment, a breaker wasn't working. In the most high-tech military facility in the world, from which all unmanned operations in Iraq and Afghanistan are coordinated, airmen were finding their way around with flashlights as they rushed to turn off the computers before their batteries died.

Finagle's law is important, as we are experiencing one of the most amazing changes in humanity's history, and yet we are completely unprepared. This is nothing new. Our forebears were likely just as unprepared for such groundbreaking new technologies as fire, printing presses, machine guns, and Pudding Pops. But they muddled through and figured it all out. That is, until some new technology came along and shook everything up again.

But today, in our overcrowded, interconnected world, the stakes are far higher than they've ever been before. Even the most ardent supporters of robotics, AI, and the Singularity warn that we have to "get it right the first time," as there is little room for "non-recoverable errors."

Unfortunately, we should expect errors. It is not merely Finagle's law at work in our machines. Mistakes are not just in robot nature, but also in human nature. As Albert Einstein advised, "Never attribute to malice that which can be adequately explained by stupidity. Only two things are infinite, the universe and human stupidity, and I'm not sure about the former."

Compounding the challenge is the fact that we have less time to react and adjust to these immense changes than ever before. Our concern shouldn't be merely change itself. Rather, as Admiral Michael Mullen, the chairman of the Joint Chiefs of Staff, put it, "What has gripped me the most is the pace of all this change."

In the blink of an eye, things that were just fodder for science fiction are creeping, crawling, flying, swimming, and shooting on today's battlefields. And these machines are just the first generation of these new technologies, some of which may already be antiquated as you read these lines.

One army officer captured well what happens when you combine an incredible pace of change with a lack of preparation: "We will only be able to react, and by the time we have responded we will be even further behind the next wave of change and very quickly left in the dust of accelerating change.... Change is coming, it is coming faster than nearly everyone expects, and nothing can be done to stop it."

## CREATIVITY CONCERNS

This all heightens the need to start discussing the issues that come as unmanned technologies are increasingly used in our society and, even more so, in our wars.

Part of the reason is to take some of the shock and sting out of these transitions, which will feel overwhelming to many, and might even spur some to violence. As terrorism expert Richard Clarke explained, "We need to have discussion of issues before they are on us. Violence comes if people feel surprises." Or as the aptly named band Army of Me put it in their aptly named song "Going Through Changes": "It's hard to accept what you don't understand, And it's hard to launch without knowing how to land."

But most of all, the reason for having these discussions is that our scientists, our military, and our political and business leaders are making decisions now that will matter for all of human society in decades to come. We are not merely building machines that will be with us for years, but setting in motion potentially irreversible research and development on what these machines can and cannot do. Even more so, we are just now creating the frameworks that will fill the current vacuum of policy, law, doctrine, and ethics. That is, how we frame an issue now will shape how we will understand and respond to the challenges that will pop up years from now.

Yet for the most part, we are deciding such important matters from a position of ignorance. "Ignorance" actually has two meanings. The first is the one we tend to think of, "the state of being uninformed." But it can also mean the "willful neglect or refusal to acquire knowledge which one may acquire and it is his duty to have." This latter definition may be more apt when it comes to robotics today. We fund, research, and use these new technologies more and more, especially in war. Yet we willfully refuse to acknowledge that the reality of robotics is now upon us. "We are already in *A Brave New World,* but just don't want to admit it," says one military consultant on unmanned systems. "We refuse to take our blinders off."

The result is that leaders are ill equipped to handle all the emerging complications and dilemmas. Tells one worker in the military's robotics test programs, "The people higher up, who are making the decisions that matter, do not have a good understanding of this technology. They are older and more mature, but they don't get it. People fear what they don't know."

Most of all, we have to start questioning into what exactly we want to invest our society's collective intellect, energy, drive, and resources. These are exciting, thrilling times, but I cannot think about them without a bit of disappointment. There is an inherent sadness in the fact that war remains one of those things that humankind is especially good at. As Eisenhower once said, "Every gun that is made, every warship launched, every rocket fired, signifies in the final sense a theft from those who hunger and are not fed, those who are cold and are not clothed. The world in arms is not spending money alone. It is spending the sweat of its laborers, the genius of its scientists and the hopes of its children."

Humans have long been distinguished from other animals by our ability to create. Our distant ancestors learned how to tame the wild, reach the top of the food chain, and build civilization. Our more recent forebears figured out how to crack the codes of science, and even escape the bonds of gravity, taking our species beyond our home planet. Through our art, literature, poetry, music, architecture, and culture, we have fashioned awe-inspiring ways to express ourselves and our love for one another.

And now we are creating something exciting and new, a technology that might just transform humans' role in their world, perhaps even create a new species. But this revolution is mainly driven by our inability to move beyond the conflicts that have shaped human history from the very start. Sadly, our machines may not be the only thing wired for war.

## [ACKNOWLEDGMENTS]

No one could write a book like this without an immense amount of help. Ralph Wipfli of Brookings and Elina Noor of ISIS-Malaysia tirelessly assisted in gathering the immense amount of research that made the project possible. They are two of the smartest young political analysts out there, so I am sure many a time they wondered just how they had ended up researching robots and editing chapters that name-dropped Paris Hilton. The Brookings Institution sponsored and hosted the project through the Stephen and Barbara Friedman Fellowship, the Sydney J. Stein Jr. Chair, and the President's Special Initiative Fund. Brookings is the world's first "think tank," and thus is perhaps the only place confident enough in the rigor of its research to take risks on such a nontraditional but important project. Strobe Talbott, Carlos Pascual, and my colleagues in the 21st Century Defense Initiative have my deepest thanks for providing an atmosphere in which truly revolutionary scholarship can thrive.

Various organizations and conferences kindly hosted me along the way, including several military bases and operations that must remain unidentified, but still have my gratitude. But some I can thank include the Association for Unmanned Vehicle Systems International, Foster-Miller, the Harry Frank Guggenheim Foundation, the Institute for Defense and Government Advancement, the Industrial College of the Armed Forces, iRobot, Marine Corps University, the Air Force Institute of Technology, National Defense University, and the Office of Naval Research.

I am in deep appreciation of the scores of interviewees who took time out of their busy schedules to talk with me about their work and share their perspectives. They had far better things to do, be it inventing robots or fighting wars, and I appreciate their generosity. I only regret that I could not tell all their stories. A special thanks goes to Noah Shachtman, who hosts *Danger Room*, the best site for data and discussion on defense technology on the entire Internet, and someone with a wicked sense of humor that infected the research. Dan Mandel of Sanford J. Greenburger Associates shepherded the project to publishers and found it a home. Eamon Dolan showed immense faith and patience in a young author and guided it to completion with thoughtful edits and comments. And, finally, the reviewers and academic referees provided many useful suggestions that turned inferior early drafts into what you see here. If you like the book, all credit goes to them. If you do not, all bile and corrections go to me.

Finally, I wish to thank the people who bought me the toys and gadgets, gave me the books, took me to the movies, told me the stories, and shared the times, for better or for worse, that shaped me as a person and this book as a whole. More recently, they suffered through countless oddball stories of robots or war, as if either was an appropriate conversation topic at dinners, parties, and even on vacations. Who could ask for more than good friends and family?

# [NOTES]

## AUTHOR'S NOTE: WHY A BOOK ON ROBOTS AND WAR?

1   *Because robots are frakin' cool* Frak is a made-up expletive that originated in the computer science research world. It then made its way into video gaming, ultimately becoming the title of a game designed for the BBC Micro and the Commodore 64. The main character, a caveman called Trogg, would say "Frak!" in a little speech bubble whenever he was "killed." It soon spread into science fiction, appearing in such games as Cyberpunk 2020 the Warhammer 40,000 novels, and the TV series *Battlestar Galactica*. It crossed over into the mainstream when it was used in the new 2003 reboot of *Battlestar Galactica*. That the characters in the updated version of the TV show frequently cursed, albeit with a made-up word, was part of the grimier, more serious feel of the show. Soon thereafter, the word was used in such popular teen shows as *The OC* and *Veronica Mars*, completing its crossover to pop culture.

1   *"It is not a phrase to be written"* John Keegan, *Six Armies at Normandy* (New York: Random House, 2004), 30.

5   *From war sprung the very first specializations* Elizabeth Arkush and Mark W. Allen, *The Archaeology of Warfare: Prehistories of Raiding and Conquest* (Gainesville: University Press of Florida, 2006).

5   *"War is a sign of disobedience"* Jean Martensen, Director for Peace Education for the Evangelical Lutheran Church, as quoted in David R. Smock, ed., *Religious Perspectives on War: Christian, Muslim, and Jewish Attitudes Toward Force*, Perspectives Series (Washington, DC: United States Institute of Peace Press, 2002), 42.

5   *It is our arrogance chastised* Christopher Coker, *Waging War Without Warriors? The Changing Culture of Military Conflict*, IISS Studies in International Security (Boulder, CO: Lynne Rienner Publishers, 2002), 20.

5   *we sure do seem to be obsessed with war* John Keegan, *The Face of Battle* (New York: Viking Press, 1976), x.

6   *"fighting is where man will win glory"* Coker, *Waging War Without Warriors?* as quoted on 33.

6   *war is described as a test* Ibid., 21–32.

6   *a cruel teacher who reveals* Ibid., 30.

6   *Democracy came from the phalanx* Ibid., 24.

6   *"to end all wars"* Ibid., 30, 99.

6   *all find their definitive expressions* Christopher Coker, *Humane Warfare* (London, New York: Routledge, 2001).

6   *Yet the reality is "ever again"* Yael Danieli, "What Determines How Social Scientists and Psychologists Try to Understand the Next War," paper presented at Imagining the Next War, Guggenheim Conference, New York City, March 25, 2006.

7   *"The Future Ain't What It Used to Be"* Ray Kurzweil, *The Singularity Is Near: When Humans Transcend Biology* (New York: Viking, 2005).

7   *"producing more history"* Vago Muradian, "Interview with John Hillen, Assistant U.S. Secretary of State, Political-Military Affairs," *Defense News*, October 9, 2006, 110.

7   *"As I look at the trends"* Bill Gates, "A Robot in Every Home," ScientificAmerican.com, December 16, 2006 (cited December 17, 2006); at http://www.sciam.com/article.cfm?-a-robot-in-every-home.

8   *there were more robots* "iRobot Co-Founder Comes Clean: Roomba Vacuum Cleaner a Worldwide Success," CNN.com, May 30, 2005 (cited May 30, 2005); available at http://www.cnn.com/2005/TECH/ptech/05/30/techprofile.irobot.ap/index.html.

8   *Another 7 million more* Gates, "A Robot in Every Home."

8   *a robot in every home by 2013* Ibid.

8   *One industry leader projects* "The Robots Are Coming!" Gizmag.com, January 19, 2004 (cited July 6, 2005); available at http://www.gizmag.com/go/2801.

8   *assembly-line factory robotics* "Born Again Robots," *Fortune Small Business*, October 2007, 57.

8    *Roughly one of every ten workers* Gates, "A Robot in Every Home."

8    *"the electronics industry is on the cusp"* Ian Rowley et al., "Ready to Buy a Home Robot?" *Business-Week*, July 19, 2004; available at http://www.businessweek.com/magazine/content/04_29/b3892141_mz070.htm.

8    *"A Robotics Gold Mine"* "A Robotics Gold Mine," *BusinessWeek*, July 19, 2004; available at http://www.businessweek.com/magazine/content/04_29/b3892145_mz070.htm.

9    *"the flying machine"* Sean Price, "When Man Took to the Skies: One Hundred Years Ago This Month, in Kitty Hawk, N.C., the Wright Brothers Gave the World Powered Flight" (Scholastic, Inc., 2005), http://www.thefreelibrary.com/When+man+took+to+the+skies:+one+hundred+years+ago+this+month,+in...-a0112585040 (accessed August 13, 2006).

10   *"Genetics, nanotechnology, and robotics"* Steve Martini, *The Jury* (New York: Putnam's, 2001), 112.

11   *"science fiction and futurism"* Michael E. O'Hanlon, *Technological Change and the Future of Warfare* (Washington, DC: Brookings Institution Press, 2000), 33.

11   *"The true watersheds"* Daniel Boorstin, "History's Hidden Turning Points," *U.S. News & World Report*, April 22, 1991, 52.

13   *"These robots are extensions of us"* Lee Gutkind, *Almost Human: Making Robots Think*, 1st ed. (New York: W.W. Norton & Co., 2006), 32.

14   *Both groups tend to disregard* K. Eric Drexler, *Engines of Creation*, 1st ed. (Garden City, NY: Anchor Press/Doubleday, 1986).

14   *these three problems can be diminished* Ibid.

14   *but also much continuity* Stephen D. Biddle, *Military Power: Explaining Victory and Defeat in Modern Battle* (Princeton, NJ: Princeton University Press, 2004).

## 1. INTRODUCTION: SCENES FROM A ROBOT WAR

19   *"We are building the bridge to the future"* U.S. Army colonel, interview, Peter W. Singer, November 16, 2006.

19   *averaging nearly 2,500 a month* Greg Grant, "U.S. Army: Active Protection Not Needed in Iraq," *Defense News*, September 25, 2006, 30. See also Michael O'Hanlon and Jason H. Campbell, "The Iraq Index: Tracking Reconstruction and Security in Post-Saddam Iraq" (Washington, DC: Brookings Institution, 2008); available at http://www.brookings.edu/iraqindex.

20   *"IEDs are my number one threat"* General John Philip Abizaid, Commander CENTCOM, as quoted in Thomas E. Ricks, *Fiasco: The American Military Adventure in Iraq* (New York: Penguin Press, 2006), 219.

20   *spending more than $6.1 billion* Whitney Terrell, "The Bomb Squad," *Washington Post*, October 29, 2006, W20.

20   *"one of the most important assignments"* Noah Shachtman, "The Baghdad Bomb Squad," *Wired* 13.11 (2005); http://www.wired.com/wired/archive/13.11/bomb.html.

20   *In a typical tour in Iraq* U.S. Navy EOD technician, interview, Peter W. Singer, Washington, DC, May 26, 2006.

20   *the pressure from the blast* Renae Merle, "Fighting Roadside Bombs: Low-Tech, High-Tech, Toy Box," *Washington Post*, July 29, 2006, A1.

20   *they loaded the remains* Noncommissioned officer, interview at the Military Robotics Conference in Washington, DC, Peter W. Singer, April 10–12, 2006.

22   *"We were the longest overnight success"* David Whelan, "Fights Wars, Lint," *Forbes* 178, no. 4 (2006): 96.

23   *"rescuers [that] are unaffected by the carnage"* Jennifer 8. Lee, "Agile in a Crisis, Robots Show Their Mettle," *New York Times*, September 27, 2001, G7.

23   *"We began to run out of Afghans"* Andrew Bennett, interview, Peter W. Singer, November 16, 2006.

23   *PackBots made their debut in a cave* Justin Pope, "Looking to Iraq, Military Robots Focus on Lessons of Afghanistan," *Detroit News*, January 12, 2003, http://www.detnews.com/2003/technology/0301/12/technology-57614.htm.

23   *the war robot business grew* Associated Press, "Robots Sniff Out Bombs," CNN.com, March 30, 2007 (cited March 30, 2007); available at http://www.cnn.com/2007/TECH/03/30/robot.warriors.ap/index.html.

24   *"really excited by it"* Patrick Seitz, "With Scoob, iRobot Looks to Clean Up for the Second Time," *Investor's Business Daily*, January 17, 2006, A5.

24   *"I have four boys and two cats"* iRobot Corporation, "iRobot Customer Quotes" (cited November 16, 2005); available at http://www.irobot.com/sp.cfm?pageid=155.

24   *"You've done enough"* iRobot Corporation, "iRobot® Dirt Dog™ Workshop Robot" (cited October 13, 2006); available at http://store.irobot.com/product/index.jsp?productId=2475131.

24   *"This is all very new stuff"* Colin Angle, as quoted in Seitz, "With Scoob, iRobot Looks to Clean Up for the Second Time."

24   *"the demographics of our purchasers"* Ibid.

24   *"there are no clear buyers yet"* iRobot designer, interview, Peter W. Singer, November 16, 2006.

25   *"A robot may not harm"* Isaac Asimov, *I, Robot* (New York: Doubleday & Company, 1950).

25   *Asimov would definitely not approve* Dave White, "War Robots Dominate iRobot Show" *Mobile*

*Magazine*, October 12, 2006 (cited October 12, 2006); available at http://www.mobilemag.com/content/100/313/C10030/.

25 *"I think he would think it's cool as hell"* Helen Greiner, interview, Peter W. Singer, November 16, 2006.

26 *Foster-Miller has boomed* Associated Press, "Robots Sniff Out Bombs."

26 *an additional $20 million repair* "U.S. Navy Orders Talon Robots," *Defense News,* October 23, 2006, 46.

27 *"The soldiers have started taping"* Edward Godere, interview, Peter W. Singer, November 17, 2006.

27 *"is all about robotics"* Andrew Bennett, interview, Peter W. Singer, November 16, 2006.

27 *"We don't build Buicks"* Whelan, "Fights Wars, Lint," 96.

27 *"These robots are on a mission"* Helen Greiner, as quoted in "iRobot Co-Founder Comes Clean: Roomba Vacuum Cleaner a Worldwide Success," CNN.com, May 30, 2005 (cited May 30, 2005); available at http://www.cnn.com/2005/TECH/ptech/05/30/techprofile.irobot.ap/index.html.

27 *"a defense firm at heart"* Bob Quinn, interview, Peter W. Singer, November 17, 2006.

27 *"We're industrialists looking for needs"* Foster-Miller executive, interview, Peter W. Singer, November 17, 2006.

28 *the family got 40 percent annual returns* Daniel Golden, James Bandler, and Marcus Walker, "Bin Laden Family Has Intricate Ties with Washington," *Wall Street Journal Europe,* September 28–29, 2001, 4.

28 *"We hear that robots are trendy"* iRobot executive, interview, Peter W. Singer, November 16, 2006.

28 *"We don't just do robots"* Foster-Miller executive, interview, Peter W. Singer, November 17, 2006.

29 *"I wouldn't use anything else"* Foster-Miller, Inc., *The Soldier's Choice—Talon Robots. Talon E-mails from Iraq,* brochure, 2005.

29 *Another Talon serving* Foster-Miller employee, interview, Peter W. Singer, November 17, 2006.

29 *"This little guy saved our butts"* Peter W. Singer, "Research Visit to Foster-Miller," 2006.

29 *"most amazing inventions"* "The Most Amazing Inventions of the Year," Time.com, November 21, 2004, http://www.time.com/time/press_releases/article/0,8599,785326,00.html.

30 *"with this increased firepower"* Discovery Channel Pictures, "Smart Weapons," in *Future Weapons,* Discovery Channel, broadcast on May 17, 2006.

30 *"You can read people's nametags"* Frank Colucci, "Explosive Ordnance Disposal Robots Outfitted with Weapons," *National Defense* 88, no. 597 (2003): 44.

31 *"It's small. It's quiet"* Ibid., 44.

31 *"bootstrap development process"* Singer, "Research Visit to Foster-Miller."

31 *"not everything has to be super high tech"* Anthony Sebasto, as quoted in Michael Regan, "Armed 'Robo-Soldier' Set for Iraq" *Sydney Morning Herald,* February 4, 2005, http://www.smh.com.au/articles/2005/02/03/1107409974357.html.

31 *gunslingers cost just $230,000* "Hi, Robot," *Time* 164, no. 22 (2004): 81.

31 *In a test of its antitank rockets* Regan, "Armed 'Robo-Soldier' Set for Iraq."

31 *"pinpoint precision" as "nasty"* Foster-Miller employee, interview, Peter W. Singer, November 17, 2006.

31 *"It eliminates the majority"* Regan, "Armed 'Robo-Soldier' Set for Iraq."

31 *"The SWORDS doesn't care"* David Platt, as quoted in Regan, "Armed 'Robo-Soldier' Set for Iraq."

31 *"G.I. of the 21st century"* Associated Press, "Robots Sniff Out Bombs."

32 *"They have been a hit"* Eric Lenkowitz, "Robots Roll into Iraq War Zone," *New York Post,* August 4, 2007.

32 *reach as high as 12,000* Robert S. Boyd, "They're Very Expensive, but They Save Lives: U.S. Enlisting Smart Robots for War's Dirty, Deadly Jobs," *Philadelphia Inquirer,* February 20, 2006, E2.

32 *some twenty-two different robot systems* Jefferson Morris, "Military Projects 4,000 Robots in Theatre in FY'06," *Aerospace Daily & Defense Report* 217, no. 26 (2006): 4.

32 *"The Army of the Grand Robotic"* Charles Dean, "Unmanned Ground Vehicles for Armed Reconnaissance," paper presented at the Military Robotics Conference, Institute for Defense and Government Advancement, Washington, DC, April 10–12, 2006. Charles Dean is a former lieutenant colonel with the United States Army who now is project manager at Foster-Miller.

32 *"looks like a baby plane"* Susan B. Glasser and Vernon Loeb, "A War of Bridges: 225,000 U.S. and British Troops Are Now Within Striking Distance," *Washington Post Foreign Service,* March 2, 2003, A1.

33 *"a flying meat fork"* Max Boot, *War Made New: Technology, Warfare, and the Course of History, 1500 to Today* (New York: Gotham Books, 2006), 362.

34 *in the first two months of operations* Ibid., 367.

34 *"The Predator is my most capable sensor"* Elizabeth Bone and Christopher Bolkcom, *Unmanned Aerial Vehicles: Background and Issues* (Congressional Research Service, Library of Congress, 2003), 24.

34 *"Our major role"* Glasser and Loeb, "A War of Bridges: 225,000 U.S. and British Troops Are Now Within Striking Distance," A1.

34 *He had been an F-16 pilot* Eric Schmitt, "Remotely Controlled Aircraft Crowd Dangerous Iraqi and Afghan Skies," *New York Times,* April 5, 2005, A9.

34 *The idea then arose* Air force general, interview, Peter W. Singer, March 22, 2007.

35  *"it was a big problem"* Ibid.

35  *"It saddens me to know"* Ibid.

35  *In the words of one U.S. officer* Boot, *War Made New*, 383.

35  *Predators carried out 2,073 missions* Bill Sweetman, "USAF Predators Come of Age in Iraq and Afghanistan as Reaper Waits in the Wings," *Jane's International Defence Review* 39, no. 6 (2006): 52.

36  *Global Hawk can fly* "RQ-4 Global Hawk," Wikipedia, March 24, 2007 (cited March 30, 2007); available at http://en.wikipedia.org/wiki/Global_Hawk.

36  *"you basically hit the land button"* Air force officer, interview at Pentagon, Peter W. Singer, March 31, 2008.

36  *The plane itself costs some $35 million* Renae Merle, "Price of Global Hawk Surveillance Program Rises," *Washington Post*, 2004, A17.

36  *the U.S. Air Force plans to spend* Bill Sweetman, "Long Range Endurance UAS Targets the Adversary," *Jane's International Defence Review* 39, no. 8 (2006): 41.

37  *"It is more of a rush"* Kevin Maurer, "Pilotless Plane Guides 82nd," *Fayetteville (NC) Observer*, August 13, 2004.

37  *"You throw the bird up"* Noah Shachtman, "Attack of the Drones," *Wired* 13.06 (2005), http://www.wired.com/wired/archive/13.06/drones.html.

37  *the number of Ravens in service* Ibid.

37  *"reconnaissance with firepower"* Owen West and Bing West, "Lessons from Iraq," *Popular Mechanics* 182, no. 8 (2005): 50.

37  *there were 5,331 drones* Tom Vanden Brook, "Report: Insurgents Benefit from Drone Shortage," *USA Today*, March 25, 2008.

37  *"given the growth trends"* David A. Deptula, "Unmanned Aircraft Systems: Taking Strategy to Task," *Joint Force Quarterly*, no. 49 (2008): 50.

39  *CRAM required a congressional earmark* David Wichner, "Army Eyes Raytheon's High-Tech, Seagoing Gatling Gun (Mortar Defense)," *Arizona Daily Star*, May 19, 2005.

39  *The business of protecting* " 'Boys with Toys' Expo Hawks Security Goods," CNN.com, April 28, 2008, available at http://www.cnn.com/2008/TECH/04/25/security.expo/index.html.

39  *"Thank you, Osama bin Laden!"* Ibid.

39  *"reached heights not seen"* Stephen Handelman, "Technology vs. Terrorism," *Popular Science* 269, no. 3 (2006): 33.

39  *some one thousand robots* Jim Pinto, "Intelligent Robots Will Be Everywhere," Automation.com (cited August 22, 2005); available at http://www.automation.com/sitepages/pid1014.php.

39  *"we will sell tens of thousands"* Business executive, interview at the Military Robotics Conference in Washington, DC, Peter W. Singer, April 20, 2006.

40  *The robot border-cop helped* Sweetman, "USAF Predators Come of Age in Iraq and Afghanistan as Reaper Waits in the Wings," 56.

40  *"But the acceptability of using these systems"* Bruce V. Bigelow, "Robot Planes' New Role Won't Fly with Some," *San Diego Union-Tribune*, April 19, 2004.

40  *One example is the "Border Hawk"* The group's online site is http://www.americanborderpatrol.com/.

40  *"The Second Mexican-American War"* Max Blumenthal, "Vigilante Injustice," Salon.com, May 22, 2003,http://dir.salon.com/story/news/feature/2003/05/22/vigilante/index.html. See also Glenn Spencer, "The Second Mexican-American War," DVD of the 2002 American Renaissance conference, Herndon, Virginia, February 22–24, 2002; available at http://www.amren.com/estore/catalog/product_1672_2002_AR_Conference_Samuel_Francis_and_Glenn_Spencer_cat_94.html.

40  *The drones are launched* "Pictures of American Border Patrol UAV on the Arizona Border," *Desert Invasion—U.S.*, 2004 (cited March 22, 2007); available at http://www.desertinvasion.us/invasion_pictures/pics_american_border_patrol.html.

40  *"broadcasting the invasion live"* Noah Shachtman, " 'Vigilantes' Use Drones on Border Patrol," Defensetech.org, May 14, 2003 (cited July 21, 2006); available at http://www.defensetech.org/archives/000418.html.

40  *Silver Fox UAVs searched for survivors* Correspondents in Baton Rouge, "Drones Aid Katrina Rescue," Australian IT, September 5, 2005 (cited September 9, 2005); available at australianit.news.com.au/articles/0,7204,16494558%5E26199%5E%5Enbv%5E15306-15319,00.html.

41  *"aerial cell tower"* Larry Dickerson, "UAV's on the Rise," *Aviation Week & Space Technology*, January 15, 2007, 116.

## 2. SMART BOMBS, NORMA JEANE, AND DEFECATING DUCKS: A SHORT HISTORY OF ROBOTICS

42  *"The further backward you look"* As quoted in Ray Kurzweil, *The Singularity Is Near: When Humans Transcend Biology* (New York: Viking, 2005), 35.

42  *"Perhaps the most wonderful piece"* David Brewster, as quoted in Jay Richards, *Are We Spiritual Machines? Ray Kurzweil vs. the Critics of Strong AI*, 1st ed. (Seattle: Discovery Institute Press, 2002).

42  *called it "most deplorable"* Rony Gelman, "Gallery of Automata," 1996 (cited November 17, 2006); available at http://www.nyu.edu/pages/linguistics/courses/v610051/gelmanr/ling.html.

42  *"the Defecating Duck"* Jessika Riskin, "The Defecating Duck, or, the Ambiguous Origins of Artificial Life," *Critical Inquiry* 29, no. 4 (2003).

43  *"getting assistance by producing some machines"* Gelman, "Gallery of Automata."

43 *these punch cards would inspire* George Dyson, "The Undead: The Little Secret That Haunts Corporate America . . . A Technology That Won't Go Away," *Wired* 7.03 (1999), http://www.wired.com/wired/archive/7.03/punchcards.html.

43 *container of "artificial excrement"* Etienne Benson, "Science Historian Examines the 18th-Century Quest for 'Artificial Life,'" *Stanford Report*, October 19, 2001, http://news-service.stanford.edu/news/2001/october24/riskinprofile-1024.html.

44 *"If every tool, when ordered"* Alan L. Mackay and Maurice Ebison, *Scientific Quotations: The Harvest of a Quiet Eye* (New York: Crane, Russak, 1977).

44 *Archytas used it to carry* "A Brief History of Robotics," Megagiant.com (cited November 25, 2005); available at http://robotics.megagiant.com/history.html.

45 *they were automated* Rodney Brooks, *Flesh and Machines: How Robots Will Change Us* (New York: Pantheon, 2002).

45 *von Kempelen had hidden a dwarf* This is much like the fighting robot Homer built for Bart in *The Simpsons*, which turned out just to have Homer hidden inside, getting beat up by the real robots. T. H. Tarnóczy and H. Dudley, "The Speaking Mashine of Wolfgang von Kempelen," *Journal of Acoustical Society of America* 22, no. 2 (1950). See also Robert Capps, "The 50 Best Robots Ever," *Wired* 14.01 (2006), http://www.wired.com/wired/archive/14.01/robots.html?pg=2&topic=robots&topic_set=.

45 *the field of modern chemistry* J. Boone Bartholomees Jr., "The Heirs of Archimedes: Science and the Art of War through the Age of Enlightenment," *Parameters* 35, no. 4 (2005): 136.

46 *"to see what would happen"* "Charles Babbage," Wikipedia, April 20, 2007 (cited April 20, 2007); available at http://en.wikipedia.org/wiki/Charles_Babbage.

46 *"I called an official"* Robert Finkelstein, "Military Robotics: Malignant Machines or the Path to Peace," paper presented at the Military Robotics Conference, Institute for Defense and Government Advancement, Washington, DC, April 10–12, 2006.

47 *the Germans protected their coast* Steven M. Shaker and Alan R. Wise, *War Without Men: Robot on the Future Battlefield* (Washington, DC: Pergamon Brassey's International Defense Publishers Inc., 1988).

48 *load them up with twenty-two thousand pounds of Torpex* Anthony J. Lazarski, "Legal Implications of the Uninhabited Combat Aerial Vehicle," *Aerospace Power Journal* 16, no. 2 (2002), http://www.airpower.maxwell.af.mil/airchronicles/apj/apj02/sum02/lazarski.html.

49 *"bombing without knowledge"* Chris Gray, *Postmodern War: The New Politics of Conflict* (New York: Guilford Press, 1997)

50 *"put a bomb in a pickle barrel"* Max Boot, *War Made New: Technology, Warfare, and the Course of History, 1500 to Today* (New York: Gotham Books, 2006), 278.

51 *Besides the automated bombsight* Ibid.

51 *"Giant Electronic Brain"* David Hambling, *Weapons Grade: How Modern Warfare Gave Birth to Our High-Tech World* (New York: Carroll and Graf, 2005), 90.

52 *"Why don't we just have a network"* Ibid., 99.

53 *a "formal set of instructions"* Ibid., 103.

53 *"how to get the driver out"* Robert Finkelstein, interview, Peter W. Singer, July 7, 2006.

53 *"certainly in the range of 2015–2030"* Ibid.

54 *"The sad thing"* Finkelstein, "Military Robotics: Malignant Machines or the Path to Peace?"

54 *"Then Nixon pulled the plug"* Finkelstein, interview, Peter W. Singer, July 7; Finkelstein, "Military Robotics: Malignant Machines or the Path to Peace?"

54 *"the gift wrap industry was larger"* Finkelstein, "Military Robotics: Malignant Machines or the Path to Peace?"

54 *"One decision criterion of mine"* Finkelstein, interview, Peter W. Singer, July 7, 2006.

54 *the Fire Fly flew 3,435 missions* Kit Lavell, "Defending America in the 21st Century: Unmanned Aerial Vehicles Are Coming of Age," *San Diego Union-Tribune*, February 16, 2003, http://www.signonsandiego.com/news/op-ed/techwar/20030216-9999_main2.html.

55 *"It took decades for UAVs to recover"* Ibid.

55 *The Aquila was to be* General Accounting Office, *Unmanned Aerial Vehicles: Outrider Demonstrations Will Be Inadequate to Justify Further Production* (Washington, DC: General Accounting Office, 1997).

56 *what is called "customer pull"* Finkelstein, "Military Robotics: Malignant Machines or the Path to Peace?"

56 *While they reloaded* Ralph Sanders, "An Israeli Military Innovation: UAVs," *Joint Force Quarterly*, no. 33 (2002).

56 *"The Iraqis came to learn"* Dina El Boghdady, "Small Firms Turn to Drones: Demand Grows for Unmanned Craft," *Washington Post*, October 31, 2005, D1.

57 *"'smart bombs' are really only"* "Notes, 8 June 2004," in *National Security in the 21st Century: Rethinking the Principles of War* (Arlington, VA: Johns Hopkins University Applied Physics Lab, 2004); available at http://www.jhuapl.edu/POW/notes/notes_8Jun.htm.

57 *"dropping a Cadillac"* Hambling, *Weapons Grade*, 125.

58 *Only 7 percent of all the bombs* James Dunigan, "The Air Campaign in Iraq," *Strategy Page*, May 21, 2003, http://www.strategypage.com/dls/articles/20030522.asp

58  *the U.S. military had bought into the idea* Gray, *Postmodern War,* 52.

58  *"I couldn't have done it all"* Ibid., 36.

58  *"That's when it really came together"* Tom Erhard, interview, Peter W. Singer, January 31, 2007.

58  *"no longer . . . spend money the way"* General Ronald R. Fogleman, as quoted in John A. Tirpak, "The Robotic Air Force," *Air Force* 80, no. 9 (1997), http://www.afa.org/magazine/sept1997/0997robot.asp.

59  *"it was threatened"* Finkelstein, "Military Robotics: Malignant Machines or the Path to Peace?"

59  *"running an after-retirement jobs program"* Robotics firm executive, interview, Peter W. Singer, March 13, 2004.

59  *shrinking by more than 30 percent* George C. Wilson, "Tough Choices Loom for the Services," *Air Force Times,* January 20, 1997, 14.

59  *"dead soldiers are America's most vulnerable center"* Robert Scales, "Urban Warfare," *Military Review,* February 2005, 9; available at http://www.au.af.mil/au/awc/awcgate/milreview/scales.pdf.

59  *an added reason for investing* Edward Luttwak, "Post-Heroic Armies," *Foreign Affairs* 75, no. 4 (1996): 33.

60  *"So what do you do?"* Senator John Warner, as quoted in George C. Wilson, "A Chairman Pushes Unmanned Warfare," *National Journal* 32, no. 10 (2000): 718.

60  *"Every now and then"* Ibid.

60  *"The Robot Is Our Answer"* H. R. Everett, interview, Peter W. Singer, October 20, 2006.

61  *the total Pentagon budget* Steven Kosiak, Classified Funding in the FY 2009 Defense Funding Request, Center for Strategic and Budgetary Analysis, June 17, 2008, www.csbaonline.org.

61  *"unchecked growth"* Jeffrey M. Tebbs, "Smelting the Triangle: Constraining Congress, Defense Contractors, and the Military Brass to Restore a Fiscally Prudent Defense Budget" (Washington, DC: Brookings Institution, 2006), 3.

61  *the black budget is not released to the public* John Bennett, "CSBA: 'Black' Spending Doubled Since 1995," *Defense News,* July 30, 2007, 22.

61  *"Prior to 9/11"* Larry Dickerson, "UAV's on the Rise," *Aviation Week & Space Technology,* January 15, 2007, 115.

61  *"Make 'em as fast as you can"* Stayne Hoff, interview, Peter W. Singer, December 5, 2006.

61  *93 percent of the bombs and missiles dropped* Sean J. A. Edwards, "Swarming and the Future of Warfare," doctoral thesis, Pardee Rand Graduate School, 2005, 36–37.

61  *"The undertaking has attracted"* Renae Merle, "Fighting Roadside Bombs: Low-Tech, High-Tech, Toy Box," *Washington Post,* July 29, 2006, A1.

62  *"Just as World War I accelerated"* Robert Finkelstein and James Albus, "Technology Assessment of Autonomous Intelligent Bipedal and Other Legged Robots" (DARPA, 2004), 104.

62  *"the most dynamic growth sector"* "Teal Group: UAV Spending to Triple Within Decade," *Aerospace Daily & Defense Report,* September 1, 2006; available at http://www.aviationweek.com/aw/.

62  *global spending on unmanned planes* Ibid.

62  *"ground vehicles are just now on the edge"* Finkelstein, "Military Robotics: Malignant Machines or the Path to Peace?"

62  *"One, the technology has finally matured"* H. R. Everett, interview, Peter W. Singer, October 20, 2006.

63  *"They're not afraid"* Tim Weiner, "Pentagon Has Sights on Robot Soldiers," *New York Times News Service,* February 16, 2005.

63  *"Can you keep your eyes open"* Hambling, *Weapons Grade,* 324.

63  *Even using the same mine-detecting* Jerry Harbor, "Assessing Unmanned System Performance," paper presented at the Military Robotics Conference, Institute for Defense and Government Advancement, Washington, DC, April 10–12, 2006.

63  *"inclement weather, smog, and smoke"* Patrick Eberle, "To UAV or Not to UAV: That Is the Question; Here Is One Answer," *Air & Space Power Journal—Chronicles Online Journal,* October 9, 2001, http://www.airpower.au.af.mil/airchronicles/cc/eberle.html.

64  *"The airplane was too good"* As quoted in George Friedman and Meredith Friedman, *The Future of War: Power, Technology, and American World Dominance in the Twenty-first Century* (New York: Crown, 1996), 296.

64  *"the human is becoming the weakest link"* Cheryl Seal, "Frankensteins in the Pentagon: DARPA's Creepy Bioengineering Program," *Information Clearing House,* August 25, 2003, http://www.informationclearinghouse.info/article4572.htm.

64  *"the UCAV [the unmanned fighter jet]"* iRobot designer, interview, Peter W. Singer, November 16, 2006.

64  *"The trend towards the future"* Ibid.

64  *can share that skill or knowledge with another computer* Jay Richards, *Are We Spiritual Machines? Ray Kurzweil vs. the Critics of Strong AI,* 1st ed. (Seattle: Discovery Institute Press, 2002).

65  *"robots don't participate"* Robotics company executive, interview, Peter W. Singer, Acton, MA, November 18, 2006.

65  *"The military is deciding"* Eliot Cohen, interview, Peter W. Singer, Washington, DC, November 15, 2006.

65  *"preference for joint unmanned systems"* Christian Lowe, "Senators Love Robots," Defensetech.org, May 17, 2006 (cited November 1, 2006); available at http://www.defensetech.org/archives/002419.html.

65  *"We're entering an era"* As quoted in John J. Klein,

"The Problematic Nexus: Where Unmanned Combat Air Vehicles and the Law of Armed Conflict Meet," *Air & Space Power Journal—Chronicles Online Journal*, July 22, 2003, http://www.airpower.maxwell.af.mil/airchronicles/cc/klein.html.

## 3. ROBOTICS FOR DUMMIES

66 *Like a robot, sometimes I just know not* Eminem, *8 Mile*. Soundtrack. Shady Records, CDC 493508, 2002.

66 *"The ROBOTs are dressed like people"* Karel Čapek, *R.U.R. (Rossum's Universal Robots)*, in *Toward the Radical Center: A Karel Čapek Reader*, ed. Peter Kussi (Highland Park, NJ: Catbird Press, 1990), 35.

66 *first mention of the word "robot"* Ibid., 33.

66 *"And God said unto them"* Genesis 1:28.

67 *man-made devices with three key components:* Robert Finkelstein, "Military Robotics: Malignant Machines or the Path to Peace," paper presented at the Military Robotics Conference, Institute for Defense and Government Advancement, Washington, DC, April 10–12, 2006.

68 *"too much technology"* U.S. Army soldier, interview, Peter W. Singer, Washington, DC, November 2, 2006.

68 *"the TV episode of* I Love Lucy*"* Sandra Erwin, "More Eyes in the Sky May Not Generate Better intelligence," *National Defense*, June 2008, http://www.nationaldefensemagazine.org/issues/2008/June/MoreEye.htm

68 *"User interface is a big, big problem"* Andrew Bennett, interview, Peter W. Singer, November 16, 2006.

68 *"playing to the soldiers' preconceptions"* Ibid.

68 *"We modeled the controller"* Stephen Graham, "America's Robot Army," *New Statesman*, June 12, 2006, http://www.newstatesman.com/200606120018.

69 *"As a military person,"* CBS News, "Gesture Glove Not Science Fiction," CBSNews.com, August 23, 2005 (cited February 3, 2007); available at http://www.cbsnews.com/stories/2005/08/23/eveningnews/main792311.shtml.

69 *"if it takes more than two clicks"* As quoted in Giles Ebbutt, "Knowledge Is Power," *Jane's International Defence Review* 40, no. 1 (2007): 29.

69 *The new controller programs* M. O'Madharain and B. Gillespie, "The Moose: A Haptic User Interface for Blind Persons" (Stanford University, 1995), 131.

70 *"It will really make a complete fusional relation"* Jeff Wise, "Bertrand Piccard's Solar-Powered Flight Around the World," *Popular Mechanics*, September 2005, http://www.popularmechanics.com/science/air_space/1701581.html?page=2.

70 *felt just like "Pop Rock candies"* Associated Press, "Creating Superhuman Troops of Future Starts at the Tongue, April 22, 2006 (cited August 14, 2007); available at http://www.gainesville.com/apps/pbcs.dll/article?AID=/20060422/LOCAL/204220325/1078/news.

70 *"In terms of evolution"* Julian Jones, director, *How William Shatner Changed the World*, produced by the History Channel, broadcast on October 21, 2006.

71 *"We know that machines...phenomenal memory"* Andrew Smith, "Science 2001: Net Prophets," *Observer*, December 31, 2000, 18.

71 *An EEG wearer* Peter Schwartz, Chris Taylor, and Rita Koselka, "Quantum Leap," *Fortune* 154, no. 3 (2006).

71 *"It's a blurry vision"* Emily Gold Boutilier, "Thinking the World into Motion," *Brown Alumni Magazine*, January 2005, http://www.brownalumnimagazine.com/storyDetail.cfm?ID=2521.

71 *"like watching a high-definition plasma screen"* Ibid.

71 *"Every other day I wanted to die"* Discovery Science Channel, *Robosapiens: The Secret (R)evolution*, broadcast on June 18, 2006.

72 *"I do feel like it was a part of me"* Ibid.

72 *"the most lavishly funded"* Cheryl Seal, "Frankensteins in the Pentagon: DARPA's Creepy Bioengineering Program," *Information Clearing House*, August 25, 2003, http://www.informationclearinghouse.info/article4572.htm.

72 *"It's as if the first flight at Kitty Hawk"* John Fauber, "Think, Think, Shoot, Score!" *Milwaukee Journal Sentinel*, December 4, 2004, http://www.jsonline.com/alive/news/dec04/281287.asp.

73 *it wasn't so much that he could "feel"* Tim Usborne, *Stargate SG-1: True Science*, produced by Paul Sen and Rosie Kingham, Sci Fi Channel, broadcast on July 18, 2006.

73 *you can squeeze it into tight parking spaces* Rodney Brooks, *Flesh and Machines: How Robots Will Change Us* (New York: Pantheon, 2002), 227.

73 *"is going to be my mental prosthesis"* Schwartz, Taylor, and Koselka, "Quantum Leap."

73 *"We would all share information"* Robert Finkelstein, interview, Peter W. Singer, July 7, 2006.

74 *"network-enabled telepathy"* Schwartz, Taylor, and Koselka, "Quantum Leap."

74 *National Science Foundation envisions* Mihail C. Roco and William Sims Bainbridge, "Converging Technologies for Improving Human Health: Nanotechnology, Biotechnology, Information Technology and Cognitive Science" (National Science Foundation, 2002), 19.

75 *"Having a dedicated operator"* Robert Finkelstein and James Albus, "Technology Assessment of Autonomous Intelligent Bipedal and Other Legged Robots" (DARPA, 2004).

75 *"The autonomy thing is f'ing hard"* Noah Shachtman, interview, Peter W. Singer, Washington, DC, March 25, 2006.

75   *"Forget about whether the intelligence"* John Arquilla, as quoted in *Warbots*, produced by Dan Saxton Company, History Channel, broadcast on August 8, 2006.

75   *"an ability to act appropriately"* George A. Miller, "WordNet Search—3.0" (Cognitive Science Laboratory, Princeton University, 2006).

76   *They argue that a machine* Chris Gray, *Postmodern War: The New Politics of Conflict* (New York: Guilford Press, 1997), 71.

76   *"calculate faster than any human being"* David Hambling, *Weapons Grade: How Modern Warfare Gave Birth to Our High-Tech World* (New York: Carroll and Graf, 2005), 205.

77   *which is 95 percent of what we ask* David Hambling discusses this analogy in *Weapons Grade*, 205.

77   *"If you think it's easy"* Ian Rowley et al., "Ready to Buy a Home Robot?" *BusinessWeek*, no. 3892 (2004): 84.

77   *"perceive something complex and make"* Sebastian Thrun, interview, Peter W. Singer, March 18, 2007.

77   *"Simply put, we can't know"* "Interview with Lynne E. Parker," *International Journal of Advanced Robotic Systems* 2, no. 2 (2004).

78   *Starting with the acclaimed battles* Rodney Brooks, *Flesh and Machines: How Robots Will Change Us* (New York: Pantheon, 2002), 103.

78   *the size of the AI market* Ray Kurzweil, *The Singularity Is Near: When Humans Transcend Biology* (New York: Viking, 2005), 279.

78   *the U.S. military funds as much as 80 percent* Ibid., 205.

78   *"We are not close to having AI"* Neal Conan, "Interview with Helen Greiner, Chairman and Cofounder of iRobot," on *Talk of the Nation*, National Public Radio, broadcast on June 23, 2006.

78   *"the UAV is able to learn"* "UAV Learns to Think for Itself—Now Technology Will Transition to Military," Gizmag.com, February 22, 2005 (cited July 6, 2005); available at http://www.gizmag.com/go/3745/.

78   *GT Max has been able* Ibid.

79   *Thaler has created* Tina Hesman, "Stephen Thaler's Computer Creativity Machine Simulates the Human Brain," *St. Louis Post-Dispatch*, January 24, 2004.

79   *the air force lab contracted Thaler* David Hambling, "Experimental AI Powers Robot Army," *Wired News*, September 14, 2006, http://www.wired.com/news/technology/software/coolapps/news/2006/09/71779.

79   *There was even one robot* More about the Reading University experiments at http://cirg.reading.ac.uk/home.htm.

79   *the computer might learn so much by* Richards, *Are We Spiritual Machines? Ray Kurzweil vs. the Critics of Strong AI*, 1st ed. (Seattle: Discovery Institute Press, 2002).

80   *"If it's a child, you want to stop"* Preston Lerner, "Robots Go to War: Within 10 Years, Infantry Soldiers Will Go into Battle with Autonomous Robots Close Behind Them," *Popular Science* 268, no. 1 (2006): 42.

81   *"understanding the environment is the Holy Grail"* Sebastian Thrun, interview, Peter W. Singer, March 18, 2007.

81   *"black pick-up truck driven by two men"* Stephen Trimble, "US Eyes Hyperspectral Technology for UAVs," *Jane's Defence Weekly*, September 6, 2006, 31.

81   *Its multispectral sensors detect* Richards, *Are We Spiritual Machines?*, 110.

82   *"As opposed to a computer"* David Bruemmer, "Intelligent Autonomy for Unmanned Vehicles," paper presented at the Military Robotics Conference, Institute for Defense and Government Advancement, Washington, DC, April 10–12, 2006.

82   *but it has since sold only six thousand units* Associated Press, "U.S. Considers Turning Scooters into War Robots," Ctv.ca, November 28, 2003 (cited August 18, 2006); available at http://www.ctv.ca/servlet/ArticleNews/story/CTVNews/1070032823376_233//. See also Segway, "About the Robotic Mobility Platform," 2005 (cited November 16, 2005); available at http://www.segway.com/products/rmp/.

82   *recent U.S. Navy work on a system* Jennifer Bails, "Water Bug Robot," *Pittsburgh Tribune-Review*, April 6, 2006.

82   *a "diamond-tipped" buzz saw* John Canning et al., "A Concept for the Operation of Armed Autonomous Systems on the Battlefield," Dahlgren Division, Naval Surface Warfare Center, 2008.

83   *The switch from chemical to electric* Geoff Hiscock, "Gun Whips Up a Metal Storm," CNN.com, June 27, 2003 (cited September 14, 2006); available at http://www.cnn.com/2003/BUSINESS/06/26/australia.metalstorm/.

83   *is good for "crowd control"* http://www.metalstorm.com/content/view/82/166/.

83   *"The combination of robotics"* Steven Metz, interview, Peter W. Singer, September 19, 2006.

83   *The devices have a range* General Dynamics, *Long Range Acoustic Device*, 2002. Product Information Sheet.

83   *the crew used LRAD sonic blasters* John Pain, "Cruise Ship Attacked by Pirates Used Sonic Weapon," USAToday.com, July 11, 2005, http://www.usatoday.com/tech/news/techinnovations/2005-11-07-cruise-blast_x.htm.

84   *shoots blobs of compressed glue* Lothar Ibruegger, "Special Report: Emerging Technologies and Their Impact on Arms Control and Non Proliferation" (NATO Parliamentary Assembly, International Secretariat, 2001).

84   *If the ray is turned off* Associated Press, "Ray Gun Makes Targets Feel Like They're on Fire," MSNBC

.com, January 25, 2007, http://www.msnbc.msn
.com/id/16794717/.

84 *being off just a few degrees could kill* Hambling,
*Weapons Grade*, 346.

84 *Depending on the tuning* David Hambling, "Air
Force Plan: Hack Your Nervous System,"
Defensetech.org, February 13, 2006 (cited December 18, 2006); available at http://www.defensetech
.org/archives/002152.html.

84 *Another prototype is the tetanizing beam weapon*
Hambling, *Weapons Grade*, 237.

85 *The idea of lasers first came* H. G. Wells, *The War of
the Worlds* (New York: Tor/Forge, 1993 [1898]), 25.

85 *George Lucas sued the U.S. government* Hambling,
*Weapons Grade*, 119. See court case. *Lucasfilm Ltd.
v. High Frontier*, 622 F. Supp. 931 (D.D.C. 1985).

85 *called it the "Holy Grail" of weapons* Doug Beason,
*The E-Bomb* (New York: Da Capo, 2001), 188.

85 *a useful defense against "terrorists on Jet-Skis"*
Dan Wildt, as quoted in Bill Sweetman, "Fact or
Fiction," *Jane's Defence Weekly*, February 22, 2006.

86 *"instantaneous burst-combustion of insurgent
clothing"* The request is available at http://blog
.wired.com/defense/files/PASDEW.pdf.

86 *the Tactical Relay Mirror System* Joshua Kucera,
"US Eyes Fast Fielding of Attack Laser," `Jane's
Defence Weekly*, July 6, 2005, 6.

86 *A novel program at Tel Aviv University* "Introducing the Nano Battery, as Thick as a Strand of Hair,"
WorldTribune.com, November 17, 2006, http://
www.worldtribune.com/worldtribune/06/front
2454057.073611111.html.

86 *"It didn't like carbonated beer"* As quoted in Kurzweil, *The Singularity Is Near*, 248.

87 *The U.S. Air Force is presently exploring* Hambling, *Weapons Grade*, 152.

87 *Another UAV, the Global Observer* Stephen Trimble, "Multi-UAV Approach Proposed for BAMS,"
*Jane's Defence Weekly*, September 13, 2006, 10.

87 *There are all sorts of projects* Lonnie D. Henley,
"The RMA After Next," *Parameters* 29, no. 4 (1999).

87 *Chew-Chew, the "gastrobot"* Reuters, "Meat-Eating Robot Has (G)astronomic Potential,"
CNN.com, July 21, 2000 (cited February 10,
2006); available at http://archives.cnn.com/2000/
NATURE/07/21/carnivorous.robot.reut/index
.html.

87 *A contemporary of Chew-Chew's* Ibid.

87 *It is called a "vampire-bot"* Kurzweil, *The Singularity Is Near*, 248.

87 *"grass, wood, broken furniture, [and] dead bodies"*
Finkelstein, "Military Robotics: Malignant
Machines or the Path to Peace?"

88 *"Although a few of the robots"* Bill Gates, "A Robot
in Every Home," ScientificAmerican.com, December 16, 2006 (cited December 17, 2006); available at
http://www.sciam.com/article.cfm?id=a-robot-in-every-home.

88 *"The tool has to fit the task"* Mark Barber, "Force

Protection Robotics," paper presented at the Military Robotics Conference, Institute for Defense
and Government Advancement, Washington, DC,
April 10–12, 2006.

88 *"Every vehicle is a robot"* Daniel H. Wilson, *How
to Survive a Robot Uprising: Tips on Defending
Yourself Against the Coming Rebellion*, 1st U.S. ed.
(New York: Bloomsbury, 2005), 26.

88 *the "No Hands Across America" drive* Todd
Jochem, "No Hands Across America Journal," 1995
(cited November 16, 2006); available at http://
www.cs.cmu.edu/afs/cs/user/tjochem/www/
nhaa/Journal.html.

88 *Its journey ended* Wilson, *How to Survive a Robot
Uprising*, 26.

89 *it costs just $70,000 to convert* Globes Correspondent, "InRob Tech Completes Remote-Controlled
Hummer Trials," Globes.co.il, January 9, 2006,
http://www.globes.co.il/serveen/globes/docview
.asp?did=1000048585&fid=942.

89 *Wheeled vehicles can only operate* Finkelstein and
Albus, "Technology Assessment of Autonomous
Intelligent Bipedal and Other Legged Robots."

89 *"humanoid robots should be fielded"* Finkelstein,
"Military Robotics: Malignant Machines or the
Path to Peace?"

90 *describe human eyes as "badly designed"* Brooks,
*Flesh and Machines*.

90 *"In the next 10–20 years"* Rodney Brooks, "Technology Impacts on Military Robotics over the
Coming Decades," paper presented at the Military
Robotics Conference, Institute for Defense and
Government Advancement, Washington, DC,
April 10–12, 2006.

90 *"Every aspect of robotics is touched by biology"*
Ronald Arkin, as quoted in Eric Smalley, "Georgia
Tech's Ronald Arkin," 2005, http://www/trnmag
.com/stories/2005/091205/View_Ronald_Arkin_
091205.html.

91 *"getting robots to jump"* Joel Garreau, *Radical Evolution: The Promise and Peril of Enhancing Our
Minds, Our Bodies—And What It Means to Be
Human* (New York: Doubleday, 2005), 35.

91 *"detection tracking algorithm"* David Hambling,
"Selective Focus May Give Drone Aircraft Eagle
Eyes," *New Scientist*, September 25, 2006, http://
www.newscientisttech.com/article.ns?id=dn10156
&feedId=tech_rss20.

91 *the research was also used by the Pixar* Elizabeth
Corcoran, "The Stickybot," *Forbes* 178, no. 4
(2006): 106.

91 *"Fact of nature"* Finkelstein and Albus, "Technology Assessment of Autonomous Intelligent Bipedal
and Other Legged Robots," 158.

91 *Designs that find their inspiration* David Hambling,
"A Breed Apart," *Guardian* (UK), February 25, 2005
(cited December 18, 2006); available at http://www
.guardian.co.uk/technology/2005/feb/24/
onlinesupplement.insideit3.

92  *Big Dog will be "unleashed"* Preston Lerner, "The Army's Robot Sherpa from the Backcountry to the Rubble-Strewn Back Alleys of a War-Torn City, This Mechanized Pack Animal Will Follow Soldiers Wherever Duty Calls Them," *Popular Science* 268, no. 4 (2006): 72.

92  *DARPA's survey on robotics futures* Finkelstein and Albus, "Technology Assessment of Autonomous Intelligent Bipedal and Other Legged Robots."

92  *"I have so many dreams"* "Future Dreams," BBC News.com, December 21, 2006 (cited May 30, 2007); available at http://news.bbc.co.uk/1/shared/spl/hi/picture_gallery/06/technology_robot_menagerie/html/10.stm.

93  *"self-folding origami"* Henry S. Kenyon, "Programmable Matter Research Solidifies," *Signal*, June 2009, http://www.afcea.org/signal/articles/templates/Signal_Article_Template.asp?articleid=1964&zoneid=263.

93  *researchers are at work on "claytronic" robots* Tom Simonite, "Shape-Shifting Robot Forms from Magnetic Swarm," *New Scientist*, January 29, 2008.

93  *"it may be increasingly difficult to say"* Gates, "A Robot in Every Home."

## 4. TO INFINITY AND BEYOND: THE POWER OF EXPONENTIAL TRENDS

94  *"I decided I would be an inventor"* Ray Kurzweil on Discovery Science Channel, *Robosapiens: The Secret (R)evolution*, broadcast on June 18, 2006.

95  *inducted into the National Inventors Hall of Fame* "Ray Kurzweil," singularity.com (cited May 29, 2007); available at http://singularity.com/aboutray.html.

95  *"About thirty years ago"* Ray Kurzweil, interview via phone, Peter W. Singer, Washington, DC, December 7, 2006.

95  *"We use predictions"* Ibid.

95  *"I've slowed down aging to a crawl"* Ibid.

96  *this is a guy whom Bill Gates described* Brian O'Keefe, "The Smartest (or the Nuttiest) Futurist on Earth," CNNMoney.com, May 2, 2007 (cited May 2, 2007); available at http://money.cnn.com/magazines/fortune/fortune_archive/2007/05/14/100008848/.

96  *Kurzweil gets a reported $25,000* Ibid.

96  *He is also one of five members* Ibid.

96  *"only an early harbinger"* Kurzweil, interview, Peter W. Singer, December 7, 2006.

96  *will "create qualitative change"* Ibid.

96  *"In just 20 years"* Kurzweil, as quoted in Joel Garreau, *Radical Evolution: The Promise and Peril of Enhancing Our Minds, Our Bodies—And What It Means to Be Human* (New York: Doubleday, 2005), 6.

96  *"very much at the mainstream"* Kurzweil, interview, Peter W. Singer, December 7, 2006.

97  *"the pace of change"* Ray Kurzweil, *The Singularity Is Near: When Humans Transcend Biology* (New York: Viking, 2005), 35.

97  *"Skeptics said there's no way"* Kurzweil, interview, Peter W. Singer, December 7, 2006.

97  *"If you double from 1 percent every year"* Ibid.

97  *some two billion people around the world* Max Boot, *War Made New: Technology, Warfare, and the Course of History, 1500 to Today* (New York: Gotham Books, 2006), 312.

98  *Moore predicted* Gordon E. Moore, "Cramming More Components onto Integrated Circuits," *Electronics* 38, no. 8 (1965), available at http://download.intel.com/research/silicon/moorespaper.pdf.

98  *Tradic, the first computer* Mihail C. Roco and William Sims Bainbridge, "Converging Technologies for Improving Human Health: Nanotechnology, Biotechnology, Information Technology and Cognitive Science" (National Science Foundation, 2002).

98  *Moore's old company Intel* "Higher Levels of Design Abstraction," Intel.com (cited August 14, 2006); available at http://www.intel.com/technology/silicon/scl/abstraction.htm.

98  *it has roughly the same capacity* Thomas Homer-Dixon, "The Rise of Complex Terrorism," *Foreign Policy*, no. 128 (2002): 54.

98  *a present-day supercomputer* Peggy Mihelich, "Supercomputers Crunching Potato Chips, Proteins and Nuclear Bombs," CNN.com, December 5, 2006 (cited December 5, 2006); available at http://www.cnn.com/2006/TECH/12/05/supercomputers/index.html.

98  *to build a next-generation supercomputer* Ibid.

98  *Only four years later* Garreau, *Radical Evolution*, 59.

99  *refrigerator magnets that play Christmas jingles* Ibid.

99  *"riding someone else's exponentials"* Ibid.

99  *"The Law of Accelerating Returns"* Kurzweil, interview, Peter W. Singer, December 7, 2006.

99  *Internet bandwidth backbone is doubling* Roco and Bainbridge, "Converging Technologies for Improving Human Health."

99  *the number of personal and service robots* Unless otherwise noted, all figures are from Garreau, *Radical Evolution*, 59.

99  *The modern-day bomber jet* Sean J. A. Edwards, "Swarming and the Future of Warfare" (doctoral thesis, Pardee Rand Graduate School, 2005), 136.

100  *the range and effectiveness of artillery fire* Stephen D. Biddle, *Military Power: Explaining Victory and Defeat in Modern Battle* (Princeton, NJ: Princeton University Press, 2004), 30.

100  *exponential "stretching" of the battlefield* Michael E. O'Hanlon, *Technological Change and the Future of Warfare* (Washington, DC: Brookings Institution Press, 2000), 121.

100 *each plane was destroying 4.07 targets* Edwards, "Swarming and the Future of Warfare," 137.

100 *the agricultural revolution* Rodney Brooks, *Flesh and Machines: How Robots Will Change Us* (New York: Pantheon, 2002).

100 *launching the Industrial Age* Richard R. Nelson, *Technology, Institutions, and Economic Growth* (Cambridge, MA: Harvard University Press, 2005), 135.

100 *The Internet took roughly a decade* "Internet Usage Statistics—The Big Picture," *Internet World Stats*, 2007 (cited May 30, 2007); available at http://www.internetworldstats.com/stats.htm.

100 *In less than a decade, over a billion people* Chuck Klosterman, *Sex, Drugs, and Cocoa Puffs: A Low Culture Manifesto* (New York: Scribner, 2003), 112.

101 *the aggregate of technologic change* Ray Kurweil in an interview with Kip P. Nygren, "Emerging Technologies and Exponential Change: Implications for Army Transformation," *Parameters* 32, no. 2 (2002): 91.

101 *"Golden Age of Invention"* "Golden Age of Invention," Sparknotes.com, 2006 (cited May 30, 2007); available at http://www.sparknotes.com/biography/edison/section4.rhtml.

101 *"The Singularity Is Near"* Kurzweil, *The Singularity Is Near: When Humans Transcend Biology.*

101 *"We often say things like"* Kurzweil, interview, Peter W. Singer, December 7, 2006.

102 *"Within a single human generation"* Hugo de Garis, "Building Gods or Building Our Potential Exterminators?" KurzweilAI.net, February 26, 2001 (cited June 27, 2006); available at http://www.kurzweilai.net/meme/frame.html?main=/articles/art0131.html?.

102 *the human brain is created* Jay Richards, *Are We Spiritual Machines? Ray Kurzweil vs. the Critics of Strong AI*, 1st ed. (Seattle: Discovery Institute Press, 2002), 206.

102 *"about twenty thousand years of progress"* Kurzweil, interview, Peter W. Singer, December 7, 2006.

103 *"the laws of science and our ability"* As quoted in Garreau, *Radical Evolution*, 72.

103 *"Google all the time"* Peter Moon, "AI Will Surpass Human Intelligence After 2020," TTworld.com, May 3, 2007 (cited May 30, 2007); available at http://www.itworld.com/Tech/3494/070503ai2020/.

103 *"the Internet-based cognitive tools"* Vernor Vinge, *Rainbows End* (New York: Tor Books, 2006), 5.

103 *"The Coming Technological Singularity"* Vernor Vinge, "The Coming Technological Singularity: How to Survive in the Post-Human Era" (paper presented at the VISION-21 Symposium, March 30–31, 1993).

103 *"within thirty years"* Ibid.

103 *"point where our old models"* Ibid.

104 *"We are on the edge of change"* Vinge, as quoted in Garreau, *Radical Evolution*, 71–72.

104 *"It's a future period"* Kurzweil, *The Singularity Is Near*, 7.

104 *"It's not merely a technology"* Robert Epstein, interview, Peter W. Singer, Washington, DC, October 25, 2006.

104 *"fits many of our happiest dreams"* Vinge, "The Coming Technological Singularity: How to Survive in the Post-Human Era."

104 *"physical extinction of the human race"* Ibid.

104 *"the very nature of what it means to be human"* "About the Book," Singularity.com (cited May 29, 2007); available at http://singularity.com/about thebook.html.

105 *"the non-biological intelligence"* Kurzweil, *The Singularity Is Near*, 136.

105 *"The Rapture for Nerds"* Charles Stross, "Singularity: A Tough Guide to the Rapture of the Nerds," 2005 (cited January 28, 2008); available at http://www.antipope.org/charlie/toughguide.html.

105 *"By 2030 we are likely to"* Bill Joy, "Forfeiting the Future," *Resurgence*, no. 208 (2001), http://www.resurgence.org/resurgence/issues/joy208.htm.

105 *"By the way, Joy's thesis is spot-on"* Special forces officer, interview, Peter W. Singer, Washington, DC, September 7, 2006.

105 *"Never before in history"* As quoted in Frank Schirrmacher, "Beyond 2001: HAL's Legacy for the Enterprise Generation," *Frankfurter Allgemeine Zeitung*, August 31, 2000.

105 *"The Future Is Coming Sooner Than You Think"* Jim Saxton, "Nanotechnology: The Future Is Coming Sooner Than You Think" (Washington, DC: Joint Economic Committee, U.S. Congress, 2007), available at http://www.house.gov/jec/publications/110/nanotechnology_03-22-07.pdf.

106 *"some people, smart people"* Robert Epstein, interview, Peter W. Singer, October 25, 2006.

107 *IBM and Intel found a way* Reuters, "Intel, IBM Unveil New Chip Technology," Speedguide.net, 2007 (cited May 30, 2007); available at http://www.speedguide.net/read_news.php?id=2240.

107 *sixteen times more operations than a normal computer transistor* Charles Choi, "Tiny Brainlike Computer Created," *Live Science*, March 10, 2008.

107 *That is, while an electric charge* Peter Schwartz, Chris Taylor, and Rita Koselka, "Quantum Leap," *Fortune* 154, no. 3 (2006).

107 *"The challenges facing the robotics industry"* Ibid.

107 *"doesn't require us to try"* Robert Finkelstein, interview, Peter W. Singer, July 7, 2006.

107 *"If this war keeps going"* Personal communication at the Military Robotics Conference, Institute for Defense and Government Advancement, Washington, DC, April 10–12, 2006.

107 *"When you marry all that up"* Robert Epstein, interview, Peter W. Singer, October 25, 2006.

## 5. COMING SOON TO A BATTLEFIELD NEAR YOU: THE NEXT WAVE OF WARBOTS

109 *"They're going to sneak up on us"* John Pike, as quoted in Preston Lerner, "Robots Go to War: Within 10 Years, Infantry Soldiers Will Go into Battle with Autonomous Robots Close Behind Them," *Popular Science* 268, no. 1 (2006).

109 *" Instead of us telling machines where to go"* Noah Shachtman, interview, Peter W. Singer, Washington, DC, March 25, 2006.

109 *In the course of his reporting* Noah Shachtman, "The Baghdad Bomb Squad," *Wired* 13.11 (2005), http://www.wired.com/wired/archive/13.11/bomb .html.

110 *"The robots you are seeing here"* Patrick Rowe as quoted in *Warbots*, produced by Dan Saxton Company, History Channel, broadcast on August 8, 2006.

110 *"But I'm convinced"* Lerner, "Robots Go to War."

110 *twenty-two different prototypes of intelligent ground vehicles* Robert Finkelstein, interview, Peter W. Singer, July 7, 2006.

110 *"It's already been done"* Ibid.

111 *fire a variety of ammunition* John Dyer, "Robotics in Urban Warfare," paper presented at the Military Robotics Conference, Institute for Defense and Government Advancement, Washington, DC, April 10–12, 2006.

111 *"You'll actually see the sniper"* Hiawatha Bray, "Robotic-Vacuum Maker, BU Team Up on Anti-sniper Device," *Boston Globe*, October 4, 2005, E3.

111 *"It is not an insurmountable problem"* Bob Quinn, interview, Peter W. Singer, November 17, 2006.

111 *"world's first multipurpose combat robot"* Dennis Sorenson, "Technological Development of Unmanned Systems to Support the Naval Warfighters," paper presented at the Military Robotics Conference, Institute for Defense and Government Advancement, Washington, DC, April 10–12, 2006.

112 *"It is just fucking nasty"* Noah Shachtman, interview, Peter W. Singer, March 25, 2006.

112 *"If you can avoid unnecessary situations"* Ibid.

112 *"could be operational on the battlefield"* Bill Christensen, "Trauma Pod Battlefield Medical Treatment System," Technology.com, April 5, 2005 (cited July 31, 2006); available at http://www.technovelgy.com/ ct/Science-Fiction-News.asp?NewsNum=364.

113 *"The average surgeon will become"* Robert Langreth, "Robo-Docs," *Forbes* 178, no. 4 (2006): 100.

113 *"The last thing I want to see"* History Channel, *Warbots*.

113 *Future Combat Systems (FCS) program* United States Congressional Budget Office, *The Army's Future Combat Systems Program and Alternatives* (Washington, DC, 2006).

114 *"it would have saved an NCO's life"* Fred Baker III,

"Soldiers Like FCS Test Systems So Much, They Don't Want to Return Them," *Army News Service*, February 13, 2007.

114 *cost as much as $16 billion a year* Jeffrey M. Tebbs, "Smelting the Triangle: Constraining Congress, Defense Contractors, and the Military Brass to Restore a Fiscally Prudent Defense Budget," (Washington, DC: Brookings Institution, 2006), 12.

114 *"the largest weapons procurement in history"* Robert Finkelstein, interview, Peter W. Singer, July 7, 2006.

114 *"it's the system that ate the army"* Ralph Peters, interview, Peter W. Singer, Washington, DC, March 29, 2007.

114 *the majority of technical hurdles* United States Congressional Budget Office, *The Army's Future Combat Systems Program and Alternatives*, 82, 39.

114 *"Everything's working against you"* Carl Posey, "Robot Submarines Go to War," *Popular Science* 262, no. 4 (2003).

115 *"The civilian sailors were"* James F. Dunnigan, "Robotic Ship Talks to Startled Sailors," *Strategy Page*, June 14, 2005 (cited June 14, 2005); available at http://www.strategypage.com/dls/articles/ 200561415554.asp.

115 *the Fire Scout can fly more than* Nick Brown, "Fire Scout Takes Over Landing Control," *Jane's Defence Weekly*, February 1, 2006, 30.

116 *It lands in the water* Bill Sweetman, "The Navy's Swimming Spy Plane: It Floats, It Flies, It Eliminates Enemy Targets—Meet the Water-Launched Unmanned Enforcer," *Popular Science* 268, no. 3 (2006).

116 *"I want to see a Predator"* David A. Fulghum, "Predator's Progress," *Aviation Week & Space Technology* 158, no. 9 (2003): 48.

116 *"standing alert somewhere"* Bill Sweetman, "USAF Predators Come of Age in Iraq and Afghanistan as Reaper Waits in the Wings," *Jane's International Defence Review* 39, no. 6 (2006): 52.

116 *The air force sees at least 45 percent* Robert S. Boyd, "They're Very Expensive, but They Save Lives: U.S. Enlisting Smart Robots for War's Dirty, Deadly Jobs," *Philadelphia Inquirer*, February 20, 2006, E2.

117 *looking like "a B-2 bomber's chick"* Bill Sweetman, "The Top-Secret Warplanes of Area 51," *Popular Science*, October 2006, http://www.popsci.com/ popsci/aviationspace/95e16f096bd8do10vgnvcm 1000004eecbccdrcrd/7.html.

117 *"a fully autonomous flight control"* Nick Cook, "Skunk Works Unveils Secret Polecat UAV," *Jane's Defence Weekly*, July 19, 2006, http://www .janes.com/regional_news/americas/news/jdw/ jdw060719_1_n.shtml.

117 *It would have a wingspan* Bill Sweetman, "Boeing Working on New Large UAV," *Jane's Defence Weekly*, July 5, 2006.

117  *The next step is DARPA's plan* Ramon Lopez, "Five-Year Plan," *Defense Technology International* 1, no. 7 (2007): 16.

117  *airships could literally be "parked"* David A. Fulghum, "Space-RAAM: AIM-120 Recast as Ballistic Missile Interceptor," *Aviation Week & Space Technology* 166, no. 19 (2007): 32.

117  *"itty-bitty, teeny-weeny UAVs"* Christian Lowe, "High-Flying, Secret Drone Unveiled," Defense tech.org, July 24, 2006 (cited December 18, 2006); available at http://www.defensetech.org/archives/002598.html.

117  *small, pilotless planes* Dina El Boghdady, "Small Firms Turn to Drones: Demand Grows for Unmanned Craft," *Washington Post*, October 31, 2005, D1.

117  *"It was tough to track on film"* David Hart, "Nano-Air Vehicle Program," paper presented at the Military Robotics Conference, Institute for Defense and Government Advancement, Washington, DC, April 10–12, 2006.

118  *"similar in size and shape to a maple tree seed"* Lowe, "High-Flying, Secret Drone Unveiled."

118  *"A lot of the three-letter agencies"* Robert Finkelstein, interview, Peter W. Singer, July 7, 2006.

118  *"perch and stare" into windows* Jim Pinto, "Intelligent Robots Will Be Everywhere," Automation.com (cited August 22, 2005); available at http://www.automation.com/sitepages/pid1014.php.

118  *recharge themselves off electrical outlets* Ibid.

118  *Boston College researchers* Ibid., 234.

118  *"It is a bit like when stone-age man"* Reuters, "1867 Nanomachine Now Reality," CNN.com, February 1, 2007 (cited February 3, 2007); available at http://www.cnn.com/2007/TECH/02/01/nanomachine.reut/index.html.

119  *"flat as a pancake"* Matthew Brzezinski, "The Unmanned Army," *New York Times Magazine*, April 20, 2003.

120  *"Can you see him now, sir?!?"* interview with U.S. soldier, Peter W. Singer, September 18, 2007.

120  *"space Pearl Harbor"* Shephard W. Hill, "A Legacy of Support to the Warfighter," *High Frontier Journal* 2, no. 3 (2006): 22.

121  *to "crush someone anywhere in the world"* As quoted in Walter Pincus, "Pentagon Has Far-Reaching Defense Spacecraft in Works," *Washington Post*, March 16, 2005, A3.

121  *"the single dumbest thing I have heard so far"* Ibruegger, "Special Report: Emerging Technologies and Their Impact on Arms Control and Non Proliferation," NATO Parliamentary Assembly, International Secretariat, 2001, 41.

121  *open up the floodgates for others* Bruce M. DeBlois, "Space Sanctuary: A Viable National Strategy," *Airpower Journal* 12, no. 4 (1998).

121  *"a space superpower, it is not going to be alone"* Richard Fisher, "Space to Manoeuvre," *Jane's Intelligence Review*, March 2007, 63.

121  *the Tamil Tiger group of Sri Lanka* Peter de Selding, "Intelsat Vows to Stop Piracy by Sri Lankan Separatist Group," *Space News*, April 17, 2007, 1, 4.

122  *"Once safely in orbit"* "Bots Will Battle to in Space," *New Scientist,* April 12, 2006 (cited January 21, 2007); available at http://www.newscientist.com/blog/technology/2006/04/bots-will-to-battle-in-space.html.

## 6. ALWAYS IN THE LOOP? THE ARMING AND AUTONOMY OF ROBOTS

123  *"Wars are a human phenomenon"* Thomas K. Adams, "Future Warfare and the Decline of Human Decisionmaking," *Parameters* 31, no. 4 (2001): 57.

123  *"people will always want humans"* Eliot Cohen, interview, Peter W. Singer, Washington, DC, November 15, 2006.

123  *"In some cases, the potential exists"* Patrick Eberle, "To UAV or Not to UAV: That Is the Question; Here Is One Answer," *Air & Space Power Journal—Chronicles Online Journal*, October 9, 2001, http://www.airpower.au.af.mil/airchronicles/cc/eberle.html.

124  *"It's far away enough"* Helen Greiner, interview, Peter W. Singer, November 16, 2006.

124  *"ever be able to autonomously fire"* Bob Quinn, interview, Peter W. Singer, November 17, 2006.

124  *"It helps keep people calm"* Noah Shachtman, interview, Peter W. Singer, Washington, DC, July 2, 2007.

124  *"The navigation computer"* Eberle, "To UAV or Not to UAV."

124  *The system came with four modes* George Friedman and Meredith Friedman, *The Future of War: Power, Technology, and American World Dominance in the Twenty-first Century,* 1st ed. (New York: Crown, 1996), 196.

125  *nicknamed "Robo-cruiser"* "Iran Air Flight 655," Wikipedia, July 7, 2007 (cited July 8, 2007); available at http://en.wikipedia.org/wiki/Iran_Air_Flight_655.

125  *the computer was trusted* Chris Gray, *Postmodern War: The New Politics of Conflict* (New York: Guilford Press, 1997), 69.

125  *"In ten to twenty years"* Andrew Bennett, interview, Peter W. Singer, November 16, 2006.

125  *"just a political description"* Ray Kurzweil, interview via phone, Peter W. Singer, Washington, DC, December 7, 2006.

125  *U.S. Patriot missile batteries* Defense Science Board, "Report of the Defense Science Board Task Force on Patriot System Performance" (Washington, DC, 2005).

126  *"The irony"* Robert Epstein, interview, Peter W. Singer, Washington, DC, October 25, 2006.

126  *"you cannot take the human out"* Predator pilot, interview, Peter W. Singer, August 28, 2006.

126  *"currently, at best, very ambitious"* Michael J. Barnes et al., "Soldier Interactions with Aerial and Ground Robots in Future Military Environments" (NATO, 2006).

126  *"Even if the tactical commander"* Sean J. A. Edwards, "Swarming and the Future of Warfare" (doctoral thesis, Pardee Rand Graduate School, 2005), 139.

127  *"By making them autonomous"* Jim Rymarcsuk, interview with Ralph Wipfli, Washington, DC, October 20, 2006.

127  *needs at least .3 seconds* Robert Kosinski, "A Literature Review on Reaction Time," *Clemson University* (2006), available at: http://biae.clemson.edu/bpc/bp/Lab/110/reaction.htm.

127  *"If you can automatically hit it"* U.S. military officer, interview, Peter W. Singer, October 17, 2006.

127  *"Anyone who would shoot"* Stephen Graham, "America's Robot Army," *New Statesman*, June 12, 2006, http://www.newstatesman.com/200606120018.

127  *"establish a track record of reliability"* Ibid.

127  *"supervisor who serves"* Adams, "Future Warfare and the Decline of Human Decisionmaking," 58.

128  *"are rapidly taking us to a place"* Ibid., 57.

128  *"humans will always be in"* Randall Steeb, *Examining the Army's Future Warrior: Force-on-Force Simulation of Candidate Technologies* (Santa Monica, CA: RAND, 2004), 44.

128  *"Let's design our armed unmanned systems"* Stephen Trimble, "DoD Group Seeks to Give Autonomy to Armed Drones," *Jane's Defence Weekly*, October 11, 2006, 10.

128  *"Concept for the Operation of Armed Autonomous Systems"* Ronald C. Arkin, "Governing Legal Behavior: Embedding Ethics in a Hybrid Deliberative/Reactive Robot Architecture" (Georgia Institute of Technology/U.S. Army Research Office, 2007); John S. Canning, "Concept for the Operation of Armed Autonomous Systems on the Battlefield," Dahlgren Division, Naval Surface Warfare Center, 2008.

128  *"Unmanned Effects: Taking the Human Out of the Loop"* U.S. Joint Forces Command, "Military Robots of the Future" (U.S. Joint Forces Command, 2003).

128  *"all the lip service paid"* Preston Lerner, "Robots Go to War: Within 10 Years, Infantry Soldiers Will Go into Battle with Autonomous Robots Close Behind Them," *Popular Science* 268, no. 1 (2006).

128  *"That's exactly the kind"* Special forces officer, interview, Peter W. Singer, Washington, DC, September 7, 2006.

129  *"There is a lot of fear"* Retired air force officer, interview, Peter W. Singer, January 28, 2007.

129  *"Soon you will just"* James Lasswell, interview, Peter W. Singer, Washington, DC, November 7, 2006.

129  *"We all joke about it"* As quoted in John Barry and Evan Thomas, "Up in the Sky, an Unblinking Eye," *Newsweek*, June 4, 2008.

129  *officers describe unmanned systems* Elizabeth Bone and Christopher Bolkcom, *Unmanned Aerial Vehicles: Background and Issues* (Congressional Research Service, Library of Congress, 2003).

130  *"Maybe you don't need fighter pilots at all"* Noah Shachtman, "Attack of the Drones," *Wired* 13.06 (2005), http://www.wired.com/wired/archive/13.06/drones.html.

130  *Wilkerson is not some groundpounder* Ibid.

130  *"We clearly see an evolution"* Bob Quinn, interview, Peter W. Singer, November 17, 2006.

130  *Special forces roles were felt* Robert Finkelstein and James Albus, "Technology Assessment of Autonomous Intelligent Bipedal and Other Legged Robots" (DARPA, 2004).

130  *"As technology advances"* Ibid.

130  *"2035 [that] we will have robots"* Ibid.

131  *"Common sense is not a simple thing"* Ray Kurzweil, *The Singularity Is Near: When Humans Transcend Biology* (New York: Viking, 2005), 177.

131  *our "emotional intelligence"* Ibid., 191.

131  *Rod Brooks of MIT and iRobot predicts* Rodney Brooks, *Flesh and Machines: How Robots Will Change Us* (New York: Pantheon, 2002), 22.

131  *"My job will be eliminated"* As quoted by David Bruemmer, "Intelligent Autonomy for Unmanned Vehicles," paper presented at the Military Robotics Conference, Institute for Defense and Government Advancement, Washington, DC, April 10–12, 2006.

131  *"Asking whether robots"* Rodney Brooks, interview, Peter W. Singer, Washington, DC, October 30, 2006.

132  *Haile, a robot musician* Gil Weinberg and Scott Driscoll, "Haile," 2006 (cited July 7, 2007); available at http://www-static.cc.gatech.edu/~gilwein/Haile.htm.

132  *understand and interact with human musicians* Matthew Abshire, "Musical Robot Composes, Performs and Teaches," CNN.com, October 3, 2006 (cited October 3 2006); available at http://www.cnn.com/2006/TECH/10/03/musical.robot/index.html.

132  *"I firmly believe"* H. R. Everett, interview, Peter W. Singer, October 20, 2006.

132  *"The challenge is to create a system"* Nick Turse, "Baghdad 2025: The Pentagon Solution to a Planet of Slums," TomDispatch.com, January 7, 2007, http://www.tomdispatch.com/post/155031/nick_turse_pentagon_to_global_cities_drop_dead.

133  *the soldier would call the "play"* Susan R. Flaherty et al., "Playbook® Control of Multiple Heterogeneous Weaponized UAVs," paper presented at the Unmanned Systems North America, AUVSI's 34th Annual Symposium and Exhibition, Washington, DC, August 6–9, 2007.

133  *"Just see it and shoot it is not the future"* Thomas

McKenna, interview, Peter W. Singer, Arlington Office of Naval Research, December 12, 2006.

133 *"The robot will do what robots do best"* Ibid.

133 *"human and robot roles will evolve"* Bruemmer, "Intelligent Autonomy for Unmanned Vehicles."

133 *The military, then, doesn't expect* U.S. Joint Forces Command, "Military Robots of the Future." See also "Automated Killer Robots 'Threat to Humanity': expert," Agence France-Presse, February 26, 2008.

133 *"I believe we should think"* Finkelstein and Albus, "Technology Assessment of Autonomous Intelligent Bipedal and Other Legged Robots," 182.

133 *"The next war could be fought"* Lee Dye, "New Vehicles Will Make Own Decisions Based on Commands," ABC News, November 17, 2004 (cited July 18, 2006); available at http://www.strategicstudies institute.army.mil/about/contact-us.cfm.

133 *WT-6 is a robot in Japan* Eric Mika, "This Modern Robot," *Popular Science* 269, no. 3 (2006): 66.

134 *"trust is a huge issue"* Bruemmer, "Intelligent Autonomy for Unmanned Vehicles."

7. ROBOTIC GODS: OUR MACHINE CREATORS

135 *"You have to remember"* Daniel Wilson, interview, Peter W. Singer, October 19, 2006.

135 *"Each year some forty-two thousand people"* Sebastian Thrun, interview, Peter W. Singer, March 18, 2007.

135 *The Grand Challenge is a robotics road race* DARPA, "Grand Challenge 2004 Final Report," 2004 (cited May 4 2006); available at http://www .darpa.mil/body/NewsItems/pdf/DGCreport 30July2004.pdf+%22grand+challenge+2004%22 +final+report+to+Congress&hl=en&gl=us&ct= clnk&cd=1; DARPA, "Grand Challenge 2004 Rules," 2004 (cited May 4, 2006); available at http://www .darpa.mil/grandchallenge05/Rules_8oct04.pdf; Anthony J. Tether, "Grand Challenge 2004 Briefing," 2005 (cited May 4, 2006); available at http:// www.darpa.mil/body/pdf/Courtyard_Event_ Briefing12_05_05.pdf; Lee Gomes, "Team of Amateurs Cuts Ahead of Experts in Computer-Car Race," *Wall Street Journal*, October 19, 2005.

136 *"it's an endurance race"* Warbots, produced by Dan Saxton Company, History Channel, broadcast on August 8, 2006.

136 *"the first Grand Challenge came off"* Preston Lerner, "Robots Go to War: Within 10 Years, Infantry Soldiers Will Go into Battle with Autonomous Robots Close Behind Them," *Popular Science* 268, no. 1 (2006).

137 *as much as $100 million in investment* Tether, "Grand Challenge 2004 Briefing"; Gomes, "Team of Amateurs Cuts Ahead of Experts in Computer-Car Race."

137 *"the best part of the Grand Challenge"* Business executive, interview at the Military Robotics Conference in Washington, DC, Peter W. Singer, April 20, 2006.

137 *"We trained Stanley"* Sebastian Thrun, interview, Peter W. Singer, March 18, 2007.

137 *"We all won"* Sebastian Thrun, as quoted in Elizabeth Corcoran, "Data Driver," *Forbes* 178, no. 4 (2006): 102.

137 *one of the ten best and brightest minds* Rena Marie Pacella, "DARPA Grand Challenge—Sebastian Thrun," *Popular Science*, October 2005 (cited May 4, 2006); available at http://www.popsci.com/ popsci/darpachallenge/5a6450f8d22d6010vgnvc m1000004eecbccdrcrd.html.

138 *Stanley was declared the number one robot* Robert Capps, "The 50 Best Robots Ever," *Wired* 14.01 (2006), http://www.wired.com/wired/archive/ 14.01/robots.html?pg=2&topic=robots&topic_ set=.

138 *research usually takes place* Daniel Richard O'Brien, "Area 51," 2003 (cited May 4 2006); available at http://www.area51show.co.uk/index.htm.

139 *"too many rules specifications"* Brian Miller, interview at the Military Robotics Conference in Washington, DC, Peter W. Singer, April 10–12, 2006.

139 *"Hands down, robots"* Daniel H. Wilson, "About the Author," 2005 (cited August 30, 2006); available at http://www.robotuprising.com/qanda.htm.

139 *"When you are deciding"* Daniel Wilson, interview, Peter W. Singer, October 19, 2006.

139 *"it left [me] with an empty feeling"* Colin Angle, as quoted in David Whelan, "Fights Wars, Lint," *Forbes* 178, no. 4 (2006).

139 *"Nothing could make it so clear"* Colin Angle, as quoted in Mike Miliard, "Deus Ex Machina," *Boston Phoenix*, February 18–24, 2005, http://www .bostonphoenix.com/boston/news_features/other_ stories/multi_3/documents/04475119.asp.

139 *"We always knew we would change"* Helen Greiner, interview, Peter W. Singer, November 16, 2006.

139 *"GIT Rockin': Government IT Rocks, Do You?"* "GIT Rockin': Government IT Rocks, Do You?" 2006 (cited May 4, 2006); available at www .gitrockin.com.

140 *"featured talent from Juniper Networks"* Federal Computer Week, "Special Report: GIT Rockin'," 2006 (cited June 8, 2007); available at http://www .fcw.com/gitrockin/.

140 *"Folks from around"* Ibid.

140 *up to a third of major university research faculty* Jonathan Moreno, "Mind Wars: Brain Research and National Defense," presentation at the Center for American Progress, Washington, DC, December 7, 2006.

140 *"accelerate the future into being"* Joel Garreau, *Radical Evolution: The Promise and Peril of Enhancing Our Minds, Our Bodies—And What It Means to Be Human* (New York: Doubleday, 2005), 24.

140 *And now it's focusing on robots* Ibid., 22.

141 *"who work at the forefront of"* Joel Garreau, "Perfecting the Human," May 30, 2005 (cited April 4, 2007); available at http://mindfully.org/Technology/2005/Perfecting-The-Human30 may05.htm.

141 *"By the time a technology"* Ibid.

141 *"it takes risks"* Sebastian Thrun, interview, Peter W. Singer, March 18, 2007.

141 *It is supposed to be a secret location* Marc Fisher, "Secret Buildings You May Not Photograph, Part 643," Washingtonpost.com, July 18, 2007, http://blog.washingtonpost.com/rawfisher/2007/07/secret_buildings_you_may_not_p.html.

141 *"challenges verging on the impossible"* Steven Wax, as quoted in Garreau, *Perfecting the Human.*

141 *the "Frankensteins in the Pentagon"* Cheryl Seal, "Frankensteins in the Pentagon: DARPA's Creepy Bioengineering Program," *Information Clearing House,* August 25, 2003, http://www.informationclearinghouse.info/article4572.htm.

141 *"a mind-numbing mix"* Defense industry expert, interview, Peter W. Singer, Washington, DC, September 28, 2006.

141 *"real madmen"* Ibid.

142 *"I spend an inordinate amount of time"* As quoted in Noah Shachtman, "Senate vs. Darpa," Defensetech.org, July 21, 2006 (cited July 21, 2006); available at http://www.defensetech.org/archives/002599.html.

142 *"I have had everyone complain"* Ibid.

142 *development of such varied programs* Robert Kavetsky and Christopher J. R. McCook, "The Technological Perfect Storm," *Proceedings,* October 2006.

142 *"ideas that literally changed the world"* Michael T. Isenberg, *Shield of the Republic: The United States Navy in an Era of Cold War and Violent Peace,* 1st ed. (New York: St. Martin's Press, 1993).

143 *"I was supporting some one hundred top graduate students"* Thomas McKenna, Arlington Office of Naval Research, interview, Peter W. Singer, December 12, 2006.

143 *"sometimes I just find them on the Web"* Ibid.

143 *"an idea and technology hummingbird"* Interview with NSA official, Peter W. Singer, Arlington–Crystal City, VA, February 29, 2008.

144 *The mud battery would refuel* Bijal P. Trivedi, "Mud Batteries: Power Cells of the Future?" *National Geographic Today,* May 20, 2004, http://72.14.209.104/search?q=cache:F6_mV5yUpDkJ:news.nationalgeographic.com/news/2002/01/0122_020122_tvmudbatteries.html+mud+battery&hl=en&gl=us&ct=clnk&cd=8.

144 *"You basically beat the snot"* Andrew Bennett, interview, Peter W. Singer, November 16, 2006.

145 *It took eleven days for the labs* Dale G. Uhler, "Technology: Force Multiplier for Special Operations," *Joint Force Quarterly,* no. 40 (2006).

145 *"Saving Ryan's Privates"* Michael Garrett, "Saving Ryan's Privates: New 'Armored' Shorts Protect Precious Arteries," Military.com, 2005 (cited September 13, 2006); available at http://www.military.com/soldiertech/0,14632,Soldiertech_Kevlar,,00.html?ESRC=soldiertech.nl.

145 *"When your butt's on the line"* Ibid.

145 *"Any one who wants to play"* David Bruemmer, "Intelligent Autonomy for Unmanned Vehicles," paper presented at the Military Robotics Conference, Institute for Defense and Government Advancement, Washington, DC, April 10–12, 2006.

146 *"my obsession with robots"* H. R. Everett, interview, Peter W. Singer, October 20, 2006.

146 *number sixteen on* Wired *magazine's list* Capps, "The 50 Best Robots Ever."

147 *the ultimate opportunity for customer feedback* Daniel Wilson, interview, Peter W. Singer, October 19, 2006.

147 *"Heat and computers don't mix well"* Noncommissioned officer, interview at the Military Robotics Conference in Washington, DC, Peter W. Singer, April 10–12, 2006.

147 *"Make it work"* Army specialist Jacob Chapman, interview at the Military Robotics conference in Washington, DC, Peter W. Singer, April 10–12, 2006.

148 *"We're going to take"* Tom Ryden, interview at the Military Robotics Conference, in Washington, DC, Peter W. Singer, April 10–12, 2006.

148 *"sometimes we get phone calls"* Mark Barber, "Force Protection Robotics," paper presented at the Military Robotics Conference, Institute for Defense and Government Advancement, Washington, DC, April 10–12, 2006.

148 *"The scientist did not need"* George Friedman and Meredith Friedman, *The Future of War: Power, Technology, and American World Dominance in the Twenty-first Century,* 1st ed. (New York: Crown, 1996), 43.

148 *"There are tons of guys"* Military analyst, e-mail, Peter W. Singer, June 12, 2007.

148 *"Our robots have logistic information"* Jim Rymarcsuk, interview with Ralph Wipfli, Washington, DC, October 20, 2006.

149 *went through some thirty-five different changes* Byron Brezina, interview at the Military Robotics Conference, Peter W. Singer, Washington, DC, April 10–12, 2006.

149 *a request for "warranty repair"* John Dyer, "Robotics in Urban Warfare," paper presented at the Military Robotics Conference, Institute for Defense and Government Advancement, Washington, DC, April 10–12, 2006.

## 8. WHAT INSPIRES THEM: SCIENCE FICTION'S IMPACT ON SCIENCE REALITY

150 *"You can never tell"* Donna Shirley, interview, Peter W. Singer, Washington, DC, October 2, 2006.

150 *"ridiculous . . . monstrosity"* Klosterman, *Sex, Drugs, and Cocoa Puffs: A Low Culture Manifesto* (New York: Scribner, 2003), 220.

151 *"like admitting that you masturbate"* Ibid., 149.

151 *the navy's "Professional Reading" program* Admiral Mike Mullen, "The Means of Knowledge: The Navy's New Professional Reading Program," *Proceedings* 132, no. 10 (2006): 22–23.

152 *Museum director Donna Shirley* Donna Shirley, interview, Peter W. Singer, October 2, 2006. See also Science Fiction Museum, "Donna Shirley Named Director of Science Fiction Museum," February 11, 2004 (cited October 16, 2006); available at http://www.sfhomeworld.org/press_room/donnashirley.pdf.

152 *"Although the guys in my classes"* Sally Richards, "Managing Martians: Donna Shirley, WITI Hall of Fame Inductee, Talks About Generating Creativity and Accomplishing Goals in the Workplace," *Women in Technology International*, 1989–2000 (cited July 7, 2007); available at http://www.witi.com/wire/feature/dshirley.shtml.

152 *"Not only were these events"* "Donna Shirley: Managing Martians," *Managing Creativity* (cited October 16, 2006); available at http://managingcreativity.com/.

152 *"but their heroes were always"* Donna Shirley, interview, Peter W. Singer, October 2, 2006.

152 *"educate people about science fiction"* Ibid.

153 *"The technology is not the interesting part"* Ibid.

153 *"the political and legal ramifications"* "Science Fiction," *Brainy Encyclopedia*, 2006 (cited August 21, 2006); available at http://www.brainyencyclopedia.com/encyclopedia/s/sc/science%5ffiction.html.

153 *science fiction is more about* Harry Turtledove and Martin Harry Greenberg, eds., *The Best Military Science Fiction of the 20th Century*, 1st ed. (New York: Ballantine, 2001).

153 *"remade and rereleased every time"* Donna Shirley, interview, Peter W. Singer, October 2, 2006.

153 *"Women writers tend to write"* Ibid.

153 *"in the hands of our depraved society"* Ibid.

153 *"I thought Ender's Game"* Orson Scott Card, interview by e-mail, Peter W. Singer, January 24, 2007.

154 *"Soldiers feel like Ender's Game"* Ibid.

155 *"The real question"* Ibid.

155 *"The conflict is obvious [in war]"* Robin Wayne Bailey, interview, Peter W. Singer, September 27, 2006.

155 *"Fiction is about character"* Turtledove and Greenberg, eds., *The Best Military Science Fiction of the 20th Century*, viii.

156 *"It only seems fitting"* Heinlein Centennial Inc., "The U.S.S. Robert A. Heinlein Campaign," Open Letter: The U.S.S. Robert A. Heinlein Campaign, Secretary of the Navy Donald C. Winter.

156 *"the Father of Science Fiction"* Adam Roberts, *The History of Science Fiction* (New York: Routledge, 2000), 48.

157 *called Wells's idea "moonshine"* Richard Rhodes, *The Making of the Atomic Bomb* (New York: Simon & Schuster, 1986).

157 *"The forecast of the writers"* Leó Szilárd, "Letter to Hugo Hirst on Forecast of Discoveries in Physics," Project of Nuclear Age Peace Foundation, 1934 (cited July 7, 2007); available at http://www.nuclearfiles.org/menu/library/correspondence/szilard-leo/corr_szilard_1934-03-17.htm.

157 *"the Man Who Invented Tomorrow"* Peggy Teeters, *Jules Verne: The Man Who Invented Tomorrow* (New York: Walker and Company, 1993).

158 *George is just a run-of-the-mill database administrator* CNN.com, "From Sci-Fi to Reality," August 1, 2006.

158 *"I think there is a world market"* David Hambling, *Weapons Grade: How Modern Warfare Gave Birth to Our High-Tech World* (New York: Carroll and Graf, 2005), 153.

158 *"We don't do well, historically"* Joseph J. Collins, "From the Ground Up," *Armed Forces Journal*, October 2006, 47.

158 *"no nation would permit it"* Tom Reiss, "Imagining the Worst," *New Yorker* 81, no. 38 (2005): 112.

159 *"Our advances in technical intelligence"* Collins, "From the Ground Up," 47.

159 *"Science fiction is not making predictions"* Robert Epstein, interview, Peter W. Singer, Washington, DC, October 25, 2006.

159 *"Science fiction at its best is about ideas"* Robin Wayne Bailey, interview, Peter W. Singer, September 27, 2006.

159 *"Science fiction did not predict"* Donna Shirley, interview, Peter W. Singer, October 2, 2006.

159 *"Science fiction is unreliable"* Ray Kurzweil, interview via telephone, Peter W. Singer, Washington, DC, December 7, 2006.

160 *"'I told you so'"* Orson Scott Card, interview via e-mail, Peter W. Singer, January 24, 2007.

160 *"It's the near future"* Greg Bear, *Quantico* (London: Harper Collins UK, 2005).

160 *"I mean, how many science fiction books"* David Sonntag, interview via e-mail, Peter W. Singer, Washington, DC, November 28, 2006.

160 *"think tank of patriotic science fiction"* Grant Slater, "Futuristic Writers Offer Ideas to Fight Terrorism," *St. Louis Post-Dispatch*, May 25, 2007.

160 *"If you don't read science fiction"* Ibid.

161 *"I say there are bad people"* Jay Cohen, as quoted in Slater, "Futuristic Writers Offer Ideas to Fight Terrorism."

161 *"If you lead the life"* Greg Bear, interview, Peter W. Singer, October 4, 2006.

161 *"Harry Truman loved science fiction"* Ibid.

161   *"They seem to be more like FDR"* Ibid.

161   *"How William Shatner Changed the World"* Julian Jones, director, *How William Shatner Changed the World,* produced by the History Channel, broadcast on October 21, 2006.

161   *"All this wiz-bangering"* Ibid.

162   *drink a "Commander Riker-Rita"* "Star Trek: The Experience" (cited July 7, 2007); available at http://www.startrekexp.com/.

162   *fan base is 250 million Trekkies strong* Sue Kovach Shuman, "Set Phasers on Stun: Fans Beaming Up for Special Events as 'Star Trek' Turns 40," *San Francisco Chronicle,* August 20, 2006, http://www.sfgate.com/cgi-bin/article.cgi?f=/c/a/2006/08/20/TRGPAKJDBK1.DTL.

162   *the origin of the transporter How William Shatner Changed the World.*

162   *"There's Captain Kirk"* Ibid.

162   *"In Silicon Valley, everyone's a* Star Trek *fan"* Ibid.

163   *Perlman is working* Ibid.

163   *an anthology of short stories* Turtledove and Greenberg, eds., *The Best Military Science Fiction of the 20th Century.*

163   *These range from exoskeleton suits* Such systems actually became a point of political debate in the Iraq war, as the Pentagon delayed developing and deploying such systems, despite their potential for protecting vehicles from ambush.

163   *"We wanted ground mobility"* James Lasswell, interview, Peter W. Singer, Washington, DC, November 7, 2006.

163   *"We got the idea from* Dick Tracy*"* Ibid. The Israelis have a similar device they call V-Rambo. See Associated Press, "Israeli Army Wrist Video," *Wired News,* March 6, 2005; http://www.wired.com/news/technology/0,66807-0.html.

163   *"But now we are finding"* Andrew Bennett, interview, Peter W. Singer, November 16, 2006.

164   *"People like Isaac Asimov"* Mihail C. Roco and William Sims Bainbridge, "Converging Technologies for Improving Human Health: Nanotechnology, Biotechnology, Information Technology and Cognitive Science" (National Science Foundation, 2002).

164   *"We picked the PHaSR name"* United Press International, "Military Develops a Star Trek-like Phaser," Physorg.com, 2005 (cited August 13, 2006); available at http://www.physorg.com/news8641.html.

164   *"The popularity of robots in fiction"* Bill Gates, "A Robot in Every Home," ScientificAmerican.com, December 16, 2006 (cited December 17, 2006); available at http://www.sciam.com/article.cfm?id=a-robot-in-every-home.

164   *"It's a way to make possibilities"* Developer, interview at the Military Robotics Conference, Peter W. Singer, Washington, DC, April 10–12, 2006.

164   *"Naval customers just assume"* Thomas McKenna, interview, Peter W. Singer, Arlington Office of Naval Research, December 12, 2006.

164   *"You have to beg for money"* Steven Metz, interview, Peter W. Singer, September 19, 2006.

165   *"Any sufficiently advanced technology"* Arthur Charles Clarke, *Profiles of the Future,* rev. ed. (New York: Harper & Row, 1973), 21.

165   *a magic box that killed almost a thousand spear-armed warriors* John Ellis, *The Social History of the Machine Gun,* Johns Hopkins paperback ed. (Baltimore: Johns Hopkins University Press, 1986), 89.

165   *what analysts call "Future Shock"* Alvin Toffler, *Future Shock* (New York: Random House, 1970).

165   *"There seems a strong tendency"* Timothy Hornyak, as quoted in Mark Jacob, "Japan's Robots Stride into Future," *Chicago Tribune,* July 15, 2006, 7.

166   *"stems from the fact that doomsday scenarios"* H. R. Everett, interview, Peter W. Singer, October 20, 2006.

166   *"To be realistic, it's going to be"* Andrew Bridges, "Scientists Aim to Duplicate Harry Potter's Invisibility Cloak," LiveScience.com, May 25, 2006 (cited May 25, 2006); available at http://www.livescience.com/scienceoffiction/060525_invisible_cloak.html.

166   *"And I have no idea"* Rod Brooks, interview, Peter W. Singer, October 30, 2006.

167   *"So, it is easy to use machines"* Jacob, "Japan's Robots Stride into Future."

167   *robots are given Shinto rites* Timothy N. Hornyak, *Loving the Machine: The Art and Science of Japanese Robots,* 1st ed. (Tokyo; New York: Kodansha International, 2006).

167   *"If you make something"* Jacob, "Japan's Robots Stride into Future."

168   *"too artificial and icky"* Rod Brooks, interview, Peter W. Singer, October 30, 2006.

168   *his lab has more collaboration with Asian companies* Sebastian Thrun, interview, Peter W. Singer, March 18, 2007.

168   *"nearly 100%" accuracy* Gregory M. Lamb, "Battle of the Bot: The Future of War?" *Christian Science Monitor,* July 27, 2005, http://www.csmonitor.com/2005/0127/p14s02-stct.html.

168   *"also has a speaker that beckons"* Louis Ramirez, "Robotic Sentry Shoots and Laughs at You," Gizmodo.com, November 3, 2006 (cited November 3, 2006); available at http://www.gizmodo.com/gadgets/tag/robotic-sentry-shoots-and-laughs-at-you-212241.php.

168   *The footage of a real-world automated machine gun* Autonomous Sentry Gun video available at http://www.dailymotion.com/video/xg078_robot-sentinella.

168   *"There's definitely a feedback"* Science Fiction Museum, "Donna Shirley Named Director of Science Fiction Museum."

169   *"The military is doing a fine job"* Robin Wayne Bailey, interview, Peter W. Singer, September 27, 2006.

169 *"The idea of trying to hit a bullet"* Donna Shirley, interview, Peter W. Singer, October 2, 2006.

169 *It not only predicts and influences the future* Bailey, interview, September 27, 2006.

169 *"Science fiction says 'what if?'"* Shirley, interview, October 2, 2006.


### 9. THE REFUSENIKS: THE ROBOTICISTS WHO JUST SAY NO

170 *learned to say, "No, thank you"* Illah Nourbakhsh, interview, Peter W. Singer, Washington, DC, October 31, 2006.

172 *"They said, 'Here's some thousands of dollars'"* Eric Baard, "Make Robots, Not War," *Village Voice* 48, no. 37 (2003). Note: Potter declined an interview.

172 *"astonishing robotic creations"* Ibid.

172 *His dad worked on projects* Ibid.

172 *"However, there is a slippery slope"* Ibid.

173 *a work stoppage by key scientists* George Friedman and Meredeth Friedman, *The Future of War: Power, Technology, and American World Dominance in the Twenty-first Century* (New York: Crown, 1996), 45–49.

174 *"It is, in some ways, responsible"* Bill Joy, "Why the Future Doesn't Need Us," in *Taking the Red Pill: Science, Philosophy and Religion in The Matrix*, ed. Glenn Yeffeth and David Gerrold (Chicago: BenBella Books, 2003), 219.

174 *"The experiences of the atomic scientists"* Ibid., 221.

174 *"a very touchy subject"* Illah Nourbakhsh, interview, Peter W. Singer, October 31, 2006.

174 *"I stay out of politics"* Brian Miller, interview, Peter W. Singer, April 10, 2006.

174 *"He didn't do it thinking"* Sebastian Thrun, interview, Peter W. Singer, March 18, 2007.

174 *"What you don't get"* Joel Garreau, *Radical Evolution: The Promise and Peril of Enhancing Our Minds, Our Bodies—And What It Means to Be Human* (New York: Doubleday, 2005), 43.

175 *"For 364 days out of the year"* Daniel Wilson, interview, Peter W. Singer, October 19, 2006.

175 *"That's above my pay grade"* As quoted in Garreau, *Radical Evolution*, 43.

175 *"You can't let the fear of the future"* Ibid.

175 *"I would probably put myself"* Eric Smalley, "Georgia Tech's Ronald Arkin," 2005, http://www.trnmag.com/Stories/2005/091205/View_Ronald_Arkin_091205.html.

175 *"I don't think"* Baard, "Make Robots, Not War."

176 *"Technology has begun to outstrip"* Paul Evans, "Dividing Lines," *UVA Alumni News*, Winter 2005, 21.


### 10. THE BIG CEBROWSKI AND THE REAL RMA: THINKING ABOUT REVOLUTIONARY TECHNOLOGIES

179 *"Guns and violence have the potential"* Pete Hegseth, "Lessons from a War," *Princeton Alumni Weekly*, November 7, 2007.

179 *"Here at the end of a millennium"* Arthur K. Cebrowski and John J. Garstka, "Network-Centric Warfare: Its Origin and Future," *United States Naval Institute Proceedings* 124, no. 1 (1998): 28.

179 *"a nervous energy and maverick streak"* Clay Risen, "War-Mart: So Long, Clausewitz. Hello, Tom Peters," *New Republic*, April 3, 2006, 20.

180 *"baby-sit the petri dish"* Adam Bernstein, "Adm. Arthur Cebrowski Dies; Led Pentagon Think Tank," *Washington Post*, November 15, 2005, B6.

180 *"For nearly 200 years"* Cebrowski and Garstka, "Network-Centric Warfare."

181 *"A Sudden Tempest Which Turns Everything Upside Down"* Francesco Guicciardini, as quoted in Max Boot, *War Made New: Technology, Warfare, and the Course of History, 1500 to Today* (New York: Gotham Books, 2006), 6.

181 *revolutionized by the world of online file sharing* Rodney Brooks, *Flesh and Machines: How Robots Will Change Us* (New York: Pantheon, 2002), 100.

182 *"must prepare to abandon everything"* David Rejeski, "The Next Small Thing," *The Environmental Forum*, March/April 2004.

182 *at least ten revolutions in military affairs since 1300* Andrew F. Krepinevich, "Cavalry to Computer: The Pattern of Military Revolutions," *National Interest*, no. 37 (1994).

182 *"a sudden tempest"* Francesco Guicciardini, as quoted in Max Boot, *War Made New*, 6.

182 *"when the first early man"* Ralph Peters, *Wars of Blood and Faith: The Conflicts That Will Shape the Twenty-first Century*, 1st ed. (Mechanicsburg, PA: Stackpole Books, 2007), 29.

182 *"How do you become a winner"* Boot, *War Made New*.

183 *"Imagine for a moment"* Murray Scott, "Battle Command, Decision Making, and the Battlefield Panopticon," *Military Review*, July–August 2002, 46.

183 *"the military equivalent of a duck-billed platypus"* Boot, *War Made New: Technology, Warfare, and the Course of History, 1500 to Today*, 175.

184 *That doesn't mean that industrialization* Ray Kurzweil, *The Singularity Is Near: When Humans Transcend Biology* (New York: Viking, 2005), 95.

184 *"dramatically increase force effectiveness"* David Albert and John J. Garstka, "Network-Centric Warfare. Report to Congress" (Department of Defense, 2001).

184 *The side that was networked* Frederick W. Kagan, "The U.S. Military's Manpower Crisis," *Foreign Affairs* 85, no. 4 (2006).

185  *"a move away from platforms to networks"* Timothy L. Thomas, "Chinese and American Network Warfare," *Joint Force Quarterly*, no. 38 (2005).

185  *"Everything in war is very simple"* Carl von Clausewitz, Michael Eliot Howard, and Peter Paret, *On War* (Princeton, NJ: Princeton University Press, 1976), 119.

185  with *"near-perfect clarity"* Michael J. Mazarr, Jeffrey Shaffer, and Benjamin Ederington, "The Military Technical Revolution: A Structural Framework" (Washington DC: Center for Strategic and International Studies, 1993), 38.

185  *"result in a quantum leap"* William A. Owens and Edward Offley, *Lifting the Fog of War*, 1st ed. (New York: Farrar, Straus and Giroux, 2000).

185  *"technological innovation"* MacGregor Knox and Williamson Murray, *The Dynamics of Military Revolution, 1300–2050* (Cambridge, UK; New York: Cambridge University Press, 2001), 178–79.

185  it would inevitably be *"a winning force"* Michael E. O'Hanlon, *Technological Change and the Future of Warfare* (Washington, DC: Brookings Institution Press, 2000), 8.

185  *"The U.S. is the only nation"* Stephen J. Cimbala, "Transformation in Concept and Policy," *Joint Force Quarterly*, no. 38 (2005).

186  *"The IT-RMA was pitched"* Richard Bitzinger, "Is the Revolution in Military Affairs Dead?" *Defense News*, October 23, 2006.

186  *"a revolution in the technology of war"* Ian Roxborough, "From Revolution to Transformation: The State of the Field," *Joint Force Quarterly*, no. 32 (2002).

186  *"Bush was (and remains) a firm believer"* Kagan, "The U.S. Military's Manpower Crisis," 98.

186  a mantra among the *"Vulcans"* Michael R. Gordon and Bernard E. Trainor, *Cobra II: The Inside Story of the Invasion and Occupation of Iraq*, 1st ed. (New York: Pantheon, 2006), 5.

186  *"harness the technological advances"* Max Boot, "The New American Way of War," *Foreign Affairs* 82, no. 4 (2003): 42.

186  *"They accepted the presumptions"* Frank Hoffman, "Challenging the Technocrats," *Armed Forces Journal*, January 2007, 33.

187  *"speed and agility and precision"* Donald Rumsfeld, secretary of defense, interview at WAPI-AM Radio, Birmingham, AL, Richard Dixon, September 28, 2004.

187  *"If 'Rummy' was the president's high priest"* Scott Truver, review of James Blaker, *Transforming Military Force: The Legacy of Arthur Cebrowski and Network Centric Warfare*, *Proceedings*, January 2008, 75.

187  *"In this position, he was responsible"* "Arthur K. Cebrowski," Wikipedia, September 20, 2007 (cited November 1, 2007); available at http://en.wikipedia.org/wiki/Arthur_K._Cebrowski.

188  *"Iraq, in turn, was set up"* Boot, "The New American Way of War," citation 44.

188  *"at a cost of 'only' 27,000 dead soldiers"* Ibid.

188  *"central to American military dominance"* Max Boot, "The Paradox of Military Technology," *New Atlantis*, no. 14 (2006); http://www.thenewatlantis.com/archive/14/boot.htm.

189  *"No one is shooting"* Loren Thompson, "Dot-Com Mania," *Defense News*, October 28, 2002, 12.

189  *"Theories and business models"* Hoffman, "Challenging the Technocrats," 32.

189  *"We will never operate"* Ralph Peters, "Progress and Peril," *Armed Forces Journal*, February 2007, 35.

189  *"We kind of lost track"* Gordon and Trainor, *Cobra II: The Inside Story of the Invasion and Occupation of Iraq*, 300.

190  *"When do we get red force trackers?"* Ibid.

190  *"What I discovered"* Joshua Davis, "If We Run Out of Batteries, This War Is Screwed," *Wired* 11.06 (2003); http://www.wired.com/wired/archive/11.06/battlefield.html.

190  *"if we run out of batteries"* Ibid.

190  *"Major combat missions during Gulf War II"* Noah Shachtman, "Battery Lack Almost Pulled Plug on Iraq War," Defensetech.org, September 3, 2003 (cited November 29, 2005); available at http://www.defensetech.org/archives/000555.html.

191  *"there is probably no conflict"* Milan Vego, "The NCW Illusion," *Armed Forces Journal*, January 2007, 17.

191  they debated back and forth R. Mike Worden, "Rethinking the U.S. Military Revolution," presentation at the Stanley Foundation Conference on Leveraging US Strength in an Uncertain World, Washington, DC, December 7, 2006.

192  *"Sir, the PackBot"* John Dyer, *Robots in Urban Warfare: The Evolving Threat Requires an Innovative, Flexible, and Persistent Response*, 2006. PowerPoint presentation.

192  *"I believe that we are the pioneers"* Robert Finkelstein and James Albus, "Technology Assessment of Autonomous Intelligent Bipedal and Other Legged Robots" (DARPA, 2004), 230.

192  *"nascent stage, set to burst"* Brooks, *Flesh and Machines*, 10–11.

192  *"tsunami that will toss our lives"* Ibid., 6.

192  I found it repeated See for example Martin van Creveld, "War and Technology," *Footnotes: The Newsletter of FPRI's Marvin Wachman Fund for International Education*, 12, no. 25 (2007), http://www.fpri.org/footnotes/1225.200710.vancreveld.wartechnology.html; Williamson Murray, "War and the West," *Footnotes: The Newsletter of FPRI's Marvin Wachman Fund for International Education*, 12, no. 26 (2007), http://www.fpri.org/footnotes/1226.200711.murray.warwest.html.

192  *"in something like the position"* Thomas K. Adams, "Future Warfare and the Decline of Human Decisionmaking," *Parameters* 31, no. 4 (2001): 57.

193  *"It is always hard to see"* Bill Joy, "Why the Future

Doesn't Need Us," in *Taking the Red Pill: Science, Philosophy and Religion in The Matrix*, ed. Glenn Yeffeth and David Gerrold (Chicago: BenBella Books, 2003), 209.

193 *While he was a huge supporter* Douglas McGray, "The Marshall Plan," *Wired* 11.02 (2003), http://www.wired.com/wired/archive/11.02/marshall.html.

193 *"There is a tendency to talk"* As quoted in Richard O. Hundley et al., *Past Revolutions, Future Transformations: What Can the History of Revolutions in Military Affairs Tell Us About Transforming the U.S. Military?* (Santa Monica, CA: RAND, 1999).

193 *"Historians will see the last decade"* Steven Metz, *Armed Conflict in the 21st Century: The Information Revolution and Post-modern Warfare* (Carlisle, PA: Strategic Studies Institute, U.S. Army War College, 2000), 93.

194 *"real transition time here."* As quoted in Megan Scully, "Gates Says Next-generation Bomber Might Fly Without Pilot," *CongressDaily*, May 14, 2009.

194 *"First, you had human beings without machines,"* John Pike, as quoted in Fred Reed, "Robotic Warfare Drawing Nearer," *Washington Times*, February 10, 2005.

194 *"We now stand on the cusp"* Christopher Coker, *Waging War Without Warriors? The Changing Culture of Military Conflict*, IISS Studies in International Security (Boulder, CO: Lynne Rienner Publishers, 2002), 171.

194 *"a very different age"* George Friedman and Meredith Friedman, *The Future of War: Power, Technology, and American World Dominance in the Twenty-first Century*, 1st ed. (New York: Crown, 1996), xi.

195 *"The robot kept operating"* Tim Kiska, "Robot Firm Liable in Death," *Oregonian*, August 11, 1983.

195 *"to be murdered by a robot"* Mel Croucher, "Killer Computers," *Crash*, no. 56 (1988), http://www.crashonline.org.uk/56/monitor.htm.

195 *"attacked by a humanoid robot"* "Trust Me, I'm a Robot," *Economist* 379, no. 8481 (2006); http://tmsuk.co.jp/artemis; "Japanese Prime Minister Koizumi Attacked by Humanoid Robot," in our media.org (2005).

195 *"Robots are very complex"* Daniel Wilson, interview, Peter W. Singer, October 19, 2006.

196 *"anything that can go wrong, will"* "Edward A. Murphy, Jr." Wikipedia (cited February 8, 2008), http://en.wikipedia.org/wiki/Major_Edward_A._Murphy, Jr.

196 *His response: "No."* Francis Harvey, private presentation, Brookings Institution, December 15, 2005.

196 *"There was nowhere to hide"* Graeme Hosken, Michael Schmidt, and Johan du Plessis, "9 Killed in Army Horror," *The Star*, October 13, 2007, http://www.iol.co.za/index.php?click_id=13&set_id=1&art_id=vn20071013080449804C939465.

196 *a "software glitch"* Leon Engelbrecht, "Did Software Kill Soldiers," ITWeb.com, October 16, 2007 (cited December 5, 2007); available at http://www.itweb.co.za/sections/business/2007/0710161034.asp?S=IT%20in%20Defence&A=DFN&O=FRGN.

197 *At its first demonstration* David Hambling, *Weapons Grade: How Modern Warfare Gave Birth to Our High-Tech World* (New York: Carroll and Graf, 2005), 314.

197 *the system "detected" a launch* Croucher, "Killer Computers."

197 *The U.S. Strategic Command* Tom Stockman, "NORAD False Alarm of Soviet Missile Attack November 9 1979," 2006 (cited December 5, 2007); available at http://www.tomstockman.com/columns/sac.shtml.

197 *"We've all had problems"* Noah Shachtman, interview, Peter W. Singer, Washington, DC, July 2, 2007.

198 *"It will drive off the road"* Noncommissioned officer, interview at the Military Robotics Conference in Washington, DC, Peter W. Singer, April 10–12, 2006.

198 *SWORDS doing "a Crazy Ivan"* iRobot engineer, interview, Peter W. Singer, November 16, 2006.

198 *The Marine Corps' Gladiator combat robot* Interview at Pentagon, March 31, 2008.

198 *"pressure to try to pass safety tests"* Jonathan Hall, interview, Peter W. Singer, Washington, DC, August 6–9, 2007.

199 *sometimes just crash when they fly* Ibid.

199 *"Rainman the robot"* Noah Shachtman, "The Baghdad Bomb Squad," *Wired* 13.11 (2005), http://www.wired.com/wired/archive/13.11/bomb.html.

199 *"In football, everything is complicated"* Steve Rushin, "Thus Spake Mountaineers," *Sports Illustrated*, January 22, 2007, 15.

199 *"The more complex any system becomes"* Ralph Peters, interview, Peter W. Singer, Washington, DC, March 29, 2007.

199 *"weak and easily jammed"* John A. Gentry, "Doomed to Fail: America's Blind Faith in Military Technology," *Parameters* 32, no. 4, (2002) 91.

199 *powered by plugging it* "Perspectives," *GPS World*, June 27, 2008, accessed at http://sidt.gpsworld.com/gpssidt/Latest+News/National-Space-Symposium-Day-3-OCX-and-GPS-III/ArticleStandard/Article/detail/525875.

200 *to fry the other side's electricity* O'Hanlon, *Technological Change and the Future of Warfare*.

200 *ongoing work on radio-frequency weapons* Ibid., 60.

200 *"The smarter the weapons"* Boot, *War Made New*, 448.

200 *radio-frequency weapons, or "e-bombs"* Michael Abrams, "The Dawn of the E-Bomb," *IEEE Spectrum* 40, no. 11 (2003).

200 *electronic "battles of conviction"* Ralph Peters, "The Future of Armored Warfare," *Parameters* 27, no. 3 (1997): 52.

200  *Ninety-five percent of its communications* Gentry, "Doomed to Fail: America's Blind Faith in Military Technology."

200  *"vulnerable to robots"* Jürgen Altmann and Mark Gubrud, "Anticipating Military Nanotechnology," *IEEE Technology and Society Magazine* 23, no. 4 (2004): 38.

201  *"The idea that they can make software unhackable"* Ralph Peters, interview, Peter W. Singer, March 29, 2007.

201  *"The parts are easily available"* Humphrey Cheung, "How To: Building a BlueSniper Rifle— Part 1," *Small Net Builder*, March 8, 2005 (cited December 18, 2006); available at http://www.smallnetbuilder.com/wireless/wireless-how-to/how_to_bluesniper_pt1.

201  *"Why is it that every time"* Richard Clarke, interview, Peter W. Singer, Washington, DC, August 8, 2007.

201  *"six-year-old kid"* Robert Young Pelton, "Licensed to Kill: Hired Guns in the War on Terror," presentation, Brookings Institution, Washington, DC, October 5, 2006.

202  *"We cannot expect the enemy"* Charles J. Dunlap Jr., "21st-Century Land Warfare: Four Dangerous Myths," *Parameters* 27, no. 3 (1997).

202  *the "bandwidth battle"* Gopal Ratnam, "Bandwidth Battle: Supply Falters as Demand Soars, Forcing U.S. to Manage Info Flow," *Defense News*, October 9, 2006.

202  *"During Gulf War I"* Harry Raduege, as quoted in Gopal Ratnam, "Bandwidth Battle," 37.

202  *"For him to deliver"* Jeffrey Smith, as quoted in Gopal Ratnam, "Bandwidth Battle," 37.

202  *"staring at the ground"* Lewis Crenshaw, as quoted in Gopal Ratnam, "Bandwidth Battle," 40.

203  *"hotspots on the battlefield"* Steven Boutelle, as quoted in Gopal Ratnam, "Bandwidth Battle," 37.

203  *"Who's overseeing all this crap?"* Air force pilot, interview, Peter W. Singer, September 9, 2006.

203  *"unleash a hurricane"* Metz, *Armed Conflict in the 21st Century*, xix.

204  *"Ultimately no one can fully predict"* Ibid., 99.

## 11. "ADVANCED" WARFARE: HOW WE MIGHT FIGHT WITH ROBOTS

205  *"Once in a while, everything about the world changes"* Chuck Klosterman, "Real Genius," *Esquire*, July 2004, http://www.thesongcorporation.com/klosterman-advancement2.htm: 223.

205  *the thesis comes not from* Ibid.

207  *"When people think about the future of technology"* Robert Bateman, interview, Peter W. Singer, October 27, 2006.

207  *"Kurzweil, while an interesting technologist"* Ibid.

207  *"The Turing test"* Ibid.

207  *"First and foremost"* Ibid.

207  *"completely bottom up right now"* Ibid.

207  *"leaders not able to think beyond"* Ibid.

208  *"U.S. Army had to rip out the radios"* Ibid.

208  *A doctrine is the central idea* J. F. C. Fuller, *The Foundations of the Science of War* (Fort Leavenworth, KS: U.S. Army Command and General Staff College Press, 1993), 254.

208  *"outline of how we fight"* Clinton J. Ancker III and Michael D. Burke, "Doctrine for Asymmetric Warfare," *Military Review* 83, no. 4 (2003): 18.

209  *"prostitution of the air force"* Max Boot, *War Made New: Technology, Warfare, and the Course of History, 1500 to Today* (New York: Gotham Books, 2006): 223.

209  *He set up fifty-seven committees* James S. Corum, *The Roots of Blitzkrieg: Hans von Seeckt and German Military Reform* (Lawrence: University Press of Kansas, 1992), 37; Murray Williamson, "Armored Warfare: The British, French, and German Experiences," in *Military Innovation in the Interwar Period*, ed. Murray Williamson and Allan Reed Millett (Cambridge, UK: Cambridge University Press, 1996).

210  *the French alone had more tanks* George Friedman and Meredith Friedman, *The Future of War: Power, Technology, and American World Dominance in the Twenty-first Century*, 1st ed. (New York: Crown, 1996), 124.

210  *"the Maginot Line of the 21st century"* John A. Gentry, "Doomed to Fail: America's Blind Faith in Military Technology," *Parameters* 32, no. 4 (2002): 88.

210  *developing a strategy and doctrine* Robert Finkelstein and James Albus, "Technology Assessment of Autonomous Intelligent Bipedal and Other Legged Robots" (DARPA, 2004).

210  *"smacked of attention deficit disorder"* Bill Sweetman, "UCAVs Offer Fast Track to Stealth, Long Range, and Carrier Operations," *Jane's International Defence Review* 40 (2007): 41.

210  *"There's got to be a better way"* Interview at U.S. military facility, February 19, 2008.

211  *they "lost it somewhere in Iraq"* U.S. Army soldier, interview, Peter W. Singer, Washington, DC, November 2, 2006; "American Drone Discovered in Baghdad Cache," *Danger Room*, June 20, 2008, http://blog.wired.com/defense/2008/06/insurgents-unma.html.

211  *"We don't have the strategy"* Robert Finkelstein, interview, Peter W. Singer, July 7, 2006.

211  *"We are just now thinking"* Ibid.

211  *"And it's been a mess for decades"* Scientist, interview, Peter W. Singer, July 17, 2006.

211  *"mainly bottom-up"* Noah Shachtman, interview, Peter W. Singer, Washington, DC, March 25, 2006.

211  *"They still think of robots"* iRobot executive, interview, Peter W. Singer, November 16, 2006.

211  *"nothing yet on logistics"* Foster-Miller executive, interview, Peter W. Singer, November 17, 2006.

211 *"It started out with people arguing"* Scientist, interview, Peter W. Singer, July 17, 2006

211 *"in all sorts of offices"* Ibid.

212 *"The Navy has programs"* U.S. Joint Forces Command, "Military Robots of the Future" (U.S. Joint Forces Command, 2003).

212 *"We were defeated by one thing only"* Arthur C. Clarke, "Superiority," in *The Best Military Science Fiction of the 20th Century*, ed. Harry Turtledove and Martin Harry Greenberg (New York: Ballantine, 2001), 129.

212 *"We now realize"* Ibid., 131.

212 *"How We Lost the High-Tech War"* Charles J. Dunlap Jr., "How We Lost the High-Tech War of 2007: A Warning from the Future," *Weekly Standard* 1, no. 19 (1996); Charles J. Dunlap Jr., "The Origins of the American Military Coup of 1912," *Parameters*, 12, no. 4 1992.

213 *"are part of the traditional U.S. military repertoire"* Jeffrey Record, "Why the Strong Lose," *Parameters* 35, no. 4 (2005): 16.

213 *"During the Cold War"* Steven Metz, *Learning from Iraq: Counter-Insurgency in American Strategy* (Carlisle, PA: U.S. Army War College, 2006), 78.

213 *not being localized battles of asymmetry* Rick Brennan et al., "Future Insurgency Threats" (RAND Corporation, 2005); David Kilcullen, "Countering Global Insurgency," *Journal of Strategic Studies* 28, no. 4 (2005).

213 *"in discussing any modernization effort"* Ann Roosevelt, "FCS Would Bring Significant Advantages to Future Insurgency-Type Operations, Harvey Says," *Defense Daily*, January 23, 2007.

213 *"We continue to focus"* Thomas X. Hammes, *The Sling and the Stone: On War in the 21st Century* (St. Paul, MN: Zenith Press, 2004), 3.

213 *"On the battlefields of the future"* Qiao Liang and Wang Xiangsui, *Unrestricted Warfare: China's Master Plan to Destroy America* (Beijing: PLA Literature and Arts Publishing House, 1999).

213 *"We have made huge leaps"* USAF lieutenant general Lance L. Smith, as quoted in Boot, *War Made New*, 394.

214 *"The success of DPhil papers"* Tom Baldwin, "Editorial Review: *Learning to Eat Soup with a Knife*," Amazon.com (cited December 13, 2007); available at http://www.amazon.com/Learning-Eat-Soup-Knife-Counterinsurgency/dp/product-description/0226567702.

214 *"Defeating an insurgency"* John Nagl, "A Better War in Iraq," *Armed Forces Journal*, August 2006, 23.

214 *"The use of force is but temporary"* Edmund Burke and Andrew Jackson George, *Burke's Speech on Conciliation with America, 1775* (Boston: D. C. Heath & Co., 1895).

215 *"When it comes to reorganizing"* Frederick W. Kagan, "The U.S. Military's Manpower Crisis," *Foreign Affairs* 85, no. 4 (2006): 107.

215 *"more effective than all the high-tech shit"* Robert D. Kaplan, *Imperial Grunts: The American Military on the Ground* (New York: Random House, 2005), 337.

215 *"Insurgents don't show up"* Boot, *War Made New*, 239.

215 *"After all the GBUs"* Todd Fredericks, "Comments and Discussion: We Have a Serious COIN Shortage," *Proceedings* 133, no. 7 (2007): 79.

215 *"I'm bothered by the old canard"* Steven Metz, interview, Peter W. Singer, September 19, 2006.

216 *finally led them to be truly accepted* Quote from an Ohio State professor at a presentation by the author on "Wired for War," October 10, 2006.

216 *"'UAVs? Yes, give me more!'"* Eliot Cohen, interview, Peter W. Singer, November 15, 2006.

216 *"In March of 2002"* Lieutenant General Walter E. Buchanan III, Commander of USAF 9th Air Force and U.S. Central Command Air Forces, during a meeting with the Defense Writers Group on October 27, 2005, in Washington, DC, as quoted in Marc V. Schanz, "A Complex and Changing Air War," *Air Force Magazine* 89, no. 1 (2006), http://www.afa.org/magazine/jan2006/0106airwar.asp.

216 *"It was a Hunter UAV"* Nathan Hodge, "Interview with Gen. William Wallace," *Jane's Defence Weekly*, October 4, 2006, 50.

216 *"It wasn't too long"* Noah Shachtman, "Robo-Planes Log 250,000 Flight Hours This Year," *Danger Room*, December 17, 2007, http://blog.wired.com/defense/2007/12/uav-conference.html.

216 *more than seven hundred Hunters* Lolita C. Baldor, "Military Use of Unmanned Aircraft Soars," *Google News*, January 1, 2008, http://ap.google.com/article/ALeqM5i_7otabxw8XLB8yCGhhlMhX7Vs7QD8TTEH400.

216 *they responded that they wanted more* Joshua Kucera, "UAV Missions in Iraq Set to Rise," *Jane's Defence Weekly*, January 19, 2005, 11.

217 *the air force retooled its pilot training program* Noah Shachtman, "Deadly 'Drone Shortage' in Iraq?" *Danger Room*, March 25, 2008, http://blog.wired.com/defense/2008/03/theres-a-drone.html; and Noah Shachtman, "Gates, Air Force Battle over Robot Planes," *Danger Room*, March 21, 2008, http://blog.wired.com/defense/2008/03/gates-vs-usa-f-0.html.

217 *the army flew 54 percent of all drone flights* Jeffrey Kappenman, "Army Unmanned Aircraft Systems: Decisive in Battle," *Joint Force Quarterly*, no. 49 (2008): 23.

217 *"a power grab"* Amy Butler, "Let the Race Begin," *Aviation Week & Space Technology* 166, no. 13 (2007): 52.

217 *"Robots were only used"* Edward Godere, interview, Peter W. Singer, November 17, 2006.

217 *"After five years of trying"* Anthony Aponick, interview, Peter W. Singer, November 17, 2006.

217 *more than thirty thousand missions* Kris Osborn, "U.S. Wants 3,000 New Robots for War," *Defense News*, August 13, 2007, 1.

217   *"For a long time"* Charles Duhigg, "The Pilotless Plane That Only Looks Like Child's Play," *New York Times*, April 15, 2007.

218   *"We adapt, they adapt"* As quoted in Boot, *War Made New*, 411.

218   *"There is a huge intellectual battle"* Foster-Miller employee, interview, Peter W. Singer, November 17, 2006.

218   more than ninety ways of triggering IEDs John Bokel, "IEDs in Asymmetric Warfare," *Military Technology* 31, no. 10 (2007).

219   *"The enemy realizes"* As quoted in Byron Spice, "Battlefield Robots Saving Lives, Proving their Worth in Iraq," *Pittsburgh Gazette*, June 9, 2006.

219   *"They're always trying to outsmart us"* Stew Magnuson, "Bomb Disposal Teams Deliver Blunt Talk on Robots," *National Defense* 91, no. 632 (2006).

219   *"Insurgents have been intensifying"* "U.S. Navy Orders Talon Robots," *Defense News*, October 23, 2006, 46.

219   *"We figured it out"* Noncommissioned officer, interview at the Military Robotics Conference in Washington, DC, Peter W. Singer, April 10–12, 2006.

219   *"Jihadis are also concerned"* Insurgent, interview, Peter W. Singer, August 17, 2006.

219   They ranged from jury-rigged Thomas E. Ricks, *Fiasco: The American Military Adventure in Iraq* (New York: Penguin Press, 2006), 219.

220   *"like the wind was pushing it"* Kenneth Dahl, interview, Peter W. Singer, Washington, DC, March 16, 2006.

220   *"It's basically a game"* H. R. Everett, interview, Peter W. Singer, October 20, 2006.

220   A Quick Reaction Force of marines Scene re-created from Bing West, "Streetwise," *Atlantic Monthly*, Jan.–Feb. 2007, http://www.theatlantic.com/doc/200701/west-iraq.

221   *"Americans have the watches"* David Barno, "Briefing" (presentation, Brookings Institution, Washington, DC, October 4, 2007).

221   *"we can use robots"* Foster-Miller executive, interview, Peter W. Singer, November 17, 2006.

221   *"Robotics also hold great promise"* Steven Metz, *Armed Conflict in the 21st Century: The Information Revolution and Post-modern Warfare* (Carlisle, PA: Strategic Studies Institute, U.S. Army War College, 2000).

221   *"winning hearts and minds"* Charles J. Dunlap Jr., "We Still Need the Big Guns," *New York Times*, January 9, 2008, http://www.nytimes.com/2008/01/09/opinion/09dunlap.html.

221   *"Solving root causes"* Metz, *Armed Conflict in the 21st Century.*

221   eleven were killed via drones strike Sami Yousafzai, "Predators on the Hunt in Pakistan," *Newsweek*, Feb. 9, 2009.

221   *"Very frankly, it's the only"* Leon Panetta, as quoted in "CIA Chief: Drones 'Only Game in Town,'" *Danger Room*, May 19, 2009, available at wired.com.

222   *"see things develop over time"* Predator pilot, interview, Peter W. Singer, August 28, 2006.

222   *"we can spot the bad guys"* Owen West and Bing West, "Lessons from Iraq," *Popular Mechanics* 182, no. 8 (2005).

222   has *"TiVo-like capabilities"* Tom Vanden Brook, "U.S. Spy Technology Caught in Military Turf Battle," *Defense News*, October 8, 2007, 54.

222   The Odin team was able Kris Osborn, "U.S. Aviators, UAVs Team Up Against IEDs," *Defense News*, January 21, 2008.

222   *"provide persistent staring"* Brian Newberry, "The Air Force in the Urban Fight," *Armed Forces Journal*, September 2006, 29.

223   *"The driver had a perfect ID"* Bing West, "Nowhere to Hide," *Popular Mechanics* 182, no. 2 (2005).

223   *"It's a comforting sound"* Thomas E. Ricks, "Beaming the Battlefield Home: Live Video of Afghan Fighting Had Questionable Effect," *Washington Post*, March 22, 2002, 1.

223   *"Situational awareness ain't deterrence,"* Sam Mundy, interview, Peter W. Singer, March 3, 2004.

223   guided in by lasers and GPS coordinates Zoran Kusovac, "Joint Intel Located Al-Qaeda Leader," *Jane's Defence Weekly*, June 4, 2006, 24.

223   *"While technology is not the sole answer"* John Bellflower, "The Indirect Approach," *Armed Forces Journal*, January 2007, 16.

223   *"an era of 'oh gee' technology"* James Lasswell, interview, Peter W. Singer, Washington, DC, November 7, 2006.

224   *"UCAVs are the answer"* As quoted in Sweetman, "UCAVs Offer Fast Track to Stealth, Long Range, and Carrier Operations," 41.

224   *"the most significant threat"* Max Boot, "The Paradox of Military Technology," *New Atlantis*, no. 14 (2006), http://www.thenewatlantis.com/archive/14/boot.htm.

224   *"faster speeds, greater stealth"* Joris Janssen Lok, "Navies Look for Ways to Tackle the Ever-Changing Close-in Threat," *Jane's International Defence Review* 37 (2004).

225   *"we are just beginning to understand"* F. W. LaCroix and Irving N. Blickstein, *Forks in the Road for the U.S. Navy* (Santa Monica, CA: RAND, 2003), ix.

225   *"is like playing"* The Military Channel, *Creating the X Craft*, broadcast on June 13, 2006.

225   *"Sometimes computers are better"* Ibid.

225   sitting at control module stations Scott Truver, "Mix and Match," *Jane's Defence Weekly*, March 16, 2005, 24.

226   *"spot on, almost visionary"* Bill Sweetman, "US Finally Looks Beyond the B-2 for Long-Range Strike Capability," *Jane's International Defence Review* 39 (2006): 44.

226 *tested out Wasp Micro Air Vehicles* Christian Lowe, "Itsy-bitsy Drone," Defensetech.org, April 5, 2005 (cited February 9, 2006); available at http://www.defensetech.org/archives/001467.html.

226 *"a critical next step"* Boeing, "Boeing Achieves First Submerged Unmanned Undersea Vehicle Recovery by a Submarine," November 26, 2007 (cited January 11, 2008); available at http://www.boeing.com/news/releases/2007/q4/071126b_nr.html.

226 *"can sit at the bottom"* Andrew Bennett, interview, Peter W. Singer, November 16, 2006.

227 *"They would act as 'force multipliers'"* Carl Posey, "Robot Submarines Go to War, Part 2: The Navy's AUVs," *Popular Science*, March 2003, http://www.popsci.com/popsci/science/6327359b9fa84010vgnvcm1000004eecbccdrcrd.html.

227 *"the under-sea fiber-optic cables"* Bill Sweetman, "Exposing the Spy Sub of the Future," *Popular Science* 267, no. 2 (2005): 81.

227 *the navy is developing a plan* Michael Fetsch, Chris Mailey, and Sara Wallace, "UV Sentry," paper presented at the Unmanned Systems North America, AUVSI's 34th Annual Symposium and Exhibition, Washington, DC, August 6–9, 2007.

227 *Similar plans are being developed* Lok, "Navies Look for Ways to Tackle the Ever-Changing Close-in Threat."

227 *The drones fly in and out* David Pugliese, "Launch and Recover UAV System Tested," *Defense News*, February 19, 2007, 14.

227 *"fork in the road"* LaCroix and Blickstein, *Forks in the Road for the U.S. Navy*, ix.

228 *"cataclysmic clashes"* Ibid.

228 *"the touchstone for U.S. naval force planning"* Frank Hoffman, "The Fleet We Need," *Armed Forces Journal*, August 2006, 29.

228 *"a truism—no one would dispute it"* "Julian Corbett," Wikipedia, January 13, 2007 (cited January 15, 2008); available at http://en.wikipedia.org/wiki/Julian_Corbett. Quote from Williamson Murray in "Corbett, Julian," *Reader's Companion to Military History* (Boston: Houghton Mifflin, 2004).

229 *"Well before it was fashionable"* Hoffman, "The Fleet We Need," 49.

230 *"An [enemy] air defense system"* David A. Fulghum and Michael J. Fabey, "F-22: Unseen and Lethal," *Aviation Week & Space Technology* 166, no. 2 (2007): 46.

230 *"If you look at nature's most efficient predators"* Noah Shachtman, as quoted in *Warbots*, History Channel, broadcast on August 8, 2006.

231 *40 percent of these victories* Sean J. A. Edwards, "Swarming and the Future of Warfare" (doctoral thesis, Pardee Rand Graduate School, 2005), 83.

231 *"an influence, a thing invulnerable, intangible"* Ibid., 64.

231 *"Obviously the birds lack"* Thomas K. Adams, "The Real Military Revolution," *Parameters* 30, no. 3 (2000).

231 *"boids," artificial birds* Craig W. Reynolds, "An Evolved, Vision-Based Model of Obstacle Avoidance Behavior," in *Proceedings*, ed. C. Langton (Redwood City, CA: Addison-Wesley, 1994).

231 *follow three simple rules* Adams, "The Real Military Revolution."

232 *"the wisdom of crowds"* James Surowiecki, *The Wisdom of Crowds: Why the Many Are Smarter than the Few and How Collective Wisdom Shapes Business, Economies, Societies, and Nations*, 1st ed. (New York: Doubleday, 2004).

232 *"We don't want to copy"* Tobey Grumet, "Robots Clean House," *Popular Mechanics* 180, no. 11 (2003): 30.

232 *"an unassailable wireless 'Internet in the sky'"* Lakshmi Sandhana, "The Drone Armies Are Coming," *Wired News*, August 30, 2002, http://www.wired.com/science/discoveries/news/2002/08/54728.

232 *"They should just go ahead"* Scientist, interview, Peter W. Singer, July 17, 2006.

232 *The Santa Fe Institute* "UCAR—The Next Generation of Unmanned Aerial Vehicles," Gizmag.com, August 17, 2003 (cited July 6, 2005); available at http://www.gizmag.com/go/2118/.

233 *PRAWNs might also carry different weapons* Dave Frelinger et al., *Proliferated Autonomous Weapons: An Example of Cooperative Behavior*, Documented Briefing (Santa Monica, CA: RAND, 1998), 6.

234 *"a dark and menacing cloud"* Carl von Clausewitz, Michael Eliot Howard, and Peter Paret, *On War* (Princeton, NJ: Princeton University Press, 1976), 581.

234 *"My vision of the future"* History Channel, *Warbots*.

234 *They might even draw inspiration* Edwards, "Swarming and the Future of Warfare," 99.

234 *"When you see one robot coming"* Justin Pope, "Looking to Iraq, Military Robots Focus on Lessons of Afghanistan," *Detroit News*, January 12, 2003, http://www.detnews.com/2003/technology/0301/12/technology-57614.htm.

234 *"zillions and zillions of robots"* As quoted in James D. McLurkin, "Stupid Robot Tricks: A Behavior-Based Distributed Algorithm Library for Programming Swarms of Robots" (Cambridge: Massachusetts Institute of Technology, 2004); available at http://people.csail.mit.edu/jamesm/McLurkin-SM-MIT-2004(72dpi).pdf.

235 *"not all novelty is desirable"* Adams, "The Real Military Revolution."

235 *"There go my people"* Gregory A. Jackson, "'Follow the Money' and Other Unsolicited Advice for CIOs," *Cause and Effect* 22, no. 1 (1999).

235 *"Basically stay the hell out of the way"* As quoted

in Joel Garreau, *Radical Evolution: The Promise and Peril of Enhancing Our Minds, Our Bodies—And What It Means to Be Human* (New York: Doubleday, 2005), 217.

236 *"decentralized decision making"* United States Marine Corps general, interview, January 16, 2007.


12. ROBOTS THAT DON'T LIKE APPLE PI:
HOW THE U.S. COULD LOSE THE UNMANNED
REVOLUTION

237 *"Technology is a double-edged sword"* George Michael Casey, "Maintaining Quality in the Force" (presentation, Brookings Institution, Washington, DC, December 4, 2007).

237 *"Sorry, sir, but we can't export"* Peter Pae, "Arms Dealers Fight It Out for Sales in Booming Asia," *Los Angeles Times*, February 27, 2006.

238 *"[Air] shows are nice"* Stayne Hoff, interview, Peter W. Singer, December 5, 2006.

238 *"knowledge, more than ever before"* Joseph S. Nye Jr. and William A. Owens, "America's Information Edge," *Foreign Affairs* 75, no. 2 (1996).

238 *"The ability to accept and capitalize"* Steven Metz, *Armed Conflict in the 21st Century: The Information Revolution and Post-modern Warfare* (Carlisle, PA: Strategic Studies Institute, U.S. Army War College, 2000), xviii.

238 *"America is by its nature"* George Friedman and Meredith Friedman, *The Future of War: Power, Technology and American World Dominance in the Twenty-first Century*, 1st ed. (New York: Crown, 1996), 1.

239 *"Technology is part of how Americans"* Metz, *Armed Conflict in the 21st Century*, 69.

239 *the first to invent or take advantage* Richard R. Nelson, *Technology, Institutions, and Economic Growth* (Cambridge, MA: Harvard University Press, 2005).

239 *"The longer you are on top"* Max Boot, *War Made New: Technology, Warfare, and the Course of History, 1500 to Today* (New York: Gotham Books, 2006), 455.

239 *video games and computers made the point* U.S. Naval Academy, interviews, Peter W. Singer, November 20, 2007.

240 *"most of the things we do"* James Lasswell, interview, Peter W. Singer, Washington, DC, November 7, 2006.

240 *"These major advances"* Talat Masood, "Shackling Shock and Awe: American and Muslim World Views on the Laws of High Tech Warfare," presentation, U.S. Islamic World Forum, Doha, January 10–12, 2004.

240 *"We have not abolished"* Orson Scott Card, interview by e-mail, Peter W. Singer, January 24, 2007.

240 *"We will see if not identical technologies"* Steven Metz, interview, Peter W. Singer, September 19, 2006.

241 *fourteen hundred corporate members in fifty nations* See Unmanned Vehicle Systems International's Web site, available at http://www.auvsi.org/about/.

241 *forty-two countries were at work* David Hughes, "A Second Kitty Hawk," *Aviation Week & Space Technology*, February 12, 2007: 49.

241 *"programmed for blasting"* "Iran Devises Robot-Soldier," IRINN, June 8, 2008, 06.08.08 13:04, available at http://news.trendaz.com/index.shtml?show=news&newsid=1263671&lang=EN.

241 *"The small U.S. humanoid robot community"* Robert Finkelstein and James Albus, "Technology Assessment of Autonomous Intelligent Bipedal and Other Legged Robots" (DARPA, 2004), 11.

241 *"failed in its assigned mission"* Ibid., 52.

241 *Dave Sonntag's job is* David Sonntag, e-mail interview, Peter W. Singer, Washington, DC, November 28, 2006.

242 *a third of all the world's industrial robots* Tim Kelly, "Rise of the Cyborg," *Forbes* 178, no. 4 (2006): 94.

242 *Japanese Ministry of International Trade and Industry* K. Eric Drexler, *Engines of Creation*, 1st ed. (Garden City, NY: Anchor Press/Doubleday, 1986), 75; "Robots Enter Japan's Daily Life," Associated Press, March 3, 2008.

243 *One of the most vehement* Prabhu Guptara, "Why the Next Decade Will Be Neither Chinese Nor Indian," *Globalist*, March 15, 2006, http://www.theglobalist.com/printStoryId.aspx?StoryId=5083.

243 *"My choice may be"* Office of the Secretary of Defense, "Airspace Integration Plan for Unmanned Aviation" (Department of Defense, 2004), 40.

243 *"put a robot in every household"* "Robot Love: South Korea to Build Robot Theme Parks," *Network World*, November 13, 2007, http://www.networkworld.com/community/node/21867.

244 *Korean robotics research* "Korea to Invest $14 Billion in Biotech," *Korea Times*, November 14, 2006, http://times.hankooki.com/lpage/biz/200611/kt2006111519261411900.htm.

244 *"The two cities"* "Robot Love: South Korea to Build Robot Theme Parks."

244 *"This means"* Chas W. Freeman, "China's Real Three Challenges to the United States," *Globalist*, December 12, 2006, http://www.theglobalist.com/DBWeb/StoryId.aspx?StoryId=5770.

244 *"much of the momentum"* Ibid.

245 *a cost of only $37,500* Jason Chen, "Chinese Beauty Robot Needs More Beauty," Gizmodo.com, August 10, 2006 (cited October 30, 2006); available at http://www.gizmodo.net/gadgets/robots/chinese-beauty-robot-needs-more-beauty-193496.php.

245 *a robot waiter to a robot chimpanzee* Jason Chen, "Chinese Robotic Gallery," Gizmodo.com, August 14, 2006 (cited October 30, 2006); available at

http://www.gizmodo.com/gadgets/robots/chinese-robot-gallery-194102.php.

245 *A five-foot-long robot* "China Develops Fish-Shaped Robot for Underwater Archeological Research," *People's Daily Online*, December 7, 2004, http://english.peopledaily.com.cn/200412/07/eng20041207_166401.html.

245 *"the perfect artificial limb"* Lisa Egan, "Intelligent Software Helps Build Perfect Robotic Hand," *Innovations Report*, November 29, 2007, http://www.innovations-report.com/html/reports/information_technology/report-99200.html.

245 *China's growing Internet presence* Wendell Minnick, "Taiwan: Chinese Virus Stole Secret Files," *Defense News*, April 16, 2007, 1.

245 *"retired fighter aircraft"* Office of the Secretary of Defense, "Annual Report on the Military Power of the People's Republic of China" (Department of Defense, 2005), 4.

246 *"the U.S. and its military"* Roger Cliff, *The Military Potential of China's Commercial Technology* (Santa Monica, CA: RAND, 2001), xv.

246 *"Technology is like 'magic shoes'"* Qiao Liang and Wang Xiangsui, *Unrestricted Warfare: China's Master Plan to Destroy America* (Beijing: PLA Literature and Arts Publishing House, 1999).

247 *"The new concept of weapons"* Ibid.

247 *"We believe that some morning"* Qiao and Wang, *Unrestricted Warfare*.

247 *the United States only has 4 percent of the world's population* Boot, *War Made New*, 322.

247 *"the United States is headed"* Robert Kavetsky and Christopher J. R. McCook, "The Technological Perfect Storm," *Proceedings,* October 2006.

247 *Only 54 percent of America's high school students* Shirley Tilghman, "'Rising Above the Gathering Storm' Through Science and Engineering Education," *Princeton Alumni Weekly*, January 24, 2007, 3. The report is available at http://www.nsf.gov/attachments/105652/public/NAS-Gathering-Storm-11463.pdf.

248 *"The longer students are exposed"* Norman R. Augustine, "Learning to Compete," *Princeton Alumni Weekly*, March 7, 2007, 36.

248 *"When I compare our high schools"* Ibid.

248 *create a "futile cycle"* Tilghman, "'Rising Above the Gathering Storm' Through Science and Engineering Education."

248 *"In the past four years"* As quoted in Augustine, "Learning to Compete."

248 *"This research, in turn"* Bruce Alberts, William A. Wulf, and Harvey Fineberg, *Current Visa Restrictions Interfere with U.S. Science and Engineering Contributions to Important National Needs,* National Academies, 2003 (cited January 8, 2007); available at http://www8.nationalacademies.org/onpinews/newsitem.aspx?RecordID=s12132002.

249 *"If action is not taken"* National Science Board, "A Companion to Science and Engineering Indicators

2004: An Emerging and Critical Problem of the Science and Engineering Labor Force" (National Science Board, 2004).

249 *In India, six engineers* Augustine, "Learning to Compete."

249 *Starbucks has to spend more* Ibid., 35.

249 *America's trade balance in high-tech goods* Ibid.

249 *three-fourths of the new R&D facilities* Ibid.

249 *"There is massive industrial espionage"* Richard Clarke, interview, Peter W. Singer, Washington, DC, August 8, 2007.

250 *angrily confronted a group* Peter W. Singer, "Research Visit to iRobot Corporation," 2006.

250 *"when they only want to buy one"* Stayne Hoff, interview, Peter W. Singer, December 5, 2007.

250 *"If the U.S. doesn't wake up"* Tina Hesman, "Stephen Thaler's Computer Creativity Machine Simulates the Human Brain," *St. Louis Post-Dispatch*, January 24, 2004.

251 *new ideas still have trouble* Credit goes to James Surowiecki for this insight.

251 *QWERTY is the way* Jared M. Diamond, *Guns, Germs, and Steel: The Fates of Human Societies*, 1st ed. (New York: W. W. Norton & Co., 1997), 248.

251 *"In no profession"* Stephen Peter Rosen, *Winning the Next War: Innovation and the Modern Military* (Ithaca, NY: Cornell University Press, 1991), 2.

252 *good enough for their heroes* J. E. Lendon, *Soldiers and Ghosts: A History of Battle in Classical Antiquity* (New Haven: Yale University Press, 2005).

252 *"foolish and unjustified discarding of horses"* David E. Johnson, *Fast Tanks and Heavy Bombers: Innovation in the U.S. Army, 1917–1945* (Ithaca, NY: Cornell University Press, 1998), 136.

252 *"The greatest hurdle"* U.S. Joint Forces Command, "Military Robots of the Future" (U.S. Joint Forces Command, 2003).

252 *"no fighter pilot is ever"* Charles Duhigg, "The Pilotless Plane That Only Looks Like Child's Play," *New York Times*, April 15, 2007.

253 *"It's like being a pilot for nerds"* Sig Christenson, "Cutting Edge of Military Aviation Has Steep Price Tag," *San Antonio Express-News*, September 18, 2007.

253 *"I was happy when drones came in"* Greg Lengyel, interview, Peter W. Singer, April 13, 2006.

253 *"Today's Air Force clings"* Ralph Peters, *Never Quit the Fight*, 1st ed. (Mechanicsburg, PA: Stackpole Books, 2006), 61.

253 *"It's seen as this geeky thing to do"* Interview at U.S. military facility, Peter W. Singer, February 19, 2008.

253 *"The reason that was given"* David Axe, "Who Killed the Killer Drone—And Why?" Defensetech.org, May 8, 2005 (cited May 9, 2005); available at http://www.defensetech.org/archives/002386.html.

254 *"perform high-speed aerobatics"* George C. Wilson, "A Chairman Pushes Unmanned Warfare," *National Journal* 32, no. 10 (2000): 718.

254 *"If you dislike change"* As quoted in P. H. Liotta, "Chaos as Strategy," *Parameters* 32, no. 2 (2002): 55.

254 *"cavorting with headhunters"* Christopher Palmeri, "A Predator That Preys on Hawks," *Business-Week*, no. 3820 (2003).

255 *"It's like a California speed shop"* Ibid.

255 *"The development of the smaller, cheaper plane"* Ibid.

256 *"We're number 1 in the world"* Boot, *War Made New*, 435.

256 *The Department of Justice* Dawn Kopecki, "On the Hunt for Fraud," BusinessWeek.com, October 10, 2006 (cited October 10, 2006); available at http://www.businessweek.com/bwdaily/dnflash/content/oct2006/db20061011_184367.htm.

256 *"routinely broken"* William Matthews, "Pentagon Inspector General: Procurement Laws Are Routinely Broken," *Defense News*, January 22, 2007, 4.

256 *being "hierarchical and top down"* Former army colonel, interview, Peter W. Singer, April 11, 2007.

256 *"The thing is 30 pounds and electric!"* Bruce Jette, "Robotics Development: An Overview of the Work of the Rapid Equipping Force," paper presented at the Military Robotics Conference, Institute for Defense and Government Advancement, Washington, DC, April 10–12, 2006.

257 *"We become prisoners"* Peters, *Never Quit the Fight*, 36.

257 *"quantitative incompetence"* Ralph Peters, "COIN of the Realm," presentation, Brookings Institution, Washington, DC, October 22, 2007.

257 *"In the year 2054"* Friedman and Friedman, *The Future of War*, 248.

257 *"If you think it is a young technology"* Demetri Sevastopulo, "US Military in Dogfight over Drones," *Financial Times*, August 19, 2007, http://www.ft.com/cms/s/0/78317cc4-4e93-11dc-85e7-0000779fd2ac.html.

257 *the number of Pentagon prime contractors* Defense Acquisition Performance Assessment Project, "Defense Acquisition Performance Assessment" (Washington, DC, 2006).

257 *"Only the dinosaurs were allowed"* Scientist, interview, Peter W. Singer, July 17, 2006.

258 *"The future belongs to those people"* "Interview: Neal Blue Chairman-CEO, General Atomics," *Defense News*, February 11, 2008, p. 26.

258 *"We just work on what the Pentagon"* Defense executive, interview, Peter W. Singer, October 4, 2007.

258 *a combined $295 billion over budget* GAO Report, "Defense Acquisitions: Assessments of Selected Weapon Programs," March 2008.

258 *91 percent of the performance bonus* This is not an isolated incident. Contractors for the F-35 Joint Strike Fighter also received their full bonus of nearly $500 million from 1999 to 2003, despite the fact that it overran the budget by $10 billion and was almost a year behind schedule.

258 *"Larger companies trend towards larger vehicles"* Mark Barber, "Force Protection Robotics," paper presented at the Military Robotics Conference, Institute for Defense and Government Advancement, Washington, DC, April 10–12, 2006. Interview with author.

259 *"There is no comparison"* Bing West, interview, Peter W. Singer, August 23, 2006.

259 *"more often smart than stupid"* Richard Szafranski, "When Waves Collide: Future Conflict," *Joint Force Quarterly*, no. 7 (1995): 82.

260 *"The ability to learn faster"* As quoted in Joel Garreau, *Radical Evolution: The Promise and Peril of Enhancing Our Minds, Our Bodies—And What It Means to Be Human* (New York: Doubleday, 2005), 257.

260 *"While learning from experience"* As quoted in Thomas Ricks, "America's Adventure," *Armed Forces Journal*, August (2006): 19.

## 13. OPEN-SOURCE WARFARE: COLLEGE KIDS, TERRORISTS, AND OTHER NEW USERS OF ROBOTS AT WAR

261 *"If I can imagine it"* Greg Bear, interview, Peter W. Singer, October 4, 2006.

261 *"It was an unusual shopping expedition"* Jason Zengerle, "Raising Money to Save Darfur," *New Republic*, March 20, 2006.

263 *"technology is both"* Max Boot, "The Paradox of Military Technology," *New Atlantis*, 14 (2006), http://www.thenewatlantis.com/archive/14/boot.htm.

263 *It is simultaneously* United Nations Office for the Coordination of Humanitarian Affairs, *Lebanon: The Many Hands and Faces of Hezbollah* (IRIN, 2006 [cited August 18 2006]); available at http://www.irinnews.org/report.aspx?reportid=26242.

263 *"There are many who belittled"* Barbara Opall-Rome, "Combating the Hizbollah Network: Israel Army Lessons from War in Lebanon," *Defense News*, October 9, 2006, 6.

264 *"a divine and strategic victory"* CNN.com, "Hezbollah Leader: Militants 'Won't Surrender Arms,'" September 22, 2006 (cited October 8, 2007); available at http://www.cnn.com/2006/WORLD/meast/09/22/lebanon.rally/index.html.

264 *a "hybrid war"* United States Marine Corps general James Mattis, presentation at the Brookings Institution, January 16, 2007. See also Frank Hoffman, "Lessons from Lebanon: Hezbollah and Hybrid Wars," Foreign Policy Research Institute, August 24, 2006 (cited August 26, 2006); available at http://www.fpri.org/enotes/20060824.military.hoffman,hezbollahhybridwars.html.

264 *Hezbollah also flew at least three* Alon Ben-David, "Israel Shoots Down Hezbollah UAV," *Jane's Defence Weekly*, August 16, 2006, 6.

264  *"able to hack into"* Noah Shachtman, "Arabs to Hezbollah: Up Yours," Defensetech.org, July 14, 2006 (cited July 14, 2006); available at http://www .defensetech.org/archives/002584.html.

264  *"hijacked" by Hezbollah hackers* Hilary Hylton, "How Hizballah Hijacks the Internet," Time.com, August 8, 2006, http://www.time.com/time/world/ article/0,8599,1224273,00.html.

264  *"In the cyberterrorism trade"* Ibid.

265  *The group even infiltrated* Opall-Rome, "Combating the Hizbollah Network: Israel Army Lessons from War in Lebanon."

265  *"The intelligence data provided"* David A. Fulghum, "Insurgents' New Tools," *Aviation Week & Space Technology* 165, no. 16 (2006).

265  *"All contempt for terrorists"* Ralph Peters, "Lessons from Lebanon," *Armed Forces Journal,* October 2006, 43.

265  *"robot mercenaries"* Special forces officer, interview, Peter W. Singer, Washington, DC, September 7, 2006.

265  *"I read an article in* Popular Science*"* Lee Gomes, "Team of Amateurs Cuts Ahead of Experts in Computer-Car Race," *Wall Street Journal,* October 19, 2005, B1.

266  *"It's a beautiful thing"* Ibid.

266  *"War made the state"* Charles Tilly, "Reflections on the History of European State-Making," in *The Formation of National States in Western Europe,* ed. Charles Tilly (Princeton, NJ: Princeton Unviersity Press, 1975), 42.

266  *past RMAs were associated* Max Boot, *War Made New: Technology, Warfare, and the Course of History, 1500 to Today* (New York: Gotham Books, 2006), 464.

267  *some sixty thousand multinational companies* For more information see New America Foundation, "Privatization of Foreign Policy Initiative" (cited January 12, 2008); available at http://www .newamerica.net/programs/american_strategy/ privatization_of_foreign_policy_initiative.

267  *so too is conflict being* John Robb, *Global Guerrillas* (cited January 10, 2008); available at http:// globalguerrillas.typepad.com.

267  *"The actual physical hardware"* Noah Shachtman, interview, Peter W. Singer, Washington, DC, March 25, 2006.

267  *"There are no friends"* Al J. Venter, *War Dog: Fighting Other People's Wars—The Modern Mercenary in Combat,* 1st ed. (Philadelphia: Casemate, 2006), 230.

267  *"A robot out of sight"* Bruce Jette, "Robotics Development: An Overview of the Work of the Rapid Equipping Force," paper presented at the Military Robotics Conference, Institute for Defense and Government Advancement, Washington, DC, April 10–12, 2006.

268  *the tiny country had hired* "Des Combats terrestres ont opposé l'armée ivoirienne et les militaires français," *Le Monde,* November 15, 2004, www .lemonde.fr.

268  *"State, nonstate, air, land, sea"* Noah Shachtman, interview, Peter W. Singer, March 25, 2006.

268  *"One of the reasons"* Logan Ward et al., "America 2025," *Popular Mechanics* 182, no. 5 (2005).

268  *"High-Tech Terror: Al-Qaeda's Use of New Technology"* Jarret M. Brachman, "High-Tech Terror: Al-Qaeda's Use of New Technology," *Fletcher Forum of World Affairs* 30, no. 2 (2006).

268  *"offer news on Iraq"* Anton LaGuardia, "Al-Qaeda Places Recruiting Ads," *Telegraph London,* August 10, 2005.

269  *There is even a site* "Now Online: Swear Loyalty to Al-Qaeda Leaders," Middle East Media Research Institute Special Dispatch 1027, 2005 (cited November 14, 2006); available at http://memri .org/bin/articles.cgi?Page=archives&Area=sd &ID=SP102705.

269  *"killer robots"* As quoted in Brachman, "High-Tech Terror: Al-Qaeda's Use of New Technology," 157.

269  *al-Qaeda explored the use of a UAV* David Hambling, "Terrorists' Unmanned Airforce," Defensetech .org, May 1, 2006 (cited July 14, 2006); available at http://www.noahshachtman.com/archives/ 002369.html.

269  *"Sooner or later"* "Quote of the Day," Time.com, February 28, 2008, at http://www.time.com/time/ quotes/0,26174,1718148,00.html?xid=feed -quoteswidget.

269  *"intersection of robotics and terrorist groups"* Robert Finkelstein, interview, Peter W. Singer, July 7, 2006.

269  *"You can be a wimp"* Noah Shachtman, interview, Peter W. Singer, Washington, DC, July 2, 2007.

270  *"a suicide bomber on steroids"* Ibid.

270  *"Robots could be very attractive"* Robert Finkelstein, interview, Peter W. Singer, July 7, 2006.

270  *"an ideal platform"* USAF Scientific Advisory Board, *Air Defense Against Unmanned Aerial Vehicles (UAVs)* (2006).

270  *costs only $1,000* http://diydrones.com, accessed April 28, 2008. See also "Build Your Own War Bot," at http://howto.wired.com/wiki/Build_ Your_Own_War_Bot, accessed March 20, 2008.

271  *The footage was so detailed* Thomas Claburn, "Terrorists Take Over Google Earth," *InformationWeek,* January 17, 2007, http://www.informationweek .com/showArticle.jhtml;jsessionid=CYKV3P1NN DZPWQSNDLPSKH0CJUNN2JVN?articleID=1 96901827.

271  *"They can make a lone actor"* Robert Finkelstein, interview, Peter W. Singer, July 7, 2006.

271  *"a few amateurs"* Ibid.

271  *"One bright but embittered loner"* Joel Garreau, *Radical Evolution: The Promise and Peril of Enhancing Our Minds, Our Bodies—and What It Means to Be Human* (New York: Doubleday, 2005), 139.

271 *"The obligation of subjects"* Christopher Coker, *Humane Warfare* (London, New York: Routledge, 2001), 18. Coker is quoting Thomas Hobbes, *Leviathan* (Oxford: J. Thornton, 1881), 170.

272 *Information on how to build* Ray Kurzweil, interview via phone, Peter W. Singer, Washington, DC, December 7, 2006.

272 *"It feels like all ten billion of us"* Garreau, *Radical Evolution*, 101.

272 *"It is no exaggeration"* Ibid., 207.

272 *"Historically, warfare"* Vernor Vinge, "Shaun Farrell Interviews Vernor Vinge," Shaun Farrell, April 2006; available at http://www.farsector.com/quadrant/interview-vinge.htm.

272 *"The future is manhunting"* Special forces officer, interview, Peter W. Singer, September 7, 2006.

273 *Skyshield, an automated machine-gun system* Alon Ben-David, "New Model Army," *Jane's Defence Weekly*, October 11, 2006, 26.

273 *the automated scanners can spot* Tom Simonite, "Scanner Recognises Hidden Knives and Guns," *New Scientist*, September 26, 2006, http://www.newscientisttech.com/article.ns?id=dn10160&feedId=tech_rss20.

273 *as many as three hundred times a day* Daniel H. Wilson, *How to Survive a Robot Uprising: Tips on Defending Yourself Against the Coming Rebellion*, 1st U.S. ed. (New York: Bloomsbury, 2005), 88.

273 *"They're the next best thing"* Stephen Kinzer, "Chicago Moving to 'Smart' Surveillance Cameras," *New York Times*, September 21, 2004, http://www.nytimes.com/2004/09/21/national/21cameras.html?_r=2&oref=slogin&oref=slogin.

274 *"They could only use"* Finkelstein, interview, Peter W. Singer, July 7, 2006.

274 *They can even scan crowds* Ibid.

274 *identify faces from as far as two hundred feet away* Noah Shachtman, "Cameras to Comb Crowds," Defensetech.org, October 24, 2006 (cited November 1, 2006); available at http://www.defensetech.org/archives/002887.html.

275 *One of the biggest data-mining efforts* Constance L. Hays, "What They Know About You," *New York Times*, November 14, 2004.

275 *"Very quietly, the core of TIA survives"* Noah Shachtman, "TIA Reboots," Defensetech.org, February 9, 2006 (cited February 9, 2006); available at http://www.noahshachtman.com/archives/002165.html.

276 *can then be matched against* Ibid.

276 *"identifying that it is a needle"* As quoted in Giles Ebbutt, "Knowledge Is Power," *Jane's International Defence Review* 40 (2007): 33.

276 *"track leads, form hypotheses"* Duncan Graham-Rowe, "Intelligence Analysis Software to Predict Terrorist Attacks in the Future," *New Scientist*, July 14, 2001, http://www.newscientist.com/article.ns?id=dn1368; Michael P. Tremoglie, "Terrorist Tracking Technology," *American Daily*, September 3, 2004, http://www.americandaily.com/article/2048.

277 *"cannot guarantee the software"* Graham-Rowe, "Intelligence Analysis Software to Predict Terrorist Attacks in the Future"; Applied Systems Intelligence, *ASI Continues Growth by Putting Brains in Army's Robots*; Eng, *Digital Warriors Artificial Intelligence May Help Spot Future Terrorism Attacks*; Tremoglie, "Terrorist Tracking Technology."

277 *"is more terrifying than losing one's privacy"* Graham-Rowe, "Intelligence Analysis Software to Predict Terrorist Attacks in the Future"; "Orwellian" quote from Wikipedia, "Information Awareness Office," December 25, 2007 (cited January 11, 2008); available at http://en.wikipedia.org/wiki/Information_Awareness_Office.

## 14. LOSERS AND LUDDITES: THE CHANGING BATTLEFIELDS ROBOTS WILL FIGHT ON AND THE NEW ELECTRONIC SPARKS OF WAR

279 *"Technological progress"* "Albert Einstein Quotes," Brainy Quote, 2008 (cited January 31, 2008); available at http://www.brainyquote.com/quotes/quotes/a/alberteins164554.html.

279 *"Increasingly, we live in a world"* Ralph Peters, "The Culture of Future Conflict," *Parameters* 25, no. 4 (1995).

279 *"I am a miner's son"* "Ralph Peters," Wikipedia, August 3, 2007 (cited August 3, 2007); available at http://en.wikipedia.org/wiki/Ralph_Peters.

280 *"simply one of the most creative"* Ralph Peters, *Beyond Baghdad: Postmodern War and Peace*, 1st ed. (Mechanicsburg, PA: Stackpole Books, 2003), back cover quote.

280 *"Ours is the age of barbarians"* Ralph Peters, *Never Quit the Fight*, 1st ed. (Mechanicsburg, PA: Stackpole Books, 2006), xv.

280 *"The soldiers of the United States Army"* Ralph Peters, "The New Warrior Class," *Parameters* 24, no. 2 (1994).

281 *"Unlike soldiers, warriors do not play"* Ibid.

281 *"postmodern warriors"* As quoted in Christopher Coker, *Waging War Without Warriors? The Changing Culture of Military Conflict*, IISS Studies in International Security (Boulder, CO: Lynne Rienner Publishers, 2002), 9.

281 *"These are people"* As quoted in Steven Metz, *Armed Conflict in the 21st Century: The Information Revolution and Post-modern Warfare* (Carlisle, PA: Strategic Studies Institute, U.S. Army War College, 2000), 48.

281 *"The archetype of the new warrior"* Ibid.

281 *"The longer the fighting continues"* Peters, "The New Warrior Class."

282 *"rightful place in the sun"* Peters, "The Culture of Future Conflict."

282 *"We live in the most dynamic age"* Ralph Peters, e-mail, Peter W. Singer, Washington, DC, March 9, 2007.

282 *"The Internet is the greatest tool"* Ibid.

282 *"The root causes of conflict"* Ibid.

283 *Americans spend over half a trillion dollars* Emily Lambert, "The Odd Couple," *Forbes* 178, no. 4 (2006).

283 *1.3 billion people in the developing world* Figures from United Nations Human Settlements Programme, *The Challenge of Slums: Global Report on Human Settlements* (London and Sterling, VA: Earthscan Publications Ltd., 2003); Paul Collier, "A Worldwide Scourge—How to Stem Civil Wars: It's the Economy, Stupid," *International Herald Tribune*, May 21, 2003; Michael Renner, "The Global Divide: Socioeconomic Disparities and International Security," in *World Security: Challenges for a New Century*, ed. Michael Klare and Yogesh Chandrani (New York: St. Martin's Press, 1998), 275.

283 *One hundred twenty-seven million Americans are "obese"* Lambert, "The Odd Couple."

284 *90 percent of the world's youth* Petersen, "A Strategy for the Future of Humanity."

284 *99 percent of the world's population growth* Jennifer Dabbs Sciubba, "The Defense Implications of Demographic Trends," *Joint Force Quarterly*, no. 48 (2008): 121.

284 *too many young males* Christian Mesquida and Neil I. Warner, "Male Age Composition and Severity of Conflicts," *Politics and Life Sciences* 18, no. 2 (1999); Richard Morin, "Boy Trouble," *Washington Post*, June 24, 2001; "Natural Born Killers: Does Biology Drive Our Need to Wage War," *Profile*, May 1999, http://www.yorku.ca/ycom/profiles/past/may99/current/dept/dispatch/dsp6.htm.

284 *the trend is particularly pronounced* Senate Select Committee on Intelligence, *The Worldwide Threat 2004: Challenges in a Changing Global Context*, Testimony by George J. Tenet, Director of Central Intelligence, February 24, 2004.

285 *As many as 250 million children* Figures from the U.S. Department of Labor, Bureau of International Affairs (2003), and Thoraya Ahmed Obaid, "State of World Population 2003: Making 1 Billion Count" (New York: United Nations Population Fund, 2003).

285 *"These poor, young billions"* Petersen, "A Strategy for the Future of Humanity."

285 *Population growth in Sudan* "Water Find 'May End Darfur War'" BBC News, July 18, 2007 (cited July 18, 2007); available at http://news.bbc.co.uk/2/hi/africa/6904318.stm.

285 *"the century of 'not enough'"* Peters, "The Culture of Future Conflict."

285 *global warming will bring water scarcity* Rob Taylor, "Millions to Go Hungry by 2080: Report," Truthout, January 30, 2007, http://www.truthout.org/issues_06/013007EA.shtml.

285 *100 million people* Michael Casey, "Report: Millions Face Hunger from Climate Change," *Christian Post*, April 10, 2007, http://www.christianpost.com/article/20070410/26802_Report:_Millions_Face_Hunger_from_Climate_Change.htm.

285 *"Unchecked climate change"* Ibid.

286 *"Now the ignorant know"* Ralph Peters, interview, Peter W. Singer, March 29, 2007.

286 *the more people are connected* Richard O. Hundley and RAND Corporation, *The Global Course of the Information Revolution: Recurring Themes and Regional Variations* (Santa Monica, CA: RAND, 2003).

286 *fifty countries that have "stateless zones"* George J. Tenet, *The Worldwide Threat 2004: Challenges in a Changing Global Context.*

286 *"Extreme losers in the information revolution"* Hundley and RAND Corporation, *The Global Course of the Information Revolution.*

287 *Al-Qaeda's movement of its training camps* George J. Tenet, *The Worldwide Threat 2004.*

287 *"It has become increasingly difficult"* Syed Hamid Albar, "Remarks at the U.S. and Islamic World Forum" (presentation, Doha, Qatar, February 17, 2006).

287 *over half of humankind* Tom Standage, *A History of the World in 6 Glasses* (Walker & Company, 2006), 93; David Oliver, "Training Street Fighters," *Military Technology* 31, no. 4 (2007): 39.

287 *are about forty times larger* Joseph Grosso, review of *Monster at Our Door* by Mike Davis (2005), *Z Magazine* 19(11) (2005), available at http://zmagsite.zmag.org/Nov2006/grosso1106.html.

287 *"The city—capstone of human organization"* Ralph Peters, "Our Soldiers, Their Cities," *Parameters* 26, no. 2 (1996): 45.

287 *"the new forests"* Ralph Peters, e-mail, Peter W. Singer, Washington, DC, March 9, 2007.

287 *"Cities are now the center of rebellion"* Ibid.

288 *"Habituated to violence"* As quoted in Coker, *Waging War Without Warriors? The Changing Culture of Military Conflict*, 10.

288 *"The future of warfare"* Peters, "Our Soldiers, Their Cities," 43.

288 *"shanty-towns and squatter communities"* Mike Davis, *Planet of Slums* (London: Verso, 2006).

288 *"the diverse religious"* Ibid.

288 *"stinking mountains of shit"* Ibid.

288 *a crossover with crime* Kenneth Turan, "Movie Review: Favela Rising," *Los Angeles Times*, August 4, 2006, http://www.latimes.com/entertainment/news/movies/cl-et-favela4aug04,1,1371999.story?coll=la-promo-entnews.

289 *"The 'feral, failed cities' of the Third World"* Mike Davis, as quoted in Nick Turse, "Baghdad 2025: The Pentagon Solution to a Planet of Slums," TomDispatch.com, January 7, 2007, http://www.tomdispatch.com/post/155031/nick_turse_pentagon_to_global_cities_drop_dead.

289 *"American Terminators"* Turse, "Baghdad 2025: The Pentagon Solution to a Planet of Slums."

289 *"where the fight"* Ibid.

289 *"The array of threats"* As quoted in Metz, *Armed Conflict in the 21st Century*, 44.

289 *"gallant struggles in green fields"* Peters, "Our Soldiers, Their Cities," 43.

290 *"it's a no-brainer for the enemy"* Ralph Peters, interview, Peter W. Singer, March 29, 2007.

290 *"A host of unmanned vehicles"* Turse, "Baghdad 2025: The Pentagon Solution to a Planet of Slums."

290 *"to make the foreign city"* Ibid.

290 *A similar program* David Hughes, "Street-Smart Maps," *Aviation Week & Space Technology* 165, no. 21 (2006): 77.

290 *"pressing need in urban warfare"* As quoted in Turse, "Baghdad 2025: The Pentagon Solution to a Planet of Slums."

291 *"You have continuous coverage"* Graham, "America's Robot Army."

291 *"unprecedented awareness"* Ibid.

291 *"There is a uniquely American pursuit"* Ralph Peters, interview, Peter W. Singer, March 29, 2007.

291 *Bush administration "urgently" needed* Richard Clarke, "Presidential Policy Initiative/Review— The Al Qida Network," Memorandum, Condoleezza Rice, Washington, DC, January 25, 2001; available at http://www2.gwu.edu/~nsarchiv/NSAEBB/NSAEBB147/clarke%20memo.pdf.

291 *"You give bin Laden too much credit"* William Douglas, "White House Tries to Discredit Counterterrorism Coordinator," CommonDreams.com, March 22, 2004, http://www.commondreams.org/headlines04/0322-10.htm.

292 *"Something very fundamental"* Richard Clarke, interview, Peter W. Singer, Washington, DC, August 8, 2007.

294 *"It goes to the last frontier"* Tom Erhard, interview, Peter W. Singer, January 31, 2007.

294 *"will be a moral battlefield"* Rodney Brooks, *Flesh and Machines: How Robots Will Change Us* (New York: Pantheon, 2002), x.

295 *neo-Luddites will also see* Antón, Silberglitt, and Schneider, *The Global Technology Revolution: Bio/Nano/Materials Trends and Their Synergies with Information Technology by 2015*.

295 *short for "Freedom Club"* "The Unabomber: A Chronology," Court TV Online (cited August 12, 2007); available at http://www.courttv.com/trials/unabomber/chronology/chron_8587.html.

295 *"As society and the problems"* "Industrial Society and Its Future," Wikipedia, August 9, 2007 (cited August 10, 2007); available at http://en.wikisource.org/wiki/Industrial_Society_and_Its_Future.

295 *"We therefore advocate a revolution"* Ibid.

296 *"huge potential for strange bedfellows"* Richard Clarke, interview, Peter W. Singer, August 8, 2007.

296 *"Since the stake is so high"* Hugo de Garis, "Building Gods or Building Our Potential Exterminators?" KurzweilAI.net, February 26, 2001 (cited June 27, 2006); available at http://www.kurzweilai.net/meme/frame.html?main=/articles/art0131.html?

296 *"The great paradox"* Ralph Peters, e-mail, Peter W. Singer, March 9, 2007.

## 15. THE PSYCHOLOGY OF WARBOTS

297 *"Warfare is about changing"* Ralph Peters, *Never Quit the Fight*, 1st ed. (Mechanicsburg, PA: Stackpole Books, 2006), 39.

297 *"Human versus robot?"* Eliot Cohen, interview, Peter W. Singer, Washington, DC, November 15, 2006.

297 *the book for civilian leaders* Ahmad Faruqui, "The Apocalyptic Vision of the Neo-Conservative Ideologues," *CounterPunch*, November 26, 2002, http://www.counterpunch.org/faruqui1126.html.

298 *"at which a majority"* John Keegan, *The Face of Battle* (New York: Viking Press, 1976), 276.

298 *"while soldiers will fight"* Fred Reed, "Robotic Warfare Drawing Nearer," *Washington Times*, February 10, 2005.

298 *"an almost helpless feeling"* Edward Godere, interview, Peter W. Singer, November 17, 2006.

298 *"without even having to fire"* Discovery Channel Pictures, "Smart Weapons," in *Future Weapons*, Discovery Channel, broadcast on May 17, 2006.

300 *But after these strange, fearsome men* Jared M. Diamond, *Guns, Germs, and Steel: The Fates of Human Societies*, 1st ed. (New York: W. W. Norton & Co., 1997), 68 and 75.

300 *"Any sufficiently advanced technology"* Christopher Coker, *The Future of War: The Re-enchantment of War in the Twenty-first Century*, Blackwell Manifestos (Malden, MA; Oxford, UK: Blackwell, 2004), 130.

300 *how American air power* Charles J. Dunlap Jr., "America's Asymmetric Advantage," *Armed Forces Journal*, September 2006.

300 *"You have a guy"* Interview with JSOC official by author, March 11, 2009, Brookings Pakistan-COIN conference in Washington, DC.

300 *"With all the dust"* Bing West, "Nowhere to Hide," *Popular Mechanics* 182, no. 2 (2005).

300 *While the hostage takers gathered* Peter W. Singer, "Research Visit to Foster-Miller," 2006.

301 *robots that "creep people out"* Seth Borenstein, "Scientists Try to Make Robots More Human," USAToday.com, November 22, 2006, http://www.usatoday.com/tech/news/robotics/2006-11-22-humanistic-robots_x.htm.

302 *"market forces will shape things"* David Hanson, interview via phone, Peter W. Singer, Washington, DC, October 12, 2007.

302 *influence the "attitudes, feelings"* United States Joint Chiefs of Staff, Joint Publication 3-35, *Doctrine for Joint Psychological Operations*, September 5, 2003, 102.

302 *evokes a state of terror* From Sigmund Freud, *Beyond the Pleasure Principle (1920)*, as discussed in Christopher Coker, *Waging War Without Warriors? The Changing Culture of Military Conflict*, IISS Studies in International Security (Boulder, CO: Lynne Rienner Publishers, 2002), 19.

302 *"makes Robocop look like"* Jonathon Keats, "The Idea Man," Popsci.com, 2004 (cited August 18, 2006); available at http://www.popsci.com/popsci/technology/generaltechnology/6b0898b0c9b84010vgnvcm1000004eecbccdrcrd.html.

303 *"That's not funny anymore"* Francis J. West, *No True Glory: A Frontline Account of the Battle for Fallujah* (New York: Bantam, 2005), 273.

304 *"might even be a profession"* Robert Finkelstein, interview, Peter W. Singer, July 7, 2006.

304 *"an object acting as a human"* David Hanson, interview, Peter W. Singer, October 12, 2007.

304 *"The keyboard and the monitor"* Robert Epstein, "My Date with a Robot," *Scientific American*, June–July 2006, 68–73.

304 *"People's empathy increases"* Mark Jacob, "Japan's Robots Stride into Future," *Chicago Tribune*, July 15, 2006.

305 *"It's not just that"* Robert Epstein, interview, Peter W. Singer, October 25, 2006.

305 *"The more familiar with technology"* Ibid.

305 *"If they are not used to robots"* David Hanson, interview, Peter W. Singer, October 12, 2007.

305 *"When my daughter first saw"* Epstein, "My Date with a Robot."

305 *"The uncanny valley is"* Andrew Bennett, interview, Peter W. Singer, November 16, 2006.

305 *The raids killed hundreds of thousands* Anthony D'Amato, "International Law, Cybernetics, and Cyberspace," *Naval War College International Law Studies* 76 (2006): 66.

305 *proved "fundamentally flawed"* H. R. McMaster, *Dereliction of Duty: Lyndon Johnson, Robert McNamara, the Joint Chiefs of Staff, and the Lies That Led to Vietnam*, 1st ed. (New York: HarperCollins, 1997), 327.

306 *"I think that it will discourage"* U.S. Army UAV pilot, interview, Peter W. Singer, November 8, 2007.

306 *"It must be daunting"* Interview at U.S. military facility, Peter W. Singer, February 19, 2008.

306 *"I didn't really imagine"* Yousif Basil, interview, Peter W. Singer, August 15, 2006.

307 *"you have to remember"* Nir Rosen, interview, Peter W. Singer, August 11, 2006.

307 *"What is Osama bin Laden's"* Peter D. Feaver, "To Maintain That Support, Show Us What Success Means," Duke University (cited August 4, 2007); available at http://www.duke.edu/web/forums/feaver.html.

307 *Showing personal bravery* United States Marine Corps general, interview, Peter W. Singer, January 16, 2007.

308 *"Victory comes from human beings"* Charles J. Dunlap Jr., "How We Lost the High-Tech War of 2007: A Warning from the Future," *Weekly Standard* 1, no. 19 (1996): 96.

308 *The future hotbed of rebellion* LTC Todd Megill, intelligence officer with 4th Infantry Division, in Max Boot, *War Made New: Technology, Warfare, and the Course of History, 1500 to Today* (New York: Gotham Books, 2006), 401.

308 *Khouri is also the editor-at-large* Rami Khouri, interview, Peter W. Singer, August 22, 2006.

309 *"This [the robotics revolution]"* Noah Shachtman, "More Robot Grunts Ready for Duty," *Wired News*, December 1, 2004, http://www.wired.com/news/technology/0,65885-0.html.

310 *"The optics of the situation"* Noah Shachtman, interview, Peter W. Singer, Washington, DC, March 25, 2006.

310 *17 percent of the global average* Herwig Schopper, "Islam and Science," *Nature*, November 1, 2006, http://www.nature.com/news/2006/061030/full/444035a.html.

310 *"encourages heresy"* For a transcript see "Samir Ubeid, an Iraqi Researcher Living in Europe: The Nobel Prize Is Racist and Stems from the Protocols of the Elders of Zion," Middle East Media Research Institute, October 31, 2006 (cited August 14, 2007); available at http://www.memritv.org/clip_transcript/en/1313.htm.

310 *lacks "a cultural base"* Ibid.

310 *"This type of warfare"* Talat Masood, "Shackling Shock and Awe: American and Muslim World Views on the Laws of High Tech Warfare" (presentation, U.S. Islamic World Forum, Doha, January 10–12, 2004).

311 *"distance warfare"* Ibid.

311 *"Overreliance on the military"* Ibid.

311 *"The concept of 'shock and awe'"* Ibid.

311 *"The mythology surrounding"* Mansoor Ijaz, "An Alliance Too Vital to Jeopardize with Poor Intelligence," *Financial Times*, January 17, 2006, 13.

311 *"America's heartless terrorism"* Munish Puri, e-mail, Peter W. Singer, Washington, DC, July 20, 2006.

311 *give credence to the unintended psychological consequences* Mubashar Jawed Akbar, interview, Peter W. Singer, Washington, DC, September 12, 2006.

312 *Using robots in war* Ibid., 65.

313 *"The insurgents are defending"* Nir Rosen, interview, Peter W. Singer, August 11, 2006.

313 *"If they play by these rules"* Rami Khouri, interview, Peter W. Singer, August 22, 2006.

313 *"Against them make ready"* Koran, sura 8, verse 60.

314 *"the study of battle"* Keegan, *The Face of Battle*, 298.

314 *"vision, a dream, a nightmare"* Ibid., 294.

## 16. YOUTUBE WAR: THE PUBLIC AND ITS UNMANNED WARS

315 *"Wars are a human phenomenon"* Thomas K. Adams, "Future Warfare and the Decline of Human Decisionmaking," *Parameters* 31, no. 4 (2001): 57.

315 *"Robotics and all this unmanned stuff"* Larry Korb, interview, Peter W. Singer, September 30, 2006.

316 *"There will be more marketing"* Ibid.

317 *"The Army belongs"* Stephen J. Cimbala, "Transformation in Concept and Policy," *Joint Force Quarterly*, no. 38 (2005): 28.

317 *"War is much more than strategy"* R. D. Hooker Jr., "Beyond Vom Kriege: The Character and Conduct of Modern War," *Parameters* 35, no. 2 (2005): 8.

317 *"Society is an intimate participant"* Ibid.

318 *"Rather than summoning Americans"* Andrew J. Bacevich, "The Right Choice?" *American Conservative*, March 24, 2008, http://www.amconmag.com/2008/2008_03_24/article.html.

318 *Josiah Bunting is a former major general* Josiah Bunting, "What Determines Why People Support the Next War?," Imagining the Next War, Guggenheim conference, New York, March 25, 2006.

319 *"This may be a positive way"* Michael Kan, "The Evolution of Warfare," *Michigan Daily*, July 27, 2005, http://www.michigandaily/com/vnews/display.v/ART/2005/03/31/424be2fd00491.

319 *"They [unmanned systems] lower"* Tom Malinowsky, interview, Peter W. Singer, Washington, DC, January 29, 2007.

319 *"Anything that makes it morally"* Special forces officer, interview, Peter W. Singer, Washington, DC, September 7, 2006.

320 *"Taking the human factor out"* Patrick Eberle, "To UAV or Not to UAV: That Is the Question; Here Is One Answer," *Air & Space Power Journal—Chronicles Online Journal*, October 9, 2001, http://www.airpower.au.af.mil/airchronicles/cc/eberle.html.

320 *entertainment, or "war porn"* Interview with U.S. Army War College officer, Peter W. Singer, Washington, DC, February 8, 2008.

320 *"A global spectator sport"* "Notes, 8 June 2004," in *National Security in the 21st Century: Rethinking the Principles of War* (Arlington, VA: Johns Hopkins University Applied Physics Lab, 2004).

321 *"the pleasure of a spectacle"* Christopher Coker, *Humane Warfare* (London, New York: Routledge, 2001), 150.

321 *Nations often go to war* Geoffrey Blainey, *The Causes of War* (New York: Free Press, 1973).

321 *"use it or lose it" mentality* "Nanotech Arms Races," Center for Responsible Nanotechnology, June 30, 2004 (cited July 18, 2006); available at http://crnano.typepad.com/crnblog/2004/06/nanotech_arms_r.html.

321 *"if we believe the hype"* Daniel Wilson, interview, Peter W. Singer, October 19, 2006.

321 *"lower the threshold for violence"* James Der Derian, interview, Peter W. Singer, September 20, 2006.

321 *"If one can argue that such new technologies"* Ibid.

322 *"option of last resort"* A. J. Bacevich and Lawrence F. Kaplan, "The Clinton Doctrine," *Weekly Standard* 215, no. 14 (1996): 16.

322 *cruise missile diplomacy of the 1990s* John A. Gentry, "Doomed to Fail: America's Blind Faith in Military Technology," *Parameters* 32, no. 4 (2002): 100.

322 *"feel good for a time"* Ibid.

322 *"The military thinks"* Daniel Wilson, interview, Peter W. Singer, October 19, 2006.

323 *One private military company executive* Robert Young Pelton, "Licensed to Kill: Hired Guns in the War on Terror," presentation, Brookings Institution, Washington, DC, October 5, 2006.

323 *Instead of widespread engagement* Paul W. Kahn, "War and Sacrifice in Kosovo," *Philosophy & Public Diplomacy Quarterly*, 2/3 (1999), http://www.publicpolicy.umd.edu/IPPP/spring_summer99/kosovo.htm.

323 *"a reflection of the moral character"* Ibid.

324 *"The life of one NATO soldier"* Ibid.

324 *"become so intoxicated by the idea"* Coker, *Humane Warfare*, 150.

324 *"making war is the act of killing"* Francis J. West, *No True Glory: A Frontline Account of the Battle for Fallujah* (New York: Bantam, 2005), 323.

324 *"is forced travel, no good food"* Paul Fussell, "What Determines Why People Support the Next War?" (paper presented at the Imagining the Next War, Guggenheim Conference, New York City, March 25, 2006).

324 *"And so I tried to cut away"* Susanna Rustin, "Hello to All That," *Guardian* (UK), July 31, 2004, http://books.guardian.co.uk/departments/history/story/0,,1272911,00.html.

325 *"If darkness had mercifully hidden"* Ibid.

325 *"If there is no risk"* Fussell, "What Determines Why People Support the Next War?"

325 *"people will support the next war"* Ibid.


## 17. CHANGING THE EXPERIENCE OF WAR AND THE WARRIOR

326 *"The introduction of every new technology"* Illah Nourbakhsh, interview, Peter W. Singer, Washington, DC, October 31, 2006.

327 *"The 101st kicks ass"* As quoted in Rym Brahimi et al., "Pentagon: Saddam's Sons Killed in Raid," CNN.com, July 22, 2003 (cited March 30, 2007); available at http://www.cnn.com/2003/WORLD/meast/07/22/sprj.irq.sons/index.html.

327 *"It was like a Super Bowl party"* Noah Shachtman, "Drone School, a Ground's-Eye View," *Wired News*, May 27, 2005, http://www.wired.com/news/technology/0,1282,67655,00.html.

328 *Colonel Gary Fabricius graduated from* Gary Fabricius, interview, Peter W. Singer, Pentagon, August 29, 2006.

328 *"kicking, screaming, clawing, and scratching"* Ibid.

328 *"But after a month, I became a believer"* Ibid.

329 *"You could see him climbing out of the hole"* Ibid.

329 *"If you want to pull the trigger"* Robert D. Kaplan, "Hunting the Taliban in Las Vegas," *Atlantic Monthly* 298, no. 2 (2006).

329 *the action felt so intense* Joel Garreau, "Bots on the Ground: In the Field of Battle (Or Even Above It), Robots Are a Soldier's Best Friend," *Washington Post*, May 6, 2007, D1.

330 *Marshall Harrison tells* Marshall Harrison, *A Lonely Kind of War: Forward Air Controller, Vietnam* (Novato, CA: Presidio Press, 1989), 27.

330 *"a wife, three children, and a well-mortgaged home"* Ibid., 43.

330 *"At the end of the duty day"* Kaplan, "Hunting the Taliban in Las Vegas."

330 *"YOU'LL BE SCARED"* Nancy Sherman, *Stoic Warriors: The Ancient Philosophy Behind the Military Mind* (New York: Oxford University Press, 2005), 101.

330 *"Most of the time,"* Noah Shachtman, "Attack of the Drones," *Wired* 13.06 (2005), http://www.wired.com/wired/archive/13.06/drones.html.

330 *getting home in time* Predator pilot, interview, Peter W. Singer, August 28, 2006.

331 *"No, he doesn't meet my definition"* Special forces officer, interview, Peter W. Singer, Washington, DC, September 7, 2006.

331 *"If you see it through their eyes"* Ibid.

331 *"so many brave and valiant men"* Max Boot, *War Made New: Technology, Warfare, and the Course of History, 1500 to Today* (New York: Gotham Books, 2006), 22.

331 *"fighting by remote"* Robert Epstein, interview, Peter W. Singer, Washington, DC, October 25, 2006.

331 *"passive disdain"* Boot, *War Made New*, 86.

331 *"Three men and a machine gun"* Ibid., 167.

332 *"The real function of an army"* T. R. Fehrenbach, *This Kind of War: The Classic Korean War History*, 1st Brassey's ed. (Washington: Brassey's, 1994), 66.

332 *"The mysterious quality"* As quoted in Henry G. Gole, "Reflections of Courage," *Parameters* 27, no. 4 (1997): 147. See also Charles McMoran Wilson Moran, *The Anatomy of Courage*, 1st American ed. (Boston: Houghton Mifflin, 1967).

332 *"It's like a video game"* Shachtman, "Drone School, a Ground's-Eye View."

332 *"Every now and then"* Hart Seely, "Robot Plane Pilots Have Bird's Eye View of Iraq War," Newhouse News Service, November 4, 2005.

332 *"You've got to be thankful"* Ibid.

332 *"Yeah, war is hell"* Ibid.

332 *"I won the last game"* Ibid.

332 *move toward "virtueless war"* Andrew White, "Uninhabited Military Vehicles as 'Virtueless' War: A Psycho-Social Exploration of Behavioural Responses" (NATO, 2006). Air Chief Marshal Sir Brian Burridge cited on p. 11-1 and later pp. 11-8.

333 *"war is not just in transition"* Chris Gray, *Postmodern War: The New Politics of Conflict* (New York: Guilford Press, 1997), 3.

333 *"From this day to the ending of the world"* William Shakespeare, *Henry V*, 4.3.58–62.

333 *"We're a family"* Robert D. Kaplan, *Imperial Grunts: The American Military on the Ground* (New York: Random House, 2005), 281.

334 *"Cohesion requires trusting"* Ralph E. McDonald, "Cohesion: The Key to Special Operations Teamwork, Research Report No. AU-ARI-94-2" (Maxwell AFB: Air University Press, October 1994).

334 *the new experiences of warriors* Bing West, interview, Peter W. Singer, August 23, 2006.

334 *"Make no mistake"* Interview at U.S. military facility, Peter W. Singer, February 19, 2008.

334 *"founding father of the study"* Josh Hyatt, "The SOUL of a New Team," *Fortune* 153, no. 11 (2006): 134–43.

335 *"The only sound"* Mihail C. Roco and William Sims Bainbridge, "Converging Technologies for Improving Human Health: Nanotechnology, Biotechnology, Information Technology and Cognitive Science" (National Science Foundation, 2002), 164.

335 *"MySpace Generation"* Jessi Hempel, "The MySpace Generation," BusinessWeek.com, December 12, 2005 (cited November 2, 2006); available at http://www.businessweek.com/magazine/content/05_50/b3963001.htm.

335 *"Computers by their nature"* Thomas E. Ricks, *Fiasco: The American Military Adventure in Iraq* (New York: Penguin Press, 2006), 313.

335 *"skittering like water bugs"* Joel Garreau, *Radical Evolution: The Promise and Peril of Enhancing Our Minds, Our Bodies—And What It Means to Be Human* (New York: Doubleday, 2005), 219.

335 *"op-con [operational control] isn't real"* United States Marine Corps general, interview, Peter W. Singer, January 16, 2007.

336 *"Ninety percent of the time"* Gary Fabricius, interview, Peter W. Singer, Pentagon, August 29, 2006.

336 *"Staff Sergeant Smithy"* Ibid.

336 *"I hate it"* Ibid.

336 *"what's funny about using Microsoft Chat"* Joshua Davis, "If We Run Out of Batteries, This War is Screwed," *Wired* 11.06 (2003), http://www.wired.com/wired/archive/11.06/battlefield.html.

336 *"Yipee-ki-aye! Motherfucker"* Interview at U.S. military facility, Peter W. Singer, February 19, 2008.

336 *"everyone thinks they have a vote"* Predator pilot, interview, Peter W. Singer, August 28, 2006.

336 *"Textual communications accounts"* Michael Downs, interview, Peter W. Singer, Washington, DC, September 13, 2006.

337 *"and confusion led the UAV"* Special forces officer, interview, Peter W. Singer, September 7, 2006.

337 *"Can you fix it?"* Garreau, "Bots on the Ground: In the Field of Battle (Or Even Above It), Robots Are a Soldier's Best Friend," *Washington Post*, May 6, 2006, D1.

337 *"This has been a really great robot"* Ibid.

338 *"wanted Scooby-Doo back"* Peter W. Singer, "Research Visit to iRobot Corporation," 2006.

338 *"I don't get happy about robots"* As quoted in Lee Gutkind, *Almost Human: Making Robots Think*, 1st ed. (New York: W. W. Norton & Co., 2006), 36.

338 *"You start to associate personalities"* Preston Lerner, "Robots Go to War: Within 10 Years, Infantry Soldiers Will Go into Battle with Autonomous Robots Close Behind Them," *Popular Science* 268, no. 1 (2006).

338 *"award 'battlefield promotions'"* Garreau, "Bots on the Ground."

339 *"It was a big deal"* Ibid.

339 *"I wish you all could be here"* Kari Thomas, "Robotics on the Battlefield," *Robotics Update News Letter*, 2 (2004), http://www.nosc.mil/robots/newsletter/RoboticsUpdate_4_2.pdf.

339 *an EOD soldier ran fifty meters* Peter W. Singer, "Research Visit to Foster-Miller," 2006.

340 *"The Colonel could not stand"* Garreau, "Bots on the Ground."

340 *the people had much the same brain activity* Robin Marantz Henig, "The Real Transformers," *New York Times Magazine*, July 29, 2007.

340 *describe their computers as having "agency"* Betya Friedman, Peter H. Kahn, and Jennifer Hagman, "Hardware Companions? What Online AIBO Discussion Forums Reveal About the Human-Robotic Relationship," in *Conference on Human Factors in Computing Systems* (Fort Lauderdale: ACM Press, 2003), 274.

340 *"Oh yeah, I love Spaz"* Peter H. Kahn et al., "Social and Moral Relationships with Robotics Others?," paper presented at the IEEE International Workshop on Robot and Human Interaction, Okayama, Japan, September 20–22, 2004, 548.

341 *this line of what is alive or not* Mihail C. Roco and William Sims Bainbridge, "Converging Technologies for Improving Human Health: Nanotechnology, Biotechnology, Information Technology and Cognitive Science" (National Science Foundation, 2002).

341 *Ibn Sina, a robot built* "How to Make (Robot) Friends and Influence People," *Technology Review*, May 5, 2009.

341 *"Robots will need to work"* Hiroko Tabuchi, "Japan Looks to a Robot Future," Associated Press, March 2, 2008.

342 *"Hey [whatever their name was]"* Jennie J. Gallimore and Sasanka Prabhala, "Creating Collaborative Agents with Personality for Supervisory Control of Multiple UCAVs," paper presented at the Symposium on Human Factors of Uninhabited Military Vehicles as Force Multipliers, Biarritz, France, October 9–11, 2006, 14.

342 *The other would say* Ibid.

342 *"Here is the last known target"* Ibid., 16.

342 *"Let's say you design robots"* Peter H. Kahn, "Social and Moral Relationships with Personified Robots" (presentation, Navy Center for Applied Research in Artificial Intelligence, March 12, 2007).

342 *"But he was somebody"* Paul Fussell, "What Determines Why People Support the Next War?," paper presented at the Imagining the Next War, Guggenheim Conference, New York City, March 25, 2006.

342 *"Being a Marine"* Nathaniel Fick, *One Bullet Away: The Making of a Marine Officer* (Boston: Houghton Mifflin, 2005), 17.

343 *"We joined the military"* United States Marine, interview, Peter W. Singer, May 15, 2007.

## 18. COMMAND AND CONTROL . . . ALT-DELETE: NEW TECHNOLOGIES AND THEIR EFFECT ON LEADERSHIP

344 *"You are watching the most violent actions"* LTC Michael Downs, interview, Peter W. Singer, Washington, DC, September 13, 2006.

346 *"What angers me"* Interview at U.S. military facility, Peter W. Singer, February 19, 2008.

347 *"Emotional exhaustion and burnout"* as quoted in Aaron Retica, "Drone-pilot Burnout," *New York Times Sunday Magazine*, Dec. 14, 2008, available at http://www.nytimes.com/interactive/2008/12/14/magazine/2008_IDEAS.html.

348 *his role in the operation* Interview, Peter W. Singer, Brookings Institution, December 17, 2007.

348 *"the personal bond"* John Keegan, *The Face of Battle* (New York: Viking Press, 1976), 114.

348 *"towards centralization of command"* Chris Gray, *Postmodern War: The New Politics of Conflict* (New York: Guilford Press, 1997), 274.

348 *"just turned the radios off"* Ibid., 63.

348 *"GCCS—known as 'Geeks' "* Joshua Davis, "If We Run Out of Batteries, This War Is Screwed," *Wired*, 11.06 (2003), http://www.wired.com/wired/archive/11.06/battlefield.html.

349 *ordered the captain* Andrew Exum, interview, Peter W. Singer, April 28, 2008.

349 *"It's like crack for generals"* Noah Shachtman, "Attack of the Drones," *Wired* 13.06 (2005), http://www.wired.com/wired/archive/13.06/drones.html.

349 *"incentives to intervene tactically"* Robert Killebrew, "Why Doctrine Matters and How to Fix It," *Armed Forces Journal*, October 2006, 22.

350 *"5,000 mile long screwdriver"* Barry Rosenberg, "Technology and Leadership," *Armed Forces Journal*, July 2007, 18.

350 *"You get too focused"* As quoted in Thomas E. Ricks, "Beaming the Battlefield Home: Live Video

of Afghan Fighting Had Questionable Effect," *Washington Post,* March 22, 2002, A1.

351  *the commanders thought* Stephen D. Biddle, *Military Power: Explaining Victory and Defeat in Modern Battle* (Princeton, NJ: Princeton University Press, 2004), 65.

351  *eat "a shit sandwich"* Michael R. Gordon and Bernard E. Trainor, *Cobra II: The Inside Story of the Invasion and Occupation of Iraq,* 1st ed. (New York: Pantheon, 2006), 314.

351  *"Mother may I?" syndrome* The term was used in four different interviews.

351  *"it sat in someone's e-mail"* Interview at U.S. military facility, Peter W. Singer, February 19, 2008.

351  *"It's the old story"* Robert D. Kaplan, "Hunting the Taliban in Las Vegas," *Atlantic Monthly* 298, no. 2 (2006).

351  *"One bad general is better"* Nicholas Wade, "Bytes Make Might," *New York Times Magazine,* March 12, 1995, 28.

352  *"each commander thinks he's in control of you"* Susan B. Glasser and Vernon Loeb, "A War of Bridges: 225,000 U.S. and British Troops Are Now Within Striking Distance," *Washington Post Foreign Service,* March 2, 2003, A1.

352  *"power struggles galore"* Ibid.

352  *"they were in a position"* Ricks, "Beaming the Battlefield Home: Live Video of Afghan Fighting Had Questionable Effect," 1.

352  *"flattening of the chain of command"* Rosenberg, "Technology and Leadership," 17.

353  *"Don't do anything beyond patrol"* United States Marine Corps general, interview, Peter W. Singer, January 16, 2007.

353  *"You may have some general officer"* John J. Klein, "The Problematic Nexus: Where Unmanned Combat Air Vehicles and the Law of Armed Conflict Meet," *Air & Space Power Journal—Chronicles Online Journal,* July 22, 2003, http://www.air power.maxwell.af.mil/airchronicles/cc/klein .html.

353  *"in the near future"* Bing West, interview, Peter W. Singer, August 23, 2006.

353  *"it'll be like taking LBJ"* Michael Wynne, interview, Peter W. Singer, Washington, DC, January 25, 2008.

353  *"techniques of leadership"* Paul Fussell, "What Determines Why People Support the Next War?" paper presented at the Imagining the Next War, Guggenheim Conference, New York City, March 25, 2006.

354  *"the passion of command"* Bryan McCoy, *The Passion of Command: The Moral Imperative of Leadership* (Quantico, VA: Marine Corps Association, 2006).

354  *"Commanding an army of droids"* David Sherman and Dan Cragg, *Star Wars: Jedi Trial,* 1st ed. (New York: Random House, 2004), 264.

354  *"where the strategic, operational, and tactical levels"* Richard A. Chilcoat, "The 'Fourth' Army War College: Preparing Strategic Leaders for the Next Century," *Parameters* 25, no. 4 (1995).

354  *"The strategic leader best adapted"* Ibid.

355  *flexibility to lead a "learning organization"* Janine Davidson, "Learning to Lift the Fog of Peace: The U.S. Military in Stability and Reconstruction Operations" (University of South Carolina, 2005).

355  *"I speculate that the digital general"* Paul T. Harig, "The Digital General: Reflections on Leadership in the Post-Information Age," *Parameters* 26, no. 3 (1996): 134.

355  *"To the strategic commander of the Information Age"* Chilcoat, "The 'Fourth' Army War College: Preparing Strategic Leaders for the Next Century."

356  *New inventions like the radio and teletype* Credit for this point goes to Harlan Ullman.

356  *"Engage your brain"* James Mattis, presentation, Brookings Institution, January 16, 2007.

356  *they identified such qualities* Chilcoat, "The 'Fourth' Army War College: Preparing Strategic Leaders for the Next Century."

356  *"In the end, it could be argued"* Harig, "The Digital General: Reflections on Leadership in the Post-Information Age," 133.

356  *"the decision cycle of the future"* John Bennett, "DoD Struggles to Craft Offensive Cyberspace Plan," *Defense News,* February 26, 2007, 1.

357  *"The solution to this problem"* Thomas K. Adams, "Future Warfare and the Decline of Human Decisionmaking," *Parameters* 31, no. 4 (2001).

357  *AI that allows a commander* Tony Skinner, "DARPA Develops Strategic Decision Support Tools," *Jane's Defence Weekly,* January 4, 2007, 7.

357  *"battle management" systems* Gray, *Postmodern War,* 58.

357  *"provide the commander"* "Interview: Dr. Alexander Kott: RAID program manager, DARPA," *Jane's International Defence Review,* 41, March 2008, 66.

357  *"virtual battle management" AI* Barbara Opall-Rome, "Israeli Defense to Use Artificial Intelligence," *Defense News,* January 21, 2008.

358  *"three-inch-deep" folder* Interview at U.S. military facility, Peter W. Singer, February 19, 2008.

358  *Recent neurological findings* Drew Westen, *The Political Brain: The Role of Emotion in Deciding the Fate of the Nation* (Public Affairs, 2007), ix, 69–88, 417–20.

358  *two underrated factors* Stephen Peter Rosen, *War and Human Nature* (Princeton, NJ: Princeton University Press, 2005), 28.

358  *low levels of serotonin* Ibid., 87.

358  *"The history of human conflicts"* Charles J. Dunlap Jr., *Technology and the 21st Century Battlefield: Recomplicating Moral Life for the Statesman and the Soldier* (Carlisle, PA: Strategic Studies Institute, United States Army War College, 1999), 12.

359  *"In war, as in life"* Christopher Coker, *The Future of War: The Re-enchantment of War in the Twenty-first*

*Century*, Blackwell Manifestos (Malden, MA; Oxford, UK: Blackwell, 2004), 73.

359  *"automatically send and collate information"* James Lasswell, interview, Peter W. Singer, Washington, DC, November 7, 2006.

359  *with each focusing* U.S. Navy, Convergence of Sea Power and Cyber Power, Strategic Studies Group XXVI, Naval War College, Newport, RI, July 2007.

## 19. WHO LET YOU IN THE WAR? TECHNOLOGY AND THE NEW DEMOGRAPHICS OF CONFLICT

360  *"How can I be a professional"* As quoted in Don M. Snider, "Jointness, Defense Transformation, and the Need for a New Joint Warfare Profession," *Parameters* 33, no. 3 (2003).

360  *"Simplicity"* Joel Clark, interview, Peter W. Singer, Washington, DC, November 8, 2007.

360  *his high school transcript revealed* Noah Shachtman, "Attack of the Drones," *Wired* 13.06 (2005), http://www.wired.com/wired/archive/13.06/drones.html.

360  *"but the idea of running a robot spy"* Ibid.

361  *"I love my job"* Joel Clark, interview, Peter W. Singer, November 8, 2008.

361  *a "military culture clash"* Noah Shachtman, "Unmanned Culture War," e-mail, defensetech@yahoogroups.com, May 27, 2005.

362  *singular professional identity* Don M. Snider and Gayle L. Watkins, "The Future of Army Professionalism: A Need for Renewal and Redefinition," *Parameters* 30, no. 3 (2000).

362  *both working with and competing against* Ibid.

362  *"will again revolutionize"* Christopher Coker, "Biotechnology and War: The New Challenge," *Australian Army Journal* 2, no. 1 (2004): 128.

362  *"UAVs are piloted"* Shachtman, "Attack of the Drones."

363  *"It might be necessary"* Steven Metz, *Armed Conflict in the 21st Century: The Information Revolution and Post-modern Warfare* (Carlisle, PA: Strategic Studies Institute, U.S. Army War College, 2000), 83.

363  *"After weapons system qualification"* Hap Carr, brigadier general, interview, Peter W. Singer, September 13, 2007.

364  *"We'll get them a few rides"* Air force colonel, interview, Peter W. Singer, April 18, 2006.

364  *"Those PlayStation 2s really do the trick"* As quoted in Eric Fleischauer, "Robots in Combat: Remote-Control Warfare: How PlayStation 2 Saves U.S. Lives," *Decatur Daily*, September 27, 2005.

365  *"you are never far from the Madden crowd"* Jeff Macgregor, "Imitation of Life," *Sports Illustrated*, August 21, 2006, 19.

365  *"The Army will draw on a generation"* Richard Szafranski, "When Waves Collide: Future Conflict," *Joint Force Quarterly*, no. 7 (1995): 79.

365  *"The video game generation learns"* Chris Barylick, "iRobot's PackBot on the Front Lines," *Space Daily*, February 23, 2006, http://www.spacedaily.com/reports/iRobots_PackBot_On_The_Front_Lines.html.

365  *today's pop culture and video games* Steven Johnson, *Everything Bad Is Good for You: How Today's Popular Culture Is Actually Making Us Smarter* (New York: Riverhead, 2005).

366  *"It will be no surprise"* Paul T. Harig, "The Digital General: Reflections on Leadership in the Post-Information Age," *Parameters* 26, no. 3 (1996).

366  *"Battles are won"* Owen West and Bing West, "Lessons from Iraq," *Popular Mechanics* 182, no. 8 (2005).

366  *"The younger airmen and women"* Air force colonel, interview, Peter W. Singer, April 18, 2006.

367  *"If you talk to my seven-year-old son"* Interview at U.S. military facility, Peter W. Singer, February 19, 2008.

367  *"The native creativity, innovativeness"* Robert H. Scales, "Clausewitz and World War IV," *Armed Forces Journal*, July 2006.

367  *"How do you manage to train"* James Lasswell, interview, Peter W. Singer, Washington, DC, November 7, 2006.

367  *"an American tendency to think"* David R. Smock, ed., *Religious Perspectives on War: Christian, Muslim, and Jewish Attitudes Toward Force*, Perspectives Series (Washington, DC: United States Institute of Peace Press, 2002), 22.

367  *"erase the pain given and taken"* Macgregor, "Imitation of Life," 19.

368  *"When you have a radical idea"* As quoted in Jonathon Keats, "The Idea Man," Popsci.com, 2004 (cited August 18, 2006); available at http://www.popsci.com/popsci/technology/generaltechnology/6b0898b0c9b84010vgnvcm1000004eecbccdrcrd.html.

368  *the Pentagon is now the largest day-care provider* Christopher Coker, *Humane Warfare* (London, New York: Routledge, 2001), 98.

369  *"Social intelligence and diplomatic skills"* Scales, "Clausewitz and World War IV."

369  *"Sixty is the new forty"* Ralph Peters, "The Geezer Brigade: Wartime Needs and Military Retirees," *Armed Forces Journal*, July 2007, http://www.armedforcesjournal.com/2007/07/2792594.

369  *"the Old Farts"* John Scalzi, *Old Man's War*, 1st ed. (New York: A Tom Doherty Associates Book, 2005).

369  *Our image of a soldier* For more on this, see Nancy Sherman, *Stoic Warriors: The Ancient Philosophy Behind the Military Mind* (New York: Oxford University Press, 2005).

369  *"inoperable pilonidal cyst"* Michael Arkush, *Rush!* (New York: Avon, 1993).

369  *"having a strong bladder and big butt"* Noah Shachtman, interview, Peter W. Singer, Washington, DC, March 25, 2006.

369 *"it is likely that a pasty-faced scholar"* Qiao Liang and Wang Xiangsui, *Unrestricted Warfare: China's Master Plan to Destroy America* (Beijing: PLA Literature and Arts Publishing House, 1999), 32.

370 *"Over time, the proportion of soldiers"* Elliott Abrams and Andrew J. Bacevich, "A Symposium on Citizenship and Military Service," *Parameters* 31, no. 2 (2001).

370 *"If you let the geeks wage war"* Foster-Miller executive, interview, Peter W. Singer, November 17, 2006.

370 *"much of the military regimen"* Bruce M. Lawlor, "Information Corps," *Armed Forces Journal*, January 1998, 26, 28.

370 *wars are fought by a new class of military reservists* Joe W. Haldeman, *Forever Peace*, 1st ed. (New York: Ace Books, 1997).

370 *the military's monopoly on war* P. W. Singer, *Corporate Warriors: The Rise of the Privatized Military Industry* (Ithaca, NY: Cornell University Press, 2003).

371 *"While these actions are principally motivated"* Charles J. Dunlap Jr., *Technology and the 21st Century Battlefield: Recomplicating Moral Life for the Statesman and the Soldier* (Carlisle, PA: Strategic Studies Institute, United States Army War College, 1999), 13.

371 *"We take these Army guys"* As quoted in Noah Shachtman, "Drone School, a Ground's-Eye View," *Wired News*, May 27, 2005, http://www.wired.com/news/technology/0,1282,67655,00.html.

371 *"certain soldiers were not as comfortable flying"* Ibid.

371 *"government-owned, contractor-operated"* Nathan Hodge, "Viper Strike Makes Its Combat Debut," *Jane's Defence Weekly*, November 14, 2007, 7.

372 *"surrogate warriors"* Bryan Bender, "Defense Contractors Quickly Becoming Surrogate Warriors," *Defense News*, March 28, 1997, 490.

372 *"They have their heart"* Rickey Smith and Helen Lardner, interview, Ralph Wipfli, November 14, 2006.

372 *"the nature of their status"* Dunlap, *Technology and the 21st Century Battlefield*, 14.

372 *"Even though international law recognizes"* Paul Kennedy and George J. Andreopoulos, "The Laws of War: Some Concluding Reflections," in *The Laws of War: Constraints on Warfare in the Western World*, ed. Michael Eliot Howard, George J. Andreopoulos, and Mark R. Shulman (New Haven, CT: Yale University Press, 1994).

372 *UAVs in wartime* John J. Klein, "The Problematic Nexus: Where Unmanned Combat Air Vehicles and the Law of Armed Conflict Meet," *Air & Space Power Journal—Chronicles Online Journal*, July 22, 2003, http://www.airpower.maxwell.af.mil/airchronicles/cc/klein.html.

373 *"he changes not only his clothing"* Coker, *Humane Warfare*, 91.

373 *"protect and defend the Constitution"* Morris Janowitz, *The Professional Soldier: A Social and Political Portrait* (Glencoe, IL: Free Press, 1960).

373 *"However much sociologists might argue"* Richard Holmes, *Acts of War: The Behavior of Men in Battle*, 1st American ed. (New York: Free Press, 1986), 31.

373 *"At least once a day"* Patrick O'Driscoll, "Losing a Limb Doesn't Mean Losing Your Job," USAToday.com, May 5, 2004, http://www.usatoday.com/news/nation/2004-05-05-cover-fit-to-serve_x.htm.

374 *Over a thousand soldiers have lost a limb* Sheri Waldrop and Michele Wojciechowski, "The 'Bionic' Warrior: Advances in Prosthetics, Technology, and Rehabilitation," *PT Magazine* 15, no. 4 (2007).

374 *they can "feel" a temperature change* Associated Press, "Bionic Arm Provides Hope for Amputees," CNN.com, September 14, 2006 (cited September 14, 2006); available at http://www.cnn.com/2006/TECH/09/14/bionic.arm.ap/index.html/2006/TECH/09/14/bionic.arm.ap/index.html.

374 *"as close to real limbs"* Art Pine, "Military Prosthetics: The Next Generation," *Proceedings* 133, no. 2 (2007): 13.

374 *40 percent of the soldiers* Ibid., 14.

374 *"It's sort of the Luke Skywalker"* As quoted in Discovery Science Channel, *Robosapiens: The Secret (R)evolution.*

375 *"As us baby boomers get older"* As quoted in Bay and Ford, *Cybernetics: Merging Machine and Man.*

375 *French physician Ambroise Paré* Ibid.

375 *a status symbol for rappers* Meredith May, "The Gold Standard of Style," *San Francisco Chronicle*, May 1, 2005, http://sfgate.com/cgi-bin/article.cgi?f=/c/a/2005/05/01/BAG32CIDA81.DTL.

375 *more than eleven million cosmetic surgeries* Reuters, "2006: Nearly 11 Million Cosmetic Surgeries in U.S.," *NewsMax*, March 22, 2007, http://www.newsmax.com/archives/articles/2007/3/22/132047.shtml.

375 *Florida-based VeriChip company* VeriChip, "Corporate FAQ," 2006 (cited August 14, 2007); available at http://www.verichipcorp.com/content/company/corporatefaq#g5.

375 *The implants have also been used* Bruce Schneier, "The ID Chip You Don't Want in Your Passport," *Washington Post*, September 16, 2006, A21.

375 *"It makes me jealous"* Discovery Science Channel, *Robosapiens: The Secret (R)evolution.*

376 *"Essentially we'll have a PDA"* Ibid.

376 *"the equivalent of a flash drive"* Robert Finkelstein, interview, Peter W. Singer, July 7, 2006.

376 *"take a book and gulp it down"* Mihail C. Roco and William Sims Bainbridge, "Converging Technologies for Improving Human Health: Nanotechnology, Biotechnology, Information Technology and Cognitive Science" (National Science Foundation, 2002), 110.

376  *"All war presupposes human weakness"* Carl von Clausewitz, Michael Eliot Howard, and Peter Paret, *On War* (Princeton, NJ: Princeton University Press, 1976), 257.

376  *"the weakling of the battlefield"* Paul F. Gorman, "SuperTroop via I-Port: Distributed Simulation Technology for Combat Development and Training Development" (Institute for Defense Analyses, 1990).

376  *"The G.I., the stamped government issue"* George Friedman and Meredith Friedman, *The Future of War: Power, Technology, and American World Dominance in the Twenty-first Century*, 1st ed. (New York: Crown, 1996), 392.

377  *"I draw the line"* Special forces officer, interview, Peter W. Singer, Washington, DC, September 7, 2006.

377  *"Being a guinea pig"* Ibid.

378  *"We need to separate emotion"* Krystyna Rudzki, "IAAF: Pistorius' Prosthetic Legs Provide Less Air Resistance," USAToday.com, July 16, 2007, http://www.usatoday.com/sports/olympics/summer/track/2007-07-16-iaaf-prosthetics-study_N.htm.

379  *"You never quite get used to the implants"* R. J. Pinero, "Air Infantry," in *Future Wars*, ed. Martin Harry Greenberg and Larry Segriff (New York: DAW Books, 2003), 212.

379  *a new type of human species* Joel Garreau, *Radical Evolution: The Promise and Peril of Enhancing Our Minds, Our Bodies—And What It Means to Be Human* (New York: Doubleday, 2005), 11.

380  *the human body actually alters* Discovery Science Channel, *Robosapiens: The Secret (R)evolution*.

380  *a growing division of "natural" humans* Ibid., 8.

381  *"One of the reactions I had"* Andrew Smith, "Science 2001: Net Prophets," *Observer*, December 31, 2000, 18.

381  *"If you are not upgraded"* Julian Jones, director, *How William Shatner Changed the World*, produced by the History Channel, broadcast on October 21, 2006.

## 20. DIGITIZING THE LAWS OF WAR AND OTHER ISSUES OF (UN)HUMAN RIGHTS

382  *"We risk continuing to fight"* John Reid, British defence secretary, "20th-Century Rules, 21st-Century Conflict" (speech, Royal United Services Institute for Defense and Security Studies, London, April 3, 2006).

382  *War is a special kind of hell* Nancy Sherman, *Stoic Warriors: The Ancient Philosophy Behind the Military Mind* (New York: Oxford University Press, 2005).

382  *"In times of war, the laws fall mute"* "Miscellaneous Military Quotes: Oats, and Military Proverbs," MilitaryQuotes.com, 2003 (cited March 30, 2007); available at http://www.military-quotes.com/misc%20quotes.htm.

382  *"War is still a rule-governed activity"* Michael Walzer, *Just and Unjust Wars*, 4th ed. (New York: Basic Books, 1977), 36.

383  *"Whether war is waged"* David J. DiCenso, "IW Cyberlaw," *Airpower Journal* 13, no. 2 (1999).

383  *"a rule-governed activity of equals"* As quoted in Sherman, *Stoic Warriors*, 11.

384  *"reconcile the necessities of war"* Peter Herby, "The Future of Weapons, Technology and International Law" (presentation, Brookings Institution, Washington, DC, October 17, 2006).

384  *"There are four pillars"* Ibid.

385  *"We have no particular viewpoint"* Ibid.

385  *"with every major scientific revolution"* Ibid.

385  *"There is so much terrible going on"* Interview with Peter W. Singer, October 17, 2006.

385  *"idiot protection device"* iRobot executive, interview, Peter W. Singer, November 16, 2006.

385  *"Our responsibility"* iRobot engineer, interview, Peter W. Singer, November 16, 2006.

386  *the conference on RMAs* Institute for National Security and Counterterrorism (INSCT) at Syracuse University, New Battlefields, Old Laws conference, Omni Shoreham Hotel, Washington, DC, October 8, 2007.

386  *"There is nothing set for this"* Colonel Gary Fabricius, USAF, interview, Peter W. Singer, August 29, 2006.

387  *The current "legal limbo"* Sebastian Thrun, interview, Peter W. Singer, March 18, 2007.

387  *"The lawyers tell me"* Tim Weiner, "A New Model Army Soldier Rolls Closer to the Battlefield," *New York Times*, February 16, 2005.

387  *"There is no consensus"* Steven Metz, interview, Peter W. Singer, September 19, 2006.

388  *participated "in over 50 interrogations"* Mother Jones Radio, "Marc Garlasco," October 2, 2005 (cited January 30, 2007); available at http://www.motherjones.com/radio/2005/10/garlasco_bio.html.

388  *"Isaac Asimov was a bit of a dirty old man"* Marc Garlasco, Human Rights Watch, interview, Peter W. Singer, Washington, DC, January 30, 2007.

390  *"It used to be a simple thing to fight a battle"* As quoted in Charles J. Dunlap Jr., "Lawfare in Modern Conflicts," *The Reporter* USAF JAG Corps Keystone Edition (2005): 94.

390  *"My JAG doesn't like this"* John J. Klein, "The Problematic Nexus: Where Unmanned Combat Air Vehicles and the Law of Armed Conflict Meet," *Air & Space Power Journal—Chronicles Online Journal*, July 22, 2003, http://www.airpower.maxwell.af.mil/airchronicles/cc/klein.html.

390  *"The ability to backtrack"* James Lasswell, interview, Peter W. Singer, Washington, DC, November 7, 2006.

391  *"there is a legal and moral duty"* "International Law—The Conduct of Armed Conflict and Air Operations, Pamphlet 110-31," ed. Department of the Air Force (1976).

391  *"What if a commander chooses"* Charles J. Dunlap

Jr., *Technology and the 21st Century Battlefield: Recomplicating Moral Life for the Statesman and the Soldier* (Carlisle, PA: Strategic Studies Institute, United States Army War College, 1999), 11.

391 *"We'll see far more lawyers"* Lawrence Korb, interview, Peter W. Singer, Washington, DC, September 30, 2006.

391 *"They are intent on manipulating"* Dunlap, "Lawfare in Modern Conflicts," 96.

391 *"We approach war in terror of lawsuits"* Ralph Peters, *Never Quit the Fight*, 1st ed. (Mechanicsburg, PA: Stackpole Books, 2006), 36.

391 *"The truth is"* Steven Green, as quoted in Andrew Tilghman, "I Came Over Here Because I Wanted to Kill People," *Washington Post*, July 30, 2006. Steven Green's trial date is set for April 27, 2009; see Associated Press, "Ex-soldiers Indictment Dispute Rejected," September 2, 2008.

392 *"revenge for our sister"* Michael Hedges, "Killings of Two Soldiers Perhaps Retaliation for Slain Iraqi Family," *Houston Chronicle*, July 4, 2006, http://www.chron.com/disp/story.mpl/front/4022556.html.

392 *"When you put young people"* Stephen E. Ambrose, *Americans at War* (Jackson: University Press of Mississippi, 1997), 152.

392 *"It is vitally important to recognize"* David Perry, "Why Hearts and Minds Matter," *Armed Forces Journal*, September 2006.

393 *"Anger is as much a part of war"* Nancy Sherman, *Stoic Warriors: The Ancient Philosophy Behind the Military Mind*, 68.

393 *"Situations where normal, good people"* Thomas Grassey, "Worse than a Failure of Leadership," *Proceedings* 132, no. 12 (2006): 44.

393 *"and the great expence of bloud is avoyed"* as quoted in Dunlap, *Technology and the 21st Century Battlefield*.

393 *"Warfare on some levels will never be moral"* Retired army officer, interview, Peter W. Singer, Doha, Qatar, January 10, 2004.

393 *"gave the detainee a good, swift kick"* U.S. Army War College meeting, Washington, DC, February 8, 2008.

394 *45 percent of soldiers wouldn't report a fellow soldier* Ronald C. Arkin, "Governing Legal Behavior: Embedding Ethics in a Hybrid Deliberative/Reactive Robot Architecture" (Georgia Institute of Technology/U.S. Army Research Office, 2007), 7.

394 *"behaves as if it were a single personality"* As quoted in Robert D. Kaplan, *Imperial Grunts: The American Military on the Ground* (New York: Random House, 2005), 60.

394 *"In the dreadful presence of suffering"* Paul Van Riper and Robert H. Scales Jr., "Preparing for War in the 21st Century," *Parameters* 27, no. 3 (1997).

394 *"The big advantage of moving"* Bob Quinn, interview, Peter W. Singer, November 17, 2006.

394 *"slow, methodical approach"* Andrew Bennett, interview, Peter W. Singer, November 16, 2006.

395 *"can coolly pick out targets"* Michael Fumento, "Is Anything Mightier Than This Sword?" *Tech Central Station*, January 6, 2005, http://www.fumento.com/military/sword-robot.html.

395 *"are not a benevolent God"* Chuck Klosterman, *Sex, Drugs, and Cocoa Puffs: A Low Culture Manifesto* (New York: Scribner, 2003).

395 *"It's like a video game"* Interview at U.S. military facility, Peter W. Singer, February 19, 2008.

395 *"that needs to be serviced"* David Singer, "What Determines Why People Support the Next War?," paper presented at the Imagining the Next War, Guggenheim Conference, New York City, March 25, 2006.

395 *"The greater the distance"* D. Keith Shurtleff, "The Effects of Technology on Our Humanity," *Parameters* 32, no. 2 (2002): 104.

396 *"as war becomes safer"* Ibid., 103.

396 *this sort of "externalization"* Chris Gray, *Postmodern War: The New Politics of Conflict* (New York: Guilford Press, 1997).

396 *"the impersonalization of battle"* John Keegan, *The Face of Battle* (New York: Viking Press, 1976), 320.

397 *"We're convinced that it was"* Marc W. Herold, "The Problem with the Predator," *Dissident Voice*, January 15, 2003, http://www.dissidentvoice.org/Articles/Herold_Predator.htm.

397 *"Why did you do this?"* Ibid.

398 *the military target might have moved* Max Boot, *War Made New: Technology, Warfare, and the Course of History, 1500 to Today* (New York: Gotham Books, 2006), 396.

398 *If the planes could have flown lower* John F. Burns, "U.S. Leapt Before Looking, Angry Villagers Say," *New York Times*, February 17, 2002.

398 *"gives you the ability to shoot second"* Andrew Bennett, interview, Peter W. Singer, November 16, 2006.

399 *Israeli UAV opened fire* Yaakov Katz, "Disaster Averted: UAV Fires at IDF, IAF Halts Fire," *Jerusalem Post*, July 25, 2006, http://www.jpost.com/servlet/Satellite?pagename=JPost%2FJPArticle%2FShowFull&cid=1153291989822.

399 *punched in the wrong coordinates* Patrick Eberle, "To UAV or Not to UAV: That Is the Question; Here is One Answer," *Air & Space Power Journal—Chronicles Online Journal*, October 9, 2001, http://www.airpower.au.af.mil/airchronicles/cc/eberle.html.

399 *"The Predator has been used"* "Exclusive: CIA Aircraft Kills Terrorist," ABC News, May 13, 2005 (cited May 18, 2006); available at http://abcnews.go.com/WNT/Investigation/story?id=755961.

399 *"The attack killed 18 civilians"* Mansoor Ijaz, "An Alliance Too Vital to Jeopardize with Poor Intelligence," *Financial Times*, January 17, 2006.

400 *"such as mistaking a civilian"* Robert Finkelstein, interview, Peter W. Singer, July 7, 2006.

400 *"Inevitably, sooner or later"* Dennis Sorenson, "Technological Development of Unmanned Systems to Support the Naval Warfighters," paper pre-

sented at the Military Robotics Conference, Institute for Defense and Government Advancement, Washington, DC, April 10–12, 2006.

400 *"It depends on the situation"* Roboticist, interview, Peter W. Singer, July 7, 2006.

400 *"The big question for military law"* Finkelstein, interview, July 7, 2006.

400 *"collateral damage estimation methodology"* Ibid.

401 *"able to be perfectly ethical in the battlefield"* Arkin, "Governing Legal Behavior: Embedding Ethics in a Hybrid Deliberative/Reactive Robot Architecture," 1, 7.

401 *"We could reduce man's inhumanity"* Ronald Arkin, as quoted in Tom Abate, "If It Only Had a Heart: Can Robots Behave Humanely?" *San Francisco Chronicle*, January 29, 2008, B1.

401 *The target could be attacked only* Colin Kahl, "How We Fight," *Foreign Affairs* 85, no.6 (2006), 86.

402 *The Somali warrior* Sean J. A. Edwards, "Swarming and the Future of Warfare" (doctoral thesis, Pardee Rand Graduate School, 2005).

402 *any farmers who refused* Alon Ben-David, "IDF Introspective Prior to Withdrawal from Lebanon," *Jane's Defence Weekly*, September 20, 2006, 8.

403 *"What happens when things don't work out"* UAV pilot, interview, Peter W. Singer, August 29, 2006.

403 *"We're not in the business"* "Robots Could Demand Legal Rights," BBC News, December 21, 2006 (cited December 22, 2006); available at http://news.bbc.co.uk/1/hi/technology/6200005.stm.

403 *a lawyer defended the right* Ray Kurzweil, *The Singularity Is Near: When Humans Transcend Biology* (New York: Viking, 2005), 379.

404 *"If they are given human level intelligence"* Robert Finkelstein, "Military Robotics: Malignant Machines or the Path to Peace?" paper presented at the Military Robotics Conference, Institute for Defense and Government Advancement, Washington, DC, April 10–12, 2006.

404 *"Will we ever get to the point"* Marc Garlasco, interview, Peter W.Singer, January 30, 2007.

404 *whether we endow the robot* Rodney Brooks, *Flesh and Machines: How Robots Will Change Us* (New York: Pantheon, 2002).

404 *psychologists are finding that people* Betya Friedman, Peter H. Kahn, and Jennifer Hagman, "Hardware Companions? What Online AIBO Discussion Forums Reveal About the Human-Robotic Relationship," in *Conference on Human Factors in Computing Systems* (Fort Lauderdale: ACM Press, 2003); Betya Friedman and Peter H. Kahn, "Human Values, Ethics, and Design," in *The Human-Computer Interaction Handbook*, ed. J. A. Jacko and A. Sears (Mahwah, NJ: Erlbaum, forthcoming); Peter H. Kahn et al., "Robotic Pets in the Lives of Preschool Children," *Interaction Studies* 7, no. 3 (2006); Peter H. Kahn, "Social and Moral Relationships with Personified Robots," presentation, Navy Center for Applied Research in Artificial Intelligence, March

12, 2007; Peter H. Kahn et al., "Social and Moral Relationships with Robotics Others?" paper presented at the IEEE International Workshop on Robot and Human Interaction, Okayama, Japan, September 20–22, 2004; Peter H. Kahn et al., "What is a Human?" *Interaction Studies* 8, no. 3 (2007).

405 *"Humans are stupid"* Daniel Wilson, interview, Peter W. Singer, October 19, 2006.

405 *"There will be people"* Salamander Davoudi, "Future Shock as Study Backs Rights for Robots in a PC world," *Financial Times*, December 21, 2006, 1.

406 *"the right to use all necessary means"* Anthony D'Amato, "International Law, Cybernetics, and Cyberspace," *Naval War College International Law Studies* 76 (2006): 62.

407 *The interpretation of robot rights* Charles J. Dunlap Jr., "The Revolution in Military Legal Affairs: Air Force Legal Professionals in 21st Century Conflicts," *Air Force Law Review* 51 (2001).

407 *"The Revolution in Military Legal Affairs"* Ibid.

407 *when the average new house cost $7,450* The People History, "1949 History: News, Events, Technology, Prices and Popular Culture," 2007 (cited November 3, 2007); available at http://www.thepeoplehistory.com/1949.html.

407 *"revolution in military legal affairs"* Dunlap, "The Revolution in Military Legal Affairs: Air Force Legal Professionals in 21st Century Conflicts."

408 *"only military trigger pullers"* Klein, "The Problematic Nexus: Where Unmanned Combat Air Vehicles and the Law of Armed Conflict Meet."

408 *"If these same Canadian forces"* Ibid.

409 *"If a robot were programmed"* D'Amato, "International Law, Cybernetics, and Cyberspace," 62.

410 *"Scientists and technologists must take"* Joel Garreau, *Radical Evolution: The Promise and Peril of Enhancing Our Minds, Our Bodies—And What it Means to Be Human* (New York: Doubleday, 2005), 165.

411 *"Science cannot by itself establish"* Ibid., 164.

412 *we might see international agreements* Jürgen Altmann and Mark Gubrud, "Anticipating Military Nanotechnology," *IEEE Technology and Society Magazine* 23, no. 4 (2004).

412 *"You have to remember"* Steven Metz, interview, Peter W. Singer, September 19, 2006.

412 *"The more society adheres"* Dunlap, "Lawfare in Modern Conflicts," 102.

## 21. A ROBOT REVOLT?
### TALKING ABOUT ROBOT ETHICS

413 *"Any machine could rebel"* Daniel H. Wilson, *How to Survive a Robot Uprising: Tips on Defending Yourself Against the Coming Rebellion*, 1st U.S. ed. (New York: Bloomsbury, 2005), 14.

413 *"As a kid, I fell in love"* Daniel H. Wilson, "About the Author," 2005 (cited August 30, 2006); available at http://www.robotuprising.com/qanda.htm.

413 *Wilson decided to try his hand* Wilson, *How to Survive a Robot Uprising*, 10.

413 *He details the warning signs* Ibid., 32.

414 *"the chance of a Hollywood-style"* Wilson, "About the Author."

414 *"Who will be man's successor?"* Ray Kurzweil, *The Singularity Is Near: When Humans Transcend Biology* (New York: Viking, 2005), 205.

415 *"matches human capabilities"* Robert Finkelstein, "Military Robotics: Malignant Machines or the Path to Peace?" paper presented at the Military Robotics Conference, Institute for Defense and Government Advancement, Washington, DC, April 10–12, 2006.

415 *"the robots will eventually succeed us"* Andrew Smith, "Science 2001: Net Prophets," *Observer*, December 31, 2000, 19.

415 *"our machines are evolving faster"* K. Eric Drexler, *Engines of Creation*, 1st ed. (Garden City, NY: Anchor Press/Doubleday, 1986), 171.

415 *"humanity looks to me"* Smith, "Science 2001: Net Prophets," 18.

415 *"In the game of life and evolution"* George Gilder and Richard Vigilante, "Stop Everything . . . It's Techno-Horror!," *American Spectator* 34, no. 2 (2001): 40.

415 *"a 50 percent chance of survival"* Smith, "Science 2001: Net Prophets," 18.

415 *"leapingly, screamingly insane"* Joel Garreau, *Radical Evolution: The Promise and Peril of Enhancing Our Minds, Our Bodies—And What It Means to Be Human* (New York: Doubleday, 2005), 73.

415 *"Well, yeah, but I've decided"* Smith, "Science 2001: Net Prophets," 18.

416 *"In designing software and microprocessors"* Bill Joy, "Why the Future Doesn't Need Us," in *Taking the Red Pill: Science, Philosophy and Religion in The Matrix*, ed. Glenn Yeffeth and David Gerrold (Chicago: BenBella Books, 2003), 211.

416 *"As one strong AI immediately begets"* Kurzweil, *The Singularity Is Near*, 262.

417 *"Evolution, Morpheus, evolution"* Robin Handson, "Was Cypher Right? (Part I): Why We Stay in Our Matrix," in *Taking the Red Pill: Science, Philosophy and Religion in The Matrix*, ed. Glenn Yeffeth and David Gerrold (Chicago: BenBella Books, 2003), 24.

417 *a robot takeover "will never happen"* Rodney Brooks, *Flesh and Machines: How Robots Will Change Us* (New York: Pantheon, 2002), ix.

417 *They would have to have lost* Ibid., 199–204.

418 *a robot takeover rests on a massive assumption* Garreau, *Radical Evolution*, 211.

418 *You don't get machines beyond control* Brooks, *Flesh and Machines*.

418 *Why would machines ever need* Read Mercer Schuchardt, "What Is the Matrix?" in *Taking the Red Pill: Science, Philosophy and Religion in The Matrix*, ed. Glenn Yetheff and David Gerrold.

419 *just as pornography helped launch* Brooks, *Flesh and Machines*, 147.

419 *"something we all await with excitement"* Scientist, interview, Peter W. Singer, July 7, 2006.

419 *"People are [already] willing to have sex"* "Trust Me, I'm a Robot," *Economist* 379, no. 8481 (2006): 20.

419 *"Pedophiles may argue"* Ibid.

419 *Are these the sorts of "experiences"* Ed Habershon and Richard Woods, "No Sex Please, Robot, Just Clean the Floor," *Sunday Times*, June 18, 2006.

419 *What are the boundaries* Eric Smalley, "Georgia Tech's Ronald Arkin," 2005, http://www.trnmag.com/Stories/2005/091205/View_Ronald_Arkin_091205.html.

420 *"I am sure there will be new dilemmas"* Rodney Brooks, interview, Peter W. Singer, Washington, DC, October 30, 2006.

420 *"fifteen minutes of privacy"* As quoted in Garreau, *Radical Evolution*, 100.

420 *"scares the shit out of me"* Daniel Wilson, interview, Peter W. Singer, October 19, 2006.

420 *"The only realistic alternative"* As quoted in Jay Richards, *Are We Spiritual Machines? Ray Kurzweil vs. the Critics of Strong AI*, 1st ed. (Seattle: Discovery Institute Press, 2002), 223.

420 *"We aren't at war"* Ibid., 224.

421 *"We are curious as a species"* Discovery Science Channel, *Robosapiens: The Secret (R)evolution*, broadcast on June 18, 2006.

421 *"If we knew what it was"* As quoted in Richards, *Are We Spiritual Machines?*, 133.

421 *"We would have to repeal capitalism"* Ibid., 54.

421 *"We'll be chasing our fucking tails"* J. Sigger, "New 'WMD' Definition?" October 16, 2006 (cited October 20, 2006); available at http://armchairgeneralist.typepad.com/my_weblog/2006/10/new_wmd_definit.html.

421 *"No matter what we do"* Robert Epstein, interview, Peter W. Singer, Washington, DC, October 25, 2006.

421 *"You can't say it's not part"* K. Eric Drexler, "Nanotechnology: Six Lessons from Sept. 11," Open Letter, Foresight Institute, December 2001.

422 *"We've got to be pro-active"* Ibid.

422 *"We have to manage the ethics"* Habershon and Woods, "No Sex Please, Robot, Just Clean the Floor."

422 *"Ethicists have written at length"* Nick Bostrom, "Nanotechnology Perceptions: A Review of Ultraprecision," *Nanotechnology Perceptions: A Review of Ultraprecision Engineering and Nanotechnology* 2, no. 2 (2006).

423 *"People ask me about"* Rodney Brooks, interview, Peter W. Singer, October 30, 2006.

423 *"Asimov's rules are neat"* Daniel Wilson, interview, Peter W. Singer, October 19, 2006.

423 *"realized during a robot exhibition"* "Trust Me, I'm a Robot," *Economist* 379, no. 8481 (2006): 20.

423 *"You don't want to tell"* Mark Barber, "Force Protection Robotics," paper presented at the Military Robotics Conference, Institute for Defense and Government Advancement, Washington, DC, April 10–12, 2006.

423  *"There is a lot of push"* Rodney Brooks, interview, Peter W. Singer, October 30, 2006.

424  *"Businesses are notoriously uninterested"* Robert Sawyer, "On Asimov's Three Laws of Robotics," 1994 (cited November 1, 2007); available at http://www.sfwriter.com/rmasilaw.htm.

424  *"We can't simply do our science"* Bill Joy, "Forfeiting the Future," *Resurgence* no. 208 (2001), http://www.resurgence.org/resurgence/issues/joy208.htm.

424  *"I gave him a six-foot extension cord"* Roger Nygard, "Grief Counseling," *The Office*, produced by B. J. Novak et al., broadcast on October 12, 2006.

424  *an ethic of "design ahead"* Drexler, *Engines of Creation.*

424  *"There is no sense"* Daniel Wilson, interview, Peter W. Singer, October 19, 2006.

425  *"Redundancy can bring"* Drexler, *Engines of Creation,* 178.

425  *"with a 'conscience' that would reflect"* Smalley, "Georgia Tech's Ronald Arkin."

426  *"Make a habit of two things"* Hippocrates, *The Epidemics,* book 1, section V.

426  *"precautionary principle"* David Runciman, "The Precautionary Principle," *London Review of Books,* 2004, http://www.lrb.co.uk/v26/n07/print/runc01.html.

426  *"define and deal"* Julia A. Moore, "The Future Dances on a Pin's Head; Nanotechnology: Will It Be a Boon—Or Kill Us All?," *Los Angeles Times,* November 26, 2002.

426  *"We have reached a point"* Neal Pollard, "Technology and Intelligence Reform: Opportunities and Hurdles," in *The Faces of Intelligence Reform,* ed. Angela M. Sapp, Barton B. Brown II, and James T. Kirkhope (Washington, DC: CENSA, 2005), 42.

426  *scientists "don't have a seat at the table"* Policy Center director, interview, Peter W. Singer, Washington, DC, November 3, 2006.

427  *"human impact statement"* Pollard, "Technology and Intelligence Reform: Opportunities and Hurdles," 42.

427  *"think that the technology"* Robert Finkelstein, interview, Peter W. Singer, July 7, 2006.

## 22. CONCLUSION: THE DUALITY OF ROBOTS AND HUMANS

428  *a character "who prevails"* American Film Institute, "AFI's 100 Years . . . 100 Heroes & Villains," 2007 (cited January 19, 2008); available at http://www.afi.com/tvevents/100years/handv.aspx.

429  *"an Age of Transitions"* Mihail C. Roco and William Sims Bainbridge, "Converging Technologies for Improving Human Health: Nanotechnology, Biotechnology, Information Technology and Cognitive Science" (National Science Foundation, 2002).

429  *the story of five American troops* Jomana Karadsheh, "Roadside Bomb Kills 5 U.S. Soldiers in Iraq,"

CNN.com, January 28, 2008 (cited January 28, 2008); available at http://www.cnn.com/2008/WORLD/meast/01/28/iraq.main/index.html.

429  *"more ponies than panzers"* Max Boot, *War Made New: Technology, Warfare, and the Course of History, 1500 to Today* (New York: Gotham Books, 2006), 467.

431  *"the twenty-first century could end"* Roco and Bainbridge, "Converging Technologies for Improving Human Health: Nanotechnology, Biotechnology, Information Technology and Cognitive Science."

431  *"Without more kindliness"* As quoted in Nick Bostrom, "A History of Transhumanist Thought," *Journal of Evolution and Technology* 14, no. 1 (2005), http://jetpress.org/volume14/bostrom.html.

432  *"represents the single most important factor"* Martin van Creveld, *The Transformation of War* (New York: Free Press, 1991), 273.

432  *defy the normal rules of logic* I am indebted to Sebastian Kaempf of the University of Queensland for this insight.

432  *"paradox of riskless warfare"* Paul W. Kahn, "The Paradox of Riskless Warfare," *Philosophy & Public Policy Quarterly* 22, no. 2 (2002): 2–8.

432  *"propels us beyond the ethics of warfare"* Ibid., 3.

433  *the codes and values that defined them* Christopher Coker, *Waging War Without Warriors? The Changing Culture of Military Conflict,* IISS Studies in International Security (Boulder, CO: Lynne Rienner Publishers, 2002).

433  *"a defeat renowned"* Christopher Coker, *Humane Warfare* (London, New York: Routledge, 2001), 45.

433  *In making war less human* Ibid.

434  *"get it right the first time"* Ray Kurzweil, *The Singularity Is Near: When Humans Transcend Biology* (New York: Viking, 2005).

434  *"What has gripped me the most"* Michael G. Mullen, "The Nation's Navy: Beyond Iraq—Sea Power for a New Iraq" (Briefing, Brookings Institution, Washington, DC, April 3, 2007).

434  *"We will only be able to react"* Kip P. Nygren, "Emerging Technologies and Exponential Change: Implications for Army Transformation," *Parameters* 32, no. 2 (2002): 93.

435  *"We need to have discussion"* Richard Clarke, interview, Peter W. Singer, Washington, DC, August 8, 2007.

435  *"Ignorance" actually has two meanings* Webster's *Revised Unabridged Dictionary,* 1913 (cited December 5, 2007); available at http://dictionary.die.net/ignorance/ignorance.

435  *"We are already in* A Brave New World" Charles McLaughlin, interview, Peter W. Singer, Washington, DC, January 30, 2008.

435  *"The people higher up"* Jonathan Hall, interview, Peter W. Singer, Washington, DC, August 6–9, 2007.

436  *"Every gun that is made"* As quoted by David Shukman, *Tomorrow's War: The Threat of High-Technology Weapons,* 1st U.S. ed. (San Diego: Harcourt Brace, 1996).